For
Charles E. Hen,
Dedicatee Extraodinane

Dannek

HORTICULTURAL REVIEWS
VOLUME 13

edited by

Jules Janick
Purdue University

A WILEY-INTERSCIENCE PUBLICATION

John Wiley & Sons, Inc.

NEW YORK / CHICHESTER / BRISBANE / TORONTO / SINGAPORE

LC card number 79-642829
ISBN 0-471-57499-6
ISSN 0163-7851

Printed and bound in the United States of America by Braun-Brumfield, Inc.

10 9 8 7 6 5 4 3 2 1

Contents

Contributors

P. C. ANDERSEN. University of Florida, IFAS, Agricultural Research and Education Center, P.O. Route 4, P.O. Box 63, Monticello, FL 32344.

EDWARD N. ASHWORTH. Department of Horticulture, Purdue University, West Lafayette, IN 47907.

ALAN B. BENNETT. Department of Vegetable Crops, University of California, Davis, CA 95616.

RICHARD J. CHAMBERS. Horticulture Research International, Worthing Road, Littlehampton, West Sussex, BN17 6LP, United Kingdom.

R. L. DARNELL. Fruit Crops Department, University of Florida, Gainesville, FL 32611.

DEAN DELLAPENNA. Deparment of Plant Science, University of Arizona, Tucson, AZ 85721.

J. D. EARLY. Pomology Department, University of California, Davis, CA 95616.

M. FENNER. Department of Biology, The University, Southampton S09 5NH, England.

ROBERT L. FISCHER. Department of Plant Biology, University of California, Berkeley, CA 94720.

JEAN M. GERRATH. Department of Horticultural Science, University of Guelph, Guelph, Ontario N1G 2W1, Canada.

JAMES J. GIOVANNONI. Department of Plant Breeding and Biometry, Cornell University, Ithaca, NY 14853.

NEIL L. HELYER. Horticulture Research International, Worthing Road, Littlehampton, West Sussex, BN17 6LP, United Kingdom.

ANWAR A. KHAN. Department of Horticultural Sciences, New York State Agricultural Experiment Station, Cornell University, Geneva, NY 14456.

C. M. KINGSTON. Ministry of Agriculture and Fisheries, Levin Horticultural Research Centre, Private Bag, Levin, New Zealand.

G. A. LANG. Department of Horticulture, Louisiana State University, Baton Rouge, LA 70803.

G. C. MARTIN. Pomology Department, University of California, Davis, CA 95616.

R. C. PLOETZ. University of Florida, IFAS, Tropical Research and Education Center, 18905 S.W. 280 Street, Homestead, FL 33031.

BRUCE SCHAFFER. University of Florida, IFAS, Tropical Research and Education Center, 18905 S.W. 280 Street, Homestead, FL 33031.

PAUL I. SOPP. Horticulture Research International, Worthing Road, Littlehampton, West Sussex, BN17 6LP, United Kingdom.

WALTER E. SPLITTSTOESSER. Department of Horticulture, University of Illinois, Urbana, IL 61801.

G. W. STUTTE. Department of Horticulture, University of Maryland, College Park, MD 20742.

KEITH D. SUNDERLAND. Institute of Horticultural Research, Worthing Road, Littlehampton, West Sussex, BN17 6LP, United Kingdom.

GEOFFREY O. TUNYA. Egerton University, P.O. Njoro, Kenya.

BARBARA D. WEBSTER. Associate Vice Chancellor—Research, University of California, Davis.

Charles E. Hess

Dedication

This volume is dedicated with respect and affection to Charles E. Hess, Assistant Secretary of Agriculture for Science and Education. In that capacity. Charley is responsible for USDA research and education programs in the food and agricultural sciences, including planning, evaluation, and coordination of the state-federal activities through various committee structures.

However, to those of us in the horticultural sciences, Charley wears several other hats. We know him best as a genial and inspiring leader in our field, but also as a President of the American Society for Horticultural Science, as a Dean of the College of Agricultural and Environmental Sciences at the University of California at Davis, as a Department Head of Horticulture and Forestry and as a Dean of Cook College at Rutgers University, and as a Professor of Horticulture at Purdue University. In these and in other leadership roles, Charley's achievements have been of major significance to the field of horticulture and agriculture.

Charley grew up in New Jersey where his father was a nurseryman. He earned his baccalaureate with high honors from Rutgers in 1954 and received a Master's degree (1954) and a Ph.D. in Plant Physiology and Plant Pathology (1957) from Cornell University—and was elected to Phi Beta Kappa along the way. He was subsequently selected as a Fellow of the American Association for the Advancement of Science and the American Society for Horticultural Science. In 1983, Purdue University awarded him an Honorary Doctoral Degree. In 1985, the Cook College Cooperative Agricultural Extension Service Alumni presented to him the George Hammell Cook Distinguished Service Award, and one year later he was named Outstanding Educator of the Year by the California Association of Nurserymen. In 1988, he received a United States Department of Agriculture Award for Distinguished Service.

Charley came to UC Davis in 1975 to administer an extremely diverse college representing 41 majors; 33 departments, divisions, and research groups involving 400 faculty; 4000 undergraduate students; and 1000 graduate students. During his 12 years of service as Dean, Charley provided significant input into the development of interdisciplinary and multidisciplinary research programs addressing major needs in the College and the Agricultural Experiment Station. It was Charley who facilitated development of Davis programs in Consumer Research, Agricultural Issues, the McKnight Foundation Program in Plant Disease Research, and Integrated Pest Management. As a prophet in his own time, he put into place a sustainable agriculture program, a genetic resources center, and a coordinated biotechnology program long before these areas became the fashion of the day. His leadership in initiating new, broadbased, and interdisciplinary programs over a decade ago, when the future of agriculture in the United States was under intense scrutiny, helped to lay the cornerstone for a new era in agriculture, one in which biotechnology, sustainability, biodiversity, and environmental enhancement now receive increasing emphasis. It is this type of programmatic leadership which brought recognition on national and international levels to Charley, and which ultimately resulted in his present appointment as Assistant Secretary of Agriculture.

Wherever Charley has been, he is remembered by his associates as a gracious, fair, and concerned colleague; a man of great exuberance, enthusiasm, and charm; an accomplished administrator; and a dedicated horticulturist and scientist. It is an honor for us all to dedicate this volume of *Horticultural Reviews* to him.

Barbara D. Webster
Associate Vice Chancellor—Research
University of California, Davis

Integrated Pest Management of Greenhouse Crops in Northern Europe

Keith D. Sunderland, Richard J. Chambers, Neil L. Helyer, and Paul I. Sopp[1]
Horticulture Research International,
Worthing Road, Littlehampton
West Sussex BN17 6LP, United Kingdom

I. INTRODUCTION

There is a long, and increasing, list of pest (used throughout this review to mean invertebrate pest, rather than diseases or weeds) problems in northern European greenhouses (Cavalloro 1991). The availability of

[1]We thank our colleagues Paul Richardson, Norman Crook, Paul Jarrett, and Adrian Gillespie for advice on the insect pathology sections. Neil Wilding kindly provided information on Entomophthorales nomenclature. Three anonymous referees improved the manuscript with constructive criticisms. The work was funded by the U.K. Ministry of Agriculture, Fisheries and Food.

chemical pesticides is declining, and therefore the need for biological control is greater than at any previous time (Sunderland 1990). The advantages of biological control, which are considerable, include safety for operator, consumer, and environment, and the elimination of phytotoxic damage.

The opportunities for biological and integrated pest control are greater in greenhouses than in field crops because the high value of the crop makes it economically feasible to rear and release natural enemies. Once released, these enemies are contained and there is the potential to control the environment, to some extent, to suit the natural enemy. It is rarely possible to biologically control all the pests in a crop throughout the growing season; therefore, chemical insecticides are also needed (as indeed are fungicides, herbicides, and plant growth regulators). It follows that any pesticides must be compatible with the natural enemies in use, or, if this is not possible, ways must be found to separate the biological and chemical components either spatially or temporally. Many other components of control can be incorporated, such as plant resistance and the chemical modification of the behavior of pests and natural enemies. Together, these methods form the basis of integrated pest management (IPM).

Biological control agents of some of the major pests have been available for a long time. In 1926 Speyer was sent the parasite *Encarsia formosa* (Gahan) from a site in Hertfordshire, England (Speyer 1927a), and by 1935 he was mass-producing it for the control of glasshouse whitefly, *Trialeurodes vaporariorum* (Westwood) (Hussey and Scopes 1985). Interest was lost in this approach in the late 1940s with the development of synthetic insecticides, which growers perceived as being easier to use. Frequent applications of these insecticides selected for resistant strains of pests. Two-spotted spider mite, *Tetranychus urticae* (Koch), became resistant to DDT, a chlorinated hydrocarbon, in 1945 and by the early 1950s was also resistant to some organophosphorus compounds (Gould 1985). In 1960, the predatory mite *Phytoseiulus persimilis* Athias-Henriot was discovered in Germany on orchids imported from Chile. This species subsequently proved to be an efficient predator of *Tetranychus urticae*. Early research into optimal strategies for using *Encarsia formosa* and *Phytoseiulus persimilis* were carried out at the Glasshouse Crops Research Institute in England. A number of small companies in England began to produce and sell these natural enemies in the late 1960s. *Phytoseiulus persimilis* was produced commercially in Scandinavia in the early 1970s. *Encarsia formosa* was widely used in Europe by 1975 (Hussey 1985a).

Biological control of the major pests led to problems in controlling other pests, because most broad spectrum insecticides could not be used. Changes in cultural practices also resulted in a change in the status of some pests; onion thrips, *Thrips tabaci* Lindeman, for example, became more of a problem in cucumbers with the switch to rockwool culture. By the late 1970s, an integrated chemical technique was demonstrated to be successful in England. This consisted of applying a mixture of deltamethrin and polybutene to the plastic sheeting below the crop to catch and kill thrips dropping from the plants to pupate. This had the advantage of not disrupting other pest-natural enemy interactions already established in the crop. The use of predatory mites, *Amblyseius* spp., for the control of *Thrips tabaci* was investigated in the Netherlands in the early 1980s (Ramakers 1983b). Leaf-mining fly larvae also became more difficult to control, and by 1976 research into the use of a range of ecto- and endo-parasites was well developed. These parasites subsequently became available commercially for the control of tomato leafminer, *Liriomyza bryoniae* (Kaltenbach).

Progressive development of resistance to a range of insecticides has been observed in two greenhouse aphid species. Various natural enemies of aphids—including parasites, predators, and pathogens—have been investigated since resistance began to appear in the 1970s (Hussey 1985a). Predatory larvae of the midge *Aphidoletes aphidimyza* (Rondani), were researched in Finland in the early 1970s and found to have the advantage of feeding on all species of greenhouse aphid. A pathogenic fungus, *Verticillium lecanii* (Zimmerman) Viegas, isolated in England in 1956, was subsequently shown to have strains that would attack aphids and whitefly. This was marketed in the 1980s under the trade names "Vertalec" (aphid strain) and "Mycotal" (whitefly strain). The fungus requires a period of high humidity for spore germination, and there was inconsistent pest control in the drier cropping situations. *Verticillium lecanii* has recently reappeared on the market. Research is in progress to determine if short periods of high humidity in the greenhouse will potentiate *V. lecanii* without stimulating plant pathogens. Non-chemical means of controlling plant pathogens within an integrated control system are also under investigation (Cavalloro 1987).

There is a dearth of effective selective chemicals for caterpillar control, but bacterial insecticides have proved very useful. *Bacillus thuringiensis* Berliner (*B.t.*) was discovered in Germany in 1911. It was being used commercially for caterpillar control by 1973. *B.t.* products currently available are not as effective against cabbage moth, *Mamestra brassicae* (L.), as against other greenhouse caterpillars.

Black vine weevil, *Otiorhynchus sulcatus* (F.), is a serious pest of ornamentals. Biological control agents have recently been developed. Insect parasitic nematodes are sold on a small scale in the Netherlands, and two nematode products have been launched in the United Kingdom for control of the weevil on cyclamen. The fungus *Metarhizium anisopliae* (Metsch.) Sorok. also shows promise for vine weevil control.

The area of northern Europe covered by greenhouses is about 27,600 ha, *E. formosa* being used on 1400 ha and *P. persimilis* on 4590 ha. In

1985, the world area of greenhouses was approximately 150,000 ha, and the usage of biological control agents was as follows; *E. formosa* 2361 ha; *P. persimilis* 5176 ha; *B.t.* 1000 ha; leafminer parasites 460 ha; *A. aphidimyza* 81 ha; and *Amblyseius* spp. 65 ha (van Lenteren and Woets 1988). The figures suggest that there is opportunity for increased usage of biological control techniques.

The need for IPM is growing, not only because of the development of resistance to chemical pesticides, but also because an upturn in world trade in plant material is accelerating the rate at which new pests (frequently pesticide-resistant) are accidentally introduced into Europe. For example, 370 million plants belonging to 1500 genera were imported into the U.K. in 1988. Inspectors of the U.K. Ministry of Agriculture, Fisheries, and Food have intercepted more than 1000 named and 80 unnamed invertebrate species in the last decade. It is inevitable that some species evade detection and become established. These immigrant pests are often cosmopolitan species which have been subjected to intensive spraying campaigns in various parts of the world and are multiply-resistant by the time they enter Europe. Recent examples are western flower thrips, *Frankliniella occidentalis* (Pergande); cotton whitefly, *Bemisia tabaci* (Gennadius); and South American leafminer, *Liriomyza huidobrensis* (Blanchard).

The range of chemicals available for pest control is rapidly declining. Some were withdrawn by the manufacturer (e.g., cyhexatin) and others by regulatory authorities (e.g., aldrin). Regulatory authorities in England are now systematically reviewing chemicals cleared under older safety schemes. This is likely to result in more withdrawals. A high proportion of minor-use pesticides may not be re-registered. As the pressures caused by environmental concerns increase, governments are beginning to set targets for the banning of pesticides. For example, the Dutch government plans to reduce the use of pesticides by 50% over a period of five years. The rate of introduction of new pesticides is not keeping up with the rate at which they are withdrawn. In the 1960s, 22 new products were launched each year, but this has now declined to 12 largely due to increases in the cost of research, development, and registration. The average research and development cost per product is in the region of £80 million (Finney 1988), and registration can cost an extra £2 million.

Apart from avoiding hazards to users, consumers, and the environment, IPM can confer economic benefits because the crop suffers less phytotoxic effects from pesticide use. Increases in crop yield of 10% to 15% may be obtained by tomato and cucumber growers who switch to biological control of *Trialeurodes vaporariorum* and *Tetranychus urticae* (Hussey 1985b). It also becomes possible to use bees for pollination on crops where no chemical pesticides are applied; more than 100 ha of greenhouse tomato are now pollinated in this way in Belgium, the Netherlands, and the U.K. The need to develop alternatives to chemical pesticides is widely acknowledged; the extent to which alternatives are already used and the prospects for new IPM measures are reviewed below.

This review draws mainly on research carried out in northern Europe, but references from other regions of the world are included if especially apposite. Similarly, although it relates almost entirely to greenhouse crops, the occasional reference to research on field crops is included in cases where the technology described is likely to find application in protected crops. The main emphasis is on present practice, but likely future developments are also described. Each sub-section within the section "Integrated Control by Pest Type" is structured in the sequence (i) pest biology, (ii) predators and parasites, followed by pathogens, and (iii) integration with chemical pesticides.

II. INTEGRATED CONTROL BY PEST TYPE

A. Spider Mites

Tetranychus urticae, commonly known as red spider mite (diapausing form) or two-spotted spider mite (usually green-colored with a dark spot on either side of its body), is a major pest of both vegetable and ornamental crops throughout the world (Hussey and Huffaker 1976). The mite is an arrhenotokous parthenogenetic species: unmated females produce only haploid eggs which develop into males. Mated females produce both haploid and diploid eggs which develop into males and females respectively (Sabelis 1981). Male and female mites become distinguishable at the deutonymph stage, males having a more elongated body than the rounded females. Ewing (1914) noted the "premature courting" behavior of male mites around the resting deutonymph prior to the final moult. Cone et al. (1971a, 1971b) found that males are attracted to female resting deutonymphs by pheromone and induced to remain within two body lengths by tactile stimuli and web production (Penman and Cone 1974). Males, once attracted, will remain on guard and fight other males if they approach too close (Potter et al. 1976). Each adult female lays up to 120 spherical (0.14 mm), translucent eggs over a period of three weeks (Hussey et al. 1969). The small, whitish larva has three pairs of legs, whereas the protonymph and older stages have four pairs. Between the nymphal stages are both feeding and resting phases.

Temperature plays an important role in the development of spider mites, the life cycle from egg to egg at 32°, 21°, 18°, and 15.5°C takes 3.5, 14.5, 21, and 30 days, respectively (Scopes 1985). Relative humidity (r.h.), according to Sabelis (1981), is much less important, particularly under greenhouse conditions. The effect of r.h. between 45–90% at 23°C has no significant effect on fecundity, development, or mortality, possibly due to webbing and defecation or more importantly leaf evapo-transpiration acting as a buffer to changes in humidity.

Mites feed on parenchyma cells, puncturing the cell and emptying the contents, causing a fine speckling visible on the upper leaf surface (Jepson et al. 1975). Leaves filled with collapsed parenchyma cells appear devoid of chloroplasts. If left unchecked, this may cause leaf shriveling and ultimately death or at least severe disfigurement of the plant.

The protonymphs and deutonymphs are sensitive to shortening daylength, low temperatures, and unfavorable food supply which induce mated females to enter diapause (Parr and Hussey 1966). This occurs during late September to October when females assume a deep orange-red color, cease feeding, and migrate from the crop to seek winter hibernation sites, usually within the greenhouse structure. Diapause is normally terminated the following spring when favorable conditions such as higher temperature and longer daylength return.

Following widespread development of resistance to a range of chemical pesticides, the predatory mite *Phytoseiulus persimilis* is now used to control *Tetranychus urticae* on a large area of heated vegetable crops (McMurtry 1982). *Phytoseiulus persimilis* is a little larger than its prey, with a pear-shaped bright orange body. Females lay 50 or more eggs over a period of about three weeks, singly on the underside of leaves among colonies of prey. After three days at 22°C, these eggs hatch and develop into adults by the seventh day, about twice as fast as *T. urticae* at the same temperature. All stages of the prey are consumed. Once established, the predator can overcome the pest rapidly; however, under dry conditions egg laying is severely reduced, resulting in a static or even declining population of the predator. *Phytoseiulus persimilis* is also not effective at temperatures above 30°C; it prefers to stay lower in the crop while the spider mite migrates to the top of the plants. Heat-tolerant strains have been selected in the USSR by Voroshilov (1979). In Australia, a naturalized strain of *P. persimilis* has effectively controlled spider mites on roses at between 35° and 40°C, although temperatures higher than 40°C should be avoided (Gough 1988).

There are two main methods of biological control of *T. urticae* on cucumber (Anon. 1976). With the "classical" or "pest-in-first" method, 50 spider mites are introduced onto every fifth plant followed by five predators onto the same plants 1–2 weeks later (Markkula and Tiittanen 1976). This system is timed so that the predators are established before previously-diapausing mites return to the crop. This method, although successful for first-time users, is not favored by the majority of growers in subsequent years. The second method involves waiting until the first signs of spider mite damage occur and then, depending on the evenness of the infestation, either introducing predator at one per plant or larger numbers at trouble spots. Introductions based on one predator per plant are usually done pro rata to every 3rd, 5th, or 10th plant depending on the extent of infestation (Ledieu et al. 1989). During the early stages of crop growth, before the plants are touching each other, introductions of *P. per-simils* are best made at regular intervals. Once the predator has become established, any severe spider mite problems can be dealt with on an area basis either by releasing large numbers of extra predators or by chemical spot treatments with fenbutatin oxide or tetradifon, both of which are safe to the predator (Helyer et al. 1984).

Spider mite damage is often related to a damage index (Hussey and Parr 1963; revised Scopes 1985); the areas of leaf showing feeding marks are totaled and the resulting value is used to determine the required control strategy. The damage threshold for cucumber is about 30% of the total leaf area; leaf damage exceeding this will result in yield loss (Hussey and Parr 1963). On tomato, 10% of leaf area damaged by mites caused a 9% loss of yield. A threshold below which no loss occurred was not apparent (Stacey et al. 1985). The damage symptoms of leaf speckling are much less noticeable on tomato, so larger numbers of spider mite often tend to build up before control measures are taken. Detecting the outbreak and introducing sufficient predators as early as possible are the best methods for successful biological control. The introduction of five predators to each fifth plant throughout the crop is usual (Ledieu et al. 1989), but heavily infested patches may be treated with 100–200 predators per plant. As with cucumber, spot treatments with chemicals may be necessary.

On tomato a strain of "hypertoxic" spider mite has been observed which produces severe foliar damage at very low pest density (Foster and Barker 1978; Ledieu and Helyer 1981). This may occur through the injection into the plant during feeding of a non-systemic toxin (Ledieu and Helyer 1981). Although some species of spider mite are known to cause plant diseases by transmitting viruses (Robertson and Carroll 1988), viruses have not been associated with "hypertoxic" spider mite damage (Ledieu and Helyer 1981). This mite was thought to be *Tetranychus cinnabarinus* (Boisduval), but work in the Netherlands (Dupont 1979) suggests that *T. cinnabarinus* and *T. urticae* may not be taxonomically distinct species. The two strains are only clearly distinguishable by the damage they cause. Control of the "hypertoxic" strain of spider mite by *P. persimilis* is not satisfactory because severe crop damage can occur before spider mite numbers decrease. New botanic insecticides show promise for control of "hypertoxic" spider mite (see Section III A).

Spider mites occurring on sweet peppers have been successfully controlled by predators, but high numbers of predators (5–10 per plant) were needed. There may be at least two reasons for this. Firstly, the host plant may influence the quality of the mites, as *T. urticae* reared on sweet peppers do not seem to be so readily consumed by *P. persimilis*. Secondly, most sweet pepper crops are grown in such a way that there is no leaf contact between plants until late in their development. This drastically reduces the mobility of the predator. One possible solution is to place wires along the line of the plants to enable the predators, and spider mites, to wander through the crop. Nevertheless, recent trials in the U.K. have shown successful control with between 2–5 predators per plant (Buxton et al. 1990).

While *P. persimilis* remains the principal biological control agent for *T. urticae*, the predatory midge *Theriodiplosis persicae* Keiffer may have a role to play (Wardlow and Tobin 1990). It occurs naturally throughout most of Europe and is similar in size and appearance to the aphidophagous midge *Aphidoletes aphidimyza*. It has an orange-colored larva which kills spider mites by sucking out their body fluids. The late instar larva spins a silken cocoon in which it pupates on the underside of the leaf, often close to a leaf vein. *Theriodiplosis persicae* was mentioned as a potentially useful predator by Speyer (Speyer 1926; Hussey 1985a), although relatively little has been done so far to develop its potential. Where biological control and reduced pesticide usage is practiced, the midge has survived and offers a useful addition to the growers armory. It has been studied by Vacante (1981) who found it to be very efficient in various crops. Due to the presence of the hyperparasite *Aphanogmus parvulus* Roberti, its use is restricted to the winter months, the midge being almost decimated from May onwards (Vacante 1984).

The fungus *Beauveria bassiana* (Balsamo) Vuillemin is sometimes found infecting *T. urticae* in the greenhouse (Rombach and Gillespie 1988). It is claimed that the application of conidial dusts can suppress populations of this mite (Lipa 1971), but there are few studies to date. Various other species of fungi attack *T. urticae*, including *Triplosporium tetranychi* (= *Neozygites tetranychi* [Weiser]) (Lipa 1971), *Entomophthora fresenii* (= *Neozygites fresenii* [Nowakowski]) (Carner and Canerday 1968), *Entomophthora thaxteriana* (Petch) (= *Conidiobolus obscurus* Hall and Dunn) (Lipa 1985), *Neozygites adjarica* (Tsintsadze and Vartapetov) (Keller and West 1983; Lipa 1985), and *Neozygites floridana* (Weiser and Muma). Of these, *N. floridana* is unlikely to be useful in the greenhouse because of its high humidity requirements; Smitley et al. (1986) succeeded in inducing an epizootic in the greenhouse by 14 h day^{-1} periods of 100% r.h., but not by 30 min of simulated rain per day.

The fungus *Hirsutella thompsonii* Fisher will also infect *T. urticae*; this fungus was formerly available commercially as "Mycar" (Abbott Laboratories, USA) for control of citrus rust mite, *Phyllocoptruta oleivora* (Ashmead) (Bartlett and Jaronski 1988). Trials suggest that it is of limited value in the greenhouse because infection rates are poor, even when the humidity is experimentally raised (Gardner et al. 1982).

The whitefly-active strain (19-79) of the fungus *Verticillium lecanii* will also infect mites (Gillespie et al. 1982a). Gillespie (1988a) found that this strain and strain 277-85 gave 97–100% reduction of *T. urticae* 13 days after application to cucumber in a greenhouse. There has, however, been little research or practical application of *V. lecanii* for mite control, probably because control with *P. persimilis* is generally satisfactory.

The selective use of pesticides with P. *persimilis* and other natural enemies has been studied by various members of the International Organization for Biological Control (IOBC) working group "Pesticides and Beneficial Organisms." The initial results of the various screening programs (Franz et al. 1980; Hassan et al. 1983, 1987, 1988; Ledieu and Helyer 1983a) indicate the toxicity of pesticides and their persistence. Osborne and Petitt (1985) evaluated insecticidal soap and found that it could be integrated with P. *persimilis* if applied three days after the release of predators, but was harmful as a contact spray against the adults. This combination works because the soap is a short persistence pesticide with little detrimental effect on the predator eggs.

Other acaricides applied just to the top of plants have little detrimental effect on P. *persimilis*. Similarly, spot chemical treatments of whole plants may kill all the predators, but they will soon reinvade from the surrounding area. Heptenophos, dimethoate, and pirimicarb can be integrated with *Phytoseiulus persimilis* (and *Encarsia formosa*) and are used against aphids on tomato, cucumber, and sweet pepper (Richardson and Helyer 1990).

Another method of controlling spider mites, or at least reducing their numbers, is the cultivation of mite-resistant cultivars. Tomato cultivars with a high density of glandular trichomes (leaf hairs) have been noticed to trap dispersing mites in the sticky exudate (van Haren et al. 1987). Rasmy (1985) noted that leaves of *Lycopersicon hirsutum* stripped of glandular hairs were still toxic to developing mites, the leaf exudate having an acaricidal effect.

The use of *Phytoseiulus persimilis* on ornamental crops was successful until the widespread application of pesticides to control American serpentine leafminer, *Liriomyza trifolii* (Burgess), and western flower thrips, *Frankliniella occidentalis*, ruled out its use (van de Vrie 1989; Wardlow 1989). The use of predators on ornamentals is likely to increase rapidly once biological control of other pests is reliable (Wardlow 1990).

B. Whiteflies

Glasshouse whitefly, *Trialeurodes vaporariorum*, has become one of the most difficult pests to control on protected crops. Since first reported in the U.K. in 1856 (Westwood 1856) it has assumed major pest status world-wide (Russell 1977). Its success is attributable to its life cycle, which was described by Hargreaves (1915); the developing eggs, larvae and pupae are well protected under a waxy cuticle through which most pesticides cannot penetrate, leaving the adult as the main target (O'Reilly 1974). Also the adults are highly mobile, readily flying from plant to plant and greenhouse to greenhouse.

Adult females lay eggs primarily on the undersides of apical leaves, often in circles or small groups. The egg pedicel is inserted directly into the host plant tissue and not stomatally as in some other whitefly species (Paulson and Beardsley 1985). Eggs hatch after about a week at 22°C as first instar larvae or "crawlers" which are initially mobile, moving for only a few hours before settling on the leaf. Once settled they insert their mouthparts into the leaf tissue and remain static as "scales" throughout the remainder of their larval and pupal development. The temperature requirements of *T. vaporariorum* were studied by Madueke and Coaker (1984) and reviewed earlier by van Lenteren and Hulspas-Jordan (1982). Information on development times for nine different host plant species over the temperature range 7–40°C was compiled by van Lenteren and Noldus (1990). Much of this information was also used in the development of a predictive computer model for *T. vaporariorum* on tomato (Hulspas-Jordan and van Lenteren 1989). The lowest developmental threshold temperature was calculated by Osborne (1982) to be 8.3°C. Fluctuating temperatures were shown to have an influence on the developmental rate of both *T. vaporariorum* and its parasite *Encarsia formosa* (Stenseth 1971, 1976). Both insects developed faster when reared under fluctuating temperatures than when kept under constant conditions. Adults emerge from the waxy pupae 28–30 days after the eggs are laid and commence oviposition after 1–3 days.

The cotton, sweet potato, or tobacco whitefly (Gerling 1990), *Bemisia tabaci* (Genn.), is a polyphagous species widespread throughout the tropics (Anon. 1986). Since 1987 it has been reported in several northern European countries (Enkegaard 1990; Fransen 1990). Differentiation between *Trialeurodes vaporariorum* and *B. tabaci* is aided by pupal morphology; *B. tabaci* has an elliptical pupa with the dorsal surface convex (Lopes-Avila 1986a), whereas *T. vaporariorum* has raised sides to the pupal case. Unfortunately, both species are polyphagous and their pupal cases may vary in appearance depending on the form of the host plant cuticle on which they develop (Mound 1963).

Duration of the life cycle of *B. tabaci* is closely correlated with climatic conditions and is also affected by the host plant. Azab, Megahed, and El-Mirsawi (1971) stated that, in Egypt, the life cycle varied from 14–75 days under field conditions. Coudriet et al. (1985) found that the time required to complete development from egg to adult at 26.7° ± 1°C was influenced by the host plant on which it was confined. Development was completed in 30% less time on lettuce, cucumber, eggplant, and squash than on broccoli or carrot. Development times from egg to adult on a range of greenhouse pot plants were recorded by Fransen (1990). Adult female longevity on poinsettia was greatest at 16°C (30 days) and only half as long at 28°C (Enkegaard 1990), which is comparable with the findings of Sharaf and Batta (1985) on tomato, where longevity of 27.3 and 14.3 days at 14° and 25°C, respectively, were reported.

Mean fecundity at 26.7° and 32.2°C was reported to be 81 and 72, respectively (Butler et al. 1983). In contrast, Azab et al. (1972) stated that the number of eggs per female varied from 48 to 394 with a mean of 161 (temperature not given). The preoviposition period of B. tabaci ranged from about two to four days at 16°–28°C (Enkegaard 1990), with the maximum number of eggs being produced some five to 14 days after adult emergence (at 16° and 28°C, respectively).

Bemisia tabaci is a very important vector of plant viruses, and is known to transmit at least 19 viruses and numerous viruslike agents, whereas T. vaporariorum is a vector of only two viruses (Brunt 1986).

Biological control of B. tabaci on greenhouse poinsettia has been successfully achieved with E. formosa in Italy (Benuzzi et al. 1990). Weekly introductions of one parasite per 5–10 plants has also provided satisfactory control in Germany (Albert 1990). Encarsia formosa is unlikely to be used as a self-perpetuating control system when a crop is infested exclusively with B. tabaci (Boisclair et al. 1990) because the parasites become less effective when reared on this host. The prospects for biological control of B. tabaci were discussed by Osborne et al. (1990); the efficacy of a number of parasites, predators, and pathogens is currently being evaluated. A review of parasites and predators attacking B. tabaci and many alternative host species was compiled by Lopez-Avila (1986b); 26 parasites and 30 predators are listed as enemies of B. tabaci, but detailed biological information is available only for the parasites Encarsia lutea (Masi) and Eretmocerus mundus Mercet.

The most serious damage by whitefly is caused by the excretion by all stages of honeydew, which rains down on to fruit and the upper surface of leaves. Under moist conditions sooty molds (Cladosporium spp.) can develop on the honeydew, reducing photosynthesis and hindering respiration (Hussey et al. 1958). Heavy infestations may necessitate the washing of fruit before sale. Various diseases can be transmitted by T. vaporariorum, including geranium leaf spot (O'Reilly 1974) and viruses of cucumber, lettuce, and ornamentals (Duffas 1965).

Chemical control of whitefly is difficult as most pesticides are only effective against the adult and first instar crawlers. Pesticide resistance, particularly to organophosphate and organochlorine compounds, became widespread in the early 1970s (Wardlow et al. 1972; French et al. 1973). The advent of the synthetic pyrethroids was hailed as a major advance in whitefly control (Anon. 1978), but since their introduction during the late 1970s (Herve et al. 1977), very serious resistance problems have arisen. By the mid-1980s, after about eight years of widespread use, synthetic pyrethroids could no longer be relied upon to control whitefly. Wardlow (1985a) tested 17 populations of glasshouse whitefly from commercial greenhouses in the U.K. and found susceptibility to permethrin decreased by six to 4000-fold (only two populations were less than 1000-fold resistant). Since the first failures of pyrethroids, many other control measures have been applied, including high volume sprays of household detergents 3–4 times each week. This has helped to control whitefly but can also lead to severe phytotoxic damage (I. J. Wyatt, unpublished data).

Trapping of adult whitefly using yellow sticky traps for both monitoring and control has been widely evaluated (Yano and Koshihara 1984; Gillespie and Quiring 1987). The efficacy of various colors, reflectance, crop position, and adhesive were studied in detail by Webb et al. (1985). They concluded that bright yellow traps, hanging vertically with the top level with the crop canopy and coated with a polybutene adhesive, offered the best trapping mechanism. Traps of this sort are used in conjunction with the parasite E. formosa (Webb and Smith 1980; van de Veire and Vacante 1984) to provide early season integrated control. Although compatible with E. formosa (as determined by a high percentage parasitism), parasites were trapped when whitefly numbers were low.

Biological control of whitefly can be achieved by the parasite E. formosa, which was first successfully used in England at the Cheshunt Experimental Station during 1926 (Speyer 1927a). Within two years of its discovery in the U.K. more than 250,000 parasites had been sent to several counties in England and Wales and a further 20,000 to Canada (Speyer 1930). The life cycle and habits were investigated by Speyer (1927b). Each female parasite can lay 60–100 eggs over a period of 10–14 days, inserting these, singly, into 3rd or 4th instar whitefly scales (Nell et al. 1976; Parr et al. 1976; Nechols and Tauber 1977). Encarsia formosa kills whitefly scales peredominantly by parasitism but also by direct host feeding. Host feeding has been observed on all whitefly developmental stages. Scales used for feeding are not parasitized and feeding does not occur on previously parasitized scales (Nell et al. 1976; van Alphen et al. 1976). At 20°–22°C parasitized scales turn black after about 10–14 days when the parasite pupates. When fully developed (at 23°C, 10–11 days after blackening), the adult parasite cuts a hole in the top of the "black scale" to emerge (van Keymeulen and Degheele 1977). Parasite activity and rate of development are temperature-dependent, taking about 25–35 days from egg to adult at 21°C (Stenseth 1975), but almost twice as long at 12°C (van Lenteren and Hulspas-Jordan 1982). Fresh black scales can be cold-stored for up to three weeks at 12°–13°C (Scopes and Biggerstaff 1971). Encarsia formosa can be introduced to a crop as black scales either on fresh leaf material (usually tobacco), which the grower cuts into strips of about 150 black scales before distributing them around the crop, or on pieces of card which contain either a section of leaf or black scales that have been transferred to card (van Lenteren and Woets 1988). The cards are then hung on leaves around the crop, the advantage being that they remain visible on the plants, and further introductions can be systematically placed throughout the greenhouse (Helyer and Payne 1986). The parasite is used on approximately 45% of the area of heated tomatoes in the U.K. (Sly 1986).

Where whitefly are a recurring problem with ready migration from one greenhouse to another, a continuous "dribble" (multiple introduction) technique is often used with the parasite (Hussey 1985a). This involves introducing about 5000 E. formosa black scales per ha each week throughout most of the life of the crop, thus ensuring that some adult parasites are always present. Alternatively, if only a few whitefly are present early in the growing season, 6–8 weekly introductions of 15,000 E. formosa per ha should allow establishment of the parasite for the rest of the season (Ledieu et al. 1989). Where whitefly are only an occasional problem on tomato, three or four fortnightly introductions of about one parasite per plant per ha should be sufficient; however, during mid-summer when adult whitefly can enter the greenhouse from outside, further parasite introductions may be necessary. Alternatively, a spray of a short-persistence insecticide such as a pyrethrum/resmethrin mixture to the tops of the plants can help to restore the pest/parasite equilibrium.

The trend to earlier sowing dates for tomato and the transfer of whitefly in higher numbers from one season's crop to the next, combined with lower growing temperatures, has caused some failures in the establishment of E. formosa during the winter months. This may be due to the effects of low temperatures on the egg maturation of E. formosa (Kajita and van Lenteren 1982). On the other hand, the parasite was found to migrate well at low temperatures (Hulspas-Jordan et al. 1987) and produce 60–86% parasitization. Other species of Encarsia and one Eretmocerus sp. have been studied to estimate their potential in controlling whitefly on greenhouse tomato under a low-temperature regime (Vet and van Lenteren 1981).

In the Channel Islands, where the population density of whitefly can be considerable even early in the season, very high introduction rates of E. formosa (>50,000 per ha) have been evaluated (Hocart 1981).

The use of E. formosa on ornamentals has been restricted partly due to the almost zero tolerance of pests and beneficials allowed on such crops and also due to the nature of the plant itself, i.e., ornamentals tend to be densely spaced crops with hairy leaves and are grown at lower temperatures than most edible crops (Wardlow 1989). Nevertheless, the parasite has been used on gerbera (van de Vrie 1989) and poinsettia (Benuzzi et al. 1990) with some success.

A range of pathogenic fungi have been found attacking whitefly, and some have been commercialized. Initial greenhouse trials with the fungus Beauveria bassiana against T. vaporariorum were encouraging (Treifi 1984). Similarly, an aphid-active strain of Verticillium lecanii was found to give control of T. vaporariorum on cucumber, but did not spread from scale to scale, and so frequent applications of the fungus were needed (Kanagaratnam et al. 1982).

Hall (1982) isolated a strain (19-79) of V. lecanii from whitefly and demonstrated that this strain, when formulated with a carbohydrate nutrient, would spread between scales to give sustained control of *T. vaporariorum* on cucumber. Spread can be achieved by water splash and by spores being carried on the bodies of insects. The fungus was marketed for a few years as "Mycotal" (Microbial Resources Ltd, U.K.) and was generally very effective, with a few exceptions (e.g., Albert 1987). The whitefly-active strain is not very effective against aphids as it does not spread readily in the aphid population (Rombach and Gillespie 1988). Humidities are generally lower in tomato crops, and Gillespie (1988a) reported whitefly mortalities due to strain 19-79 to be consistently below 50% in this crop. *Verticillium lecanii* can, however, be integrated with the use of *E. formosa* for whitefly control (Gillespie and Moorhouse 1989). It would be useful as a supplement to *E. formosa* on tomato, providing the growing conditions favor high humidity (e.g., damp soil or low greenhouse structure or use in early summer before night-time venting [Hall 1985a]).

The fungus *Aschersonia aleyrodis* Webber is reported to have whitefly as its only host. Strains originating in Vietnam, Trinidad, China, and Bulgaria are claimed to provide good control of *T. vaporariorum* in the greenhouse (references cited in Rombach and Gillespie 1988). Infection of whitefly scales by *A. aleyrodis* is easily detected as bright orange spore masses on the cadavers. *Aschersonia placenta* Berkeley and Broome also attacks whitefly, but is not brightly colored (Ramakers et al. 1982). *Aschersonia aleyrodis* does not infect eggs (Fransen et al. 1987) or adults (Ramakers 1983a). An advantage of *A. aleyrodis* is rapid germination; Ramakers and Samson (1984) found that microclimatic conditions during the night after spraying allowed the conidia to germinate. It could, therefore, be useful on tomato, where the period of high humidity is limited, although the night-time temperatures under some tomato growing regimes would not favor the fungus (Hall 1985a). It appears to be equally effective in rainy periods and during sunny weather (Ramakers 1983a; Ramakers and Samson 1984). On the other hand, its usefulness is limited because it does not grow saprophytically on the leaf surface or spread in the greenhouse. Since most whitefly instars are sedentary, repeated and efficient applications would be necessary for control.

Ramakers and Samson (1984) found that at high dose rates of *A. aleyrodis* spores, 75% of larvae became infected on mature cucumber plants in the greenhouse, but mortality was more variable and sometimes lower on younger plants. *Aschersonia aleyrodis* will not grow at 37°C (and so, in this respect, is safe to mammals) and it does not infect *E. formosa*; it might therefore prove a useful adjunct to *E. formosa* for whitefly control. Ramakers et al. (1982) recorded 85% whitefly mortality on cucumber due to the combined use of *A. aleyrodis* and *E. formosa*, compared with 49% for *E. formosa* alone.

Small-scale culture of *A. aleyrodis* is possible (and is carried out in the

Netherlands, for use in conjunction with *E. formosa*) on solid substrates such as rice or bran (Rombach and Samson 1982), but more work will be needed before conidia of a standard high quality can be mass-produced (Ramakers 1983a; Bartlett and Jaronski 1988; Rombach and Gillespie 1988). Research aimed at improved commercial production of both *A. aleyrodis* and *V. lecanii* is also being carried out in the USSR (Filippov 1989), and *Aschersonia placenta* f. *vietnamica* B. et Br. and *Aschersonia flava* Petch are used to control *T. vaporariorum* on 10 ha of cucumber in the Ukraine (Lipa 1985).

Buprofezin, a potent whitefly insecticide, may be useful in correcting a pest/parasite imbalance and is also relatively safe to the parasite (Garrido et al. 1984; Martin and Workman 1986). Buprofezin is an insect growth regulator with a strong volatile action and relatively long persistence and as such could be used as a spot treatment in areas of high whitefly density (N. L. Helyer unpublished).

Integrated control programs using the broad-spectrum pesticide oxamyl have been developed (Helyer et al. 1983b; Martin and Workman 1984) for early season control of whitefly, followed by introductions of *E. formosa* when conditions better favor the parasite. By predicting the life cycle of whitefly at various temperatures (van Lenteren and Hulspas-Jordan 1982) a systemic pesticide (such as oxamyl) can be applied just before the first pupae are formed (usually early winter). This kills all the pre-pupae and younger stages. The parasite is introduced when the next whitefly generation develops to the third instar. This usually occurs about 6–8 weeks later when daylength and solar gain are more favorable for parasite activity.

Plant breeding programs to produce plants that confer some resistance to whitefly have been studied for sweet pepper (Laska et al. 1986) and tomato (van Gelder and de Ponti 1987). A program to produce cucumber plants with less hairy leaves has improved the host searching activity of *E. formosa* (Li et al. 1987).

C. Thrips

Many species of thrips can be found in European greenhouses, (Mantel and van de Vrie 1988a), but two are of particular concern: *Thrips tabaci* (onion thrips) and more recently, *Frankliniella occidentalis* (Pergande) (western flower thrips). Thrips in both adult and larval stages feed by puncturing plant epidermal cells and ingesting the escaping sap. This causes a flecking on lower leaf surfaces, flowers, and fruits, which may turn to silvering with more intense damage. Distortion and leaf mottling can also occur. Small black drops of frass are often visible within the damaged area. Immature thrips are paler than adults, are wingless, and after two nymphal stages pass through a prepupal and pupal stage, often in the soil or on the leaves, before emerging as adults. Females make a slit in the leaf tissues in which to lay their eggs.

Thrips tabaci is an indigenous species about 1 mm long which is variable in color, ranging from pale yellow to dark brown. Reproduction is by parthenogenetic females, and males are rare. It is very polyphagous and is a pest on cucumber, carnation, chrysanthemum, and other ornamentals.

Frankliniella occidentalis is a yellow or yellowish brown species, also widely polyphagous, and reproduction is partly sexual, partly parthenogenetic. Robb (1989) made detailed measurements of life history parameters at a range of temperatures, including development rate and intrinsic rate of increase. The arrival of *F. occidentalis* was the single most significant development in greenhouse entomology in Europe in the 1980s and has proved a serious threat to both established and developing integrated control programs. Its success has been partly attributed to its polyphagous nature, having been found living on 219 species from 59 genera of host plants. It is also at least partially predatory, capable of feeding on mite eggs (Pickett et al. 1988). Resistance to a wide range of pesticides together with the increase in trade in plant material around the world have contributed to its pest status.

The strain of *F. occidentalis* infesting greenhouses originated in the western USA and spread during the 1980s through all of the USA, Canada, and more recently throughout the world. It arrived in the Netherlands in 1983 (Mantel and van de Vrie 1988b), where it was found infesting saintpaulia. It later arrived in Denmark in 1985, probably transmitted via the Netherlands or West Germany, and by 1987 the majority of greenhouses in southern Germany were infested (Brodsgaard 1987). It appeared in France in 1986 on chrysanthemum and saintpaulia (Fougeroux 1987) and arrived in the U.K. in the same year. Since then it has spread rapidly, becoming a major problem in ornamentals in 1987. In the U.K., attempts to eradicate the thrips were unsuccessful, and in 1988 and 1989 it spread to cucumber and sweet pepper. It is now regarded as an established species in the U.K. and ceased to be a "notifiable" pest as of June 1989. In Finland, *F. occidentalis* appeared in 1987, introduced on saintpaulia and rose, and, as in the U.K., the majority of infestations appeared initially on ornamentals, with some on cucumber (Tiittanen and Markkula 1989). Scarring and curling of immature cucumber fruit is a major problem with *F. occidentalis* in cucumber.

Frankliniella occidentalis has been transmitting Tomato Spotted Wilt Virus (TSWV) in the U.K. since November 1988. The virus has been present on U.K. holdings since the 1940s, but good control of *T. tabaci* (the second thrips species recorded as transmitting the virus in Europe) has prevented the disease from becoming a serious problem. The virus is widespread throughout temperate regions, and affects almost 200 plant species worldwide, including both cultivated crops and weeds. It has not been found on commercial crops in Denmark (Brodsgaard 1987). Symptoms induced by TSWV vary greatly, and depend on a combination of factors including duration of infection, tolerance, and nutritional status of the host. In many cases, TSWV causes systemic necrosis, but in some hosts, symptoms may be much less severe. TSWV can only be acquired by larvae of thrips, and acquisition takes a minimum feeding time of 15 min (Sakimura 1962). A latent period of several days follows before transmission by adults can occur (Bald and Samuel 1931). Control of TSWV depends on a combination of good plant hygiene with control of the vector species.

Palm thrips, *Thrips palmi* Karny, is similar in appearance and life cycle to *T. tabaci*. It has a wide host range and so may pose a threat in future to the main greenhouse edibles and ornamentals. At present it is not established in northern Europe, but has been found on imported material. It is of mainly Asian distribution with a reported presence in parts of Africa and South America. It has caused serious crop losses to cucurbits in Japan. It is reported to be tolerant to organophosphorus insecticides but is believed not to be a virus vector.

Yellow sticky traps can alert a grower to the presence of thrips some time before they become obvious on the leaves (Jacobson 1989b). Moffit (1964) and Yudin et al. (1987) found that white sticky traps were the most effective in field conditions, but Brodsgaard (1989) found that a particular blue shade was the most effective for *F. occidentalis*; however, the trap catch was a poor indicator of the number of thrips on pot plants. Teulon and Ramakers (1990) reviewed the potential use of chemical attractants for trapping thrips in protected crops, either for monitoring or for control purposes. Brodsgaard (1990) showed that blue sticky traps with anisaldehyde caught more *F. occidentalis* than unscented traps, although the improvement in catch was relatively small.

Thrips reproductive rates may be reduced to a degree by damp growing conditions; high humidities and damp soil floors deter them, the latter by encouraging natural fungal disease control of the pupal stage or even by drowning. The drier conditions which prevail with plastic floor coverings tend to encourage thrips. Insecticidal soaps will depress thrips populations but are unlikely to eliminate them.

Problems with the control of *T. tabaci* arose after the ending of routine broad-spectrum pesticide treatments against two-spotted spider mite and whiteflies and their replacement with biological control agents. At this time, chemical control of thrips (using diazinon, dichlorvos, or permethrin) was damaging to the interaction of *Phytoseiulus persimilis* with *Tetranychus urticae*. This stimulated research into biological control of thrips using both arthropod predators and fungal pathogens. Further impetus to research resulted from the arrival of *F. occidentalis* in Europe.

Woets (1973) observed predatory mites attacking *T. tabaci* in greenhouses, and subsequently Ramakers (1978) observed *Amblyseius* spp.,

Orius minutus L. (Fam. Anthocoridae), and *Entomophthora* spp. as antagonists under glass. Initial trials using *Amblyseius* spp. were conducted by Ramakers (1980), and rearing methods for two species of *Amblyseius* were developed (Ramakers and van Lieburg 1982; Ramakers 1983b). Rearing companies in Europe now produce *Amblyseius cucumeris* Oudemans and *Amblyseius barkeri* (Hughes) (= *Amblyseius mckenziei* Schuster & Pritchard and *Neoseiulus barkeri* Hughes).

Adult *A. cucumeris* mites are light brown, pear-shaped, and difficult to see when stationary. Females lay white eggs which are often attached to plant hairs. The mite develops rapidly, taking 13 days from egg to adult at 20°C, and it is capable of increasing more rapidly than its prey. The mite feeds on small thrips larvae, as adult thrips are relatively large and can defend themselves from attack. *Amblyseius cucumeris* has the advantage of feeding on other food such as pollen (Dosse 1961) and spider mites (Dosse 1955) if thrips are difficult to find. In the laboratory, the mite will survive, develop, and lay eggs on a diet of pepper pollen (van Rijn and Sabelis 1990). This feature helps to maintain *A. cucumeris* at low prey densities, and they will persist on pepper plants for some while before arrival of the pest (Ramakers 1990). The biology and life cycle parameters of *A. barkeri*, including development rate, sex ratio, longevity, oviposition rate, increase rate, and feeding rates, were studied by Bonde (1989). The increase rate of *A. barkeri* feeding on *T. tabaci* was higher than that of *T. tabaci* itself. The mean feeding rate over all life stages was 3.3 thrips nymphs per day in laboratory conditions where searching for prey was minimal. Two-spotted mites, broad mite (*Polyphagotarsonemus latus* [Banks]), and pollen were all consumed in the absence of thrips prey (Bonde 1989). Because of this low kill rate compared to other arthropod predators, high predator:prey ratios are required for the mite to make an impact on the thrips population. The increase rate measured in the laboratory is probably not maintained in the greenhouse, and *Amblyseius* spp. are not necessarily in dynamic equilibrium with their prey. However, low-cost mass-rearing on storage mites, *Acarus farris* (Oudemans) (= *Acarus siro* L.) feeding on bran (Ramakers and van Lieburg 1982) allows the liberation of large numbers at affordable prices. Ramakers and van Lieburg describe a mass-production system, producing 100,000 predators per litre of bran on *A. farris* as a factitious prey. Mass-production using thrips as prey is uneconomic. The bran is heated to kill existing insects and mites, and a high humidity is maintained during rearing. A modified mass-production system is described by Hansen and Geyti (1987).

In 1985, de Klerk and Ramakers (1986) used *A. cucumeris* for thrips control in sweet peppers. The predators were able to keep thrips numbers at low levels and were also able to feed on other food, including pollen. An integrated control program was proposed which included pirimicarb for aphid control. Ravensberg and Altena (1987) reported the successful introduction of *A. cucumeris* into sweet pepper crops on commercial nurseries in the Netherlands against *T. tabaci* from 1985; satisfactory control was reported on up to 80% of the holdings by 1986. Ramakers (1988) assessed both *A. barkeri* and *A. cucumeris* for control of *T. tabaci* in sweet pepper. *Amblyseius cucumeris* established more readily and outcompeted *A. barkeri* when the two were introduced together. It is the preferred species for this crop. On sweet pepper, pollen feeding by *Amblyseius* spp. makes preventative introductions possible. Good thrips control can be achieved if 70–80% of the pepper leaves have at least one mite or mite egg before the thrips arrive. This program requires a clean crop at the start, and several treatments of dichlorvos are advisable before the first predator introduction (Altena and Ravensberg 1990).

On cucumber, Ravensburg and Altena (1987) concluded that *A. barkeri* applied against *T. tabaci* was unsuccessful on commercial holdings, despite success in small-scale trials. The mite was unreliable, and too frequent correction of the balance of predator and prey with chemicals was required. However, Hansen (1988) in Denmark confirmed that *T. tabaci* could be controlled on cucumber with *A. barkeri*, although large numbers of mites were required (300–600 per m^2). In Finland, Lindqvist and Tiittanen (1989), working with the same mite on cucumber, suggested a regime of introductions made at weekly intervals after 10% of the plants were infested, totaling 60–80 mites per m^2. Higher initial infestations required 100–180 mites per m^2 in total. The repeated inundative releases at high rates that appear necessary to establish *Amblyseius* spp. on cucumber have been attributed to low kill rates by the mite on this crop. It is known that *A. cucumeris* walks further and faster on leaves of sweet pepper than cucumber due to the hairy surface of the latter (Petersen 1989). The mites can survive on sweet pepper pollen if prey are difficult to find. In the U.K. on cucumber, Bennison et al. (1990) recommended that *Amblyseius* spp. introductions should be delayed until late March since the short days induce diapause (Jacobson 1989a; Gilkeson et al. 1990), although it is suggested that growers could raise temperatures in March to overcome this difficulty. *Amblyseius* spp. did not always establish satisfactorily on cucumber, and may need to be supplemented by an application of the fungus *Verticillium lecanii*.

Improved methods of release are needed for *Amblyseius* spp. At present, the mite is released by tipping the bran in which it is supplied onto the leaves. Good dispersal around the crop is important, but on cucumber leaves bran slides off easily and mites can be lost. "Open rearing units" consisting of small quantities of bran containing storage mites with *A. cucumeris* can be placed on the rockwool blocks beneath the plants. The bran can be applied in a small heap or in perforated plastic bags. These will then produce *A. cucumeris* over several weeks (Ramakers 1990). A similar system has been developed by a U.K. company with the bran contained in a packet suspended from a cucumber leaf petiole (Anon. 1990).

More recent research on commercial holdings suggests that *Amblyseius* spp. can be effective in controlling *T. tabaci* on cucumber, but if introduced before mid-March they may enter a reproductive diapause and fail to give good early-season control (Jacobson 1989a). Development work on *Amblyseius* spp. against *T. tabaci* has also been carried out in the USSR (Beglyarov and Suchalkin 1983) and Norway (Stenseth 1986), but more recently, attention has turned to control of *F. occidentalis* with *Amblyseius* spp. Jakobsen (1987) used *A. barkeri* to control *F. occidentalis* on pot plants. Gillespie (1989) found *A. cucumeris* to be an effective predator against both *T. tabaci* and *F. occidentalis* on cucumber. There is a need for further research on arthropod biological control against thrips on cucumber as well as ornamentals. Ramakers (1978) found *Orius* spp. (Anthocoridae) attacking thrips in greenhouses. *Orius minutus* (L.) is about the size of a thrips larva and can enter small crevices where thrips hide. Anthocorids are known to be voracious predators, but are difficult to mass-rear, with cannibalism being one restriction. They sometimes invade the greenhouse in response to a build-up of thrips or aphids and may provide some control.

Several species of fungi are known to infect *Thrips tabaci*. These include *Zoophthora radicans* (Brefeld) Batko.; *Neozygites parvispora* sp. nov. (Macleod et al. 1976; Keller and West 1983); *Entomophthora thripidum* Samson, Ramakers & Oswald (Samson et al. 1979); *Beauveria bassiana* (Dyadechko 1964; Lipa 1985); and *Verticillium lecanii* (Binns et al. 1982).

Gillespie (1984, 1986) investigated the use of *V. lecanii*, *B. bassiana*, *Metarhizium anisopliae*, and *Paecilomyces fumosoroseus* (Wize) Brown & Smith for control of *T. tabaci*. *Metarhizium anisopliae* and *B. bassiana* were very pathogenic under ideal laboratory conditions. In the USSR, two applications of "Boverin" (*B. bassiana*) to cucumber at 10-day intervals reduced *T. tabaci* populations by 83% (Lipa 1985). A strain of *V. lecanii* isolated from thrips gave control of *T. tabaci* on the underside of cucumber leaves, but not on the upper surface (Gillespie et al. 1982a). However, a whitefly-active strain of *V. lecanii* (strain 19-79) successfully suppressed *T. tabaci* under greenhouse conditions (Gillespie 1986). A single application can give excellent control for more than eight weeks (Gillespie 1989a). This fungus also causes high mortality rates of *F. occidentalis* in the laboratory, but its performance against this pest in the greenhouse remains to be determined (Gillespie 1989a).

Entomophthora thripidum only sporulates from the abdomen, and *T. tabaci* can still move about when sporulation is in progress. The thrips climbs to a high place, and spores are forcibly ejected. Although this species has been observed to cause epizootics in the greenhouse in late summer, attempts to cultivate it on artificial media have so far failed (Samson et al. 1979). *Neozygites parvispora* causes natural infections of thrips in unheated greenhouses (Samson et al. 1979), but it cannot yet be cultured in vitro (Gillespie 1988b), and commercialization is, therefore, not imminent; there are similar mass-production problems with other Entomophthorales (Wilding and Latteur 1987; Bartlett and Jaronski 1988).

Insect-parasitic rhabditid nematodes (*Steinernema* spp., *Heterorhabditis* spp.) have not yet been used against thrips, but should be investigated for control of pupae in soil and compost (Richardson 1987). Wilson and Cooley (1972) reported a tylenchid nematode, *Howardula aptini* (Sharga), parasitizing adult *F. occidentalis*. This endoparasite is believed to sterilize female thrips.

The use of "Thripstick" and similar compounds has proved effective for control of *T. tabaci* in cucumber. Thripstick is composed of a sticky grade of polybutene with water and the pyrethroid insecticide deltamethrin added (Pickford 1983). Thripstick has been in use in the U.K. since 1981 (Helyer et al. 1983a) and was registered for use in the Netherlands on cucumber and sweet pepper in 1986 (Ravensburg and Altena 1987). It is applied by spraying onto plastic surfaces under the cucumber plants. Thrips descending for molting to the pupal stage are trapped and killed. Control lasts about 10 weeks. Thripstick control of *T. tabaci* allowed the application of *E. formosa* for whitefly, which was previously impossible because of the susceptibility of *E. formosa* to chemicals (gamma HCH and diazinon) used for thrips control. Thripstick fitted in well with the practice of covering floors with plastic sheeting, the absence of soil sterilization, and the growing of plants in rockwool slabs. However, some growers and their workforce dislike Thripstick, finding its sticky nature disagreeable to work with. Thripstick appears to have little effect on *F. occidentalis* numbers on cucumber, probably because, unlike *T. tabaci*, many of the larvae pupate on the crop.

Thrips tabaci is not difficult to control with chemicals, but there can be problems if biological control of other pests is being practiced. Diazinon is commonly used for the control of *T. tabaci*. It can be persistent, but if used as a soil treatment against the pupae it will not interfere with predatory mites, although it will harm *Encarsia formosa* and *Aphidoletes aphidimyza* (Steiner and Elliott 1987). Pesticide resistance in *F. occidentalis* is commonplace (Robb and Parrella 1986), and at present this thrips can only be controlled by chemicals that are inimical to biocontrol agents. Control is made more difficult by the feeding habits of the larvae which enter flower buds and leaf growing points and are inaccessible to chemicals (Mantel and van de Vrie 1988b). Foliar sprays recommended in the U.K. against *F. occidentalis* are dichlorvos, endosulfan, and pyrazophos. Deltamethrin can be fogged. Aldicarb and gamma-HCH can be used as soil treatments, but gamma-HCH vapor will remain toxic to beneficials for many months.

In the search for compounds active against *F. occidentalis*, deltamethrin was identified as effective while minimizing phytotoxicity to ornamentals (Bohmer 1986). Heungens et al. (1989) tested 29 pesticides against the thrips on chrysanthemum. Even after six treatments with the best insecticides, mortality reached only 88% to 96%. Recommended chemicals were acephate, carbofuran, chlorpyrifos, dichlorvos, methamidophos, methomyl, and omethoate. Freuler and Benz (1988) reported resistance of *F. occidentalis* to cypermethrin.

Gamma radiation was also found to be effective in killing *F. occidentalis* during postharvest clean-up, having the advantage of being able to penetrate flowers. Irradiated second instar larvae died before reaching the pre-pupal stage (Wit and van de Vrie 1985).

High volume (HV) sprays of dichlorvos can control *F. occidentalis* but can cause phytotoxic damage as scorch or fruit loss, especially on cucumber (Staay and Uffelen 1988). Two sprays may normally be required as the chemical does not kill the eggs laid into plant tissues. Dichlorvos applied as an aerosol is less phytotoxic than the HV spray, and dichlorvos aerosols have been used successfully in the Netherlands. Pyrazophos may be less phytotoxic than dichlorvos (Helyer 1990) but may not be integratable with biological control agents.

Ramakers et al. (1989) described an integrated program for control of pests on cucumber with very high introduction rates of *Amblyseius* spp. for thrips control coupled with *Encarsia formosa* for whitefly. Cyanide was suggested for *Aphis gossypii* Glover which is pirimicarb-resistant, although the use of this compound is unlikely to be acceptable in many European countries. Thrips control was effective in the early part of the crop but easily broke down later. Good hygiene between crops and the application of dichlorvos before the first flowers appear, together with routine sticky trap monitoring and Thripstick applied to the plastic floor covering, are also to be recommended for control of *F. occidentalis*.

In sweet pepper, *A. cucumeris* used to control *T. tabaci* works well, but this predator is sensitive to chemicals used to control minor pests. Ravensberg and Altena (1987) reported that heptenophos to control *A. gossypii* on sweet peppers was not compatible with *A. cucumeris*, nor was pirimicarb used to control other aphid species; however, heptenophos has only a short residual effect. De Klerk and Ramakers (1986) also recognized the damaging side effect of pirimicarb for aphid control on *A. cucumeris*. Buxton et al. (1990), however, reported minimal effect of pirimicarb on *Amblyseius* spp. in trials on sweet pepper. Aphid control in sweet pepper is the main remaining biological control requirement for an integrated control program. A more resistant strain of *A. cucumeris* or another arthropod beneficial is required. It is to be hoped that current trials with *Aphidoletes aphidimyza* and *Aphidius* spp. might prove successful.

The current proposed integrated program for thrips control on sweet pepper in the U.K. is to apply dichlorvos fog at five-day intervals for 6–8 weeks after planting in December. *Amblyseius* spp. should be introduced at the end of February, three weeks after the final dichlorvos application. Several introductions may be necessary. Thripstick can be used to supplement control (Richardson and Helyer 1990).

D. Aphids

Under glass, aphids are generally considered to be a minor pest in contrast to two-spotted spider mite or whitefly. However, they can cause persistent and repetitive problems in commercial crops for two reasons. Firstly, repeated spray applications in the relatively isolated greenhouse environment have unintentionally selected clones of aphids resistant to insecticides. Secondly, with the increase in the use of biological control for major pests, aphids previously killed by broad-spectrum sprays have become more serious pests in their own right. Both of these factors have created a demand for biological control agents for aphids.

Aphids most commonly found under glass are polyphagous species such as *Myzus persicae* (Sulzer) and *Aphis gossypii,* both of which have developed strong pesticide resistance. Other polyphagous species frequently encountered are *Macrosiphum euphorbiae* (Thomas), *Aulacorthum solani* (Kaltenbach), *A. circumflexum* (Buckton), and *Brachycaudus helichrysii* (Kaltenbach). *Macrosiphoniella sanbornii* (Gillette) is occasionally a problem on chrysanthemum (Hall 1985b). Blackman and Eastop (1984) detail identification, host-plant relationships, and life history information on aphid species that are pests in horticulture and agriculture. Minks and Harrewijn (1988) have edited a comprehensive guide to aphids and their natural enemies.

Typically, aphids have no sexual phase in greenhouses, increasing by viviparous parthenogenetic reproduction. They are phloem feeders and can cause distortion of leaves and growing points as well as cover leaf surfaces with sticky honeydew. Damage tolerance to aphids is lowest on ornamentals where they may deposit unsightly honeydew, or cause distortion of growing points and encourage sooty mold. On food crops, greater tolerance is possible, and aphids are more likely to pass unnoticed.

Aphids are vectors of plant viruses. *Myzus persicae* is capable of transmitting Tomato Aspermy, Tomato Yellow Net, Tomato Yellow Top, and Lettuce Mosaic. *Aphis gossypii* is a vector of Cucumber Mosaic Virus among others.

Wyatt and Brown (1977) calculated increase rates for four greenhouse species, the most rapid rate recorded being over 20-fold per week for *A. gossypii* on cucumber at 24°C, and 17-fold for *M. persicae* on chrysanthemum. Increase rates depend upon host plant, aphid biotype, and environmental factors; in practice *A. gossypii* may increase at about 10-fold per week.

The predatory midge *Aphidoletes aphidimyza* is commercially available for aphid control in the U.K., Netherlands, and other (east and west) European countries. Early biological control studies were conducted by Markkula and Tiittaanen (1977) on rose and sweet pepper. *Aphidoletes aphidimyza* was first used commercially in Finland in 1978 (Tiittanen and Markkula 1989). The midge is a delicate fly, a few millimeters long, which is nocturnal and lives in shady damp habitats, such as below hedges. Female midges lay orange eggs beside aphid colonies, or even on individual aphids. The aphidophagous larva grows to approximately 3 mm in length before descending to the soil to form a cocoon just below the surface in which pupation takes place. The midge is polyphagous, having been recorded feeding on 61 aphid species (Harris 1973). The predatory Cecidomyiidae were reviewed by Nijveldt (1988) and Kulp et al. (1989).

Midge larvae are sensitive to photoperiod in the final larval instar. From two to four cycles of short days (less than 15–15.5 h of light) are sufficient to trigger a diapausing larval form within the cocoon (Havelka 1980). Diapausing larvae are paler and broader in appearance, and remain within the cocoon for a longer period. The threshold photoperiod for diapause induction (as well as other ecological and morphological features) has been found to be related to geographical latitude (Havelka and Zemek 1988). Gilkeson and Hill (1986a) selected midges for non-diapause by maintaining them over a sequence of generations in an 8 h photoperiod at constant temperatures. Selection under fluctuating temperatures was unsuccessful. Midge larvae are sensitive to low light intensities (Gilkeson and Hill 1986b), and diapause can be prevented by suspending a low wattage bulb over the plants to extend the photoperiod in spring or autumn. A 60-watt incandescent bulb will avert diapause in the majority of larvae within a 10-m radius of the light source.

Currently, prices limit the use of *A. aphidimyza* to a rate of 1–2 per m², probably only with fortnightly introductions. One U.K. supplier of the midges suggests about one cocoon released fortnightly per 8 m² to combat a low and diffuse infestation on pot plants. Another supplier suggests 0.5 cocoons per m² as a preventative measure, 1–2 per m² fortnightly as soon as aphids are seen, with release points every 40 m² and extra cocoons placed beneath aphid "hotspots." Early introduction of cocoons while aphid densities are very low is stressed by several suppliers, as is the need to keep cocoons moist after release. Releases can sometimes be made close to drip-feeds to ensure a moist release medium is maintained.

Mass-rearing of the midge on cereal aphids was described by Belousov and Popov (1989), with a system capable of producing 15,000 cocoons per day from the labor of one person. Most midge rearing now appears to use the method of collecting the larvae in water when they are ready to form cocoons (van Leiburg and Ramakers 1984). It is possible to cold-store *A. aphidimyza* for eight months at 5°C with mortality under 9%. Storage of the larvae in diapause seems to be unnecessary. However, emergence is over an extended period which may limit cold-storage to research uses (Gilkeson 1990a).

In Denmark, Hansen (1983) experimented with "open rearing units" (boxes containing plants and aphids not injurious to the crop) which were placed among sweet pepper to control M. persicae. Control on the crop was successful, but supplementary introductions of midge cocoons were necessary.

Aphidoletes aphidimyza may have a role to play in ornamentals where the aesthetic requirement for "clean" plants is important. The midge leaves little evidence of its presence after predation is over and the larva has descended to pupate. Like Verticillum lecanii, A. aphidimyza leave the dead remains of aphids on the plant, but these do not stay long. They are attached by their stylets, but fall off when the plant is moved or sprayed. Any remaining aphids or larvae might be removed with a final "clean-up" spray. In chrysanthemum, where the crop is subjected to short photoperiods to induce flowering, midge diapause can be partially averted by the use of light-transmitting thermal screens as "blackouts" (Chambers 1990).

Where A. aphidimyza is used over plastic floors, as in modern sweet pepper production, many can be lost as they leave the plant to pupate. Similar losses would be expected if Thripstick were in use. The provision of a substrate for the cocoons beneath the plants may enhance the efficiency of this predator.

The first attempts to use hymenopterous parasites to control aphids under glass were by Wyatt (1970) and Scopes (1970) who released Aphidius matricariae Haliday on chrysanthemum. Control was obtained by Wyatt, but only after aphids had reached a self-limiting density unacceptable to the grower; however, control was better on cultivars more resistant to aphids. Further investigations of this parasite were made on eggplant (Tremblay 1974; Rabasse 1980) and on sweet pepper (van Lenteren et al. 1979). Mass-rearing of A. matricariae for release against M. persicae was discussed by Shijko (1989), who reported good control possibilities on sweet pepper.

Experience suggests that A. matricariae is more effective at lower aphid densities, while A. aphidimyza is more successful when aphid numbers have built up (van Lenteren 1988). Aphidius matricariae is currently available from some commercial suppliers in Europe and is being recommended as a backup to A. aphidimyza, often at rates of one per m^2 per week. Release of both parasite and predator into sweet pepper crops is a method showing signs of promise (Gilkeson 1990b).

The effects of parasitism on an aphid population by Aphidius spp. and other aphidiid parasites were summarized by Stary (1988a). Although some aphid colonies can be reduced, elimination of the prey is unlikely by aphidiids alone, and parasitoids can themselves be attacked by hyperparasitoids, usually some time after establishment. Other aphidiids such as Ephedrus cerasicola Stary may prove useful. Ephedrus cerasicola has a more limited host range and slower development than A. matricariae, but is more fecund and may be less susceptible to hyperparasites (Hagvar and Hofsvang 1990).

The impact of aphelinid parasitoids is less well researched (Stary 1988b). Wyatt (1972) released *Aphelinus* sp. aff. *flavipes* Kurdjomov from India into U.K. cucumber crops for control of *A. gossypii*. Control was only possible if parasites were present at the start of the aphid infestation, ensuring a high initial parasite:host ratio, and if night temperatures were adjusted to tip the balance in favor of the parasite. Kuo-Sell and Kreisfeld (1987) appraised three species of cereal aphids as hosts for *Aphelinus asychis* (Walker). Female parasites were heavier if reared on *Rhopalosiphum padi* (L.), but more host-feeding took place on *Sitobion avenae* (F.).

The predatory larvae of syrphids have a greater voracity than *A. aphidimyza* (Bondarenko and Asyakin 1981). Chambers (1986) assessed predator:prey ratios of eggs and larvae of *Metasyrphus corollae* (F.) for control of *A. gossypii* on cucumber. Control could be obtained over short periods with first and second instar larvae; aphid control in a crop over a periods of weeks would require the continuous presence of gravid female flies. Mass-rearing costs would therefore need to be low enough to enable growers to make frequent releases. The aphidophagous Syrphidae were reviewed by Chambers (1988).

Chrysopa carnea Stephens is currently mass-produced for sale by one Italian biocontrol supplier. Methods of mass-production are well developed (Morrison 1985), and research on improved methods of adult feeding and egg collection continues (Karelin et al. 1989).

Anthocorids are recognized as voracious predators of aphids, thrips, and mites, and currently attempts are underway to mass-rear them, but there are difficulties with cannibalism. Niemczyk (1988) concluded that *Anthocoris nemorum* L. was an effective aphid predator when aphid numbers were low and the predator:prey ratio was relatively high. A comprehensive study of the potential impact of Anthocoridae on aphids infesting greenhouse crops has not been conducted.

Coccinellids have not been used successfully for aphid control in greenhouses; their relatively long life cycle and high cost of mass-production currently preclude it. Quilici et al. (1985) released small larvae of *Propylea 14-punctata* L. against *M. persicae* on eggplant, but found that leaf hairs disturbed the predators. Control did not occur, but aphid population development was delayed. Larger larvae and adults were more effective.

Aphids remain problematical for biological control with arthropod predators and parasites. There appears to be no single predator or parasite currently known which has sufficient voracity, fecundity, and rate of development to reliably and effectively contain aphid population growth other than temporarily. This means that repeated inundative introductions are needed. Future advances will require investigation of "new" beneficials and research into the more effective application of known species.

Entomogenous fungi are appropriate candidate natural enemies for pest control in greenhouses because there is the potential for manipulating the environmental conditions to suit the fungus (Wilding 1983). Also, growers are innovative and familiar with predictable pest problems involving regular prophylactic pest control (Quinlan 1988).

There are few cases of Entomophthorales being used in greenhouses (Zimmerman 1983), but a strain of the hyphomycete *Verticillium lecanii* has been developed for aphid control (Hall and Burges 1979). It has a wide host range and can also grow saprophytically in soil, rotten wood (Quinlan 1988), on food material (Domsch et al. 1980), on the phylloplane, and as a hyperparasite of rusts (Spencer 1980). It is safe to handle because it does not infect mammals or cause allergic reactions (Hall 1981; Quinlan 1988). If a spore comes in contact with an aphid and germinates successfully, the germ tube can penetrate the aphid's cuticle (Hughes and Gillespie 1985) by the use of enzymes (proteases, lipases, and chitinases [St. Leger et al. 1986]) and mechanical pressure. Once inside the haemocoel, growth morphology changes and becomes yeastlike, and the aphid dies in four to seven days (Payne and Lynch 1988). Sporulation follows on the external surface of the dead aphid. Conidia are produced in sticky slime heads, improving the chances of dispersal and the development of an epizootic.

Encarsia formosa and *Phytoseiulus persimilis* might aid dispersal of *V. lecanii;* the former can also be a host but mortality is usually negligible (Rombach and Gillespie 1988). The fungus is compatible with the parasite *Aphidius matricariae,* and with *P. persimilis* and leafminer parasites (Hall 1985b).

Conidia have a short half-life after drying, and so aerial dispersal is likely to be unimportant (Hall 1981; Gardner et al. 1984). For germination and growth, conidia require a relative humidity of at least 93% at temperatures between 15°C and 27°C (Milner and Lutton 1986; Rombach and Gillespie 1988).

Verticillium lecanii will sporulate and grow on a range of media and can be produced commercially in submerged culture or on semi-solid media such as husked millet (Bartlett and Jaronski 1988). Commercial formulations contain a nutrient to enable the fungus to grow saprophytically on the leaf surface. Hyphal bodies can survive on foliage for long periods only if the humidity is sufficiently high; this increases the chances of transmission of the disease to wandering aphids. Commercial products were available on the European market in the past (e.g. "Vertalec", Microbial Resources Ltd, U.K.) and others are currently on sale (e.g. "Microgermin", Christian Hansen's BioSystems, Denmark) or are under development.

Verticillium lecanii gave good control of *Myzus persicae* on chrysanthemum and eggplant (Zimmerman 1983), but was less effective against *Brachycaudus helichrysi* and *Macrosiphoniella sanborni* (Hall and Burges 1979; Gardner et al. 1984). The commercial formulation also controlled *Aphis gossypii* on cucumber (Hall 1982, 1985b). Variations in the degree of control are not related to the relative degree of susceptibility of species, but rather to differences in the aphid's behavior and location on the plants (Hall and Burges 1979). *Myzus persicae* is highly mobile, while *A. gossypii* and *M. sanborni* are sedentary; *M. persicae* and *A. gossypii* feed mainly under the leaves, but *M. sanborni* feeds preferentially on the exposed stem. The species are therefore subject to different microclimates and degrees of exposure to sprays. Mobile species are more likely to contact spores than sedentary species.

Application of *V. lecanii* followed immediately by the aphid alarm pheromone, (E)-β-farnesene, stimulated movement of *A. gossypii* and increased the level of mortality due to fungus (Hockland et al. 1986). This is not, however, observed with all clones of *A. gossypii* (P. I. Sopp, unpublished data), and response to the pheromone in aphids such as *M. persicae* can be related to insecticide resistance (Dawson et al. 1983). Another means of improving aphid control on chrysanthemum is to make frequent applications of the fungus at low dosages. Although this is effective (Helyer and Wardlow 1987), it is not favored by growers because of the increased labor costs and risk of plant diseases due to the application of large volumes of water (e.g. 1000 liters per ha). To obviate these problems, Sopp et al. (1989) applied spores in 10 liters per ha using an experimental ULV electrostatic sprayer, the APE-80 (Arnold and Pye 1980). This also had the advantage of directing a larger proportion of spores to the undersides of leaves where most aphids are found and where the humidity is higher. This technique gave an acceptable level of control of *A. gossypii*. All the experiments on chrysanthemum described above employed black polythene blackouts; it is not yet known if *V. lecanii* can be used effectively against aphids where thermal screens are used as blackouts. In these conditions, the humidity is likely to be lower.

Future commercial exploitation of *V. lecanii* for aphid control need not necessarily rely on strains that have been or are currently being mass-produced; Hall (1980) reported that three out of eight strains isolated from fungal hosts were highly pathogenic to aphids. It may also eventually prove possible to improve strains by novel recombination techniques (Payne and Jarrett 1984) or genetic manipulation (Heale 1988). Drummond and Heale (1988), for example, have used protoplast fusion and hyphal anastomosis to carry out 31 crosses between four *V. lecanii* isolates. Pathogenicity results from a number of attributes (such as adhesion to the host, germination rate, host cuticle degradation, toxin production, etc.), but progress is being made in the genetic manipulation of these traits (Heale et al. 1989).

Some fungi other than *V. lecanii* have been evaluated for aphid control. Dedryver and Rabasse (1982) carried out greenhouse trials on lettuce to assess the effect of applications of resting spores of *Conidiobolus obscurus* Hall & Dunn and mycelium of *Erynia neoaphidis* Remaud. & Henn. on populations of *Aulacorthum solani*, *Macrosiphum euphorbiae* Thomas, *Nasonovia ribis-nigri* Mosl., and *Acyrthosiphon lactucae* Pass. *Conidiobolus obscurus* spread slowly and killed a maximum of 8% of aphids. *Erynia neoaphidis* spread more rapidly and killed 80% of aphids at high aphid density, but only 5% at low aphid density. Other species of fungi could be assessed for aphid control; Hall (1985b) reported *Zoophthora erinacea* (= *Erynia erinacea* [Ben-Ze'ev & Kenneth]) providing natural control of aphids on chrysanthemum in Colombia.

In the U.K., aphids became a problem on chrysanthemum due to the advent of *M. persicae* and *A. gossypii* clones resistant to systemic carbamate and organophosphate insecticides; however, these two species are susceptible to direct contact sprays of nicotine (Cross et al. 1983). Resistance to the carbamate insecticide pirimicarb has more recently appeared in the cucumber biotype of *A. gossypii*, an aphid which has become more common than 15 years ago, possibly as a result of the switch to biological control systems. It has evolved pesticide-resistant strains very rapidly on cucumber, chrysanthemum, and other ornamentals where it is most commonly found. In the Netherlands, where resistance in *A. gossypii* is severe, cyanide is suggested for aphid control in integrated programs on cucumber (Ramakers et al. 1989).

Pirimicarb is a relatively selective systemic and contact insecticide for use against aphids which has been used successfully in integrated control programs and is a widely-used pesticide on edible greenhouse crops (Sly 1986). The non-specific insecticides dimethoate, heptenophos, and demeton-S-methyl are also used for aphid control, but OP resistance can extend to these as well. Heptenophos and pirimicarb may be difficult to integrate with *Amblyseius* spp. for thrips control, as they can be toxic to this predator (de Klerk and Ramakers 1986; Ravensberg and Altena 1987). Nevertheless, Buxton et al. (1990) reported that pirimicarb had little effect on *Amblyseius* spp. in integrated control trials on sweet pepper. It does seem likely, however, that a greater reliance on *Aphidoletes aphidimyza*, *Aphidius* spp., or other aphid antagonists will be required in integrated programs in the future. *Amblyseius* spp. are therefore biological control partners to aphidophages.

On sweet pepper, *A. aphidimyza* and *A. matricariae* introduced together is now proposed as a means of restraining aphid populations in the Netherlands (Altena and Ravensberg 1990). In Canada, Gilkeson (1990b) reported that the best control was given by a combination of these two antagonists. However, in the U.K. the control of aphids on sweet pepper by *A. aphidimyza*, whether alone or assisted by parasitoids, has been reported as variable (Buxton et al. 1990), and may have been due to the absence of pupation sites for the midge in rockwool-grown crops.

Root aphids cause occasional problems on lettuce, but are currently controlled with diazinon incorporated into compost or applied as a soil drench.

E. Leafminers

The tomato leafminer, *Liromyza bryoniae* (Kaltenbach), is the most important leafminer species on protected salad crops in the U.K. The chrysanthemum leafminer, *Chromatomyia syngenesiae* Hardy, may occasionally be found on vegetables such as lettuce, but rarely causes significant damage. The American serpentine leafminer, *Liriomyza trifolii* (Burgess), and the South American leafminer, *Liriomyza huidobrensis*, are both polyphagous species which are notifiable pests in the U.K.

Liriomyza trifolii appeared as a pest on ornamental plants in France during 1977 (Lyon 1986) and on vegetables in the Netherlands during the early 1980s (Minkenberg and van Lenteren 1986); it was first recorded during 1989 (Bartlett et al. 1989). These species are occasionally imported to the U.K. on chrysanthemum and other leafy plant material, but so far all occurrences have been contained and destroyed. *Liriomyza huidobrensis* has a wider host range than *L. trifolii* that includes eggplant, cucumber, endive, lettuce, and tomato as edible crops and chrysanthemum, carthamum, exacum, gypsophila, and many weeds (Bartlett et al. 1989). Nevertheless, Parella and Bethke (1984) concluded that *L. huidobrensis* is unlikely to attain the same pest status as *L. trifolii* on chrysanthemum and that it is easier to control with pesticides.

Liriomyza bryoniae adults are 2.5 mm long, black, with a yellow spot on the thorax between their wings. They attack both tomato and cucumber plants. The adults feed on the sap exuded after the female fly has made an oviposition mark. These are visible as pale spots just less than 1 mm across which can be invaded by fungi and bacteria. Price and Harbaugh (1981) noted an increase of bacterial leafspot in chrysanthemum infested with *L. trifolii*. Only about 10% of these leaf punctures contain eggs (Helyer, unpublished data), the remainder being feeding punctures. The eggs hatch after about one week into small larvae which begin to mine within the leaf, causing the characteristic leaf mines. These mines reduce the photosynthetic capacity of the plant, and heavy infestations may cause desiccation and even death of seedlings and young plants. Yield losses can also occur as a result of leafminer infestation. The economic injury level was assessed to be 15 mines of *L. bryoniae* per leaf (tomato), particularly if the leaves were adjacent to fruit at an early-to-mid stage of swelling (Ledieu and Helyer 1982). If damage on leaves adjacent to a truss reached 30 mines per leaf when the fruit was half swollen, a 10% loss of yield resulted (Ledieu and Helyer 1985).

Mature larvae of *L. bryoniae* and *L. trifolii* usually vacate the leaf to pupate on the ground, whereas L. huidobrensis may either remain stuck to the leaf or pupate on the ground. Thus, sticky trap materials applied to the floor covering below a crop have significantly reduced leafminer populations (Helyer et al. 1983a). The emergent larvae of L. trifolii are negatively phototactic (Leibee 1984). Pupation can be delayed for a limited time by continuous lighting forcing the larva to find sites of low light intensity where temperature and humidity are more likely to be suitable (Leibee 1986). Oatman and Michelbacher (1959) noted that a related species, Liriomyza pictella (Thompson) (the melon leafminer), pupated more rapidly under dull conditions than in full light. In natural light they moved up to 40 mm (at a rate of 30 mm h^{-1}) before pupating under objects or in the soil (Oatman and Michelbacher 1959). During the summer months adults of L. trifolii emerge from the puparium after about one week, but diapause, induced in the leafminer population by shortening daylength, delays and extends the period of adult emergence from August onwards (Helyer and Ledieu 1990). Leafminer should be controlled before the onset of diapause in August to reduce leafminer incidence on early-sown vegetable crops.

Liriomyza bryoniae and Chromatomyia syngenesiae may be controlled biologically by the same parasites. Diglyphus isaea (Walker) is a polyphagous parasite which has been reared from more than 25 species of Agromyzidae (Minkenberg and van Lenteren 1986). This species is an ectoparasite which commonly occurs naturally from June onwards in the U.K. and is commercially available throughout the year from several insect rearing companies. The adult female locates the host larvae by drumming her antennae along the mines, then punctures the leaf and larva to paralyze the latter before laying one or more eggs. Host feeding by the female, in laboratory studies by Ibrahim and Madge (1979), accounted for almost half the number of C. syngenesiae larvae killed. The developing parasite has three larval instars; the last instar characteristically builds meconial pillars of frass as structural support for its pupal chamber. Dacnusa sibirica Telenga is an endoparasite which is often supplied commercially mixed with another endoparasite, Opius pallipes Wesmael. Both parasites lay their eggs inside the leafminer larvae. Opius pallipes is ineffective against L. trifolii because the parasite eggs are encapsulated within the host larvae (Woets and van der Linden 1982). The parasites are introduced at rates from 1000 to 10,000 per ha depending on the severity of the leafminer attack and the time of year (Wardlow 1985b).

The susceptibility of Liriomyza trifolii pupae to 11 strains of entomopathogenic hyphomycetes (Beauveria bassiana, Metarhizium anisopliae, Paecilomyces farinosus, and P. fumosoroseus) was tested in the laboratory (Bordat et al. 1988). Strains of P. fumosoroseus and P. farinosus proved to be the most pathogenic of those tested. The rhabditid nematode Steinernema feltiae Filipjev will attack L. trifolii and can be sprayed onto foliage. If this is done with a large volume of water in the evening (to prevent desiccation of the nematode), leafminer infestation can be prevented (Harris and Warkentin 1988).

The efficiency of parasites may be impaired by low light intensity and short days in the winter months (Scopes and Biggerstaff 1973). The low light intensity may affect the gain in body temperature by solar radiation and thus reduce parasite efficiency (Minkenberg and van Lenteren 1986). Therefore, leafminers attacking young plants during winter or early spring are usually controlled chemically as this offers a relatively quick method. Young plants are susceptible to phytotoxic damage by many pesticides when reared under poor winter light conditions; however, diazinon or pyrazophos sprays repeated at 10–14-day intervals have given good control with minimal phytotoxic damage (Ledieu and Helyer 1983b). Both are organophosphorous (OP) compounds and should have relatively little impact on the OP-tolerant *Phytoseiulus persimilis* widely used in Europe (Hassan 1982). The aphicide, heptenophos, is also a useful product against leafminers, but limited data are available for crop safety during the propagating stage. Heptenophos is a systemic OP insecticide of very short persistence and has shown no adverse effects on established populations of *Encarsia formosa* and *Phytoseiulus persimilis* when repeatedly sprayed at 10–14-day intervals (Helyer and Ledieu 1986). At least three spray applications at 10–14-day intervals are necessary for satisfactory leafminer control. This extended period is necessary to kill the progeny derived from those individuals which are either egg or pupa at the time of the first chemical application.

Cultural control of leafminers may take the form of sticky traps as used for whitefly (van de Veire and Vacante 1984) which will reduce the number of flying adults. The effectiveness of different positions, heights, and directions of yellow sticky traps was evaluated for *L. trifolii* by Yathom et al. (1988). For gerbera, they found free-hanging vertical traps 30 cm above the plants to be the best position. In a study by Jones and Parella (1986) on chrysanthemum, female *L. trifolii* were found to fly further than males and, although both sexes were trapped on yellow sticky cards, predominantly males were caught. Traps placed up to 26 m apart gave adequate information to determine population trends.

F. Leafhoppers

The greenhouse leafhopper *Hauptidia maroccana* (Melichar) (= *Zygaena pallidifrons* Edw.) has become more evident in recent years. This is likely to be a result of the decreased usage of pesticides to control the more important pest species. These insects are now reaching damaging levels and require control measures which will integrate with the use of natural enemies against the major greenhouse pests. The leafhopper, *H. maroccana*, feeds by sucking out the cell contents of leaves, causing blanched areas to appear on the upper surface that become confluent as feeding continues. They have a wide host range including eggplant, cucumber, french bean, tomato, chrysanthemum, fuchsia, geranium, and many other ornamental plants (Copland and Soeprapto 1985). When seedlings are damaged, plant growth may be severely checked. The adult is a slender, pale yellow insect about 3 mm long with two dark, V-shaped bands across the back, formed when the wings are folded. Mature hoppers are long-lived (MacGill 1934) and extremely active in sunny weather, leaping into flight when disturbed. Eggs are laid into secondary veins on the undersides of leaves and hatch after about one week to produce inactive whitish nymphs which feed on the undersides of leaves.

Biological control of leafhoppers has been recorded (Hussey et al. 1969) using the tiny mymarid wasp *Anagrus atomus* (L.). This egg parasite readily locates and attacks the host, but the percentage of parasitized leafhoppers remains low (Copland and Soeprapto 1985). Parasitized eggs may be clearly seen as they develop a reddish color which is also the color of the adult; however, the wasp is not commercially produced as it is relatively short-lived and easily desiccated in detached leaves, making distribution difficult.

Although various entomogenous fungi, including *Metarhizium anisopliae* (Gillespie et al. 1982b), have been shown to be effective in the laboratory against species such as *H. maroccana,* they failed to control the pest in the greenhouse (Gillespie 1984).

Leafhoppers may be controlled by heptenophos which is also effective against aphids and leafminers; however, the chemical does not appear to be effective against the eggs of leafhopper, so more than one application is necessary to ensure that the pesticide is present at a susceptible stage of the life cycle. In greenhouse trials on cucumber and tomato (Helyer and Ledieu 1986), weekly sprays of heptenophos gave excellent control but were too damaging to the natural enemies for practical use. Fortnightly sprays were much less disruptive to *Encarsia formosa* and *Phytoseiulus persimilis* and were reasonably effective against leafhopper but less so against thrips. A practical compromise of three applications, each at 10-day intervals, was proposed; this should control both pests adequately without undue adverse effects on beneficial insects and mites.

G. Tarsonemid Mites

The broad mite *Polyphagotarsonemus latus* Banks is becoming an increasingly common pest on greenhouse crops, mainly due to the trend toward decreased usage of pesticides (Bassett 1985). These mites are very small (150 μm long) and have been found on eggplant, cucumber, tomato, and sweet pepper crops grown in the U.K. (Bassett 1981; Cross and Bassett 1982). Damage is normally noticeable only when large numbers of mites are present, and takes the form of distorted or russetted fruit and brittle down-curled leaves with a shiny undersurface. During the first recorded outbreak of P. latus on cucumber in the U.K. in 1980 (Bassett 1981), growers mistakenly took the observed damage to be that of hormone weedkiller spray drift. Adult females live for about 10 days and produce about five eggs daily. The mite life cycle is completed in less than one week at 20°C, allowing a rapid population increase.

The potential of biological control of P. latus by phytoseiid mites has been studied by McMurtry et al. (1984). Wardlow (1989) has suggested that Amblyseius spp. mites, as used to control thrips, would probably control P. latus.

Acaricides used to control P. latus need to be applied repeatedly at short intervals, but this frequent use may be harmful to natural enemies used to control other pests. Fortunately, the acaricides fenbutatin oxide and tetradifon applied as HV sprays control P. latus and are relatively safe to the predatory mite Phytoseiulus persimilis (Hassan et al. 1987); however, these acaricides must still be used carefully—otherwise they will have detrimental effects by reducing numbers of the prey, Tetranychus urticae, to levels below those at which P. persimilis can survive.

H. Darkwinged Fungus Gnats

The larvae of sciarids or darkwinged fungus gnats are mostly saprophagous, feeding on decaying plant tissues, but a few are primary pests. The most commonly occurring greenhouse sciarids belong to the genus Bradysia and are important pests of seedlings and young plants. Species within the family Sciaridae are extremely difficult to identify, so identification relies almost entirely on the characteristics of the male external genitalia (Freeman 1983). They are now also becoming a pest of rockwool-grown plants, the larvae feeding on surface-growing algae and on root hairs. They are also known to transmit root pathogens (Binns 1977); however, Fagg and Fletcher (1987) demonstrated that although sciarid flies could transmit the pathogen of tomato stem rot (Didymella lycopersici Kleb.), this was probably an infrequent occurrence. The adults are often seen around the bases of plants where they are attracted by the presence of fungi and the production of ammonia released by the breakdown of nitrogenous fertilizers. Adult females lay over 100 eggs which hatch after four days at 22°C. The larvae pass through four instars in just over three weeks, pupate in the soil or rockwool, and emerge as adults four days later.

There is potential for developing entomopathogenic nematodes and bacteria for the control of sciarid flies. The tylenchid nematode, Tripius sciarae Bovien, reduced populations of Bradysia paupera Tuomikoski within four weeks of introduction to a greenhouse; larvae were killed and infected adults sterilized (Poinar 1971); however, since it has no resistant free-living stage and infective females remain viable for only two weeks

(Richardson 1987), this nematode species is unlikely to be commercialized. The mermithid nematode *Tetradonema plicans* Cobb, on the other hand, is more promising because it kills or sterilizes the host, has a high fecundity, and can be cheaply mass-produced and stored for up to a year at 10°C in the egg stage or for several years as the larval stage in aerated water containing formalin (Hudson 1974). Timing of introductions need not be crucial because *T. plicans* lays its eggs over a two-week period, and these hatch in sequence. The prospects for developing nematode control of these pests appear to be good, especially since *Bradysia paupera* larvae are also attacked by rhabditid nematodes (Richardson 1987) such as *Steinernema bibionis,* which are already mass-produced for the control of other pests.

Laboratory bioassays demonstrated 92% mortality of *Bradysia coprophila* (Linter) larvae exposed to *Bacillus thuringiensis* var. *israelensis* (H-14) (Osborne et al. 1985). Young larvae were 23 times more susceptible than nine-day-old larvae. Drenches of this bacterium were effective in reducing *B. coprophila* populations in the greenhouse, three weeks after application.

Sciarid larvae may be controlled by drenching potting composts with diazinon or diflubenzuron; the latter may be incorporated into the potting medium before planting. Adult sciarid populations may be reduced by placing yellow sticky traps amongst the plants and by spraying around the bases of plants with Thripstick. Both of the latter methods must be regarded as aids to control rather than as total control strategies (Richardson and Helyer 1990).

I. Vine Weevil

The most important of these wingless weevils to horticultural crops is the black vine weevil, *Otiorhynchus sulcatus* (F.). This is a pest of a range of pot-grown ornamentals including cyclamen, begonia, impatiens, gloxinia, rose, and other slow-growing plants and shrubs. The adult female is about 9 mm long and black, with roughened wing cases on which there is a fine yellow speckling. Reproduction is parthenogenetic, no males being known, and begins about 10 weeks after maturation from the pupa (Hussey et al. 1969). Individuals oviposit for at least four months, producing an average of seven eggs per day. Small (0.7 mm) eggs are laid in the surface soil near plants, these are white at first but melanize to a light brown after a few days. The eggs hatch in about three weeks to produce legless, yellowish white grubs which have brown heads and lie in a characteristically curved manner when disturbed. They feed for about 60–80 days causing considerable damage to the roots and corms of their host plants, which are often severely stunted or killed.

The insect-pathogenic bacterium *Bacillus thuringiensis* does not appear to be effective against *O. sulcatus* (Blackshaw 1984). Although *O.*

sulcatus is susceptible to some of the new coleopteran-active strains (Herrnstadt et al. 1986), it is slow-acting and a large dose is needed.

Known fungal antagonists of *O. sulcatus* include *Metarhizium flavoviride* W. Gams & J. Roszypal var. *flavoviride* Rombach, Humber & Roberts (Poprawski et al. 1985); *Metarhizium anisopliae* (Zimmerman 1981; Soares et al. 1983); *Paecilomyces fumosoroseus; Paecilomyces farinosus* (Holm ex S. F. Gray) Brown & Smith (Poprawski et al. 1985); *Beauveria bassiana* (Saito and Ikeda 1983); *Beauveria brongniartii* (Saccardo) Petch (Tillemans and Coremans-Pelseneer 1987; Coremans-Pelseneer and Tillemans 1989; Gillespie and Moorhouse 1989); and *Verticillium lecanii* (Marchal 1977).

Gillespie (1988a, 1989c) bioassayed a range of fungal isolates in pot plant tests and found that *Beauveria* spp. and *Paecilomyces* spp. were generally less effective than *Metarhizium* spp. Of 13 *Metarhizium* isolates, four reduced vine weevil populations by 80–100%, and the best was a U.K. isolate of *M. anisopliae* from *O. sulcatus,* which gave up to 100% control on pot-grown begonia and impatiens (Gillespie 1989b). Soares et al. (1983) also found large differences among strains in the performance of *M. anisopliae;* the best strains were superior to any of the strains of *M. flavoviride* var. *flavoviride* tested.

Vine weevil eggs are susceptible to infection by a range of fungi, excluding *B. bassiana* (Poprawski et al. 1985), but, with *M. anisopliae* at least, infection rate is reduced as the chorion becomes tanned (Zimmerman 1982; Gillespie 1989b). Fungi can infect larvae hatching more than 20 days after contamination of the eggs (Poprawski et al. 1985). Prophylactic applications of *M. anisopliae* spore suspensions to the soil surface were more effective (in terms of number of larvae killed per unit quantity of spores) than either mixing spores with soil or a curative application to the soil surface (Zimmerman 1981, 1984; Gillespie 1989b). This contrasts with *B. brongniartii,* where curative applications were more effective than prophylactic treatments (Coremans-Pelseneer and Tillemans 1989). *Metarhizium anisopliae* spores survive well in compost for several months (Gillespie 1988a), which suggests that cost-effective strategy of making a single prophylactic application may result in prolonged weevil control. The infection can also be spread by passive transport of conidia on the bodies of microarthropods in the soil or compost (Zimmerman and Bode 1983). Larval mortality on cyclamen was 30–40% compared with 80–100% on potted strawberry, azalea, and kalanchoe, and this was thought to be due to the production of fungistatic substances (such as the saponin, cyclamin) in the root zone (Zimmerman 1984).

The research carried out by Zimmerman (1981, 1984) and by Gillespie and Moorhouse (1989) suggests that a product based on a formulation of *M. anisopliae* may be available for greenhouse growers in the near future. Gillespie (1989c) estimated that enough fungi to treat one hectare of crop could be produced on 2 kg barley in 10 days.

Vine weevil larvae and pupae are susceptible to parasitism by rhabditid nematodes such as *Steinernema* (= *Neoaplectana*) *bibionis* Bovien and *Heterorhabditis heliothidis* (Khan, Brooks & Hirschmann) (Bedding and Miller 1981). These nematodes, in common with entomogenous fungi, will not be effective if desiccated, although some species can survive at least four days at 48% r.h. (Simons and Poinar 1973). They are therefore particularly suitable for use against vine weevils and other pests in soil and compost where these nematodes occur naturally (Blackshaw 1988). When sprayed onto foliage in chrysanthemum houses there was a decline in numbers of live *Steinernema feltiae* Filipjev of 12% per hour, due to death from desiccation (Begley 1987). They are mutualistically associated with insect-pathogenic bacteria (genus *Xenorhabdus*), which are held in an oesophageal vesicle. Infective nematodes are attracted to hosts (Kaya 1985), e.g., to carbon dioxide (Gaugler et al. 1980) and enter through the mouth or anus, or also, in the case of *Heterorhabditis* spp., by penetration of the cuticle. *Xenorhabdus* spp. are released and multiply, killing the insect host by septicaemia usually within two days of invasion. The nematode feeds on *Xenorhabdus* spp. and host tissue and reproduces; thousands of infective larval nematodes are released from the cadaver within two weeks (Payne 1988), equivalent to one million larvae per gram of insect host (Richardson 1987). Hermaphroditic females of *Heterorhabditis* spp. release up to 1000 eggs (Wouts 1984).

Mass-production in vitro (Bedding 1981, 1984; Poinar 1986), using axenic cultures of nematodes and *Xenorhabdus* spp. added to crumbed polyurethane sponge enriched with homogenate of pig's kidneys and beef fat, has achieved yields of 2000 million nematodes per 5 kg of homogenate. Occasionally, during production, the primary (and naturally-occurring) phase of *Xenorhabdus* spp. shifts to a secondary phase, reducing the capacity for nematode reproduction (Akhurst 1986; Boemare 1989); this problem is being tackled by research programs as part of the development of large-scale liquid fermentation culture of nematodes (Miller 1987).

Attention is also being given to efficient methods of storage, transport (Westerman and Simons 1988), and application. Nematodes can withstand 70.3 kg cm^{-2} (= 1000 psi) (Kaya 1985) and can be delivered through conventional spray booms (Loosbroek and Theunissen 1985), trickle irrigation systems (Reed et al. 1986), and mist blowers without ill effect. Previous rhabditid nematode products (e.g., "Seek", "Spear", and "Bionem") have had short commercial lives and it is important to ensure product reliability so that this form of biological control will continue to be available to the greenhouse industry (Richardson 1989a).

Levels of control in the region of 90–100% have been demonstrated on potted ornamentals and soft fruit using a Dutch strain of *Heterorhabditis* sp. (Simons 1981; Zimmerman and Simons 1986) and Tasmanian strains of *S. bibionis* and *H. heliothidis* (Bedding and Miller 1981). Although *Steinernema glaseri* Steiner has the capacity to search a larger area than other nematode species (Georgis and Poinar 1984; Zimmerman and Simons 1986; Schroeder and Beavers 1987), it is not necessarily the best species for vine weevil control (Rovesti and Deseo 1989). Zimmerman and Simons (1986) reviewed the potential of a range of species, including *S. feltiae* and *S. glaseri*, and concluded that *H. heliothidis* was the most promising species because it is effective and stores well. Stimman et al. (1985) and Georgis and Poinar (1984) also reported higher mortality with *H. heliothidis*. Dolmans (1983) found that *Heterorhabditis* sp. gave good control on rose and astilbe in the greenhouse. Richardson (1989b) showed that larvae on greenhouse impatiens were highly susceptible to four species of *Steinernema* and were eradicated both by *S. bibionis* and *Steinernema kraussei* Steiner, with some control being exerted for up to eight months after application to the compost. Nematodes could migrate from pot to pot via horticultural capillary matting (Richardson and Godliman 1989). Viable infective nematodes can also be passively disseminated through a phoretic association with microarthropods; Epsky et al. (1988), for example, observed *S. feltiae* infectives forming tightly packed rafts on the dorsum of large soil mites. Plant condition, soil moisture, and the composition of soil (Kaya 1985) or compost will also have an important impact on the control results, e.g., by restricting nematode movement (Simons 1981).

Some of these nematodes (e.g., *S. bibionis* and *Heterorhabditis* sp.) are now commercially available for the control of vine weevil in greenhouses in a number of European countries, including the U.K., the Netherlands, Switzerland, Germany, Denmark, and Italy. In Switzerland 60,000 million nematodes were sold by a Cooperative Society in 1986 (Klingler 1989). In the U.K., a research program supported by the Agricultural Genetics Company, at Horticulture Research International, resulted in the launch in 1989 of "Nemasys", a nematode product for the control of *O. sulcatus* on U.K. greenhouse cyclamen. It is applied as a drench to the compost surface three and nine weeks after potting (Richardson 1989c).

Nematodes also have potential for controlling other weevil species. Rutherford et al. (1987) reported some success with *Heterorhabditis* sp. against the strawberry root weevil, *Otiorhynchus ovatus* L., on potted fir seedlings in the greenhouse, and Deseo et al. (1986) obtained 92% mortality of *Otiorhynchus salicicola* Heyden on azalea using *H. heliothidis*. In greenhouse trials, Klingler (1986) reported 90% control of *O. salicicola* with *Heterorhabditis* sp. Under laboratory conditions, *S. glaseri* gave 100% parasitism of the weevil *Neoplinthus tigratus* Boh. on *Ruscus* sp. rhizomes (Deseo et al. 1986).

Remedial chemical control for vine weevil was often difficult due to the depth and inaccessibility of the larvae, so persistent pesticides such as aldrin were incorporated at potting. However, aldrin has now been withdrawn from use in the U.K. (Anon. 1989).

J. Caterpillars

The caterpillars of several moths can cause severe damage to many protected crops. The principal pests are members of the Noctuidae. The young caterpillars usually feed on the undersides of leaves for several days without causing damage visible from above. As they grow, the damage becomes more noticeable as either "windows" or holes in the leaves. They may also damage fruit. The tomato moth, *Lacanobia oleracea* (L.), is the major caterpillar pest on protected crops, being found on tomato, chrysanthemum, and sweet pepper (Jarrett 1985a). The color of the caterpillars varies from pale green to yellow-brown, and they have a prominent, lateral, deep yellow line below the spiracles. When fully grown, they reach a length of 45 mm. The adult female lays batches of between 50 and 300 eggs on the underside of leaves. After 6–7 days at 20°C the eggs hatch, and the young caterpillars feed gregariously for two days before dispersing. Mature caterpillars feed on leaves, fruit, and stems before pupating in the soil or under plant debris. The life cycle during early summer can be 9–10 weeks, but later generations may overwinter as pupae, generally in the soil. Overwintering pupae may be killed by soil sterilization; Foster (1979) indicated that steam was more effective than the use of chemical sterilants generating methyl isothiocyanate.

The cabbage moth, *Mamestra brassicae* (L.), is predominantly a brassica pest but may also cause severe damage to chrysanthemum, lettuce, tomato, and sweet pepper (Jarrett 1985a). Adults emerging outdoors during May and June invade greenhouses where females each lay up to 500 eggs on the undersides of leaves and even on the greenhouse structure. The caterpillars feed voraciously for several weeks before pupating in the soil. Other important species include the silver-y moth, *Autographa gamma* (L.), damaging protected lettuce and sweet pepper crops, and *Noctua pronuba* (L.), the large yellow underwing, an occasional pest on sweet pepper.

Control of *L. oleracea* was initially by the use of hand-picking and sprays of lead arsenate (Lloyd 1920) followed by sprays and dusts of DDT after World War II. During this time further work was undertaken on the evaluation of potential chemical pesticides including sodium fluoride in beer and molasses to trap adult moths, sodium arsenate prepared and used as above, and cuprous cyanide to kill caterpillars. The use of cuprous cyanide soon ended due to its capacity to liberate hydrogen cyanide (Speyer and Parr 1948). Pesticides continue to be used particularly where biological control is not practiced. Ledieu (1978) evaluated several pesticides for use on greenhouse crops. Where biological control of other pests is being practiced, the selective agent diflubenzuron or the HD-1 strain of the microbial insecticide *Bacillus thuringiensis* may be used with safety to control most of the common caterpillar pests. Diflubenzuron is effective as a stomach and contact poison, acting by inhibition of chitin synthesis.

The potential of natural enemies was also investigated by Speyer and Parr (1948), including pupal predators such as mice and several caterpillar parasites. *Trichogramma evanescens* Westwood, an egg parasite, has been evaluated for use against *M. brassicae* on cabbage crops by Pak and DeJong (1987), and may find a use under glass. Similarly, *Chrysoperla carnea* (Stephens) has been found to be a useful egg predator against *M. brassicae* (Sengonca et al. 1987). However, little other work appears to have been done with arthropod natural enemies for use on greenhouse crops due to the widespread adoption of *B. thuringiensis*.

Bacteria and viruses are currently the best prospects for the biological control of caterpillars using pathogens. There are, however, some initial studies with fungi, e.g., *Nomuraea rileyi* (Farlow) Samson against *M. brassicae* eggs (Rodriguez-Rueda and Fargues 1980); and nematodes, e.g., *Steinernema feltiae* against *Spodoptera exigua* (Hubner) (Timper et al. 1988). *Steinernema feltiae* can be sprayed onto foliage and will control second instar *Spodoptera exigua* on chrysanthemum cuttings and reduce numbers of fourth instar *S. exigua,* but not to a commercially acceptable level (Begley 1987). Nematodes also show promise against the cutworm, *Agrotis segetum* (Denis & Schiffermuller), on outdoor lettuce (Theunissen and Fransen 1984; Lossbroek and Theunissen 1985). When nematodes are used against *S. exigua,* they are compatible with the use of baculoviruses; Kaya and Burlando (1989) showed that *S. feltiae* reproduced successfully in 68% of nuclear polyhedrosis virus-infected hosts and that the nematodes retained sufficient active nuclear polyhedrosis virus (see below) to infect larvae.

The insect pathogenic bacterium *Bacillus thuringiensis* is used for the control of some species of Lepidoptera in greenhouses. It is an aerobic bacterium that produces at sporulation a large proteinaceous protoxin crystal. When bacteria coating the surface of foliage are eaten by a caterpillar the crystals are dissolved by the caterpillar's alkaline gut fluid and degraded by gut proteases, releasing toxic polypeptides. As a consequence, the muscles of gut and mouthparts are paralyzed, feeding stops, the gut epithelium lyses, and death occurs 30 minutes to three days after ingestion (Payne and Lynch 1988). *Bacillus thuringiensis* was the first microbial pathogen to be commercialized, and the *B. thuringiensis* var. *kurstaki* HD-1 isolate is now sold under at least eight trademarks and is quite safe to humans and other vertebrates (Heimpel 1971). It can be applied as a conventional high-volume spray (Helyer and Payne 1986). A pulse-jet fogging machine can also be used, because the high temperature is too transitory to harm the bacterium (Jarrett 1985b). Tomato moth,

L. oleracea, is the most susceptible of the greenhouse caterpillars and can be controlled with a relatively low dose (Jarrett and Burges 1982). The angleshades moth, Phlogophora meticulosa (L.); the silver-y moth, Autographa gamma; and the carnation tortrix moth, Cacoecimorpha pronubana (Hubner), on chrysanthemum, sweet pepper, and lettuce require a higher dose (especially against the later instars) and repeat applications, because they are very mobile. Thermal fogs are not effective on these crops because the dense foliage prevents penetration of droplets. Later instars of M. brassicae, the large yellow underwing; Noctua pronuba, the turnip moth; Agrotis segetum (Jarrett 1985b); and the beet armyworm, S. exigua (Moar and Trumble 1987), are, unfortunately, less susceptible.

Experimental use of "thuringiensin" (a β-exotoxin produced by some strains of Bacillus thuringiensis) was found to potentiate the effect of commercial formulations of B. thuringiensis var. kurstaki against S. exigua (Moar et al. 1986), but registration of this exotoxin seems unlikely. Programs of strain selection (Jarrett and Burges 1986) and strain recombination (Payne and Jarrett 1984) are developing new strains with good activity against Mamestra brassicae and other noctuid species. Genetic manipulation can also be used to insert B. thuringiensis toxin genes into plants such as tomato to control caterpillar pests (Fischhoff et al. 1987). In general, B. thuringiensis is compatible with other biological control agents, but under conditions of stress, natural enemies such as the lacewing Chrysoperla carnea can sometimes succumb (Donegan and Lighthart 1989).

Baculoviruses cause disease in a small number of insect orders, but predominantly in Lepidoptera. They are amongst the most specific of all microbial insecticides (Heimpel 1971). The virus particles can be packaged singly as granulosis viruses (GV) or in large numbers as nuclear polyhedrosis viruses (NPV) within large proteinaceous crystals or occlusion bodies (Payne and Lynch 1988). Infection occurs when susceptible caterpillars eat food contaminated with the virus; the occlusion body dissolves, releasing virus particles which multiply in gut epithelial cells and spread rapidly to other tissues. Death usually occurs more than four days after infection. A GV has been observed to attack Lacanobia oleracea and, at a high concentration, to give good control of moths introduced into a greenhouse (Crook et al. 1982). The GV suppressed the pest on three nurseries in Scotland, without re-introduction, for five years (Foster and Crook 1983). Virus persistence is improved in greenhouses because levels of ultraviolet radiation are lower under glass, and this results in reduced rates of inactivation of virus particles (Smits et al. 1987) (the same principle also applies to other microbial pesticides). Even so, inactivation of virus may be up to 70% on a sunny day (Smits 1986). Mamestra brassicae NPV is highly infective for the key caterpillar pests of protected vegetables (Allaway and Payne 1984) and is produced commercially ("Mamestrin"—Calliope, France); registration is being sought for greenhouse use (Guillon 1987). Another virus which is close to commercialization is *S. exigua* NPV (Gelertner and Federici 1986; Smits and Vlak 1988a). Trials on chrysanthemum, gerbera, kalanchoe, and tomato showed that more than 95% control of *S. exigua* can be achieved if the virus is applied against the early larval instars (Smits et al. 1987), the virus being as effective as some chemical pesticides (Gelertner et al. 1986). Accurate timing of virus application is important, and this is probably best achieved by monitoring moth populations with light traps, because pheromone traps (Persoons et al. 1981) are not always effective in the greenhouse (van den Bos 1983). Application of virus with an electrostatic sprayer could also improve control, because the young larvae feed underneath leaves without consuming the upper epidermis (Smits et al. 1987), and charged droplets of microbial pesticides are more likely to impact on the undersurface of leaves (Sopp et al. 1989). The virus is transmitted vertically to some extent (adults developing from virus-infected larvae transmit the virus to their progeny, 10–28% becoming infected [Smits and Vlak 1988b]) and can be horizontally transmitted by predators (such as the heteropteran, *Podisus* sp.) consuming dead larvae and spreading active virus via their feces (Smits 1987).

III. NOVEL METHODS AND SPRAY APPLICATION

A. Semiochemicals

A semiochemical is a chemical involved in an interaction between organisms. Examples are pheromones, allomones, and kairomones (Nordlund et al. 1981). The use of such chemicals in pest management in protected environments is still at the early research stage.

Pheromones are substances secreted by an organism which cause a specific reaction in a receiving organism of the same species; an example is the sex pheromone of Lepidoptera. Pheromones are extensively used in monitoring and trapping of many pests of field crops and orchards. The use of pheromones under glass is complicated by the enclosed environment affecting the dispersal of the chemical. Van den Bos (1983) summarized the results of a number of Dutch workers on pheromone trapping of members of the Tortricidae (*Clepsis spectrana* Tr., *Adoxophyes orana* F. V. R.), the Noctuidae (*Spodoptera exigua, S. littoralis* Boids, *Chrysodeixis chalcites* Esper.), and the Gelechiidae (*Phthorimaea operculella* Zell) in greenhouses. Van den Bos concluded that pheromone-mediated communication was less effective in greenhouses than in the open air and unlikely to be effective for monitoring or mass-trapping lepidopteran pests; however, he also concluded that greenhouses may be suitable for using pheromones to disrupt mating. In Scotland, Foster (1987) attempted to use pheromones to trap the tomato moth, *Lacanobia*

1. IPM OF GREENHOUSE CROPS IN NORTHERN EUROPE

oleracea, in tomato crops. He found that the traps were unable to prevent damage from occurring, and further work was abandoned, although he concluded that the traps were useful for monitoring. Sex pheromones have also been found in aphids (Dawson et al. 1987) and the glasshouse whitefly, Trialeurodes vaporariorum (Yin and Maschwitz, 1983); however, no attempts have been made to use them in pest management.

The aphid alarm pheromone, (E)-β-farnesene has potential for use with contact insecticides or fungal pathogens for aphid control by increasing dispersal of the aphid and hence contact with the control agent (Pickett et al. 1986). Gibson and Pickett (1983) found that the wild potato Solanum berthaultii Hawkes actually secretes the aphid alarm pheromone as a defense mechanism, and if this ability could be transferred into other cultivated plants it may provide valuable protection against aphid colonization and virus transmission.

Allomones are chemicals, such as plant-derived antifeedants, which evoke a behavioral or physiological reaction in the receiver which is adaptively favorable to the emitter. Very little work has been published on the use of allomones against greenhouse pests. One of the best known is a botanic insecticide extracted from the neem tree, Azadirachta indica A. Juss., which contains limonoid tetranortriterpenoids including azadirachtin which is active against lepidopteran larvae, phytophagous mites, whitefly, and leafminers (Schmutterer and Ascher 1987). Neem has a very wide range of activities including antifeedant, repellent, growth-disruption, fecundity-reductant, and some insecticidal activity.

Three international conferences have been held on the effects of neem (Schmutterer et al. 1981; Schmutterer and Ascher 1984, 1987). Although no neem-based product is available in Europe, an extract is available in the USA for the control of leafminer on chrysanthemum crops (Larew 1987; Larson 1987). The product is used as a drench and acts as a systemic insecticide against the larvae (antiecdysial activity), but has no effect on oviposition or the initial development of mines. Significant repellency of adult leafminers has been reported on neem-treated foliage (Webb et al. 1983).

In experiments with the sweetpotato whitefly, Bemisia tabaci, foliage treated with neem extract resulted in a reduction in oviposition (through adult repellency) and egg viability, and in prolonged larval development and mortality (Coudriet et al. 1985). In Europe, studies have concentrated on spider mites (Tetranychus cinnabarinus and T. urticae) (Schauer and Schmutterer 1981; Mansour et al. 1987). Mansour et al. (1987) found that solvent extracts caused high mortality in T. cinnabarinus and significant reduction in the fecundity of the survivors. At certain concentrations and with certain solvents, mortality and fecundity reduction of the predacious mite Phytoseiulus persimilis was much lower than for the pest mite.

The active molecule of neem is probably too complicated for economic synthesis, but other potent antifeedants have much less complex structures and are more readily synthesized, e.g., polygodial (Nakanishi and Kubo 1977). Treated plants are much less likely to be colonized by aphids, including those highly resistant to conventional insecticides. Polygodial has also been shown to significantly reduce virus transmission by aphids on some field crops (Gibson et al. 1982). Jones and Klocke (1987) reported feeding deterrent activity from several phenolic compounds incorporated into artificial diets for several species of aphid, including *Myzus persicae*. However, no published accounts of the activity of these compounds in the greenhouse have yet appeared.

Kairomones are chemicals which evoke a behavioral or physiological reaction in the receiver which is adaptively favorable to the receiver. The two-spotted spider mite, *T. urticae*, has been shown to produce a kairomone which is used by the predatory mite *P. persimilis* to locate its prey (Sabelis et al. 1984; de Moraes and McMurtry 1985). Ledieu (1976) showed that *Encarsia formosa* was able to detect plants infested with *T. vaporariorum* at a distance and that it rarely landed on uninfested plants, suggesting the presence of a volatile kairomone. The honeydew produced by *T. vaporariorum* is known to act as a contact kairomone for *E. formosa* (van Vianen and van de Veire 1988). There has been no commercial development of such kairomones, and none appears imminent. Dicke et al. (1986) showed that the diet of the predatory mite *Amblyseius potentillae* (Garman) affects the response of the mite to kairomones produced by one of its prey species, *T. urticae*. This may have implications for the searching efficiencies of captive-bred predators intended for release in greenhouses.

B. Biological Aspects of Spray Application

Application of chemical sprays in European greenhouses is still almost exclusively carried out using traditional HV hydraulic systems (Gould and Kingham 1964). The efficiency of such sprays, in terms of active ingredient usage, is generally recognized as being very low. Few studies on the biological efficiency of application systems in greenhouses have been published, and most that have relate to low volume (LV) applications. The design and physical aspects of sprayers are outside the scope of this review, which concentrates on the biological aspects of chemical application.

Chemical fogs are the most common form of LV application. Jarrett et al. (1978) compared the spray coverage of a Pulsefog application with an HV spray and an LV rotary atomizer application of *Bacillus thuringiensis* to mature chrysanthemum. Almost no deposits were recovered from the undersides of leaves on the fogged treatment, whereas about 30% of the total HV deposit was on the undersides and just 10% of the rotary atomizer deposit; however, the total deposit on the fog and rotary atomizer treatments was twice that on the HV, due to extensive run-off on the HV-treated plants. The effectiveness of the deposits was not assessed, although with most insect pests inhabiting the undersides of leaves, the HV spray would probably be the most effective.

LV-controlled droplet application (CDA) systems may be more efficient in the use of active ingredient (a.i.). Small uniform spray droplets are the most efficient way to apply chemicals for the control of *Tetranychus urticae* (Munthali 1984; Munthali and Wyatt 1986) and *Trialeurodes vaporariorum* (Palmer et al. 1983; Wyatt et al. 1984; Adams et al. 1987). Adams and Palmer (1986) and Abdelbagi and Adams (1987) compared the use of several LV sprayers, including electrostatic, in applications to tomato. They found that the electrostatic sprayers gave approximately equal deposits on both upper and lower leaf surfaces; however, air-assistance in the form of a small fan behind the atomizer was necessary to obtain good penetration into the crop. Furthermore, Abdelbagi and Adams (1987) demonstrated that the combination of small droplets (efficient use of a.i.) and electrostatics (underleaf coverage, minimal drift) could, potentially, lead to dramatic reductions in the amount of active ingredient applied.

Electrostatics were also used by Sopp et al. (1989) to increase the underleaf deposits of spores of the insect pathogenic fungus *Verticillium lecanii* on chrysanthemum. This led to improved spore survival because of the sheltered environment, and closer spore-aphid contact. These factors in turn led to earlier infection of the aphid population and improved control.

The integration of natural enemies and pesticides relies upon the natural enemy being resistant to the chemical or being separated from it in space or time. Adams and Palmer (1989) demonstrated that using an LV electrostatic sprayer, spatial separation could be achieved. They applied permethrin, for *T. vaporariorum* control, to the upper foliage of tomato whilst minimizing contamination of lower foliage inhabited by the parasitic wasp *Encarsia formosa*.

IV. THE FUTURE

In northern Europe, the practical implementation of biological control is more advanced in protected environments than on outdoor crops. This is due to several factors, most notably because resistance is widespread and provides strong motivation for developing alternatives to chemical control. In addition, crop value per unit area is very high and can bear the cost of rearing natural enemies, and greenhouse ecosystems are relatively simple and natural enemies are to some extend confined. Finally, environmental conditions can be modified to benefit biological control.

Systems of integrated control must be devised which provide economic control of all pests and diseases throughout the life of the crop. All control measures must be compatible or the system fails. Problems can be expected from pests spanning a wide range of taxonomic groups, but equally, the grower's armory contains a great diversity of natural enemies. Many of these are now proven to be effective, both in experimental trials and in commercial practice. After effective IPM systems have been developed and implemented, they are easily disrupted by changes in cultural practice or by the arrival of new pests for which no compatible control measure currently exists. The latter problem is becoming more frequent because of the acceleration in world trade in plant material. Therefore, if integrated control is to be adopted as a long-term strategy, it is necessary to have an underpinning research base that includes investigations into a wide range of natural enemies and other control options. In Europe, this research effort is shared through collaboration chiefly under two bodies: the Commission of the European Communities and the International Organisation for Biological Control. In the latter, working groups on "Integrated Control in Glasshouses" and "Pesticides and Beneficial Organisms" play important roles in coordinating research and exchanging information.

The statistics show that biological control under glass is employed much less frequently on ornamentals than on vegetables. This is probably because fewer pests can be tolerated on ornamental crops, and acceptable residue levels are not so severe as for edible produce. Nevertheless, resistance problems and "green politics" are now combining to create a greater demand for IPM on ornamentals. On some of the major ornamental crops it is now possible to control all the pests with a combination of biological control agents, narrow-spectrum chemical pesticides and other techniques, and an increase in the acreage employing IPM can be anticipated.

The trend towards reduced pesticide usage on greenhouse crops is likely to continue. In the medium-term, chemical pesticides are likely to be eliminated on some crops and reduced on others by the adoption of more sophisticated application techniques and the introduction of new natural enemies. In the longer-term, research goals are: (1) improved biological control agents; (2) computerized microclimate management to optimize natural enemy efficiency without adverse effects on the crop; (3) semi-resistant varieties of crop; and (4) the use of semiochemicals to concentrate pests in zones of natural enemy abundance. If these research programs are successful, it should be possible to greatly reduce the use of chemical pesticides on all greenhouse crops with no reduction in yield or quality.

LITERATURE CITED

Abdelbagi, H. A., and A. J. Adams. 1987. The influence of droplet size, air-assistance and electrostatic charge upon the distribution of ultra-low volume sprays on tomatoes. *Crop Prot.* 6:226–233.

Adams, A. J., M. R. Abdulla, I. J. Wyatt, and A. Palmer. 1987. The relative influence of the factors which determine the spray droplet density required to control the glasshouse whitefly, *Trialeurodes vaporariorum. Aspects Applied Biol.* 14:257–266.

Adams, A. J., and A. Palmer. 1986. Deposition patterns of small droplets applied to a tomato crop using the Ulvafan and two prototype electrostatic sprayers. *Crop Prot.* 5:358–364.

_____ . 1989. Air-assisted electrostatic application of permethrin to greenhouse tomatoes: Droplet distribution and its effect upon whiteflies (*Trialeurodes vaporariorum*) in the presence of *Encarsia formosa. Crop Prot.* 8:40–48.

Akhurst, R. J. 1986. Recent advances in *Xenorhabdus* research. p. 304–307. In: R. A. Samson and J. M. Vlak (eds.), Fundamental and Applied Aspects of Invertebrate Pathology, Foundation of the 4th Int. Coll. Inv. Path., Wageningen, Netherlands.

Albert, R. 1987. Experiences with the introduction of biological control methods in greenhouses in south-west Germany. *Bull. SROP/WPRS* 1987/X/2:13–17.

_____ . 1990. Experiences with biological control measures in glasshouses in southwest Germany. *Bull. SROP/WPRS* XIII/5:1–5.

Allaway, G., and C. C. Payne. 1984. Host range and virulence of five baculoviruses from lepidopterous hosts. *Ann. Appl. Biol.* 105:29–37.

Altena, K., and W. J. Ravensberg. 1990. Integrated pest management in the Netherlands in sweet peppers from 1985 to 1989. *Bull. SROP/WPRS* XIII/5:10–13.

Anon. 1976. The biological control of cucumber pests (2nd edition). *GCRI growers bulletin 1,* Littlehampton, U.K.

_____ . 1978. A teaspoon per acre for top fruit pest control. *The Grower.* 90:710.

_____ . 1986. Chapter 2; Distribution. p. 13–16. In: *Bemisia tabaci:* A Literature Survey on the Cotton Whitefly with an Annotated Bibliography. (Ed M. J. W. Cock). CAB/FAO; International Institute of Biological Control, Silwood Park, England.

_____ . 1989. Ministry puts immediate ban on use of aldrin. *Grower* 111:5.

_____ . 1990. Buntings better WFT control with bags of predators. *Grower* 114 (3):29.

Arnold, A. J., and B. J. Pye. 1980. Spray application with charged rotary atomizers— Systems for the 1990s. *Proc. BCPC Applications Symposium.* p. 109–117.

Azab, A. K., Megahed, M. M., and H. D. El-Mirsawi. 1971. On the range of host-plants of *Bemisia tabaci* (Genn.) (Homoptera: Aleyrodidae). *Bull de la Soc. Ent. d'Egypt.* 54:319–326.

_____ . 1972. On the biology of *Bemisia tabaci* (Genn.) (Homoptera: Aleyrodidae). *Bull. de la Soc. Ent. d'Egypt.* 55:305–315.

Bald, J. G., and G. Samuel. 1931. Investigations on spotted wilt of tomatoes II. *Austr. Counc. Sci. Ind. Res. Bull.* no. 54.

Bartlett, M. C., and S. T. Jaronsky. 1988. Mass production of entomogenous fungi for biological control of insects. p. 61–85. In: M. Burge (ed.). Fungi in Biological Control Systems. Manchester University Press, U.K.

Bartlett, P., J. Pryse, and S. Halliwell. 1989. Undesirable aliens. *Grower* 112(9):15–17.

Bassett, P. 1981. Observations on broad mite (*Polyphagotarsonemus latus*) (Arcarina: Tarsonemidae) attacking cucumber. *Proceedings of the 1981 British Crop Protection Conference—Pests and Diseases* 1:99–103.

_____ . 1985. Tarsonemid mites. p. 93–96. In: N. W. Hussey and N. E. A. Scopes (eds.). Biological Pest Control: The Glasshouse Experience. Blandford Press, Poole, U. K.

Bedding, R. A. 1981. Low cost, in vitro, mass production of *Neoaplectana* and *Heterorhabditis* species (Nematoda) for field control of insect pests. *Nematologica* 27:109–114.

_____. 1984. Large scale production, storage and transport of the insect-parasitic nematodes *Neoaplectana* spp. and *Heterorhabditis* spp. *Ann. Appl. Biol.* 104:117–120.

Bedding. R. A., and L. A. Miller. 1981. Use of a nematode, *Heterorhabditis heliothidis*, to control black vine weevil, *Otiorhynchus sulcatus*, in potted plants. *Ann. Appl. Biol.* 99:211–216.

Begley, J. W. 1987. Control of foliar feeding lepidopterous pests with entomophagous nematodes. *Abstr. Soc. Inv. Path.: XXth Annual Meeting*, Florida 20–24th July. p. 65–67.

Beglyarov, G. A., and F. A. Suchalkin. 1983. A predacious mite—a potential natural enemy of the tobacco thrips [Russian]. *Zashch. Rast.* 9:24–25.

Belousov, Yu. V., and N. A. Popov. 1989. Rearing of *Aphidoletes aphidimyza* (Diptera, Cecidomyiidae) on greenbugs. *Acta Entomol. Fenn.* 53:3–5.

Bennison, J. A., S. Hockland, and R. Jacobson. 1990. Recent developments with integrated control of thrips on cucumber in the United Kingdom. *Bull. SROP/WPRS*. XIII/5:19–26.

Benuzzi, M., G. Nicoli, and G. Manzaroli. 1990. Biological control of *Bemisia tabaci* (Genn.) and *Trialeurodes vaporariorum* (Westw.) by *Encarsia formosa* Gahan on poinsettia. *Bull. SROP/WPRS* XIII/5:27–31.

Binns, E. S. 1977. Fungus gnats. *The Garden: Royal Hort. Soc.* 102:76–78.

Binns, E. S., R. A. Hall, and R. J.J. Pickford. 1982. *Thrips tabaci* Lind. (Thysanoptera—Thripidae) distribution and behaviour on glasshouse cucumbers in relation to chemical and integrated control. *Entomol. Mon. Mag.* 118:55–68.

Blackman, R. L., and V. F. Eastop. 1984. Aphids on the World's Crops: An Identification and Information Guide. Wiley, Chichester, U.K.

Blackshaw, R. P. 1984. Studies on the chemical control of vine weevil larvae. *Proc. BCPC Conf.-Pests and Diseases.* Croydon, U. K. 3:1093–1098.

_____. 1988. A survey of insect-parasitic nematodes in Northern Ireland. *Ann. Appl. Biol.* 113:561–565.

Boemare, N. 1989. Importance of both variants in *Xenorhabdus* spp., bacteria associated with entomopathogenic nematodes, *Steinernematidae* and *Heterorhabditidae*. *Bull. SROP/WPRS* 1989/XII/4:3–4.

Bohmer, B. 1986. Pyrethroids are the best help against the new thrips. *Pflanzenschutz* 22:833–834.

Boisclair, J., G. J. Brueren, and J. C. van Lenteren. 1990. Can *Bemisia tabaci* be controlled with *Encarsia formosa*? *Bull. SROP/WPRS* XIII/5:32–35.

Bondarenko, N. V., and B. P. Asyakin. 1981. Behaviour of the predatory midge [*Aphidoletes aphidimyza* Rond.] and other aphidivorous insects in relation to population density of the prey. p. 6–14. In: V. P. Pristavko (ed). Insect Behaviour as a Basis for Developing Control Measures Against Pests of Field Crops and Forests. Oxonian Press, New Delhi, India.

Bonde, J. 1989. Biological studies including population growth parameters of the predatory mite *Amblyseius barkeri* [Acarina: Phytoseiidae] at 25°C in the laboratory. *Entomophaga* 34:275–287.

Bordat, D., P. Robert, and M. Renaud. 1988. Susceptibility of *Liriomyza trifolii* (Burgess) and *Liriomyza sativae* (Blanchard) (Diptera: Agromyzidae) to eleven strains of entomopathogenic fungi. *Agron. Trop.* 43:68–73.

Brodsgaard, H. F. 1987. *Frankliniella occidentalis* (Thysanoptera; Thripidae)—a new pest in Danish greenhouses, a review. *Tidsskr. Planteavl.* 93:83–91.

_____. 1989. Coloured sticky traps for *Frankliniella occidentalis* (Pergande) (Thysanoptera, Thripidae) in glasshouses. *J. Appl. Entomol.* 107:136–140.

_____. 1990. The effect of anisaldehyde as a scent attractant for *Frankliniella occidentalis* (Thysanoptera: Thripidae) and the response mechanism involved. *Bull. SROP/WPRS* XIII/5:36–38.

Brunt, A. A. 1986. Chapter 6. Transmission of diseases. p. 43–49. In: M. J. W. Cock, (ed.). *Bemisia tabaci*: A Literature Survey on the Cotton Whitefly with an Annotated Bibliography. CAB/FAO; International Institute of Biological Control, Silwood Park, England.

Butler, G. D., T. J. Henneberry, and T. E. Clayton. 1983. *Bemisia tabaci* (Homoptera: Aleyrodidae): development, oviposition and longevity in relation to temperature. *Ann. Ent. Soc. Am.* 76:310–313.

Buxton, J. H., R. Jacobson, M. Saynor, R. Storer, and L. R. Wardlow. 1990. An integrated pest management programme for peppers; three years trials experience. *Bull. SROP/WPRS* XIII/5:45–50.

Carner, G. R., and T. D. Canerday. 1968. Field and laboratory investigations with *Entomophthora fresenii*, a pathogen of *Tetranychus* spp. *J. Econ. Entomol.* 61:956–959.

Cavalloro, R. (Ed.) 1987. Integrated and Biological Control in Protected Crops. A. A. Balkema, Rotterdam.

_____. (Ed.) 1991. Practical Application of Integrated Control in Protected Crops. EEC Joint Expert's Meeting, Antibes, France 16–19 October 1989 (in press).

Chambers, R. J. 1986. Preliminary experiments on the potential of hoverflies [Dipt., Syrphidae] for the control of aphids under glass. *Entomophaga* 31:197–204.

_____. 1988. Syrphidae. p. 259–270. In: A. K. Minks and P. Harrewijn (eds.). Aphids, Their Biology, Natural Enemies and Control, Vol. 2B. Elsevier Science Publishers B. V., Amsterdam, the Netherlands.

_____. 1990. The use of *Aphidoletes aphidimyza* for aphid control under glass. *Bull. SROP/WPRS* XIII/5:51–54.

Cone, W. W., L. M. Mcdonough, J. C. Maitlen, and S. Burdajewicz. 1971a. Pheromone studies of the two-spotted spider mite. 1. Evidence of a sex pheromone. *J. Econ. Entomol.* 64:355–358.

Cone, W. W., S. Predki, and E. C. Klostermeyer. 1971b. Pheromone studies of the two-spotted spider mite. 2. Behavioral responses of males to quiescent deutonymphs. *J. Econ. Entomol.* 64:379–382.

Copland, M. J. W., and W. Soeprapto. 1985. Biology of the glasshouse leaf-hopper and its parasite. p. 58–61. In: N. W. Hussey and N. E. A. Scopes (eds). Biological Pest Control: The Glasshouse Experience. Blandford Press, Poole, Dorset, U.K.

Coremans-Pelseneer, J., and F. Tillemans. 1989. *Beauveria brongniartii* as biological control agent against *Otiorhynchus sulcatus*. *Bull. SROP/WPRS* 1989/XII/4:37–38.

Coudreit, D. L., N. Prabhaker, A. N. Kishara, and D. E. Meyerdirk. 1985. Variation in the development rate on different hosts and overwintering of the sweet potato whitefly *Bemisia tabaci* (Homoptera: Aleyrodidae). *Env. Ent.* 14:516–519.

Coudriet. D. L., N. Prabhaker, and D. E. Meyerdirk. 1985. Sweetpotato whitefly (Homoptera: Aleyrodidae): Effects of neem-seed extract on oviposition and immature stages. *Environ. Entomol.* 14:776–779.

Crook, N. E., J. D. Brown, and G. N. Foster. 1982. Isolation and characterisation of a granulosis virus from the tomato moth, *Lacanobia oleracea*, and its potential as a control agent. *J. Invertebr. Pathol.* 40:221–227.

Cross, J. V., L. R. Wardlow, R. Hall, M. Saynor, and P. Bassett. 1983. Integrated control of chrysanthemum pests. *Bull. SROP/WPRS* 1983/VI/6:181–185.

Cross, J. V., and P. Bassett. 1982. Damage to tomato and aubergine by broad mite, *Polyphagotarsonemus latus* (Banks). *Plant Path.* 31:391–393.

Dawson, G. W., D. C. Griffiths, N. F. Janes, A. Mudd, J. A. Pickett, L. J. Wadhams, and C. M. Woodcock. 1987. Identification of an aphid sex pheromone. *Nature* 325:614–616.

Dawson, G. W., D. C. Griffiths, J. A. Pickett, and C. M. Woodcock. 1983. Decreased response to alarm pheromone by insecticide resistant aphids. *Naturwissenschaften* 70:254–255.

Dedryver, C. A., and J. M. Rabasse. 1982. Attempt at biological control of lettuce aphids in glasshouses with resting spores of *Conidiobolus obscurus* (Hall and Dunn), and mycelium of *Erynia neoaphidis* (Renaud and Henn). *Proc. 3rd Int. Coll. Inv. Path.* p. 103.

de Klerk, M-L. J., and P. M. J. Ramakers. 1986. Monitoring population densities of the phytoseiid predator *Amblyseius cucumeris* and its prey after large scale introductions to control *Thrips tabaci* on sweet pepper. *Meded. Fac. Landbouww. Rijksuniv. Gent* 51:1045–1048.

de Moraes, G. J., and J. A. McMurtrey. 1985. Chemically mediated arrestment of the predaceous mite Phytoseiulus persimilis by extracts of Tetranychus evansi and Tetranhchus urticae. Expt. Appl. Acarol. 1:127–138.

Deseo, K. V., M. Costanzi, and E. Orsi. 1986. Control of pests with insect-parasitic nematodes in protected crops. In: R. A. Samson, J. M. Vlak, and D. Peters (eds.). Fundamental and Applied Aspects of Invertebrate Pathology. Proc. 4th Int. Coll. Inv. Path. p. 327.

Dicke, M., M. W. Sabelis, and A. Groeneveld. 1986. Vitamin A deficiency modifies response of predatory mite Amblyseius potentillae to volatile kairomone of two-spotted mite, Tetranychus urticae. J. Chem. Ecol. 12:1389–1396.

Dolmans, N. G. M. 1983. Biological control of the black vine weevil (Otiorhynchus sulcatus) with a nematode (Heterorhabditis sp.). Meded. Fac. Landbouww. Rijksuniv. Gent. 48 (2):417–420.

Domsch, K. H., W. Gams, and T. Anderson. 1980. 1980 Compendium of Soil Fungi, Vol. I and II:895 + 405 pp. Academic Press, London, U.K.

Donegan, K., and B. Lighthart. 1989. Effect of several stress factors on the susceptibility of the predatory insect, Chrysoperla carnea (Neuroptera: Chrysopidae), to the fungal pathogen Beauveria bassiana. J. Invertebr. Pathol. 54:79–84.

Dosse, G. 1955. Aus der Biologie der Raubmilbe Typhlodromus cucumeris Oud. (Acar., Phytoseiidae). Z. Pflanzenkr. Pflanzenschutz 62:593–598.

———. 1961. Uber die Bedeutung der Pollennahrung fur Typhlodromus pyri Scheuten (= tiliae Oud.) (Acari: Phytoseiidae). Entomol. Exp. Appl. 4:191–195.

Drummond, J., and J. B. Heale. 1988. Genetic studies on the inheritance of pathogenicity in Verticillium lecanii against Trialeurodes vaporariorum. J. Invertebr. Pathol. 52:57–65.

Duffas, J. E. 1965. Beet pseudo-yellows virus, transmitted by the glasshouse whitefly (Trialeurodes vaporariorum). Phytopathology 55:450–453.

Dupont, L. M. 1979. On gene flow between Tetranychus urticae Koch 1936, and Tetranychus cinnabarinus (Boisduval) Boudreaux 1956, (Acari:Tetranychidae): Synonymy between the two species. Entomol. Expt. Appl. 25:297–303.

Dyadechko, W. P. 1964. Thrips or Fringe-winged Insects (Thysanoptera) of the European Part of the USSR. Vroshai Publishers, Kiev, USSR.

Enkegaard, A. 1990. Age-specific fecundity and adult longevity of the cotton whitefly, Bemisia tabaci (Genn.) (Homoptera: Aleyrodidae) on Poinsettia (Euphorbia pulcherrima) at different temperatures. Bull. SROP/WPRS XIII/5:55–60.

Epsky, N. D., D. E. Walter, and J. L. Capinera. 1988. Potential role of nematophagous microarthropods as biotic mortality factors of entomogenous nematodes (Rhabditidae: Steinernematidae, Heterorhabditidae). J. Econ. Entomol. 81:821–825.

Ewing, H. E. 1914. The common red spider or spider mite. Oreg. Agr. Expt. Sta. Bull. 121.

Fagg, J., and J. T. Fletcher. 1987. Studies the epidemiology and control of tomato stem rot caused by Didymella lycopersici. Plant Pathol. 36:361–366.

Filippov, N. A. 1989. The present state and future outlook of biological control in the USSR. Acta Entomol. Fenn. 53:11–18.

Finney, J. R. 1988. World crop protection prospects: Demisting the crystal ball. Proc. 1988 Brighton BCPC Conference, Pests and Diseases 1:3–14.

Fischoff, D. A., K. S. Bowdish, F. J. Perlak, P. G. Marrone, S. M. McCormick, J. G. Niedermeyer, D. A. Dean, K. Kusano-Kretzmer, E. J. Mayer, et al. 1987. Insect tolerant transgenic tomato plants. Biotechnology 5:807–814.

Foster, G. N. 1979. Cultural factors affecting pest incidence in glasshouse tomato crops. Proceedings 1979 British Crop Protection Conference—Pests and Diseases, p. 441–445.

———. 1987. Control of the tomato moth (Lacanobia oleracea). The West of Scotland Agricultural College Report 1987, p. 111–118.

Foster, G. N., and J. Barker. 1978. A new biotype of red spider mite (Tetranychus urticae (Koch)) causing atypical damage to tomatoes. Plant Pathol. 27:47–48.

Foster, G. N., and N. E. Crook. 1983. A granulosis disease of the tomato moth. Lacanobia oleracea L.. Bull. SROP/WPRS 1983/VI/6:163–166.

Fougeroux, S. 1987. Un nouveau ravageur en France, le thrips du saintpaulia. Biologie du thrips. Quelles sont les possibilities de lutte? In: Conference Internationale sur les ravageurs en agriculture/International Conference on pests in Agriculture (Paris, December 1987). Annales ANPP 6, (I/III):474–482.

Fransen, J. J. 1990. Development of Bemisia tabaci (Gennadius) (Homoptera: Aleyrodidae) on Poinsettia and other potplants grown under glass. Bull. SROP/WPRS XIII/5:61–63.

Fransen, J. J., C. Winkelman, and J. C. van Lenteren. 1987. The differential mortality of stages of glasshouse whitefly, Trialeurodes vaporariorum, by infection of the entomopathogenic fungus Aschersonia aleyrodes. J. Invertebr. Pathol. 50:158–165.

Franz, J. M., H. Bogenschutz, S. A. Hassan, P. Huang, E. Naton, H. Suter, and G. Viaggiani. 1980. Results of a joint pesticide test programme by the Working Group "Pesticides and Beneficial Arthropods." Entomophaga 25:231–236.

Freeman, P. 1983. Sciarid flies (Diptera, Sciaridae). Handbooks for the Identification of British Insects no. 9(6), 68 pp. Royal Entomological Society of London, UK.

French, N., F. A. B. Ludlam, and L. R. Wardlow. 1973. Observations on the effects of insecticides on glasshouse whitefly (Trialeurodes vaporariorum (Westw.)). Plant Pathol. 22:99–107.

Freuler, J., and M. Benz. 1988. La sensibilite en laboratorie du thrips de l'oignon, Thrips tabaci Lind., et du thrips de Californie, Frankliniella occidentalis Pergande a l'egard de l'etrimfos, du furathiocarbe et de la cypermethrine. Rev. Suisse Vitic. Arboric. Hortic. 20:335–336.

Gardner, W. A., R. D. Oetting, and G. K. Storey. 1982. Susceptibility of the two-spotted spider mite, Tetranychus urticae Koch, to the fungal pathogen, Hirsutella thompsonii Fischer. Flor. Entomol. 65:458–465.

————. 1984. Scheduling of Verticillium lecanii and benomyl applications to maintain aphid (Homoptera: Aphidae) control on chrysanthemums in glasshouses. J. Econ. Entomol. 77:514–518.

Garrido, A., F. Beitia, and P. Gruenholz. 1984. Effects of PP618 on immature stages of Encarsia formosa and Cales noacki. Proc. of BCPC Pests and Diseases 1984 Conference, Brighton, UK, p. 305–310.

Gaugler, R., L. Lebeck, B. Nakagaki, and G. M. Boush. 1980. Orientation of the entomogenous nematode, Neoaplectana carpocapsae, to carbon dioxide. Environ. Entomol. 9:649–652.

Gelertner, W. D., and B. A. Federici. 1986. Isolation, identification and determination of a nuclear polyhedrosis virus from the beet armyworm, Spodoptera exigua, (Lepidoptera: Noctuidae). Environ. Entomol. 15:240–245.

Gelertner, W. D., N. C. Toscano, K. Kidd, and B. A. Federici. 1986. Comparison of a nuclear polyhedrosis virus and chemical insecticides for control of beet armyworm, Spodoptera exigua, (Lepidoptera: Noctuidae) on head lettuce. J. Econ. Entomol. 79:714–717.

Georgis, R., and G. O. Poinar Jr. 1984. Greenhouse control of the black vine weevil, Otiorhynchus sulcatus (Coleoptera: Curculionidae) by Heterorhabditid and Steinernematid nematodes. Environ. Entomol. 13:1138–1140.

Gerling, D. 1990. Whiteflies: their bionomics, pest status and management. In: Gerling, D. (ed.). Intercept Ltd., Andover, Hants, UK.

Gibson, R. W., and J. A. Pickett. 1983. Wild potato repels aphids by release of aphid alarm pheromone. Nature 302: 608–609.

Gibson, R. W., A. D. Rice, J. A. Pickett, M. C. Smith, and R. M. Sawicki. 1982. The effects of the repellants dodecanoic acid and polygodial on the acquisition of non-, semi- and persistent plant viruses by the aphid Myzus persicae. Ann. Appl. Biol. 100:55–59.

Gilkeson, L. A. 1990a. Cold storage of the predatory midge, Aphidoletes aphidimyza (Diptera: Cecidomyiidae). J. Econ. Entomol. 83:965–970.

_____ .1990b. Biological control of aphids in greenhouse sweet peppers and tomatoes. *Bull. SROP/WPRS.* XIII/5:64–70.

Gilkeson, L. A., and S. B. Hill. 1986a. Genetic selection for and evaluation of nondiapause lines of predatory midge, *Aphidoletes aphidimyza* (Rondani) (Diptera: Cecidomyiidae). *Can. Entomol.* 118:869–879.

_____ . 1986b. Diapause prevention in *Aphidoletes aphidimya* (Diptera: Cecidomyiidae) by low-intensity light. *Environ. Entomol.* 15:1067–1069.

Gilkeson, L. A., W. D. Morewood, and D. E. Elliott. 1990. Current status of biological control of thrips in Canadian greenhouses with *Amblyseius cucumeris* and *Orius tristicolor.* *Bull. SROP/WPRS.* XIII/5:71–75.

Gillespie, A. T. 1984. The potential of entomogenous fungi to control glasshouse pests and brown planthopper of rice. PhD Thesis, University of Southampton, UK.

_____ . 1986. The potential of entomogenous fungi as control agent for onion thrips, *Thrips tabaci.* BCPC Monograph no. 34: pp. 237–243. Biotechnology, Crop Improvement and Protection.

_____ . 1988a. The use of insect-pathogenic fungi for pest control. *Glasshouse Crops Res. Inst. Annu. Rpt. 1987.* p. 84–90.

_____ . 1988b. Use of fungi to control pests of agricultural importance. p. 37–60. In: M. Burge (ed.). Fungi in Biological Control Systems. Manchester Univ. Press, UK.

_____ . 1989a. Potential of entomogenous fungi for the control of Western Flower Thrips and other glasshouse pests. *Inst. Hort. Res. Annu. Rep. 1988.* p. 13.

_____ . 1989b. Death by fungus. *Grower* 111 (7):24–27.

_____ . 1989c. The use of fungi to control the black vine weevil, *Otiorhynchus sulcatus,* on ornamentals. *Bull. SROP/WPRS* 1989/XII/4:36.

Gillespie, A. T., R. A. Hall, and H. D. Burges. 1982a. Control of onion thrips, *Thrips tabaci,* and the red spider mite, *Tetranychus urticae,* by *Verticillium lecanii.* Proc. Int. Coll. Inv. Path. and Microbial Control, Brighton, UK, 1982., p. 100.

_____ . 1982b. Entomogenous fungi as control agents for the glasshouse leaf hopper, *Zygaena pallidifrons.* Proc. Int. Coll. Inv. Path. Microbial Control, Brighton, UK 1982., p. 108.

Gillespie, A. T., and E. R. Moorhouse. 1989. The use of fungi to control pests of agricultural and horticultural importance. p. 55–84. In: J. M. Whipps and R. D. Lumsden (eds.). The biotechnology of fungi for improving plant growth. Cambridge Univ. Press, UK.

Gillespie, D. R., 1989. Biological control of thrips [Thysanoptera: Thripidae] on greenhouse cucumber by *Amblyseius cucumeris.* *Entomophaga* 34:185–192.

Gillespie, D. R., and D. Quiring. 1987. Yellow sticky traps for detecting and monitoring glasshouse whitefly (Homoptera: Aleyrodidae) adults on greenhouse tomato crops.*J. Econ. Entomol.* 80:675–679.

Gough, N. 1988. Mites: Their Biology and Control on Ornamental Plants. Queensland Department of Primary Industries.

Gould, H. J. 1985. The advisory problem. p. 219–223. In: N. W. Hussey and N. E. A. Scopes (eds.). Biological Pest Control: The Glasshouse Experience. Blandford Press, Poole, Dorset, UK.

Gould, H. J., and H. G. Kingham. 1964. The efficiency of commercial high volume spraying with acaricides on cucumbers under glass. *Plant Pathol.* 13:60–64.

Guillon, M. 1987. Marketing of baculovirus-based biological insecticide applications to *Mamestra brassicae* and *Spodoptera littoralis* NPV. *Meded. Fac. Landbouww. Rijksuniv. Gent* 52 (2a):147–153.

Hagvar, E. B. and T. Hofsvang. 1990. The aphid parasitoid *Ephedrus cerasicola,* a possible candidate for biological control in glasshouses? *Bull. SROP/WPRS.* XIII/5:87–90.

Hall, R. A. 1980. Laboratory infection of insects by *Verticillium lecanii* strains isolated from phytopathogenic fungi. *Trans. Br. Mycol. Soc.* 74:445–446.

_____ . 1981. The fungus, *Verticillium lecanii,* as a microbial insecticide against aphids and scales. p. 483–498. In: H. D. Burges (ed.). Microbial Control of Pests and Diseases,

1970–1980. Academic Press, London, U. K.

———. 1982. Control of whitefly, *Trialeurodes vaporariorum*, and cotton aphid, *Aphis gossypii*, in glasshouses by the fungus *Verticillium lecanii*. *Ann. Appl. Biol.* 101:1–11.

———. 1985a. Whitefly control by fungi. p. 116–118. In: N. W. Hussey and N. E. A. Scopes (eds.). Biological Pest Control: The Glasshouse Experience. Blandford Press, Poole, Dorset, UK.

———. 1985b. Aphid control by fungi. p. 138–141. In: N. W. Hussey and N. E. A. Scopes (eds.). Biological Pest Control: The Glasshouse Experience. Blandford Press, Poole, Dorset, UK.

Hall, R. A., and H. D. Burges. 1979. Control of aphids in glasshouses with the fungus *Verticillium lecanii*. *Ann. Appl. Biol.* 93:235–246.

Hansen, L. S. 1983. Introduction of *Aphidoletes aphidimyza* (Rond.) (Diptera: Cecidomyiidae) from an open rearing unit for the control of aphids in glasshouses. *Bull. SROP/WPRS* 1983/VI/3:146–150.

———. 1988. Control of *Thrips tabaci* [Thysanoptera: Thripidae] on glasshouse cucumber using large introductions of predatory mites *Amblyseius barkeri* [Acarina: Phytoseiidae] *Entomophaga* 33:33–42.

Hansen, L. S., and J. Geyti. 1987. Possibilities and limitations of the use of *Amblyseius mckenziei* Sch. and Pr. for biological control of thrips (*Thrips tabaci* Lind.) on glasshouse crops of cucumber. p. 145–150. In: R. Cavalloro (ed.). Integrated and Biological Control in Protected Crops. A. A. Balkema, Rotterdam.

Hargreaves, E. 1915. The life-history and habits of the greenhouse white fly (*Aleyrodes vaporariorum* Westw.). *Ann. App. Biol.* 1:303–334.

Harris, K. M. 1973. Aphidophagous Cecidomyiidae (Diptera): taxonomy, biology and assessments of field populations. *Bull. Entomol. Res.* 63:305–325.

Harris, M., and D. Warkentin. 1988. Leafminers? Send in the nematodes. *Grower Talks* 52:74–78.

Hassan, S. A. 1982. Relative tolerance of three different strains of the predatory mite *Phytoseiulus persimilis* A.-H. (Acari, Phytoseiidae) to 11 pesticides used on glasshouse crops. *Z. Ang. Ent.* 93:55–63.

Hassan, S. A., R. Albert, F. Bigler, P. Blaisinger, H. Bogenschutz, E. Boller, J. Brun, P. Chiverton, P. Edwards, W. D. Englert, P. Huang, C. Inglesfield, E. Naton, P. A. Oomen, W. P. J. Overmeer, W. Rieckmann, L. Samsoe-Petersen, A. L. Staubli, J. J. Tuset, G. Viggiani, and G. van Wetswinkel. 1987. Results of the third joint pesticide testing programme by the IOBC/WPRS- Working Group "Pesticides and beneficial organisms." *J. Appl. Entomol.* 103:92–107.

Hassan, S. A., F. Bigler, H. Bogenschutz, E. Boller, J. Brun, P. Chiverton, P. Edwards, F. Mansour, E. Naton et al. 1988. Results of the fourth joint pesticide testing programme carried out by the IOBC/WPRS Working Group "Pesticides and Beneficial Organisms". *J. Appl. Entomol.* 105:321–329.

Hassan, S. A., F. Bigler, H. Bogenschutz, J. U. Brown, S. I. Firth, P. Huang, M. S. Ledieu, E. Naton, P. A. Oomen, W. P. J. Overmeer, W. Rieckmann, L. Samsoe-Petersen, G. Giaggiani, and A. Q. van Zon. 1983. Results of the second joint pesticide testing programme by the IOBC/WPRS Working Group "Pesticides and Beneficial Arthropods." *J. Appl. Entomol.* 95:151–158.

Havelka, J. 1980. Photoperiodism of the carnivorous midge *Aphidoletes aphidimyza* (Diptera, Cecidomyiidae). *Ent. Rev.* 59:1–8.

Havelka, J., and R. Zemek. 1988. Intraspecific variability of the aphidophagous gall midge *Aphidoletes aphidimyza* (Rondani) (Dipt., Cecidomyiidae) and its importance for biological control of aphids. *J. Appl. Entomol.* 105:280–288.

Heale, J. B. 1988. The potential impact of fungal genetics and molecular biology on biological control with particular reference to entomopathogens. p. 211–235 In: M. Burge (ed.). Fungi in Biological Control Systems. Manchester Univ. Press, UK.

Heale, J. B., J. E. Isaac, and D. Chandler. 1989. Prospects for strain improvement in entomopathogenic fungi. *Pestic. Sci.* 26:79–92.

Heimpel, A. M. 1971. Safety of insect pathogens for man and vertebrates. p. 469–489 In: H. D. Burges and N. W. Hussey (eds.). Microbial Control of Insects and Mites. Academic Press, U. K.

Helyer, N. L. 1990. Evaluation of phytotoxicity of pesticides to protected edible crops. *Tests Agrochem. Cultiv.,* 11:82–83.

Helyer, N. L., P. Bassett, and J. V. Cross. 1984. Spider mites on protected crops. *ADAS Leaflet 224.*

Helyer, N. L., G. Grimmett and R. J. J. Pickford. 1983a. The use of polybutenes in crop protection. *Proc. 10th Int. Congress Plant Protection 1983* 2:573.

Helyer, N. L., and M. S. Ledieu. 1986. The potential of heptenophos and Mk936 pesticides for control of minor pests in integrated pest control programmes under glass. *Agr. Ecosys. Environ.* 17:287–292.

_____. 1990. Seasonal variability in tomato leaf-miner (*Liriomyza bryoniae* (Kaltenbach)) life cycle. *Bull. SROP/WPRS* 1990/XIII/5:83-86.

Helyer, N. L., M. S. Ledieu, N. W. Hussey, and N. E. A. Scopes. 1983b. Early season integrated control of whitefly in tomatoes using oxamyl. *Proc. 1984 British Crop Protection Conference—Pests and Diseases* 1:293–298.

Helyer, N. L., and C. C. Payne. 1986. Current progress and future developments in integrated pest management on protected vegetable crops. *Aspects of Applied Biology* 12, (Crop Protection in Vegetables) p. 171–187.

Helyer, N. L., and L. R. Wardlow. 1987. Aphid control on chrysanthemum, using frequent, low dose, applications of *Verticillium lecanii. Bull. SROP/WPRS* 1987/X/2:62–65.

Herrnstadt, C., G. G. Soares, E. R. Wilcox, and P. L. Edwards. 1986. A new strain of *Bacillus thuringiensis,* with activity against coleopteran insects. *Biotechnology* 4:305–308.

Hervé, J. J., S. Smolikowski, P. Pastre, C. Piedallu, and L. Roa. 1977. NRDC 161 (RU22974): A new pyrethroid insecticide for use in agricultural crops. *Proc.Brit. Crop Prot. Conf. Pests and Diseases,* 613–621.

Heungens, A., G. Buysse, and D. Vermaerke. 1989. Control of *Frankliniella occidentalis* on *Chrysanthemum indicum* with pesticides. *Meded. Fac. Landbouww. Rijksuniv. Gent* 54 (3b):975–981.

Hocart, C. W. 1981. Whitefly control on Guernsey tomatoes by *Encarsia formosa.* Ph.D. Thesis, Univ. of Lond. (Imperial Coll.).

Hockland, S. H., G. W. Dawson, D. C. Griffiths, B. Marples, J. A. Pickett, and C. M. Woodcock. 1986. The use of aphid alarm pheromone ((E)-β-farnesene) to increase effectiveness of the entomophilic fungus *Verticillium lecanii* in controlling aphids on chrysanthemums under glass. p. 252 In: A. Samson, J. M. Vlak, and D. Peters (eds.). Fundamental and Applied Aspects of Inv. Path. *Foundation of the 4th Int. Coll. on Inv. Path.,* Wageningen, Netherlands.

Hudson, K. E. 1974. Regulation of greenhouse sciarid fly populations using *Tetradonema plicans* (Nematoda: Mermithoidea). *J. Invertebr. Pathol.* 23:85–91.

Hughes, J. C., and A. T. Gillespie. 1985. Germination and penetration of the aphid *Macrosiphoniella sanborni* by two strains of *Verticillium lecanii. Programme and Abstracts XVIII Ann. Meeting of Soc. of Inv. Pathology,* Sault St. Marie, Canada. p. 28.

Hulspas-Jordan, P. M., E. E. Christochowitz, J. Woets, and J. C. van Lenteren. 1987. The parasite-host relationship between *Encarsia formosa* (Hymenoptera: Aphelinidae) and *Trialeurodes vaporariorum* (Homoptera: Aleyrodidae) XXIV. *J. Appl. Entomol.* 103:368–378.

Hulspas-Jordan, P. M., and J. C. van Lenteren. 1989. The parasite-host relationship between *Encarsia formosa* (Hymenoptera: Aphelinidae) and *Trialeurodes vaporariorum* (Homoptera: Aleyrodidae) XXX. Modelling population growth of greenhouse whitefly on tomato. *Agricultural University Wageningen Papers* 89-2 (1989): p. 1–54.

Hussey, N. W. 1985a. History of Biological Control in Protected Culture: Western Europe.

p. 11–12 In: N. W. Hussey and N. E. A. Scopes (eds.). Biological Pest control: The Glasshouse Experience. Blandford Press, Poole, Dorset, UK.

———. 1985b. The economic equation. p. 224–228. In: N. W. Hussey and N. E. A. Scopes (eds.). Biological Pest Control: The Glasshouse Experience. Blandford Press, Poole, Dorset, UK.

Hussey, N. W., and L. B. Huffaker. 1976. Spider mites. p. 179–228 In: V. L. Delucchi (ed.). Studies in Biological Control. Cambridge Univ. Press, Cambridge, UK.

Hussey, N. W., and W. J. Parr. 1963. The effects of glasshouse red spider mite (Tetranychus urticae Koch) on the yield of cucumbers. J. Hort. Sci. 57:93–101.

Hussey, N. W., W. J. Parr, and B. Gurney. 1958. The effect of whitefly populations on the cropping of tomatoes. Glasshouse Crops Res. Inst. Annu. Rep. 1957, p. 79–86.

Hussey, N. W., W. H. Read, and J. J. Hesling. 1969. The Pests of Protected Cultivation. Edward Arnold (Publisher) Ltd., London.

Hussey, N. W., and N. E. A. Scopes. 1985. Biological Pest Control: The Glasshouse Experience. Blandford Press, Poole, Dorset, UK.

Ibrahim, A. G., and D. S. Madge. 1979. Parasitization of the chrysanthemum leafminer Phytomyza syngenesiae (Hardy) (Dipt., Agromyzidae), by Diglyphus iseae (Walker) (Hym., Eulophidae). Entomol. Mon. Mag. 114:71–81.

Jacobson, R. 1989a. WFT: Steps towards control. Grower 112 (22):16–21.

———. 1989b. Living with WFT. Grower 111(3):24–29.

Jakobsen, J. 1987. Muligheder for biologisk bekaempele af blomstertripsen Frankliniella occidentalis ved hjaelp af tripsrovmiden Amblyseius barkeri [In Danish] Danske Plantevaernskonference 1987, 73–79.

Jarrett, P. 1985a. Biology of greenhouse caterpillar pests. p. 99–103 In: N. W. Hussey and N. E. A. Scopes (eds). Biological Pest Control: The Glasshouse Experience. Blandford Press, Poole, Dorset, UK.

———. 1985b. Experience with the selective control of caterpillars using Bacillus thuringiensis. pp. 142–144. In: N. W. Hussey and N. E. A. Scopes (eds). Biological Pest Control: The Glasshouse Experience. Blandford Press, Poole, Dorset, UK.

Jarrett, P., and H. D. Burges. 1982. Control of tomato moth, Lacanobia oleracea, on glasshouse tomatoes and the influence of larval behaviour. Entomol. Expt. Appl. 31:239–244.

———. 1986. Isolates of Bacillus thuringiensis active against Mamestra brassicae and some other species: alternatives to the present commercial isolate HD1. Biol. Agr. Hort. 4:39–45.

Jarrett, P., H.D Burges, and G. A. Matthews. 1978. Penetration of controlled drop spray of Bacillus thuringiensis into chrysanthemum beds compared with high volume spray and thermal fog. BCPC Monograph no. 22, Controlled Drop Application. p. 75–82.

Jepson, L. R., H. H. Keifer, and E. W. Baker. 1975. Mites Injurious to Economic Plants. Univ. Calif. Press Berkeley.

Jones. K. C., and J. A. Klocke. 1987. Aphid feeding deterrency of ellagitannins, their phenolic hydrolysis products and related phenolic derivatives. Entomol. Expt. Appl. 44:229–234.

Jones, V. P., and M. P. Parrella. 1986. The movement and dispersal of Liriomyza trifolii (Diptera: Agromyzidae) in a chrysanthemum greenhouse. Ann. Appl. Biol. 109:33–39.

Kajita, H., and J. C. van Lenteren. 1982. The parasite-host relationship between Encarsia formosa (Homoptera: Aleyrodidae). XIII. Effect of low temperatures on egg maturation of E. formosa. J. Appl. Entomol. 93:430–439.

Kanagaratnam, P., R. A. Hall, and H. D. Burges. 1982. Control of glasshouse whitefly, Trialeurodes vaporariorum, by an 'aphid' strain of the fungus Verticillium lecanii. Ann. Appl. Biol. 100:213–219.

Karelin, V. D., T. N. Yakovchuk, and V. P. Danu. 1989. Development of techniques for commercial production of the common green lacewing, Chrysopa carnea (Neuroptera, Chrysopidae). Acta Entomol. Fenn. 53:31–35.

Kaya, H. K. 1985. Entomogenous nematodes for insect control in IPM systems. p. 283–302.

In: M. A. Hoy and D. C. Herzog (eds.). Biological Control in Agricultural IPM Systems. Academic Press, Florida.

Kaya, H. K., and T. M. Burlando. 1989. Development of Steinernema feltiae (Rhabditidae: Steinernematidae) in diseased insect hosts. J. Invert. Pathol. 53:164–168.

Keller, S., and J. West. 1983. Observations on three species of Neozygites (Zygomycetes: Entomophthoraceae). Entomophaga 28:123–124.

Klingler, J. 1986. Investigations on the parasitism of Otiorhynchus salicicola and O. sulcatus (Coleoptera: Curulionidae) by Heterorhabditis sp. (Nematoda). Entomophaga 33:325–331.

──────. 1989. Use of Heterorhabditis sp. against Otiorhynchus sulcatus in Switzerland. Bull. SROP/WPRS 1989/XII/4:7-8.

Kulp, D., M. Fortmann, M. Hommes, and H-P. Plate. 1989. Die rauberische Gallmucke Aphidoletes aphidimyza (Rondani) (Diptera: Cecidomyiidae)—Ein bedeutender Blattlauspradator—Nachschlagewerk zur Systematik, Verbreitung, Biologie, Zucht und Anwendung. Mitteilungen aus der Biologischen Bundesanstalt fur Land- und Fortwirtschaft, Berlin-Dahlem (FRG), 250:1–126.

Kuo-Sell, H. L., and K. Kreisfeld. 1987. Zur Wirtseignung verschiedener Getreideblatt- lausarten fur den Parasitoiden Aphelinus asychis (Walker). Meded. Fac. Landbouww. Rijksuniv. Gent, 52 (2a):353–362.

Larew, H. G. 1987. Use of neem seed kernal extract in a developed country: Liromyza leafminer as a model case. p. 375–385 In: H. Schmutterer and K. R. S. Ascher (eds.). Proceedings of the Third International Neem Conference. Nairobi, 1986.

Larson, R. O. 1987. Development of Margosan-OR, a pesticide from neem seed. p. 243–250 In: H. Schmutterer and K. R. S. Ascher (eds.). Proc. Third Intern. Neem Conference. Nairobi, 1986.

Laska, P., J. Betlach, and M. Havrankova. 1986. Variable resistance in sweet pepper, Capsicum annuum, to glasshouse whitefly, Trialeurodes vaporariorum (Homoptera, Aleyrodidae). Acta Entomol. Bohemoslov. 83:347–353.

Ledieu. M. S. 1976. Dispersal of the parasite Encarsia fòrmosa as influenced by its host, Trialeurodes vaporariorum. Bull. SROP/WPRS 1976/4, 121–124.

──────. 1978. Candidate insecticides for the control of larvae of Mamestra brassicae (Lepidoptera) (Noctuidae). Ann. Appl. Biol. 88:251–255.

Ledieu, M. S., and N. L. Helyer. 1981. Problems associated with a virulent strain of red spider mite (Tetranychus urticae) causing hypertoxic damage to glasshouse tomatoes. Proc 1981 British Crop Protection Conference—Pests and Diseases, p. 105–108.

──────. 1982. Effect of tomato leaf miner on yield of tomatoes. Glasshouse Crops Res. Inst. Annu. Rep. 1981. p. 108.

──────. 1983a. Integration of pesticides with biological control agents. Proc. 10th Int. Cong. Plant Protection, p. 1111.

──────. 1983b. Pyrazophos: A fungicide with insecticidal properties including activity against chrysanthemum leaf miner (Phytomyza syngenesiae) (Agromyzidae). Ann. Appl. Biol. 102:275–279.

──────. 1985. Observations on the economic importance of tomato leafminer (Liriomyza bryoniae) (Agromyzidae). Agr. Ecosys. Environ. 13:103–109.

Ledieu, M. S., N. L. Helyer, and D. M. Derbyshire. 1989. Pests and diseases of protected crops. pp. 405–511. In: Pest and Disease Control Handbook, 3rd edition, British Crop Protection Council, Bracknell, UK.

Leibee, G. L. 1984. Influence of temperature on the development of Liriomyza trifolii (Burgess) (Diptera: Agromyzidae) on celery. Environ. Ent. 13:497–501.

──────. 1986. Effect of light on the pupariation of Liriomyza trifolii (Diptera: Agromyzidae). Flo. Entomol. 49:758–759.

Li, Z. H., F. Lammes, J. C. van Lenteren, P. W. T. Huisman, A. van Vianen, and O. M. B. de Ponti. 1987. The parasite-host relationship between Encarsia formosa Gahan (Hymenoptera, Aphelinidae) and Trialeurodes vaporariorum (Westwood) (Homoptera,

Aleyrodidae). XXV. Influence of leaf structure on the searching activity of *Encarsia formosa. J. Appl. Entomol.* 104:297–304.

Lindqvist, I., and K. Tiittanen. 1989. Biological control of *Thrips tabaci* (Thysanoptera, Thripidae) on greenhouse cucumber. *Acta Entomol. Fenn.* 53:37–42.

Lipa, J. J. 1971. Microbial control of mites and ticks. p. 357–373. In: H. D. Burges and N. W. Hussey (eds.). Microbial Control of Insects and Mites. Academic Press, UK.

_____. 1985. History of biological control in protected culture: Eastern Europe. p. 23–29. In: N. W. Hussey and N. E. A. Scopes (eds.). Biological Pest Control: The Glasshouse Experience. Blandford Press, Poole, Dorset, UK.

Lloyd, L. L. 1920. The habits of the glasshouse tomato moth, *Hadena (Polia) oleracea* and its control. *Ann. Appl. Biol.* 7:66–102.

Lopez-Avila, A. 1986a. Chapter 1; Taxonomy and biology. p. 3–11. In: *Bemisia tabaci*: A Literature Survey on the Cotton Whitefly with an Annotated Bibliography. (Ed M. J. W. Cock). CAB/FAO; International Institute of Biological Control, Silwood Park, England.

_____. 1986b. Chapter 4; Natural enemies. p 27–35. In: M. J. W. Cock (ed.) *Bemisia tabaci*: A Literature Survey on the Cotton Whitefly with an Annotated Bibliography. CAB/FAO; International Institute of Biological Control, Silwood Park, England.

Lossbroek, T. G., and J. Theunissen. 1985. The entomogenous nematode *Neoaplectana bibionis* as a biological control agent of *Agrotis segetum* in lettuce. *Entomol. Expt. Appl.* 39:261–264.

Lyon, J. P. 1986. Specific problems posed by *Liriomyza trifolii* Burgess (Diptera: Agromyzidae) and biological control of this new pest of protected crops. *Colloq. INRA* 34, pp. 85–97.

Macgill, E. I. 1934. The biology of *Erythroneura pallidifrons* Edw. *Bull. Entomol. Res.* 23:43–43.

Macleod, D. M., D. Tyrell and K. P. Carl. 1976. *Entomophthora parvispora* sp. nov., a pathogen of *Thrips tabaci. Entomophaga* 21:307–312.

Madueke, E-D. N., and T. H. Coaker. 1984. Temperature requirements of the whitefly *Trialeurodes vaporariorum* (Homoptera: Aleyrodidae) and its parasitoid *Encarsia formosa* (Hymenoptera: Aphelinidae). *Entomol. Gen.* 9:149–154.

Mansour, F., K. R. S. Ascher, and N. Omari. 1987. Effects of neem (*Azadirachta indica*) seed kernal extracts from different solvents on the predacious mite *Phytoseiulus persimilis* and the phytophagous mite *Tetranychus cinnabarinus. Phytoparasitica* 15:125–130.

Mantel, W. P., and M. van de Vrie. 1988a. A contribution to the knowledge of Thysanoptera in ornamental and bulbous crops in the Netherlands. *Acta Phytopathol. Entomol. Hung.* 23:301–311.

_____. 1988b. De Californische trips, *Frankliniella occidentalis*, een nieuwe schadelijke tripssoort in de tuinbouw onder glas in Nederland. *Entomol. Ber. Amst.* 48:140–144.

Marchal, M. 1977. Fungi imperfecti isoles d'une population naturelle d'*Otiorhynchus sulcatus* F. (Coleoptera: Curculionidae). *Rev. Zool. Agr. Pathol. Veg.* 76:101–108.

Markkula, M., and K. Tiittanen. 1976. "Pest in first" and "natural infestation" methods in the control of *Tetranychus urticae* Koch with *Phytoseiulus persimilis* A.-H. on glasshouse cucumbers. *Ann. Agric. Fenn.* 15:81–85.

_____. 1977. Use of the predatory midge *Aphidoletes aphidimyza* (Rond.) (Diptera, Cecidomyiidae) against aphids on glasshouse cultures. p. 43–44. In: *Proc. XV Symposium Int. Congr. Entomol.*, Washington USDA-ARS-NE-85.

Martin, N. A., and P. Workman. 1984. Control of greenhouse whitefly with oxamyl granules. *Proc. 37th New Zeal. Weed and Pest Control Conference,* p. 265–267.

_____. 1986. Buprofezin: A selective pesticide for glasshouse whitefly control. *Proc. 39th New Zeal. Weed and Pest Control Conference.* p. 234–236.

McMurtry, J. A. 1982. The use of Phytoseiids for biological control: Progress and future prospects. p. 23–48. In: M. A. Hoy (ed.). Recent Advances in Knowledge of the Phytoseiidae. *Proc. Formal Conf. Acarology Soc. Am.; Ent. Soc. Am. meeting San Diego,*

Dec. 1981. 3284.

McMurtry, J. A., M. H. Badii, and H. G. Johnson. 1984. The broad mite, *Polyphagotarsonemus latus*, as a potential prey for Phytoseiid mites in California. *Entomophaga* 29:83–86.

Miller, J. P. 1987. Bug-eating nematodes hold promise as natural pesticide. *Wall Street* August 28, 1987.

Milner, R. J., and G. G. Lutton. 1986. Dependence of *Verticillium lecanii* (Fungi: Hyphomycetes) on high humidities for infection and sporulation using *Myzus persicae* (Homoptera: Aphididae) as host. *Environ. Entomol.* 15:380–382.

Minkenberg, O. P. J. M., and J. C. van Lenteren. 1986. The leafminers *Liriomyza bryoniae* and *L. trifolii* (Diptera: Agromyzidae), their parasites and host plants: A review. Agricultural University Wageningen Papers 86-2.

Minks, A. K., and P. Harrewijn. 1988. Aphids, Their Biology, Natural Enemies and Control. World Crop Pests Vols 2A and 2B. Elsevier Science Publishers B. V., Amsterdam, the Netherlands.

Moar, W. J., W. L. A. Osbrink, and J. T. Trumble. 1986. Potentiation of *Bacillus thuringiensis* var. *kurstaki* with Thuringiensin on beet armyworm (Lepidoptera: Noctudidae). *J. Econ. Entomol.* 79:1443–1446.

Moar, W. J., and J. T. Trumble. 1987. Biologically derived insecticide against beet armyworm. *Calif. Agr.* 41 (Nov–Dec 1987):13–15.

Moffit, H. R. 1964. A colour preference of the Western Flower Thrips, *Frankliniella occidentalis*. *J. Econ. Entomol.* 57:604–605.

Morrison, R. K. 1985. *Chrysopa carnea*. p. 419–426 In: P. Singh and R. F. Moore (eds.). Handbook of Insect Rearing, Vol. I. Elsevier Science Publishers B. V., Amsterdam, the Netherlands.

Mound, L. A. 1963. Host-correlated variation in *Bemisia tabaci* (Gennadius.) (Homoptera: Aleyrodidae). *Proc. Royal Ent. Soc. London.* (A) 38:171–180.

Munthali, D. C. 1984. Biological efficiency of small dicofol droplets against *Tetranychus urticae* (Koch) eggs, larvae and protonymphs. *Crop Prot.* 3:327–334.

Munthali, D. C., and I. J. Wyatt. 1986. Factors affecting the biological efficiency of small pesticide droplets against *Tetranychus urticae* eggs. *Pestic. Sci.* 17:155–164.

Nakanishi, K., and I. Kubo. 1977. Studies on the warburganal muzigadial and related compounds. *Isr. J. Chem.* 16:28–31.

Nechols, J. R., and M. J. Tauber. 1977. Age specific interaction between the glasshouse whitefly and *Encarsia formosa*: Influence of host on the parasite's oviposition and development. *Environ. Entomol.* 6:143–149.

Nell, H. W., L. A. Sevenster-van der Lelie, J. C. van Lenteren, and J. Woets. 1976. The parasite-host relationship between *Encarsia formosa* (Hymenoptera: Aphelinidae) and *Trialeurodes vaporariorum* (Homoptera: Aleyrodidae). II. Selection of host stages for oviposition and feeding by the parasite. *J. Appl. Entomol.* 81:372–376.

Niemczyk, E. 1988. Effectiveness of the bark bug (*Anthocoris nemorum* L.) as a predator of green apple aphid (*Aphis pomi* De Geer). pp. 279–283. In: E. Niemczyk and A. F. G. Dixon (eds.). Ecology and Effectiveness of Aphidophaga. SPB Academic Publishing, The Hague, the Netherlands.

Nijveldt, W. 1988. Cecidomyiidae. p. 271–278 In: A. K. Minks and P. Harrewijn (eds.). Aphids, Their Biology, Natural Enemies and Control, Vol. 2B: Elsevier Science Publishers B. V., Amsterdam, the Netherlands.

Nordlund, D. A., R. L. Jones, and W. J. Lewis, 1981. Semiochemicals, Their Role in Pest Control. Wiley, New York.

Oatman, E. R., and A. E. Michelbacher. 1959. The melon leaf miner *Liriomyza pictella* (Thomson) (Diptera: Agromyzidae) II. Ecological studies. *Ann. Ent. Soc. Am.* 52:83–89.

O'Reilly, C. J. 1974. Investigations on the biology and biological control of the glasshouse whitefly *Trialeurodes vaporariorum* (Westwood). PhD. thesis, National Univ. Ireland, Dublin.

Osborne, L. S. 1982. Temperature-dependent development of glasshouse whitefly and its parasite *Encarsia formosa*. *Environ. Entomol.* 11:483–485.

Osborne, L. S., D. G. Boucias, and R. K. Lindquist. 1985. Activity of *Bacillus thuringiensis* var. *israelensis* on *Bradysia coprophila* (Diptera: Sciaridae). *J. Econ. Entomol.* 78:922–925.

Osborne, L. S., K. Hoelmer, and D. Gerling. 1990. Prospects for control of *Bemisia tabaci*. *Bull. SROP/WPRS* XIII/5:153–160.

Osborne, L. S., and R. L. Petitt. 1985. Insecticidal soap and the predatory mite, *Phytoseiulus persimilis* (Acari: Phytoseiidae), used in management of the two-spotted spider mite (Acari: Tetranychidae) on greenhouse grown foliage plants. *J. Econ. Entomol.* 78:687–691.

Pak, G. A., and E. J. Dejong. 1987. Behavioural variations among strains of *Trichogramma* spp.: host recognition. *Neth. J. Zool.* 37:137–166.

Palmer, A., I. J. Wyatt, and N. E. A. Scopes. 1983. The toxicity of ULV permethrin to glasshouse whitefly. *Proc. 10th Int. Congr. Plant Protect.* 2:512.

Parr, W. J., H. J. Gould, N. H. Jessop, and F. A. B. Ludlam. 1976. Progress towards a biological control programme for glasshouse whitefly (*Trialeurodes vaporariorum*) on tomatoes. *Ann. Appl. Biol.* 83:349–363.

Parr, W. J., and N. W. Hussey. 1966. Diapause in the glasshouse redspider (*Tetranychus urticae* Koch) a synthesis of present knowledge. *Hort. Res.* 6:1–21.

Parrella, M. P., and J. A. Bethke. 1984. Biological studies of *Liriomyza huidobrensis* (Diptera: Agromyzidae) on chrysanthemum, aster and pea. *J. Econ. Entomol.* 77:342–345.

Paulson, G. S., and J. W. Beardsley. 1985. Whitefly (Hemiptera: Aleyrodidae) egg pedicel insertion into host plant stomata. *Ann. Entomol. Soc. Am.* 78:506–508.

Payne, C. C. 1988. Pathogens for the control of insects: Where next? *Philos. Trans. R. Soc. Lond. B. Biol. Sci.* 318:225–248.

Payne, C. C., and P. Jarrett. 1984. Microbial pesticides: Selection and genetic improvement. *BCPC Conference, Pests and Diseases 1984* 1:231–238.

Payne, C. C., and J. M. Lynch. 1988. Biological control. p. 261–287. In: J. M. Lynch and J. E. Hobbie (eds.). Micro-organisms in Action: Concepts and Applications in Microbial Ecology. Blackwell Scientific Publications, Oxford, UK.

Penman, D. R., and W. W. Cone. 1974. Role of web, tactile stimuli, and female sex pheromone in attraction of male two-spotted spider mites to quiescent female deutonymphs. *Ann. Entomol. Soc. Am.* 67:179–182.

Persoons, C. J., C. van der Kraan, W. J. Nooijen, F. J. Ritter, S. Voerman, and T. C. Baker. 1981. Sex pheromone of the beet armyworm, *Spodoptera exigua*: Isolation, identification and preliminary field evaluation. *Entomol. Expt. Appl.* 30:98–99.

Petersen, B. 1989. The effect of leaf surface characters on the foraging behaviour of the predatory mite, *Amblyseius cucumeris*: implications for the biological control of Western Flower Thrips, *Frankliniella occidentalis*, in greenhouses. *Proc. Int. Vedalia Symposium on Biological Control*, Riverside California, March 27–30 1989, p. 49.

Pickett, C. H., L. T. Wilson, and D. Gonzalez. 1988. Population dynamics and within-plant distribution of the Western Flower Thrips (Thysanoptera: Thripidae), an early-season predator of spider mites infesting cotton. *Environ. Entomol.* 17(3):551–559.

Pickett, J.A, G. R. Cayley, G. W. Dawson, D. C. Griffiths, S. H. Hockland, B. Marples, R. T. Plumb, and C. M. Woodcock. 1986. Use of alarm pheromone and derivatives against aphid mediated damage. *Abs. 6th International Congress on Pesticide Chemistry IUPAC*, Ottawa 1986.

Pickford, R. J.J. 1983. A selective method of thrip control on cucumber. *Bull. SROP/WPRS* 1983/VI/3:177–180.

Poinar, G. O. 1971. Use of nematodes for microbial control of insects. pp. 181–203. In: H. D. Burges and N. W. Hussey (eds.). Microbial Control of Insects and Mites. Academic Press, London, UK.

_____. 1986. Entomophagous nematodes. In: J. M. Frantz (ed.). Biological Plant and Health Protection. Fortschr. Zool. 32:95–121.

Poprawski, T. T., M. Marchal, and P. H. Robert. 1985. Comparative susceptibility of Otiorhynchus sulcatus and Sitona lineatus (Coleoptera: Curulionidae). Environ. Entomol. 14:247–253.

Potter, D. A., D. L. Wrensch, and D. E. Johnston. 1976. Guarding, aggressive behaviour and mating success in male two-spotted spider mites. Ann. Entomol. Soc. Am. 69:707–711.

Price, J. F., and B. K. Harbaugh. 1981. Effect of cultural practices on Liriomyza. p. 156–167 In: D. J. Schuster (ed.). Proc. IFAS—Ind. Conf. Biol. Cont. of Liriomyza Leafminers, Lake Buena Vista, Florida.

Quilici, S., G. Iperti, and J. M. Rabasse. 1985. Essais de lutte biologique en serre d'aubergine a l'aide d'un predateur aphidiphage: Propylea quatuordecimpunctata L. (Coleoptera, Coccinellidae). Frustula Entomol. 8:9–25.

Quinlan, R. J. 1988. Use of fungi to control insects in glasshouses. p. 19–36 In: M. Burge (ed.). Fungi in Biological Control Systems. Manchester University Press, UK.

Rabasse, J. M. 1980. Implantation d'Aphidius matricariae dans les populations de Myzus persicae en culture d'aubergines sous serre. Bull. SROP/WPRS 1980/III/3:175–186.

Ramakers, P. M. J. 1978. Possibilities for biological control of Thrips tabaci Lind. (Thysanoptera: Thripidae) in glasshouses. Meded. Fac. Landbouww. Rijksuniv. Gent 43(2):463–469.

_____. 1980. Biological control of Thrips tabaci (Thysanoptera: Thripidae) with Amblyseius spp. (Acari: Phytoseiidae). Bull. SROP/WPRS 1980/III/3:203–208.

_____. 1983a. Aschersonia aleyrodis, a selective biological insecticide. Bull. SROP/WPRS 1983/VI/3:167–171.

_____. 1983b. Mass production and introduction of Amblyseius mckenziei and A. cucumeris. Bull. SROP/WPRS 1983/VI/3:203–206.

_____. 1988. Population dynamics of the thrips predators Amblyseius mckenziei and Amblyseius cucumeris (Acarina: Phytoseiidae) on sweet pepper. Neth. J. Agr. Sci. 36:247–252.

_____. 1990. Manipulation of phytoseiid thrips predators in the absence of thrips. Bull. SROP/WPRS. XIII/5, 169–172.

Ramakers, P. M. J., M. Dissevelt, and K. Peeters. 1989. Large scale introductions of phytoseiid predators to control thrips on cucumber. Meded. Fac. Landbouww. Rijksuniv. Gent 54(3a):923–929.

Ramakers, P. M. J., M. C. Rombach, and R. A. Samson. 1982. Application of the entomopathogenic fungus, Aschersonia aleyrodis, in an integrated control programme against the glasshouse whitefly, Trialeurodes vaporariorum. Proc. Int. Coll. Inv. Path. and Microbial Conf., Brighton, Sept. 1982. p. 99.

Ramakers, P. M. J., and R. A. Samson. 1984. Aschersonia aleyrodis, a fungal pathogen of whitefly II. Application as a biological insecticide in glasshouses. J. Appl. Entomol. 97:1–8.

Ramakers, P. M. J., and M. J. van Lieburg. 1982. Start of commercial production and introduction of Amblyseius mckenziei Sch. & Pr. (Acarina: Phytoseiidae) for the control of Thrips tabaci Lind. (Thysanoptera: Thripidae) in glasshouses. Meded. Fac. Landbouww. Rijksuniv. Gent 47:541–546.

Rasmy, A. H. 1985. The biology of the two-spotted spider mite Tetranychus urticae as affected by resistant solanaceous plants. Agriculture, Ecosystems and Environment 13:325–328.

Ravensburg, W. J., and K. Altena. 1987. Recent developments in the control of thrips in sweet pepper and cucumber. Bull. SROP/WPRS 1987/X/2:160–164.

Reed, D. K., G. L. Reed, and C. S. Creighton. 1986. Introduction of entomogenous nematodes into trickle irrigation systems to control striped cucumber beetle (Coleoptera: Chrysomelidae). J. Econ. Entomol. 79:1330–1333.

Richardson, P. N. 1987. The use of insect-parasitic nematodes in integrated pest control programmes on protected crops. p. 163–174. In: R. Cavalloro (ed.). Integrated and Biological Control in Protected Crops. A. A. Balkema, Rotterdam, Netherlands.

_____. 1989a. Mass production and quality control of insect-parasitic rhabtidid nematodes. In: R. Cavalloro and V. Delucchi (eds.) Proc. of Parasitis, 25–28 Oct. 1988, Barcelona, Spain. p. 459–470. Ministerio de Agricultura Pesca y Alimentacion, Madrid, Spain.

_____. 1989b. Insect-parasitic nematodes. Inst. Hort. Res. Annu. Rep. 1987. p. 24–25.

_____. 1989c. Worms of death for vine weevils. Grower 112(12):28–30.

Richardson, P. N., and J. Godliman. 1989. The use of rhabditid nematodes to control black vine weevil, Otiorhynchus sulcatus, in glasshouse ornamentals. Bull. SROP/WPRS 1989/XII/4:10–11.

Richardson, P. N., and N. L. Helyer. 1991. The current state of integrated pest control in protected crops in the United Kingdom. Proc. CEC Experts' Group Meeting, 'Practical Application of Integrated Control in Protected Crops', Antibes, France, 16–19 October 1989, 8 pp (In press).

Robb, K. L. 1989. Analysis of Frankliniella occidentalis (Pergande) as a pest of floricultural crops in California greenhouses. PhD thesis, University of California, Riverside.

Robb, K. L., and M. P. Parella. 1986. Western Flower Thrips control. Greenhouse Grower 4(9):94–98.

Robertson, N. L., and T. W. Carroll. 1988. Virus-like particles and a spider mite intimately associated with a new disease of barley. Science 240:1188–1190.

Rodriguez-Rueda, D., and J. Fargues. 1980. Pathogenicity of entomopathogenic hyphomycetes Paecilomyces fumosoroseus and Nomuraea rileyi, to eggs of noctuids Mamestra brassicae and Spodoptera littoralis. J. Invertebr. Pathol. 36:399–408.

Rombach, M. C., and A. T. Gillespie. 1988. Entomogenous hyphomycetes for insect and mite control on greenhouse crops. Biocontrol News and Information 9:7–18.

Rombach, M. C., and R. A. Samson. 1982. Small scale production of the entomopathogenic fungus, Aschersonia aleyrodis. Proc. Int. Coll. Inv. Path. and Microbial Control, Brighton, Sept. 1982. p. 229.

Rovesti, L., and K. J. Deseo. 1989. Mobility and persistence of entomoparasitic nematodes in field conditions. Bull. SROP/WPRS 1989/XII/4:88.

Russell, L. M. 1977. Hosts and distribution of the greenhouse whitefly, Trialeurodes vaporariorum (Westwood) (Hemiptera: Homoptera: Aleyrodidae). USDA Coop. Plant Pest Rep. 2:449–458.

Rutherford, T. A., D. Trotter, and J. M. Webster. 1987. The potential of Heterorhabditid nematodes as control agents of root weevils. Can. Ent. 119: 67–73.

Sabelis, M. W. 1981. Biological control of two-spotted spider mites using phytoseiid predators. Part 1. Modelling the predator–prey interaction at the individual level. Agr. Res. Rep., Wageningen, Netherlands, 910 pp.

Sabelis, M. W., J. E. Vermaat, and A. Groenveld. 1984. Arrestment responses of the predatory mite, Phytoseiulus persimilis, to steep odour gradients of a kairomone. Physiol. Entomol. 9:437–446.

Saito, T., and F. Ikeda. 1983. Beauveria bassiana isolated from the black vine weevil Otiorhynchus sulcatus (Coleoptera: Curculionidae). Japan. J. Entomol. Zool. 27:71–74.

Sakimura, K. 1962. In: K. Maramorosch (ed.). Biological Transmission of Disease Agents, Academic Press, New York. p. 33.

Samson, R. A., P. M. J. Ramakers, and T. Oswald. 1979. Entomophthora thripidum, a new fungal pathogen of Thrips tabaci. Can. J. Bot. 57:1317–1323.

Schauer, M., and H. Schmutterer. 1981. Effects of neem kernal extracts on the two-spotted spider mite, Tetranychus urticae. p. 259–266. In: H. Schmutterer, K. R. S. Ascher and H. Rembold (eds.). Proc. 1st International Neem Conference (Rottach-Egern, 1980).

Schmutterer, H., and K. R. S. Ascher (Eds.). 1984. Natural pesticides from the neem tree (Azadirachta indica A. Juss) and other tropical plants. Proc. of the Second International

Neem Conference, Rauischholzhausen, 1983.

―――― . (Eds.). 1987. Natural pesticides from the neem tree (*Azadirachta indica* A. Juss) and other tropical plants. *Proc. Third International Neem Conference*, Nairobi, Kenya, 1986.

Schmutterer, H., K. R. S. Ascher, and H. Rembold. (Eds.). 1981. Natural pesticides from the neem tree (*Azadirachta indica* A. Juss). *Proc. First International Neem Conference*, Rottach-Egern, 1980.

Schroeder, W. J., and J. B. Beavers. 1987. Movement of the entomogenous nematodes of the families Heterorhabditidae and Steinernematidae in soil. *J. Nematol.* 19:257–259.

Scopes, N. E. A. 1970. Control of *Myzus persicae* on year-round chrysanthemums by introducing aphids parasitized by *Aphidius matricariae* into boxes of rooted cuttings. *Ann. Appl. Biol.* 66:323–327.

―――― . 1985. Red spider mite and the predator *Phytoseiulus persimilis*. p. 43–52 In: N. W. Hussey and N. E. A. Scopes (eds.). Biological Pest Control: The Glasshouse Experience. Blandford Press, Dorset, UK.

Scopes, N. E. A., and S. M. Biggerstaff. 1971. The production, handling and distribution of the whitefly *Trialeurodes vaporariorum* and its parasite *Encarsia formosa* for use in biological control programmes in glasshouses. *Plant Path.* 20:111–116.

―――― . 1973. Progress towards integrated pest control on year-round chrysanthemums. *Proc. 7th British Insecticide Fungicide Conference*, Brighton, UK. p. 227–234.

Sengonca, C., S. Gerlach, and G. Melzer. 1987. Effect of different prey on *Chrysoperla carnea* (Stephens) (Neuroptera: Chrysopidae). *Z. Pflanzenkr. Pflanzenschutz* 94:197–205.

Sharaf, N., and Y. Batta, 1985. Effect of some factors on the relationship between the whitefly *Bemisia tabaci* (Genn.) (Homopt: Aleyrodidae) and the parasitoid *Eretmocerus mundus* Mercet (Hymenopt: Aphelinidae). *Z. Ang. Ent.* 99:267–276.

Shijko, E. S. 1989. Rearing and application of the peach aphid parasite, *Aphidius matricariae* (Hymenoptera, Aphidiidae). *Acta Entomol. Fenn.* 53:53–56.

Simons, W. R. 1981. Biological control of *Otiorhynchus sulcatus* with heterorhabditid nematodes in the glasshouse. *Neth. J. Plant Path.* 87:149–158.

Simons, W. R., and G. O. Poinar Jr. 1973. The ability of *Neoaplectana carpocapsae* (Steinernematidae: Nematodea) to survive extended periods of desiccation. *J. Invert. Pathol.* 22:228–230.

Sly, J. M. A. 1986. Review of usage of pesticides in agriculture, horticulture and animal husbandry in England and Wales 1980–1983. Pesticide Usage Survey Report no. 41. ADAS, MAFF, UK.

Smitley, D. R., W. M. Brooks, and G. G. Kennedy. 1986. Environmental effects on production of primary and secondary conidia, infection and pathogenesis of *Neozygites floridiana*, a pathogen of the two-spotted spider mite, *Tetranychus urticae*. *J. Invert. Path.* 47:325–332.

Smits, P. H. 1986. Calculations on the polyhedra intake by beet armyworm larvae feeding on virus sprayed chrysanthemums. p. 616–619. In: A. Samson, J. H. Vlak and D. Peters (eds.). Fundamental and Applied Aspects of Invertebrate Pathology. *Foundation 4th Int. Coll. Invert. Path.*

―――― . 1987. Nuclear polyhedrosis virus as a biological control agent of *Spodoptera exigua*. PhD Thesis, Wageningen Agricultural University, The Netherlands. 127 pp.

Smits, P. H., M. van de Vrie, and J. M. Vlak. 1987. Nuclear polyhedrosis virus for control of *Spodoptera exigua* larvae on glasshouse crops. *Entomol. Expt. Appl.* 43:73–80.

Smits, P. H., M. C. van Helden, M. van de Vrie, and J. M. Vlak. 1987. Feeding and dispersion of *Spodoptera exigua* larvae and its relevance for control with a nuclear polyhedrosis virus. *Entomol. Expt. Appl.* 43:67–72.

Smits, P. H., and J. M. Vlak. 1988a. Quantitative and qualitative aspects in the production of a nuclear polyhedrosis virus in *Spodoptera exigua* larvae. *Ann. Appl. Biol.* 112:249–257.

―――― . 1988b. Biological activity of *Spodoptera exigua* nuclear polyhedrosis virus against

S. exigua larvae. J. Invert. Pathol. 51:107–114.

Soares, G. G., M. Marchal, and P. Ferron. 1983. Susceptibility of Otiorhynchus sulcatus (Coleoptera: Curculionidae) larvae to Metarhizium anisopliae and Metarhizium flavoviride (Deuteromycotina: Hyphomycetes) at two different temperatures. Environ. Entomol. 12:1886–1890.

Sopp. P. I., A. T. Gillespie, and A. Palmer. 1989. Applicaton of Verticillium lecanii for the control of Aphis gossypii by a low-volume electrostatic rotary atomizer and a high-volume hydraulic sprayer. Entomophaga 34:417–428.

Spencer, D. M. 1980. Parasitism of carnation rust (Uromyces dianthi) by Verticillium lecanii. Trans. Brit. Mycol. Soc. 74:191–194.

Speyer, E. R. 1926. Red spider: Life history and seasonal history. Ann. Rep. Exp. Res. Station Cheshunt for 1925. p. 89–93.

_____. 1927a. White Fly (Trialeurodes vaporariorum Westw.). Ann. Rep. Exp. Res. Station Cheshunt 1926, p. 55–56.

_____. 1927b. An important parasite of the greenhouse white-fly. Bull. Ent. Res. 17:301–308.

_____. 1930. Biological control of the greenhouse white-fly. Nature 126:1009–1010.

Speyer, E. R., and W. J. Parr. 1948. The tomato moth. Ann. Rep. Exp. Res. Station Cheshunt 1947. p. 41–62.

Staay, M.v.d., and J.v. Uffelen. 1988. Chemical control of Western Flower Thrips: Dichlorvos can also be used for cucumber. [in Dutch] Weekblad Groenten en Fruit, 19 February 1988, 40–43.

Stacey, D. L., I. J. Wyatt, and R. J. Chambers. 1985. The effect of glasshouse red spider mite damage on the yield of tomatoes. J. Hort. Science, 60:517–523.

Stary, P. 1988a. Aphidiidae. p. 171–184 In: A. K. Minks and P. Harrewijn (eds.). Aphids, Their Biology, Natural Enemies and Control, Vol. 2B: Elsevier Science Publishers B. V., Amsterdam, the Netherlands.

_____. 1988b. Aphelinidae. p. 185–188 In: A. K. Minks and P. Harrewijn (eds.). Aphids, Their Biology, Natural Enemies and Control, Vol. 2B: Elsevier Science Publishers B. V., Amsterdam, the Netherlands.

Steiner, M. Y., and D. P. Elliott. 1987. Biological pest management for interior plantscapes. Vegreville, AB. Alberta Environmental Centre. 32 pp.

Stenseth, C. 1971. Effect of temperature on the development of Trialeurodes vaporariorum Westwood. Forsk. Fors. Landbruket 22:493–496.

_____. 1975. Effect of temperature on the development of the parasite Encarsia formosa. Gartneryrket 65:136–139.

_____. 1976. Some aspects of the practical application of the parasite Encarsia formosa for the control of Trialeurodes vaporariorum. Bull. SROP/WPRS 1976/IV: 104–114.

_____. 1986. Bekjempelse av nelliktrips pa slangeagurk i veksthus. Forsk. Fors. Landbruket 37:15–22.

Stimman, M. W., H. K. Kaya, T. M. Burgando, and J. P. Studdert. 1985. Black vine weevil management in nursery plants. Calif. Agr. 39 (Jan-Feb 1985):25–26.

St. Leger, R. J., A. K. Charnley, and R. M. Cooper. 1986. Cuticle-degrading enzymes of entomopathogenic fungi: Synthesis in culture on cuticle. J. Invert. Pathol. 48:85–95.

Sunderland, K. D. 1990. The future for biological control. Prof. Hort. 4:11–20.

Teulon, D. A. J., and P. M. J. Ramakers. 1990. A review of attractants for trapping thrips with particular reference to glasshouses. Bull. SROP/WPRS. XIII/5:212–214.

Theunissen, J., and J. J. Fransen. 1984. Biological control of cutworms in lettuce by Neoaplectana bibonis. Meded. Fac. Landbouww. Rijksuniv. Gent 49:771–776.

Tiittanen, K., and M. Markkula. 1989. Biological control of pests on Finnish greenhouse vegetables. Acta Entomol. Fenn. 53:57–59.

Tillemans, F., and J. Coremans-Pelseneer. 1987. Beauveria brongniartii (fungus, Moniliale) as control agent against Otiorhynchus sulcatus (Coleoptera: Curculionidae). Meded. Fac. Landbouww. Rijksuniv. Gent 52 (2a):379–384.

Timper, P., H. K. Kaya, and R. Gaugler. 1988. Dispersal of the entomogenous nematode, *Steinernerma feltiae* (Rhabditida:Steinernematidae) by infected adult insects. *Environ. Entomol.* 17:546–550.

Treifi, A. H. 1984. Use of *Beauveria bassiana* (Bals.) to control the immature stages of the whitefly, *Trialeurodes vaporariorum*, (Westw.) (Homoptera: Aleyrodes) in the greenhouse. *Arab J. Plant Prot.* 2:83–86.

Tremblay, E. Z. 1974. Possibilities for utilization of *Aphidius matricariae* Hal. (Hymenoptera, Ichneumonidae) against *Myzus persicae* (Sulz.) (Homoptera, Aphidoidea, Aphididae) in small glasshouses. *Z. Pflanzenkr. Pflanzenschutz* 81:612–619.

Vacante, V. 1981. Notizie sulla presenza di *Theriodiplosis persicae* Keiffer (Diptera, Cecidomyiidae) in serra su piante orticole e floreali, attaccate da *Tetranychus urticae* Koch (Caraina, Tetranychidae). *Tech. Agr.* 5:5–14.

———. 1984. The current state of control of phytophagous mites in protected crops in Sicily. *Bull. SROP/WPRS* 1985/VIII/1:43–50.

van Alphen, J. J. M., H. W. Nell, and L. A. Sevenster van der Lelie-. 1976. The parasite-host relationship between *Encarsia formosa* Gahan (Hymenoptera: Aphelinidae) and *Trialeurodes vaporariorum* Westwood (Homoptera: Aleyrodidae). VII. The importance of host feeding as a mortality factor in greenhouse whitefly nymphs. *Bull. SROP/WPRS* 1976/4:165–169.

van den Bos, J. 1983. Experiences with pheromonal trapping of Lepidoptera in greenhouses. *Bull. SROP/WPRS* 1983/VI/3: 196–202.

van de Vrie, M. 1989. Present status of biological/integrated control of mite and insect pests in floriculture in The Netherlands. *Bull. SROP/WPRS* XII/3:18–22.

van de Veire, M., and V. Vacante. 1984. Glasshouse whitefly control through the combined use of the colour attraction system with the parasitic wasp *E. formosa* (Hym.: Aphelinidae). *Entomophaga* 29:303–310.

van Gelder, W. M. J., and O. M. B. de Ponti. 1987. Alpha-Tomatine and other steroidal glycoalkaloids in fruits of tomato lines resistant to the glasshouse whitefly (*Trialeurodes vaporariorum* Westw.). *Euphytica* 36:555–561.

van Haren, R. J. F., M. M. Steenhuis, M. W. Sabelis, and O. M. B. de Ponti. 1987. Tomato stem trichomes and dispersal success of *Phytoseiulus persimilis* relative to its prey *Tetranychus urticae*. *Expt. & Appl. Acarol.* 3:115–121.

van Keymeulen, M., and D. Degheele. 1977. The development of oocytes and the lapse of time for adult emergence of *Encarsia formosa* Gahan 1924 at a constant temperature. *Meded. Fac. Landbouww. Rijksuniv. Gent* 42:1279–1287.

van Lenteren, J. C. 1988. Evaluating the effectiveness of natural enemies. p. 175–181 In: E. Niemczyk and A. F. G. Dixon. Ecology and Effectiveness of Aphidophaga. SPB Academic Publishing, The Hague, The Netherlands.

van Lenteren, J. C., and L. P. J.J. Noldus. 1990. Whitefly-plant relationships: Behavioral and ecological aspects. p. 47–89. In: Girling, D. (ed.). Whiteflies: their bionomics, pest status and management. Intercept Ltd. Andover, Hants. UK.

van Lenteren, J. C., and P. M. Hulspas-Jordan. 1982. Influence of low temperature regimes on the capability of *Encarsia formosa* and other parasites in controlling the glasshouse whitefly, *Trialeurodes vaporariorum*. *Bull. SROP/WPRS* 1982/V/6:54–70.

van Lenteren, J. C., P. M. J. Ramakers, and J. Woets. 1979. The biological control situation in Dutch glasshouses: problems with *Trialeurodes vaporariorum* West., *Liriomyza bryoniae* Kalt. and *Myzus persicae* Sulz.. *Meded. Fac. Landbouww. Rijksuniv. Gent.* 44 (1):117–125.

van Lenteren, J. C., and J. Woets. 1988. Biological and integrated pest control in greenhouses. *Annu. Rev. Entomol.* 33:239–271.

van Lieburg, M. J., and P. M. J. Ramakers. 1984. A method for the collection of *Aphidoletes* larvae in water. *Meded. Fac. Landbouww. Rijksuniv. Gent* 49 (3a):777–779.

van Rijn, P. C. J., and M. W. Sabelis. 1990. Pollen availability and its effect on the maintenance of populations of *Amblyseius cucumeris*, a predator of thrips. *Bull. SROP/WPRS.* XIII/5:179–184.

van Vianen, A., and M. van de Veire. 1988. Honeydew of the glasshouse whitefly *Trialeurodes vaporariorum* (Westwood) as a contact kairomone for its parasite *Encarsia formosa* Gahan. *Meded. Fac. Landbouww. Rijksuniv. Gent* 53:949–954.

Vet, L. E. M., and J. C. van Lenteren. 1981. The parasite-host relationship between *Encarsia formosa* Gah. (Hymenoptera: Aphelinidae) and *Trialeurodes vaporariorum* (Westw.) (Homoptera: Aleyrodidae). *J. Appl. Entomol.* 4:327–348.

Voroshilov, N. V. 1979. Heat-resistant lines of the mite *Phytoseiulus persimilis* A–H. *Genetika* 15:70–76.

Wardlow, L. R. 1985a. Pyrethroid resistance in glasshouse whitefly (*Trialeurodes vaporariorum*, Westw.). *Meded. Fac. Landbouww. Rijksuniv. Gent* 50:555–557.

———. 1985b. Leafminers and their parasites. p. 62–65. In: N. W. Hussey and N. E. A. Scopes (eds.). Biological Pest Control: The Glasshouse Experience. Blandford Press, Poole, Dorset, UK.

———. 1989. Biological pest control on ornamentals: United Kingdom summary. *Bull. SROP/WPRS* 1989/XII/3:23–25.

———. 1990. Integrated pest management in protected ornamental crops. *Bull. SROP/WPRS* XIII/5, 222–224.

Wardlow, L. R., F. A. B. Ludlam, and N. French. 1972. Insecticide resistance in glasshouse whitefly. *Nature, Lond.* 239:164–165.

Wardlow, L. R., and A. Tobin. 1990. Potential new additions to the armoury of natural enemies for protected tomatoes. *Bull. SROP/WPRS* XIII/5: 225–227.

Webb, R. E., M. A. Hinebaugh, R. K. Lindquist, and M. Jacobson. 1983. Evaluation of aqueous solution of neem seed extract against *Liriomyza sativae* and *L. trifolii* (Diptera: Agromyzidae). *J. Econ. Entomol.* 76:357–362.

Webb, R. E., and F. F. Smith. 1980. Glasshouse whitefly control in an integrated regimen based on adult trapping and nymphal parasitism. *Bull. SROP/WPRS* 1980/III/3:235–246.

Webb, R. E., F. F. Smith, H. Affeldt, R. W. Thimijan, R. F. Dudley, and H. F. Webb. 1985. Trapping glasshouse whitefly with coloured surfaces: variables affecting efficacy. *Crop Prot.* 4:381–393.

Westerman, P. R., and W. R. Simons. 1988. Preliminary experiments with media for short-term storage and transport of the insect-parasitic nematode, *Heterorhabditis* sp.. *Meded. Fac. Landbouww. Rijksuniv. Gent* 53:919–927.

Westwood, J. O. 1856. The new Aleyrodes of the greenhouse. *Gardeners Chron.* p. 852.

Wilding, N. 1983. The current status and potential of entomogenous fungi as agents of pest control. *Proc. 10th Int. Cong. of Plant Protection*, Brighton, UK., 2:743–750.

Wilding, N., and R. G. Latteur. 1987. The Entomophtorales—problems relative to their mass production and their utilisation. *Meded. Fac. Landbouww. Rijksuniv. Gent* 52 (2a):159–164.

Wilson, T. H., and T. A. Cooley. 1972. A chalcidoid planidium and an entomophilic nematode associated with the western flower thrips. *Ann. Entomol. Soc. Am.* 65:414–418.

Wit, A. K. H., and M. van de Vrie. 1985. Gamma radiation for post harvest control of insects and mites in cut flowers. *Meded. Fac. Landbouww. Rijksuniv. Gent* 50 (2b):697–704.

Woets, J. 1973. Integrated control in vegetables under glass in the Netherlands. *Bull. SROP/WPRS* 1973/4:26–31.

Woets, J., and A. van der Linden. 1982. Serpentine leaf miner, *Liriomyza trifolii*. *Annu. Rep. GCRES 1981*, Naaldwijk, The Netherlands, p. 104.

Wouts, W. M. 1984. Nematode parasites of Lepidopterans. In: W. R. Nickle (ed.). Plant and Insect Nematodes. Marcel Dekker Inc., New York.

Wyatt, I. J. 1970. The distribution of *Myzus persicae* (Sulz.) on year-round chrysanthemums. II. Winter season. The effect of parasitism by *Aphidius matricariae*. *Ann.*

Appl. Biol. 65:31–41.

_____ . 1972. Control of *Aphis gossypii* by parasites. *Glasshouse Crops Res. Inst. Annu. Rep.* 1971. p. 100.

Wyatt, I. J., M. R. Abdulla, P. T. Atkey, and A. Palmer. 1984. Activity of discrete permethrin droplets against whitefly scales. *Proc. 1984 British Crop Protection Conference—Pests and Diseases,* pp. 1045–1048.

Wyatt, I. J., and S. J. Brown. 1977. The influence of light intensity, daylength and temperature on increase rates of four glasshouse aphids. *J. Appl. Ecol.* 14:379–399.

Yano, E., and T. Koshihara. 1984. Monitoring techniques for adults of the glasshouse whitefly, *Trialeurodes vaporariorum* (Westwood). *Bull. Veg. Ornamental Crops Res. Stn. Ser. A,* 12, December 1984, p. 85–96.

Yathom S., R. Marcus, M. Chen, and S. Tal. 1988. Comparison of different positions and heights of yellow sticky traps for sampling populations of the leafminer *Liriomyza trifolii. Phytoparasitica* 16:217–224.

Yin. L. T., and U. Maschwitz. 1983. Sexual pheromone in the greenhouse whitefly *Trialeurodes vaporariorum* Westw. *J. Appl. Entomol.* 95:439–446.

Yudin, L. S, W. C. Mitchell, and J. J. Cho. 1987. Color preference of thrips (Thysanoptera: Thripidae) with reference to aphids (Homoptera: Aphididae) and leafminers in Hawaiian lettuce farms. *J. Econ. Entomol.* 80:51–55.

Zimmerman, G. 1981. Gewachshausversuche zur Bekampfung des Gefurchten Dickmaulrusslers, *Otiorhynchus sulcatus* F., mit dem Pilz, *Metarhizium anisopiliae* (Metsch.) Sorok. *Pflanzenschutzdienstes* 33:103–108.

_____ . 1982. Unterzuchungen zur Wirkung von *Metarhizium anisopliae* (Metsch.) Sorok. auf Eier und schlupfende Eilarven von *Otiorhynchus sulcatus* F. (Coleoptera: Curculionidae). *J. Appl. Entomol.* 93:426–482.

_____ . 1983. Biological control of aphids by entomopathogenic fungi: Present state and prospects. p. 33–40. In: R. Cavalloro (ed.). Aphid Antagonists. A. A. Balkema, Rotterdam.

_____ . 1984. Weitere Versuche mit *Metarhizium anisopliae* (Fungi Imperfecti: Moniliales) zur Bekampfung des Gefurchten Dickmaulrusslers, *Otiorhynchus sulcatus* F., an Topfpflanzen in Gewachshaus. *Pflanzenschutzdienstes* 39:55–59.

Zimmerman, G., and E. Bode. 1983. Dispersal of the entomopathogenic fungus, *Metarhizium anisopliae* (Fungi Imperfecti: Moniliales) by soil arthropods. *Pedobiologia* 25:65–71.

Zimmerman, G., and W. R. Simons. 1986. Experiences with biological control of the black vine weevil, *Otiorhynchus sulcatus* F., p. 529–533. In: A. Samson, J. M. Vlak, and D. Peters (eds.). Fundamental and Applied Aspects of Invertebrate Pathology. *Foundation of the 4th Int. Coll. of Inv. Path.,* Wageningen, the Netherlands.

Polygalacturonase and Tomato Fruit Ripening

James J. Giovannoni
Department of Plant Biology, University of California
Berkeley, CA 94720

Dean DellaPenna
Department of Plant Science, University of Arizona
Tucson, AZ 85721

Alan B. Bennett
Department of Vegetable Crops, University of California
Davis, CA 95616

Robert L. Fischer
Department of Plant Biology, University of California
Berkeley, CA 94720

I. INTRODUCTION

Ripening is the final stage of fruit development and represents a complex cascade of events which eventually impart upon the fruit organ certain qualities rendering it attractive and desirable for consumption. As such the fruit serves as a vehicle for seed dispersal and the resulting propagation of the plant species. Fruit ripening is often a dramatic change, and thus it has received much attention from plant biologists looking for model systems to study fundamental principles of plant development. Because the cultivated tomato (*Lycopersicon esculentum* Mill) has been exceptionally well characterized at the physiological and genetic levels, and because of its importance as an agricultural commodity, it has emerged as an intensively studied model system of fruit maturation.

One of the most thoroughly studied aspects of tomato fruit maturation in the last decade has been the involvement of polygalacturonase, a fruit ripening specific cell wall pectinase, in the ripening and softening process (Grierson 1985, 1986; Bennett and DellaPenna 1987a; Brady et al. 1987). The activity of this enzyme has been strongly implicated as a key determinant of softening during tomato fruit ripening and hypothesized to have a more general role in the regulation of maturation of the fruit as a whole (Hobson 1964; Tigchelaar et al. 1978b; Brady et al. 1982; Huber 1983a; Bennett and DellaPenna 1987b; Baldwin and Pressey 1988). The purpose of this review will be to (1) describe the process of tomato fruit ripening; (2) provide an overview of the data leading to such interest in polygalacturonase; (3) review the literature describing the isolation of the polygalacturonase gene and the regulation of its expression; and (4) focus on recent experiments designed to identify the function of polygalacturonase in the ripening process.

A. Tomato Fruit Ripening

Tomato ripening is characterized by a variety of rapid and conspicuous changes in tissue morphology, biochemistry, and physiology (Rhodes 1980). One of the most dramatic changes associated with the ripening of fruit from many different plant species is the accumulation and/or loss of pigments, resulting in distinct changes in color. An excellent example of such a change occurs in ripening tomato fruit which display loss of chlorophyll, and subsequent changes in levels of a variety of carotenoids, including accumulation of the red pigment lycopene. This color change occurs within membrane bound plastid vesicles (Grierson 1985), resulting in the conversion of chloroplasts to chromoplasts as the highly organized photosynthetic apperati of the plastids is disrupted and lycopene accumulates in crystalline form (Grierson et al. 1987).

The first visible signs of color change during tomato ripening are preceded by the respiratory climacteric which constitutes an easily monitored hallmark of the beginning of the ripening process (Grierson 1985). The climacteric response is characterized by a large increase in respiration of the fruit and a dramatic burst of ethylene production. Ethylene (C_2H_4) is a gaseous plant hormone produced in many different plant tissues and involved in a variety of developmental programs and environmental responses, including fruit ripening (Abeles 1973; Yang 1985). In addition, ripening associated ethylene production is autocatalytic in nature; ethylene produced during tomato fruit ripening promotes its own biosynthesis in a positive manner.

The biochemical pathway of ethylene production has been elucidated by Yang (1985). The rate limiting step in this pathway is the conversion of s-adenosyl methionine (SAM) to 1-aminocyclopropane-1-carboxylic acid (ACC) by the enzyme ACC synthase. ACC is subsequently converted to ethylene by the ethylene forming enzyme (EFE) as the remainder of the modified amino acid skeleton is recycled back into SAM, thus recycling methionine. ACC synthase activity, and therefore ethylene formation, increase dramatically in ripening tomato fruit (Su et al. 1984).

Experiments employing inhibitors of ethylene action and biosynthesis have demonstrated that this hormone is necessary for normal ripening of tomato fruit, while application of exogenous ethylene to mature green tomato fruit results in hastening of the ripening process (Lyons and Pratt 1964; Rhodes 1980; Baile and Young 1981; Hobson et al. 1984). Using cloned cDNA probes several laboratories have demonstrated that ethylene acts, at least in part, to regulate the expression of specific nuclear genes during the ripening phase of tomato fruit development (Lincoln et al. 1987; Maunders et al. 1987; Deikman and Fischer 1988; Margossian et al. 1988; Cordes et al. 1989). In particular, Lincoln and Fischer (1988a) have demonstrated that ethylene can regulate gene expression during fruit ripening at both transcriptional and post-transcriptional levels. Although the mechanism(s) of ethylene action on gene expression and fruit ripening are not fully understood, this hormone clearly plays an extremely important role in the control and synchronization of maturation events in tomato fruit. Consequently, comprehension of the developmental cues which initiate ripening specific ethylene production, and insight into the mode of action of this hormone on fruit physiology, are paramount for an understanding of fruit ripening. The fact that polygalacturonase activity has been implicated in the control of ripening related ethylene evolution (Tigchelaar et al. 1978a; Baldwin and Pressey 1988) makes regulation of the biosynthesis of ethylene particularly interesting.

Coincident with autocatalytic ethylene production and lycopene accumulation during tomato fruit maturation are a variety of other ripening phenomena, including (1) the conversion of starch to sugars; (2) a general increase in susceptibility to pathogenic and saphrophytic microorganisms; (3) changes in fruit cell wall ultrastructure; (4) a significant accumulation of polygalacturonase mRNA, protein, and activity; and (5) a striking change in texture of the fruit that results in softening (Grierson 1985; Brady et al. 1987; DellaPenna and Giovannoni 1990). The last three parameters in particular are thought to be interrelated, and are the focus of this review.

B. The Cell Wall and Softening

The tomato fruit cell wall is similar to other plant cell walls in that it consists of cellulose microfibrils embedded in a matrix of crosslinking molecules (Huber 1983a). These matrix components include glycoproteins, hemicelluloses, and polyuronides (Huber 1983b). Polyuronides, or pectins, are particularly abundant in the middle lamella region joining adjacent plant cell walls (Crookes and Grierson 1983). Galacturonic acid polymers are stabilized into a gelatinous compound via electrostatic interaction with ionic calcium, resulting in a structurally sound matrix (Poovaiah et al. 1988). Thus, because of their predominance in the middle lamella, polyuronides are thought to be important contributors to structural support of plant tissues. It has been proposed that middle lamella polyuronides provide adhesion to juxtaposed cell walls thereby imparting firmness to plant organs, including unripe fruit (Crookes and Grierson 1983).

A dramatic change in cell wall ultrastructure occurring during tomato fruit ripening is the degradation and solubilization of polyuronides (Crookes and Grierson 1983). Because of the hypothesized structural role of pectin in plant tissue rigidity, the loss of pectin integrity during tomato fruit ripening has been implicated as the predominant component of ripening associated softening (Brady et al. 1982; Crookes and Grierson 1983; Huber 1983b). Polyuronide degradation has been assessed by a variety of means, in particular as increased levels of both water and chelator extractable polyuronides, and via depolymerization or decrease in polymer length of extractable polyuronides as tomato fruit ripen (Huber 1983a). Electron micrographic analysis has demonstrated the disruption of the darkly staining polyuronide rich middle lamella between fruit cell walls during ripening (Crookes and Grierson 1983). Because of the temporal correlation between polyuronide degradation and fruit softening, pectin degradation historically has been considered as the primary determinant of tomato fruit softening (Grierson 1985; Brady et al. 1987; Bennett and DellaPenna 1987a).

The enzymic activity responsible for ripening related polyuronide degradation has been attributed to endo-polygalacturonase (poly[1,4-B-D-galacturonide]glycanohydrolase, EC 3.2.1.15) which accumulates in parallel with pectin degradation and fruit softening (Huber 1983a; Grierson 1985; Brady et al. 1987; Giovannoni et al. 1989). Although exo-polygalacturonase activity has been described and purified from firm green tomato fruit (Pressey 1987), it comprises only a small fraction of total polygalacturonase activity found in ripening fruit and is not thought to play a significant role in softening. Consequently, endo-polygalacturonase activity has been implicated as the primary determinant of polyuronide degradation and softening of tomato fruit during ripening, and is therefore the subject of considerable scientific investigation. Polygalacturonase activity has also been associated with other horticulturally relevant fruit ripening parameters, including cell wall metabolism (Huber 1983a, 1983b) and general regulation of fruit ripening (Tigchelaar et al. 1978b; Baldwin and Pressey 1988).

C. Tomato as a Model System

Tomato is an excellent model system for studying fruit ripening and softening. Tomato fruit ripen uniformly, especially the small or cherry types, and can be easily and consistently staged using morphological markers (Lincoln et al. 1987). The plant itself is easy to grow and maintain and is extremely amenable to clonal propagation. The latter characteristic, in conjunction with the fact that the cultivated tomato is largely self-pollinating, allows for easy maintenance of genetic resources. Also, the crop has been well-studied genetically, providing an extensive battery of compatible germplasm in addition to that available from wild species for the introduction of novel traits (notably disease resistance markers) and the facilitation of genetic analysis (Rick 1956, 1986). In addition, the generation of a high density restriction fragment length polymorphism (RFLP) map of the tomato genome simplifies the mapping of morphological markers and cloned tomato DNA sequences (Bernatzky and Tanksley 1986). Tomato is also amenable to *Agrobacterium* mediated DNA transfer and tissue culture regeneration, allowing for molecular genetic analysis of plant processes, including fruit ripening (McCormick et al. 1986; Fillatti et al. 1987). As a well-studied and important agricultural crop in the United States and other regions throughout the world, tomato represents a model crop for the study of genetic and developmental control of key developmental events.

D. Polygalacturonase and Fruit Softening

Hobson (1965) described the correlation between fruit softening and polygalacturonase activity as determined by the ability of crude cell wall protein extracts to degrade a purified pectin substrate in vitro. As fruit matured from green to red ripe, the level of polygalacturonase activity increased dramatically. This initial observation ultimately led to an expanded effort in many laboratories to identify the enzyme responsible for this activity and assess its physiological function during ripening.

The correlation between increased polygalacturonase activity and fruit softening was the first of several convincing observations leading researchers to attribute changes in texture during ripening to polygalacturonase activity (Hobson 1965; Grierson 1985). As yet, no significant polygalacturonase activity has been observed in other tomato organs such as leaves, roots, or stems, though an endopolygalacturonase activity has been reported in flower abscission zones and germinating pollen grains (Tucker et al. 1984; Brown and Crouch 1990). Also, a variety of fruit ripening mutants exist in which fruit ripening is inhibited or delayed and which exhibit corresponding decreased levels of polygalacturonase activity during fruit development (Tigchelaar et al. 1978b). It is important to note that none of these mutations map to the chromosomal location of the single ripening specific polygalacturonase gene which is located on chromosome 10 (Bird et al. 1988). Furthermore, a rough correlation has been demonstrated between levels of polygalacturonase activity and degree of softening among different lines of cultivated tomato (Tigchelaar et al. 1978b; Brady et al. 1983). Also, Yoshida et al. (1984) have shown that storage of mature green tomato fruit at 33°c resulted in inhibition of ripening, polygalacturonase activity, and softening. Picton and Grierson (1988) demonstrated that this loss of activity is due to a decrease in polygalacturonase mRNA accumulation. Return of fruit to 22°c led to partial ripening, softening, and correspondingly low polygalacturonase activity (Yoshide et al. 1984). Finally, there is little correlation between the activities of several other cell wall degrading enzymes (cellulase, β-1,3-glucanase, and pectinesterase) and rate of tomato fruit softening (Hobson 1968; Wallner and Walker 1975; Tigchelaar et al. 1978b). Thus, there are several lines of evidence supporting the role of polygalacturonase in tomato fruit softening.

The evidence supporting polygalacturonase involvement in fruit softening is strictly correlative in nature. In fact, other results indicate that softening may not be regulated exclusively by the action of polygalacturonase. First, there are exceptions to the correlation between polygalacturonase activity and the rate of tomato fruit softening among different cultivars (C. Brady, personal communication). Hemicellulose degradation also occurs during fruit ripening and may contribute to softening by disrupting a component of the cell wall other than polyuronides (Huber 1983a). In addition, active biosynthesis of a variety of cell wall components has been demonstrated to correlate with the onset of fruit ripening and softening (Mitcham et al. 1989). Finally, other fruits, such as strawberry, have been shown to soften in the absence of dramatic increases in polygalacturonase activity (Knee et al. 1977). Consequently, the action of other enzymes or factors, such as substrate accessibility and fruit structural differences, may contribute individually or in concert to influence tomato fruit texture and ripening related softening.

In addition to the proposed role of polygalacturonase in fruit softening, it has been speculated that endo-polygalacturonase dependent poly-uronide hydrolysis may generate oligosaccharide molecules capable of influencing other aspects of the ripening process (Brady et al. 1987; Bennett and DellaPenna 1987a; Baldwin and Pressey 1988). This concept is supported in a report by Brecht and Huber (1986) demonstrating that cell wall fragments are capable of stimulating ethylene biosynthesis when applied exogenously to tomato pericarp tissue. Similarly, it has been reported that infiltration of purified polygalacturonase protein into mature green tomato fruit stimulates ethylene production (Baldwin and Pressey 1988). Finally, pectic polysaccharides isolated from plant cell walls have been shown to activate phytoalexin (Davis et al. 1986) and proteinase inhibitor 1 biosynthesis (Walker-Simmons and Ryan 1986). The fact that polygalacturonase is present in vivo at developmental periods associated with elevated ethylene biosynthesis (e.g., ripening and abscission) supports the notion that polygalacturonase activity or its resulting products may represent at least one mechanism for promoting ethylene production.

Nevertheless, the correlation between increased polygalacturonase activity and ethylene evolution has been somewhat controversial. Using immunological techniques, Grierson and Tucker (1983) showed that increased ethylene evolution preceded polygalacturonase synthesis by 20 h in ripening tomato fruit. Conversely, Lincoln et at. (1987) demonstrated that polygalacturonase mRNA accumulates in parallel with ethylene biosynthesis. In fact, others have demonstrated the appearance of significant polygalacturonase activity as much as six days before the onset of the respiratory climacteric (Poovaiah and Nukaya 1979). One must remember, however, that all of these correlations are restricted by the fact that all parameters are measured by different means, making sequence determinations of temporally similar events tentative at best. Also, the variation in the reported temporal sequence of ripening associated polygalacturonase accumulation and autocatalytic ethylene biosynthesis demonstrates the importance of careful and reproducible staging of experimental fruit and tissues.

In short, numerous lines of evidence associate polygalacturonase activity with polyuronide degradation, fruit softening, and possibly regulation of other components of the ripening process. Because polygalacturonase has been associated with a variety of fruit ripening parameters, it is important from both scientific and agricultural perspectives. Softening and pectin degradation represent critical components of cell wall metabolism. In addition, softening is an impediment to harvest and shipment of the tomato crop. Unfortunately, from the point of view of the consumer, the modern tomato has been subjected to breeding and harvesting practices promoting the ability to withstand the stress of mechanical harvesting and transport rather than richness of flavor and overall fruit quality. It becomes clear from the standpoint of the considerable implications of involvement in fruit development and potential as a candidate for genetic manipulation, that tomato fruit endopolygalacturonase has been a worthy recipient of the extensive scrutiny it has received in recent years.

One correlation substantiated by experimental results is the relationship between polygalacturonase activity and disruption of middle lamella polyuronides. Crookes and Grierson (1983) have used electron microscopy to examine the cell wall ultrastructure of ripening tomato fruit. Their observations of middle lamella disruption coincident with polygalacturonase accumulation were verified by the induction of similar degradation of mature green fruit cell walls by purified polygalacturonase isozymes.

During the past five years, the regulation of polygalacturonase activity, protein accumulation, and gene expression during fruit ripening have been undertaken in a variety of laboratories, including our own. Specific antibodies have been generated against purified polygalacturonase protein for the quantification of levels in ripening and non-ripening tissues, determination of kinetics of their respective accumulations, and in situ localization in ripening fruit (Ali and Brady 1982; Sato et al. 1984; DellaPenna et at 1986; Tieman and Handa 1989). cDNA and genomic polygalacturonase clones have been isolated and characterized structurally (DellaPenna et al. 1986; Grierson et al. 1986; Sheehy et al. 1987; Bird et al. 1988; Giovannoni et al 1989). Using molecular probes generated from these clones, analysis of genetic regulation and mRNA accumulation has been undertaken (DellaPenna et al. 1986, 1987, 1989; Lincoln et al. 1987; Sheehy et al. 1987; Bird et al. 1988; Knapp et al. 1989; Biggs and Handa 1989; Pear et al. 1989). In addition these clones have been utilized to analyze polygalacturonase gene structure and regulation in an assortment of ripening impaired mutants (DellaPenna et al. 1987, 1989; Lincoln and Fischer 1988b; Biggs and Handa 1989; Knapp et al. 1989). Employing biochemical techniques, a significant body of data pertaining to accumulation of activity in vitro and in vivo has also been generated (Bennett and DellaPenna 1987a and references therein). The thrust of these experiments has been to correlate polygalacturonase gene expression, activity, and localization with ripening related phenomena.

E. Understanding the Function of Polygalacturonase

Ideal for the elucidation of polygalacturonase function during fruit ripening would be a mutation in the tomato polygalacturonase structural gene resulting in a significant reduction or absence of the enzyme activity. Unfortunately, no such mutant has been identified as yet and would be a difficult one to isolate through traditional mutagenesis techniques. Large numbers of full-size tomato plants yielding mature fruit would be needed for such a screen, with no guarantee of easy identification of the mutant phenotype.

Another avenue toward the understanding of polygalacturonase and its relationship to fruit ripening is via techniques designed to inhibit production of the enzyme during normal fruit development. Heat treatment (Ogura et al. 1975a, 1975b; Yoshida et al. 1984; Picton and Grierson 1988) and incubation in the presence of a variety of inhibitors of ethylene action and biosynthesis (Hobson et al. 1984; Yang 1985; Lincoln et al. 1987; Davies et al. 1988) have been employed to reduce the levels of polygalacturonase mRNA and protein in ripening tomato fruit. However, these methods also result in the inhibition of a considerable number of other ripening parameters making derivation of polygalacturonase function extremely difficult. Recently, procedures for antisense repression of nuclear gene expression have been developed and applied to suppress the expression of specific genes (Melton 1985; Ecker and Davis 1986; Green et al. 1986 [review]; Cabrera et al. 1987; Knecht and Loomis 1987; Rothstein et al. 1987; van der Krol et al. 1988). This procedure has been successfully utilized to suppress polygalacturonase gene expression (Sheehy et al. 1988, Smith et al 1988). The absence of effects on other ripening genes, contrary to the results from the use of the less specific inhibitors described above, makes data interpretation more direct.

To gain insights into the function of fruit endo-polygalacturonase during tomato ripening and softening we have employed a mutant complementation approach (Giovannoni et al. 1989). That is, a chimeric polygalacturonase gene was constructed and integrated into the genome of the fruit ripening mutant, *ripening inhibitor (rin)*, impaired in endogenous polygalacturonase gene activation and fruit softening. Because the chimeric polygalacturonase gene was designed to be transcriptionally active in *rin* mutant fruit, we were able to analyze the effect of polygalacturonase enzyme activity on cell wall structure and fruit softening. Throughout the rest of this chapter we will: (1) review work demonstrating the regulation and expression of polygalacturonase during tomato fruit ripening; (2) review the recent advances resulting in antisense repression and mutant complementation; and (3) assess the meaning of these results.

II. THE BIOCHEMISTRY OF POLYGALACTURONASE

A. Isozymes of Polygalacturonase

Tomato fruit polygalacturonase accumulates during ripening in several forms (Pressey and Avants 1973; Rexová-Benková and Markovic 1976; Tucker et al. 1980; Ali and Brady 1982). Though reported sizes of polygalacturonase have varied somewhat in the literature, tomato fruit accumulate, early during ripening, a polygalacturonase isoform (PG1) of approximately 110 kilodaltons (kd) as determined by column chromatography (Lisker and Retig 1974; Ali and Brady 1982; Pressey 1986a). As fruit development continues, two smaller isoforms, PG2A and PG2B of approximately 42 and 46 kd respectively, accumulate (Brady et al. 1982). PG1 has been determined to have an isoelectric point (pI) of 8.6 while the two smaller forms have identical pIs of 9.4 (Ali and Brady 1982). Heat stability analysis of purified polygalacturonase isoforms has demonstrated the PG1 is considerably more thermostable than the PG2 isozymes (Knegt et al. 1988).

The relative amounts of polygalacturonase isozymes in different cultivars vary widely (Tucker et al. 1980; Pressey 1986a). However, it is generally agreed throughout the literature that PG1 accumulation precedes that of PG2A and PG2B (Tucker et al. 1980; Crookes and Grierson 1983; Moshrefi and Luh 1983; Brady et al. 1985). This variation in relative and total amounts of polygalacturonase isozymes in ripening fruit most likely represents real variations among different cultivars combined with differences in fruit sampling, extraction, and assay techniques (Pressey 1986b; Knegt et al. 1988).

Polygalacturonase isozymes PG2A and PG2B have been shown to differ in the level of glycosylation (DellaPenna and Bennet 1988). Considerable evidence has accumulated suggesting that the PG1 polygalacturonase isozyme consists of a PG2 polypeptide noncovalently bound to additional polypeptide(s). Tucker et al. (1980) showed that denaturation of polygalacturonase enzymes with sodium dodecyl sulfate (SDS) yields similar polypeptides, suggesting that the larger PG1 form is a multimer consisting at least in part of PG2 isozymes. In addition, antibodies raised against a mixture of PG2A and PG2B enzymes reacted equally well with the high molecular weight PG1 isoform (Ali and Brady 1982). Also, trypsin and chymotrypsin digestion of purified PG1 and PG2 enzymes yielded nearly identical peptide fragments (Tucker et al. 1980; Sheehy et al. 1987). Both high and low molecular weight isoforms have been shown to be glycoproteins (Moshrefi and Luh 1983, DellaPenna and Bennett 1988). We have recently reported that expression of a single polygalacturonase genomic clone in transgenic tomato fruit results in the accumulation of all three isozymes, proving their relationship at the level of DNA sequence (Giovannoni et al. 1989; DellaPenna et al. 1990).

Though PG1 is known to consist in part of PG2 polypeptides, the nature of this relationship is not well understood. Isolation of PG1 from ripening fruit cell wall protein extracts, using gel filtration techniques, yielded a PG1 fraction of nearly 200 kd (Moshrefi and Luh 1983). This size is approximately twice that previously reported for this particular isozyme (Tucker et al. 1980; Ali and Brady 1982). Subsequent SDS poly-acrylamide gel electrophoresis (PAGE) of the 200 kd fraction revealed two polypeptides of 47.5 and 41.4 kd, the larger of which comigrated with purified PG2 enzymes. In the same report, both polypeptides were shown to glycoproteins and hypothesized to exist in some combination of four molecules to form PG1.

Tucker et al. (1981) identified a heat stable, non-dialyzable factor extractable from both mature green and ripe tomatoes capable of convert-ing a mixture of purified PG2A and PG2B into PG1. Subsequently, this factor has been purified and designated PG converter (CV) by Pressey (1984). It was determined to be a protease sensitive glycoprotein of approximately 100 kd. Treatment of PG1 with base resulted in release of PG2 (which isoform is not determined), suggesting that the converter is the associated molecule (Pressey 1986a and 1986b). As yet, no relation-ship between PG converter and the 41.4 kd protein of Moshrefi and Luh (1983) has been reported. In light of the fact that the converter is recover-able from unripe fruit and vegetative tissues, and in conjunction with the finding that certain extraction conditions result in the recovery of only PG2 isozymes, it has been suggested that PG1 is an artifact of extraction technique (Pressey 1988).

Knegt et al. (1988) have proposed an alternate hypothesis for PG con-verter function and the presence of PG1 enzyme in ripening tomato fruit. They have demonstrated that purified PG converter reacts with PG2 enzymes to form an enzyme similar to, yet distinct from PG1, designated PGx. Also, heat treatment of purified PG1 yields converter activity as measured by the ability to form PGx when incubated with PG2 isozymes. In addition, they have shown unreacted PG converter to be localized in plant cell walls. Based upon these observations, it is hypothesized that PG converter is a cell wall entity responsible for the binding of polygalac-turonase, and possibly other cell wall proteins, to the plant cell wall. Should this hypothesis be correct, PG1 would most likely be the physiologically active form of the enzyme during fruit ripening. Substan-tiating this notion are recent results demonstrating that polyuronide degradation in transgenic tomato fruit results primarily in the presence of PG1 (DellaPenna et al., unpublished results).

III. TOMATO FRUIT RIPENING MUTATIONS

Over the years, several tomato mutants have been identified whose fruit do not ripen normally. Many of these mutants have been mapped using classical linkage analysis to morphological markers and shown to be located at individual unique loci (Rick 1980). Mutants such as the greenflesh (gf; Ramirez and Tomes 1964) and yellowflesh (r; Darby 1978) effect color development during ripening by inhibiting chlorophyll degradation and lycopene production respectively. Several other mutations, however, severely retard many aspects of tomato fruit ripening. Specifically, these fruit ripening specific mutants (reviewed in Grierson et al. 1987) include the ripening inhibitor (rin; Robinson and Tomes 1968), non-ripening (nor; Tigchelaar et al. 1973), Never-ripe (Nr; Rick 1956; Hobson 1968), and alcobaca allele of the nor locus (alc; Kopeliovitchet et al. 1980) mutations, all of which inhibit ripening without significantly affecting growth and development of other vegetative tissues. For a recent review on tomato ripening mutants see Grierson et al. (1987).

The ripening-impaired mutants have been utilized to further the understanding of the regulatory events involved in fruit ripening, including the function of polygalacturonase. The rin and nor mutants display the most extreme phenotypes in that their fruit fail to ripen (Tigchelaar et al. 1978b). Although rin and nor fruit become full size, they remain green and firm until falling prey to either microbial infection or dehydration. In contrast, Nr and alc fruit ripen partially, as do nor fruit in particular genetic backgrounds (Hobson 1967; Kopeliovitchet et al. 1980). Nr fruit become an orange-red color and soften moderately as compared to normal fruit (Hobson 1967), while fruit harboring the alc mutation ripen slowly to a light red color (Mutschler 1984). The fruit of all four mutants produce viable seed.

Fruit of the rin and nor mutants fail to display the expected climacteric rise in respiration (Herner and Sink 1973; Ng and Tigchelaar 1977) with only nor fruit capable of severely attenuated ethylene biosynthesis (Ng and Tigchelaar 1977). Both rin and nor fruit respond to exogenous ethylene and propylene by increased respiration, without measurable changes in endogenous ethylene biosynthesis (McGlasson et al. 1975). Fruit of rin and nor are distinctive in appearance; there are differences in the spectrophotometric profiles of carotenoids, with rin fruit accumulating relatively high levels of phytoene (Sink et al. 1974; Ng and Tigchelaar 1977). In addition, chlorophyll degradation is delayed in fruit from both mutants as compared to those which undergo normal ripening (Tigchelaar et al. 1978b). In comparison, Nr and alc fruit display reduced and delayed ethylene production and respiration coincident with delayed color formation (Tigchelaar et al. 1978b; Mutschler 1984).

A. Polygalacturonase Levels in Fruit Ripening Mutants

The correlation of low polygalacturonase enzyme levels in the ripening impaired mutants supported the hypothesized role of polygalacturonase involvement in fruit ripening. Analysis of polygalacturonase accumulation in fruit ripening mutants has been of considerable interest as a way of explaining the delayed or reduced softening observed in these fruit. Indeed, fruit of rin and nor genotypes, which soften the least, display only trace amounts of polygalacturonase activity and protein at a time corresponding to normal fruit ripening (Buescher and Tigchelaar 1975; Ng and Tigchelaar 1977; DellaPenna et al. 1987; Biggs and Handa 1989: Giovannoni et al. 1989), while partially ripening Nr and alc fruit are delayed and reduced in total polygalacturonase activity (Mutschler 1984; Biggs and Handa 1989). Finally, polygalacturonase activity staining of non-denaturing polyacrylamide gels reveals that barely detectable levels of PG1 isoform accumulate in rin fruit at a time corresponding to several weeks after the onset of normal ripening (DellaPenna et al. 1987). Tucker et al. (1980) observed only PG1 in Nr fruit while others have visualized both PG1 and PG2 isoforms several months after ripening would normally occur if the fruit were wild-type (Crookes and Grierson 1983).

As just described, the rin and nor mutants exhibit the most aberrant phenotypes of any of the tomato fruit ripening mutations reported to date. However, several reports have been published describing partial inducement of ripening parameters in both of these mutants. Treatment with ethylene or propylene of detached rin fruit hastens the normal yellowing of the fruit without noticeable effects on lycopene accumulation, softening, autocatalytic ethylene biosynthesis, or polygalacturonase activity (McGlasson et al. 1975; Giovannoni et al. 1989). However, treatment of fruit still attached to the vine with ethylene or 2-chloroethly phosphonic acid (ethephon), a chemical whose degradation products include ethylene, results in 10–20% of normal lycopene accumulation, and increased softness of the fruit (Mizrahi et al. 1975; Buescher 1977). Similar results with detached rin fruit have been reported when treatments were performed in high oxygen atmospheres (Frenkel and Garrison 1976; Tigchelaar et al. 1978b). Finally, Mizrahi et al. (1982) reported that supplementation of nutrient solutions with NaCl resulted in partial ripening of nor fruit; however, no effect was observed in similarly treated rin fruit. The partial complementation of ripening parameters coincident with partial inducement of polygalacturonase activity in hormone treated and salt stressed mutants adds to the correlative evidence implicating polygalacturonase activity in the control of tomato fruit ripening (Tigchelaar et al. 1978a, 1978b).

B. Regulation of Polygalacturonase Gene Expression

Several laboratories have reported the cloning of cDNA inserts corresponding to polygalacturonase mRNAs isolated from ripening tomato fruit (DellaPenna et al. 1986: Grierson et al. 1986; Sheehy et al. 1987; Biggs and Handa 1989). In all cases libraries were constructed from polyA+ ripening fruit RNAs isolated from a variety of tomato cultivars.

However, different approaches were employed for subsequent isolation of the polygalacturonase clones. Utilizing differential screening techniques, a variety of tomato fruit ripening related cDNA clones have been isolated, followed by identification of a polygalacturonase clone via comparison of cDNA sequence with N-terminal amino acid sequence (Grierson et at. 1986). Other investigators have constructed ripening fruit cDNA libraries in which vectors conducive to the expression of cloned sequences in bacterial hosts are employed (DellaPenna et al. 1986; Sheehy et al. 1987; Biggs and Handa 1989). In all instances, bacterial expression of tomato ripening fruit cDNAs, followed by immunodetection with antisera raised against purified PG2 enzymes, resulted in specific identification of colonies harboring polygalacturonase clones. Sheehy et al. (1987) and DellaPenna and Bennett (1988) verified their polygalacturonase sequences via codon comparison to the PG2A polypeptide sequence generated from N-terminal and C-terminal sequencing of the purified protein. Finally, nucleotide sequence analysis and cross hybridization experiments have demonstrated that the polygalacturonase cDNA clones isolated independently in all four laboratories contain nearly identical sequences (Grierson et al. 1986; Sheehy et al. 1987; DellaPenna and Bennett 1988; Biggs and Handa 1989).

C. Analysis of Polygalacturonase mRNA Accumulation

The availability of polygalacturonase cDNA clones made it possible to study the regulation of polygalacturonase gene expression during tomato fruit development. Several laboratories have demonstrated that polygalacturonase mRNA accumulation increases dramatically at the onset of ripening and continues to retain a high level of abundance throughout the remainder of fruit development (DellaPenna et al. 1986, Lincoln et al. 1987, Maunders et al. 1987, Biggs and Handa 1989). Hybridization to radiolabeled cDNA probes has demonstrated increased polygalacturonase mRNA accumulation in breaker tomato fruit (the term "breaker" refers to the stage in ripening when the first visible signs of lycopene accumulation occur) followed by a continual increase through the fully red stage (Fig. 2.1). Polygalacturonase has been reported to account for as much as 2.3% of the total mRNA in ripening fruit (DellaPenna et al. 1987). However, a significant decrease in concentration of this particular mRNA occurs as the fruit proceeded to advanced stages of ripening (DellaPenna et al. 1987; Maunders et al. 1987; Sheehy et al. 1987; Biggs and Handa 1989). Nevertheless, polygalacturonase enzyme activity increases throughout fruit development (Biggs and Handa 1989). It has also been demonstrated that polygalacturonase mRNA accumulation does not occur in roots, leaves, or stems of the tomato plant (Maunders et al. 1987; Biggs and Handa 1989).

As mentioned earlier, polygalacturonase mRNA accumulation has been suggested to be temporally associated with the autocatalytic burst of ethylene biosynthesis associated with fruit ripening. Nevertheless, Lincoln et al. (1987) have demonstrated that polygalacturonase mRNA accumulation in mature green one (MG1) stage fruit does not result as a response to an 8-h exposure to endogenous ethylene. However, others have shown that ethylene treatment for at least one day does results in considerable polygalacturonase mRNA accumulation in mature green fruit (Maunders et al. 1987). It is important to remember, however, that fruit staging methodology differs among laboratories, making comparison of results difficult at best. Nevertheless, it has been shown that polygalacturonase mRNA accumulation is severely inhibited by the inhibitors of ethylene action, norbornadiene and silver thiosulphate (Lincoln et al. 1987; Davies et al. 1988). These results suggest that polygalacturonase mRNA accumulation is not an early response of the fruit to the exogenous application of ethylene but may be influenced by events relatively late in the cascade of ethylene mediated affects on fruit ripening.

Because of the early reports of reduced or absent polygalacturonase enzyme activity in the rin, nor, and, Nr mutants, consideraable efforts were directed toward understanding the molecular basis of these

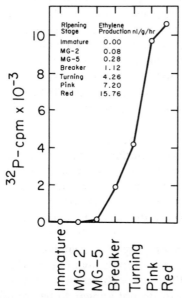

Figure 2.1. Analysis of polygalacturonase mRNA levels during wild-type tomato fruit ripening. PolyA+ RNA (1 ug) from each ripening stage was electrophoresed, blotted to nitrocellulose and probed with the ^{32}p-labeled insert of a polygalacturonase cDNA clone. After autoradiography the radioactivity associated with each stage was determined by liquid scintillation counting. Redrawn from DellaPenna et al. (1986).

phenomena utilizing the available polygalacturonase clones. Sheehy et al. (1987) reported no detectable polygalacturonase mRNA in polyA+ RNA extracted from rin fruit 50 days post anthesis. However, analysis of rin fruit development in several other laboratories revealed low but significant levels of polygalacturonase mRNA extractable from 42–49-day-old fruit (DellaPenna et al. 1987; Biggs and Handa 1989; Knapp et al. 1989). A peak of 2.5% of maximum wild-type polygalacturonase mRNA accumulation was measured in fruit of the cultivar 'Rutgers' nearly isogenic for the rin mutation (DellaPenna et al. 1987), while Knapp et al. (1989) reported a high of 0.6% when the mutation was in the 'Ailsa Craig' cultivar. No induction of polygalacturonase mRNA accumulation was observed when rin fruit were treated with exogenous ethylene (Lincoln et al. 1987; Knapp et al. 1989). Furthermore, the low levels of polygalacturonase mRNA observed in rin fruit were shown to result in a correspondingly small amount of PG1 enzyme in vivo (DellaPenna et al. 1987).

Levels up to 30% of maximum wild-type polygalacturonase mRNA concentrations were observed in fruit homozygous for the Nr mutation (DellaPenna et al. 1987; Biggs and Handa 1989; Knapp et al. 1989) which undergo a delayed and partial ripening (Tigchelaar et al. 1978b). In contrast, fruit homozygous for the nor mutation exhibit barely detectable levels of polygalacturonase mRNA (DellaPenna et al. 1987; Biggs and Handa 1989). These results suggest that the rin, nor, and Nr mutations exert their pleiotropic effects on ripening, at least in part, by restricting polygalacturonase mRNA accumulation in developing tomato fruit, and support the hypothesis that polygalacturonase activity is a prerequisite for normal ripening. It must be stressed, however, that these data are merely consistent with said hypothesis, while the question remains as to whether polygalacturonase is a primary or secondary effect of the "ripening genes."

The rin and nor mutants have been classified genetically as influencing fruit ripening in a recessive manner, while the Nr mutant is reported to be a dominant mutation (Rick 1956; Robinson and Tomes 1968; Tigchelaar et al. 1973). Analysis of polygalacturonase mRNA, protein, and activity accumulation has been performed on fruit both heterozygous and homozygous for the three fruit ripening mutations mentioned above (Biggs and Handa 1989). All three mutations were shown to reduce the accumulation of polygalacturonase mRNA, protein, and activity in a manner dependent upon copy number of the mutant alleles, with the most dramatic differences observed in rin and nor fruit. Therefore, the rin and nor alleles do not appear to be completely recessive, because in heterozygous backgrounds intermediate levels of polygalacturonase mRNA are detected, and delayed ripening is observed. In addition, all genotypes tested, including wild-type, showed similar patterns of polygalacturonase expression in that maximal mRNA, protein, and activity accumulations were separated from each other by several days (Biggs and Handa 1989).

D. Transcriptional Analysis of Polygalacturonase Gene Expression

DellaPenna et al. (1989) utilized run-on transcription analysis of isolated plant nuclei to analyze relative rates of transcription of ripening related genes, including polygalacturonase. In short, intact nuclei were isolated from fruit tissues at a variety of developmental states and allowed to continue transcription of previously initiated messages in the presence of radionucleotides (Luthe and Quatrano 1980; Lincoln and Fischer 1988a). In vitro labeled RNAs were then utilized as probes for hybridization with cDNA clones, including those containing polygalacturonase sequences. In correlation with the patterns of mRNA accumulation described above, polygalacturonase transcription is undetectable in mature green fruit (45 days post-anthesis), activated during early ripening, and exhibits a considerable decrease in activity as the fruit reaches advanced stages of maturity. In contrast, transcription of the D21-3 gene, a constitutively expressed control, is nearly constant throughout fruit development (Fig. 2.2). These results demonstrate that changes in rates of gene transcription control, at least in part, polygalacturonase mRNA concentrations in wild-type tomato fruit.

Analysis of polygalacturonase transcription in fruit containing the *rin*, *nor*, and *Nr* ripening mutations showed reduced transcription rates expected from mRNA accumulation data. Barely detectable polygalacturonase transcription in *rin* and *nor* fruit throughout development was observed, while *Nr* fruit showed both attenuated and delayed increases in both transcription and mRNA accumulation (Fig. 2.2). D21-3 transcrip-

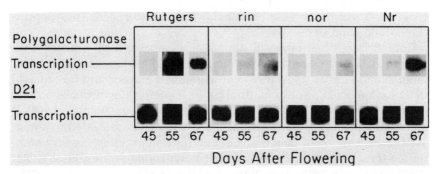

Figure 2.2. In vitro transcription rates of polygalacturonase and the constitutively expressed cDNA clone D21. Maximum cpm for polygalacturonase and D21 transcription were 379 and 507, respectively. 45, 55, and 67 days after flowering correspond to mature green, ripe, and overripe stages of ripening, respectively, for wild-type (Rutgers) fruit. Redrawn from DellaPenna et al. (1989).

tion appeared to be relatively unaffected by mutations at the fruit ripening loci (Lincoln et al. 1988b; DellaPenna et al. 1989). Therefore, the reduced levels of polygalacturonase mRNA resulted, at least in part, from reduced levels of polygalacturonase gene transcription.

E. Post-transcriptional Polygalacturonase Gene Regulation

Analysis of relative rates of gene transcription and mRNA accumulation for a variety of ripening related genes demonstrated that the relative rate of transcription initiated at the polygalacturonase promoter is low compared to other fruit ripening genes. Nevertheless, polygalacturonase mRNA concentration was at least 3-fold higher than other fruit ripening mRNAs (DellaPenna et al. 1989). These results suggest that post-transcriptional determinants, such as mRNA stability and transport into the cytoplasm, play an important role in determining polygalacturonase mRNA concentration in ripening tomato fruit.

Because of its secretion into the cell walls of ripening tomato fruit, polygalacturonase expression has been analyzed for post-transcriptional events which mediate putative activation and translocation processes (DellaPenna and Bennet 1988; Biggs and Handa 1989). In vitro transcription of a full length polygalacturonase cDNA resulted in synthesis of a 54 kd polypeptide via in vitro translation of the resultant mRNA. Subsequent processing experiments in the presence of canine pancreas microsomal membranes, shown to be competent for protein glycosylation in vitro (Walter et al. 1984), resulted in signal sequence cleavage and the appearance of two heavily glycosylated polypeptides of approximately 58 and 61 kd (DellaPenna and Bennett 1988). In contrast, chemical deglycosylation of a purified mixture of PG2A and PG2B isoenzymes yielded a single polypeptide of 42 kd. Together these results suggest that the modifications resulting in PG2A and PG2B enzymes arise via differential glycosylation of a single polygalacturonase polypeptide (DellaPenna and Bennett 1988).

Comparison of N-terminal amino acid sequence of purified PG2 enzymes to that deduced from the nucleotide sequence suggest that translational and/or post-translational processing events account for removal of 71 amino acids prior to maturation of the polygalacturonase protein (Grierson et al. 1986; Sheehy et al. 1987; DellaPenna and Bennett 1988). It has been shown that polygalacturonase mRNA is associated to a considerable degree with membrane bound ribosomes, allowing for potential co-translational processing of the polypeptide (Walter et al. 1984; Biggs and Handa 1989). DellaPenna and Bennett (1988) demonstrated the co-translational removal in vitro of the putative polygalacturonase signal sequence which consists of the 24 N-terminal amino acids. They further hypothesize that the remaining 47 N-terminal amino acids comprise a prosequence possibly involved in secretion to the cell wall, or inactivation of the precursor protein. It should be possible, utilizing molecular techniques, to dissect the prosequence function via chimeric gene constructions in which the polygalacturonase prosequence is incorporated into the N-terminal amino acid sequence of reporter proteins not normally directed to the cell wall.

F. Polygalacturonase Gene Structure and Organization

The availability of cloned polygalacturonase cDNA sequences has allowed for the isolation and characterization of corresponding genomic sequences (Bird et al. 1988; Giovannoni et al. 1989). The polygalacturonase genomic clone has been sequenced, and comparison to cDNA sequences reveals a 7 kb transcription unit containing eight introns ranging in size from 99 to 953 bp (Bird et al. 1988). Approximately 1.4 kb of the 5' upstream region has been sequenced and has been shown to contain DNA sequences sufficient to control polygalacturonase gene expression during fruit ripening (Bird et al. 1988; J. Montgomery and R. Fischer, unpublished).

Transcriptional control of polygalacturonase gene expression during tomato fruit ripening has been supported by construction of a polygalacturonase promoter-reporter gene fusion (Bird et al. 1988). Specifically, 1.4 kb of sequence 5' of the polygalacturonase gene transcription start site was ligated to protein coding sequences of a bacterial chloramphenicol acetyl transferase (CAT) gene (Colot et al. 1987) and inserted into the genomes of wild-type tomato plants via *Agrobacterium* mediated DNA transfer (McCormick et al. 1986). Subsequent analysis of CAT activity (Fig. 2.3) in a variety of tissues from transgenic plants

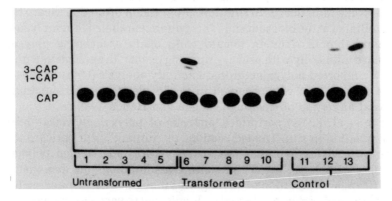

Figure 2.3. Chloramphenical acetyl transferase activity in different organs of transgenic tomato fruit containing the polygalacturonase/CAT chimeric gene construction. Protein extracts were from roots (lanes 1, 10), stems (lanes 2, 9), leaves (lanes 3, 8), green fruit (lanes 4, 7), and red fruit (lanes 5, 6). The activity of purified CAT enzyme is shown in lanes 11, 12, and 13. From Bird et al. (1988).

demonstrated significant activation of the polygalacturonase promoter only in ripening tomato fruit (Bird et al. 1988), supporting the previously described data suggesting specific transcription and accumulations of mRNA, protein, and enzyme activity during fruit ripening.

Polygalacturonase cDNA and genomic clones have simplified analysis of gene organization and structure in both normal tomato plants and ripening impaired mutants. DNA gel blot analysis, copy number reconstruction, and genomic cloning experiments suggest the presence of a single polygalacturonase gene per haploid genome in the cultivated tomato (DellaPenna et al. 1987; Bird et al. 1988; Giovannoni et al. 1989; Knapp et al. 1989). In addition, DNA gel blot hybridization analysis demonstrates no substantial differences in polygalacturonase gene structure in the genomes of ripening impaired mutants (DellaPenna et al. 1987, Knapp et al. 1989).

G. In Vivo Localization of Tomato Fruit Polygalacturonase

Employing techniques of tissue blotting to membrane filters, followed by immunodetection with purified polygalacturonase antibodies, temporal and spatial accumulation of polygalacturonase protein in ripening tomato fruit has been assessed (Tieman and Handa 1989). Tomato fruit at various ripening stages ranging from mature green to overripe were bisected parallel to the collumella (the segment of carpel running through the center of the fruit) and blotted to membranes capable of covalently binding proteins. Subsequent immunochemical localization indicated that polygalacturonase protein synthesis initiates at the collumella region and progresses to all pericarp tissues including the radial walls of pericarp. In addition, progression through pericarp tissue was initiated at the blossom end. This pattern paralleled that for lycopene accumulation in ripening tomato fruit, again associating polygalacturonase intimately with primary ripening events. In agreement with previously reported polygalacturonase activity analysis (Hobson 1964), no detectable protein was observed in the locular tissue surrounding the seeds at any stage of ripening (Tieman and Handa 1989).

Pear et al. (1989) performed analysis of polygalacturonase mRNA accumulation in situ. Treated sections of "turning" fruit pericarp tissue were utilized for hybridization analysis with radiolabeled polygalacturonase cDNA probes. The resulting audioradiograms demonstrated significant accumulation of silver grains in regions corresponding to the outer layers of pericarp cells and cells adjacent to vascular regions. These results agreed with immunocytolocalization data indicating initial polygalacturonase protein accumulation in the exocarp region of the pericarp at early stages of fruit ripening (Tieman and Handa 1989).

IV. FUNCTION OF POLYGALACTURONASE DURING TOMATO FRUIT RIPENING

A. Complementation of the Ripening-inhibitor Mutation with a Chimeric Polygalacturonase Gene

To investigate further the hypothesized roles of polygalacturonase activity in polyuronide degradation, ethylene biosynthesis, and fruit softening, we have implemented a molecular strategy to modify poly-galacturonase gene expression in vivo and assess its physiological function using the tomato fruit ripening mutant, rin (Giovannoni et al. 1989). As stated above, the rin mutation severely effects the majority of ripening phenomena including softening, polyuronide degradation, ethylene biosynthesis, lycopene accumulation and polygalacturonase activity (Tigchelaar et al. 1978b, Giovannoni et al. 1989). Because the rin mutation does not result from a lesion in the polygalacturonase gene (Bird et al. 1988), and the lack of polygalacturonase activity results from a cor-responding failure to activate gene transcription (DellaPenna et al. 1987, 1989), our goal was to induce polygalacturonase gene expression in trans-genic rin fruit and determine what aspects of fruit ripening, if any, were complemented. Thus, this is a system essentially null with respect to polygalacturonase gene expression, in which polygalacturonase activity can be induced and followed by assessment of its consequences on fruit ripening parameters.

To this end we have constructed a chimeric gene consisting of the poly-galacturonase structural gene fused to the regulatory sequences of another ripening related gene, E8 (Deikman and Fischer 1988). The func-tion of the E8 gene has not yet been determined. However, sequence com-parison demonstrates relationship to another ripening related gene, pTOM13, shown by antisense repression to be important for normal ethylene biosynthesis in ripening tomato fruit (Holdsworth et al. 1988; Grierson et al. 1990).

E8 regulatory sequences were chosen for driving polygalacturonase gene transcription for the following reasons. First, in wild-type tomato plants polygalacturonase and E8 gene expression are tightly coordinated (Lincoln et al. 1987). That is, both E8 and polygalacturonase mRNAs are abundant in ripe tomato fruit, but are not detected by hybridization analysis in other organs such as leaf, root, or stem (Lincoln and Fischer 1988a; Giovannoni 1990). Furthermore, in wild-type fruit, E8 and poly-galacturonase mRNA levels and relative rates of gene transcription increase coincidentally; however, polygalacturonase gene transcription is severely inhibited in rin fruit whereas E8 maintains 60% of the relative wild-type transcription rate (DellaPenna et al. 1989). Finally, E8, and not polygalacturonase transcription, is activated by ethylene in rin fruit (Lincoln and Fischer 1988b). In short, E8 regulatory sequences were chosen for facilitation of polygalacturonase gene expression in transgenic rin fruit because they would best reproduce wild-type patterns of fruit specific polygalacturonase gene expression, without the potentially detrimental side effects of polyuronide solubilization in non-fruit tissues of the experimental organism.

B. Construction of a Chimeric E8-Polygalacturonase Gene

A detailed description of the construction of a chimeric E8-polygalacturonase fusion gene is presented in Giovannoni et al. (1989). Initially, E8 and polygalacturonase genomic clones were isolated and characterized using a variety of conventional means (Deikman and Fischer 1988; Giovannoni 1989). The cloned polygalacturonase restriction fragments corresponded exactly to those observed in genomic DNA blot experiments, suggesting that polygalacturonase is encoded by a single-copy gene. To facilitate construction of a chimeric gene, Ncol restriction endonuclease sites were generated at the ATG translation initiation codons of both the polygalacturonase and E8 genes via oligonucleotide mediated site-directed mutagenesis. It is important to note that no nucleotides in the polygalacturonase protein coding region were altered, ensuring that the polygalacturonase enzyme resulting from expression of a chimeric gene would be identical to that produced in normal fruit development. Finally, the chimeric E8-polygalacturonase construct, consisting of 2 kb of E8 5'-flanking sequence, the E8 site of transcription initiation, and the 34 bp E8 untranslated mRNA leader fused at the ATG translation initiation site to the 7.5 kb polygalacturonase structural gene, followed by 5.5 kb of 3' polygalacturonase flanking sequence, was transferred to the rin genome by Agrobacterium mediated T-DNA transfer. Three plants were recovered and designated rin(E8/PG)-1, rin(E8/PG)-2, and rin(E8/PG)-3. In addition, a control rin plant, designated rin(C), was transformed with vector alone. Blot hybridization indicated that each transformant resulted from a single insertion event (Giovannoni et al. 1989).

C. Chimeric Gene Expression in Transgenic Fruit

Expression of the Chimeric E8-polygalacturonase gene in transgenic rin fruit was determined via analysis of accumulation of chimeric gene mRNA, protein, and extractable polygalacturonase activity. To distinguish among E8, polygalacturonase, and chimeric gene expression, mRNA was isolated from transformed and control fruits and assayed by S1-nuclease protection (Fig. 2.4). S1-nuclease protection of RNA-DNA hybrids allowed for individual assessment of mRNA accumulation resulting from the three genes just mentioned. Figure 2.4A shows that, as

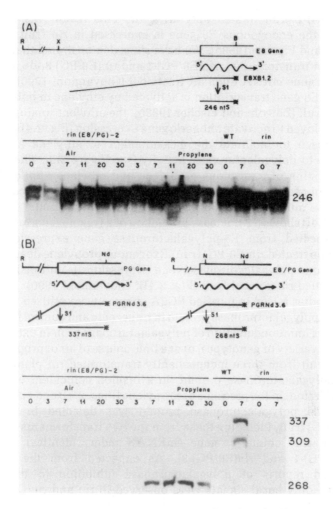

Figure 2.4. S1-nuclease protection of 25 ug samples of total RNAs extracted from rin(E8/PG)-2, wild-type (WT), and rin fruit harvested at the mature green stage (Lincoln et al. 1987) and treated with humidified air or propylene for the indicated number of days. Following hybridization to ^{32}-P end-labeled clones E8XB1.2 and PGRNd3.6 which hybridize specifically to E8 (A) and polygalacturonase (B) mRNAs, respectively, protected fragments were separated by acrylamid gel electrophoresis. The difference in sizes of protected fragments resulting from utilization of the polygalacturonase probe, PGRNd3.6 (panel B), occurs because chimeric E8/polygalacturonase fusion gene mRNA contains the E8 untranslated leader sequence which does not hybridize to the probe. From Giovannoni et al. (1989).

expected, E8 mRNA was detected in wild-type, *rin,* and rin(E8/PG)-2 fruit, whereas mRNA encoded by the endogenous polygalacturonase gene (Fig. 2.4B) was detected only in wild-type fruit. Curiously, mRNA encoded by the E8-polygalacturonase gene (Fig. 2.4B) was not detected in rin(E8/PG)-2 fruit treated only with air. This result was unexpected because the endogenous E8 gene is expressed in rin fruit (Fig. 2.4B; Lincoln and Fischer 1988b). We have since demonstrated that fruit from two other transformants, rin(E8/PG)-1 and rin(E8/PG)-3, do accumulate chimeric gene mRNA upon air treatment (Giovannoni 1990). However, because E8 gene transcription is activated by ethylene in both wild-type and rin fruit (Lincoln and Fischer 1988b), the ethylene analog propylene was employed to activate chimeric gene expression (Fig. 2.4B) Propylene was chosen to facilitate the measurement of ethylene evolution in response to polygalacturonase activity (see Figure 2.9). The effects on fruit ripening of exogenous propylene and ethylene application have been demonstrated to be very similar, except that higher concentration of propylene must be used (Burg and Burg 1967; McMurchie et al. 1972). Figure 2.4B shows S1-nuclease protection of a fragment corresponding to that expected from E8-polygalacturonase gene expression only in propylene treated rin(E8/PG) fruit. Exogenous propylene does not induce endogenous polygalacturonase gene expression in untransformed and transformed rin control fruit (Fig. 2.4B, Giovannoni 1990).

Antibodies raised to purified PG2A enzyme were utilized to assay for levels of polygalacturonase protein in transgenic and control fruit. Figure 2.5 shows immunodetection of polygalacturonase protein extracted from fruit of a variety of genotypes, over a time course of air or propylene treatment. Fruit from three independently transformed rin plants accumulated polygalacturonase protein, in a fashion coincident with mRNA accumulation, in the presence of propylene (Figs. 2.4B and 2.5). Note that very little polygalacturonase protein was detected in air treated rin(E8/PG) fruit, including those from the two transformants whose fruit accumulated chimeric gene mRNA under identical conditions (rin(E8/PG)-1 and rin(E8/PG)-3). As expected from the previously described reports of polygalacturonase inhibition in the ripening mutants, only basal signals were observed in rin and rin(C) fruit (Fig. 2.5).

To ascertain whether or not the protein resulting from expression of the chimeric gene was in fact capable of catalyzing polyuronide hydrolysis, extracts of cell wall proteins were assayed for the ability to degrade purified polygalacturonic acid in vitro. As shown in Figure 2.6A, only wild-type fruit exhibit significant levels of polygalacturonase activity during air treatment. However, exposure to exogenous propylene results in a significant increase in in vitro activity in transgenic rin fruit expressing the chimeric gene product (Fig. 2.6B; Giovannoni et al. 1989). After 30 days of propylene exposure rin(E8/PG) fruit displayed up to 60% of maximal levels found in wild-type fruit.

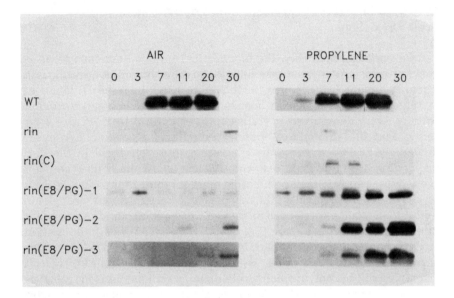

Figure 2.5. Protein gel blot analysis of cell wall protein extracts from wild-type, *rin*, and transgenic *rin* fruit. Cell wall protein extracts were isolated, fractionated by SDS-polyacrylamid gel electrophoresis, and transferred to nitrocellulose; polygalacturonase protein was detected using anti-polygalacturonase serum. From Giovannoni et al. (1989).

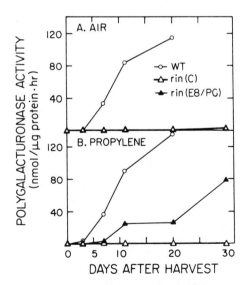

Figure 2.6. Polygalacturonase enzyme activity in wild-type, *rin*, and transgenic *rin* fruit. Mature green fruit were harvested and exposed to either air (A) or propylene (B) for the indicated period of time. Cell wall protein extracts were isolated and the levels of polygalacturonase activity determined. Redrawn from Giovannoni et al. (1989).

D. In Vivo Effects of Polygalacturonase Activity on *rin*
Fruit Physiology

As described previously, the biochemical function ascribed to poly-galacturonase protein has been the catalysis of polyuronide hydrolysis in ripening tomato fruit (Huber 1983b). To test this hypothesis, levels of extractable polyuronides, a measure of in vivo polygalacturonase activity, were determined in transgenic and control fruit. Figure 2.7A shows that the amounts of soluble uronic acid recoverable from air treated *rin*(E8/PG)-2 fruit were similar to those observed in the corresponding negative controls, *rin* and *rin*(C), while wild-type fruit accumulated approximately threefold more soluble uronic acid at 20 days of air treatment. However, in the presence of exogenous propylene, *rin*(E8/PG)-2 fruit experienced an increase in soluble uronic acids similar to that observed in wild-type (Fig. 2.7B). Subsequent size fractionation analysis has demonstrated that size distributions of soluble polyuronides change as tomato fruit ripen, with propylene treated *rin*(E8/PG) fruit displaying soluble polyuronide profiles similar to those derived from ripe wild-type tomato fruit (DellaPenna et al. 1990). It is clear from these results that polygalacturonase activity is sufficient for the dramatic increase in polyuronide solubilization characteristic of ripening tomato fruit.

The most significant results yielded by this research, however, dealt with the role of polygalacturonase in tomato fruit softening. Com-

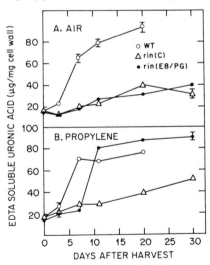

Figure 2.7.   Levels of EDTA-soluble polyuronides in wild-type, *rin*, and transgenic *rin* fruit. Fruit were harvested at the mature green stage and exposed to either air (A) or propylene (B) for the indicated number of days. Cell walls were isolated and levels of EDTA-soluble polyuronides were measured. Redrawn from Giovannoni et al. (1989).

pressibility of fruit was assayed as a measure of softening. Figure 2.8A shows compressibility of rin, rin(C)-1, and rin(E8/PG)-2 fruit changed little over the course of air treatment (a treatment which does not result in polygalacturonase protein accumulation in E8-polygalacturonase fruit), whereas wild-type fruit exhibited a fourfold increase in compressibility. Interestingly, Fig. 2.8B demonstrates that exposure of rin(E8/PG) fruit to exogenous propylene, which elicited significant levels of polygalacturonase activity and wild-type levels of polyuronide degradation after 11 days (see Fig. 6B and 7B), had no discernible effect on the softening of transgenic fruit (Giovannoni et al. 1989). This result contradicts much of the previously reported correlative evidence associating polygalacturonase activity and pectic solubilization with ripening related softening of tomato fruit.

To address hypothesized roles of polygalacturonase activity in the elicitation of other aspects of tomato fruit ripening, analysis of ethylene evolution and color development was performed. Figure 2.9 shows no significant effect of polygalacaturonase activity on ethylene biosynthesis in transgenic fruit, with the small rise in ethylene production in 25 day propylene treated transgenic fruit resulting from a single fruit (Fig. 2.9B). Finally, visual inspection of propylene treated rin(E8/PG) fruit reveals no significant differences in color formation as compared to untransformed rin controls (Giovannoni et al. 1989).

Explanations for the lack of softening and promotion of other ripening phenomena in propylene treated rin(E8/PG) fruit must include the fact

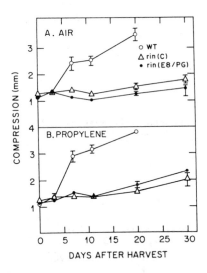

Figure 2.8. Compressibility of wild-type, rin, and transgenic rin fruit. A minimum of 5 mature green fruit were exposed to either air (A) or propylene (B) for the indicated number of days, and fruit compressibility was determined. Error bars represent standard deviations. Where error bars are not shown, the standard deviation was no greater than the size of the symbol. Redrawn from Giovannoni et al. (1989).

that little is known of how the *rin* mutation inhibits ripening, making it possible that the mutation somehow masks the effect of polygalacturonase activity in transgenic fruit. It is also possible that the physiologically appropriate isozymes are not synthesized. However, we have demonstrated that all three isozymes previously reported in wild-type ripening tomato fruit do accumulate in *rin*(E8/PG) fruit by 30 days of propylene treatment, and result in similar polyuronide degradation products as are found in non-mutant fruit ripening (Giovannoni et al. 1989; DellaPenna and Giovannoni 1990). In fact, the correlation of PG1 enzyme with maximal polyuronide degradation in *rin*(E8/PG) fruit (DellaPenna et al. unpublished) suggests that this may be the physiologically active protein. Furthermore, this result demonstrates that expression of a single polygalactauronase structural gene is sufficient for accumulation of all three isozymes, PG1, PG2A, and PG2B, found in ripening wild-type tomato fruit.

Finally, in light of reports demonstrating the partial complementation of the *rin* phenotype in fruit treated with ethylene or ethephon while still attached to the vine (Mizrahi et al. 1975; Buescher 1977; Tigchelaar et al. 1978), we initiated similar experiments with *rin*(E8/PG) fruit in the event that partial compensation of the *rin* phenotype may liberate some aspect of polygalacturonase function potentially masked by the effects of the *rin* background (Giovannoni et al. 1990). Figure 2.10 shows the induction of polygalacturonase activity in *rin*(E8/PG) fruit attached to the vine following ethephon treatment. No significant effects on texture or color

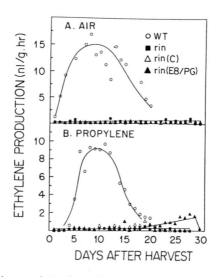

Figure 2.9.   Ethylene evolution by wild-type, *rin*, and transgenic *rin* fruit. Mature green fruit were exposed for the indicated number of days to either air (A) or propylene (B), and at the indicated period of time ethylene evolution was determined. Redrawn from Giovannoni et al. (1989).

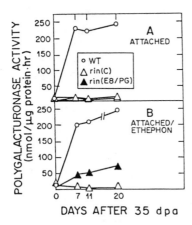

Figure 2.10. Polygalacturonase enzyme activity in wild-type, rin, and transgenic rin fruit. Mature green fruit were allowed to remain attached to the vine for the indicated number of days following either no treatment (A) or treatment with 1% ethephone (B). Redrawn from Giovannoni et al. (1990).

development, as compared to identically treated rin or rin(C) fruit, were detected. In summary, although polygalactauronase activity is necessary for polyuronide solubilization, it is insufficient to promote softening, ethylene biosynthesis, or lycopene accumulation in transgenic rin fruit.

E. Antisense Repression of Polygalacturonase Gene Expression

The mutant complementation experiments described above were designed to exploit a situation devoid of endogenous polygalacturonase activity and assay the physiological consequences resulting from addition of this protein via molecular genetic means. An alternative approach towards elucidation of the function of polygalacturonase in ripening tomato fruit, by antisense gene repression of the endogenous fruit ripening polygalacturonase gene, has been pursued independently in two laboratories (Sheehy et al. 1988; Smith et al. 1988). In both reports the constitutively active cauliflower mosaic virus 35s promoter (Guilley et al. 1982) was utilized to drive transcription of cloned polygalacturonase cDNA sequences in a reverse orientation relative to the endogenous gene. In particular, Smith et al. (1988) employed a 730 bp fragment from the 5′ end of the polygalacturonase cDNA, while Sheehy et al. (1988) used a full length cDNA clone.

Analysis of a number of transgenic antisense plants revealed similar reductions of polygalacturonase mRNA in vitro activity of 10–20% of that observed in untransformed ripening tomato fruit (Fig. 2.11). A corresponding decline in polygalacturonase protein accumulation was also observed (Smith et al. 1988). In addition, Sheehy et al. (1988) demonstrated that polygalacturonase gene transcription occurred at a rate considerably less than that observed for the inserted antisense construct; this was confirmed by another report of low polygalacturonase transcription rates (DellaPenna et al. 1989).

Both groups reported no change in lycopene accumulation during development of polygalacturonase antisense fruit as compared to untransformed controls (Fig. 2.11). In addition, Smith et al. (1988) reported no significant effect on timing nor on degree of softening of transgenic fruit. Primary transformants hemizygous for the antisense polygalacturonase gene have subsequently been selfed, and a homozygous antisense plant has been demonstrated to accumulate less than 1% normal polygalacturonase activity. In control experiments, progeny which inherit zero copies of the inserted sequence display normal polygalacturonase activity (Schuch et al. 1989). Furthermore, analysis of average molecular weight of extractable polyuronides demonstrates an inverse correlation between polyuronide polymer size and antisense gene

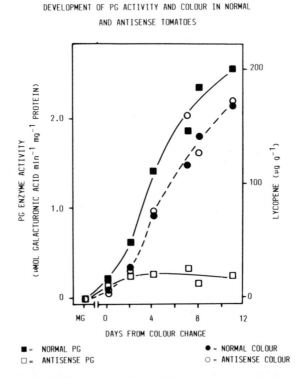

DEVELOPMENT OF PG ACTIVITY AND COLOUR IN NORMAL
AND ANTISENSE TOMATOES

■ = NORMAL PG   ● = NORMAL COLOUR
□ = ANTISENSE PG   ○ = ANTISENSE COLOUR

Figure 2.11. Polygalacturonase activity and lycopene content during the ripening of untransformed and transgenic polygalacturonase antisense tomato fruit. From Smith et al. (1988).

dosage. In agreement with results reported for transgenic *rin* fruit expressing a chimeric polygalacturonase gene (Giovannoni et al. 1989), no effects on softening, lycopene accumulation, or ethylene biosynthesis were observed in fruit containing 1% of wild-type polygalacturonase activity by antisense gene repression (Schuch et al. 1989; Grierson et al. 1990). Furthermore, invertase and pectin esterase activities were unaffected by reduction in polygalacturonase activity (Schuch et al. 1989).

Recent attempts to define the role of polygalacturonase through mutant complementation (Giovannoni et al. 1989) and antisense gene repression (Sheehy et al. 1988; Smith et al. 1988; Schuch et al. 1989) have shed considerable light on the function of this particular enzyme and the effect of polyuronide solubilization and depolymerization during tomato fruit ripening. In particular, the widely held belief that polygalacturonase activity and resulting pectin degradation are largely responsible for ripening related softening of tomato fruit has not been supported by the data described in this section.

V. ANALYSIS

A. Polygalacturonase Isozymes

Both the presence and roles of different polygalacturonase isozymes in tomato fruit ripening are area of active investigation. As described above, models exist in which PG1 is depicted as either an artifact of cell wall protein isolation techniques (Pressey 1988) or the primary in vivo component of polyuronide degradation (Kneght et al. 1988). The presence of PG converter may make resolution of this problem difficult, employing biochemical techniques because assays of in vivo activity attributable to specific isoforms will be impaired by the presence of the catalyst of PG1 production in many tissues, specifically fruit. However, a molecular genetic approach may prove useful for addressing the question of isozyme function. For example, polygalacturonase gene expression could be forced via inducible DNA constructs in tissues which do not contain endogenous converter (if any exist) in homologous and heterologous plant systems, followed by analysis of cell wall polyuronide degradation. Alternatively, antisense gene repression of PG converter in ripening tomato fruit may be utilized to generate fruit deficient in CV protein. Both examples would yield systems in which the action of PG2 isoforms could be assessed on their own. Should PG2 isozymes alone prove to be inactive in vivo, the necessity of the converter in cell wall degradation would be substantiated.

B. The Role of Polygalacturonase in Tomato Fruit Ripening

The failure of transgenic *rin* fruit expressing a chimeric polygalacturonase gene at a time corresponding to wild-type fruit ripening to soften (Giovannoni et al. 1989), in conjunction with the inability to influence wild-type tomato fruit softening by antisense repression of the fruit polygalacturonase gene (Smith et al. 1988), strongly suggests that the role of polygalacturonase during tomato fruit development is more complex than was originally thought. The correlative evidence implicating polygalacturonase involvement in fruit softening is compelling indeed; nevertheless, it has recently been demonstrated that high levels of polygalacturonase activity in ripening tomato fruit are neither necessary (Smith et al. 1988; Schuch et al. 1990) nor sufficient (Giovannoni et al. 1989) for obvious ripening related changes in fruit texture. What has been proven is that polygalacturonase activity is largely responsible for the disruption of polyuronides associated with tomato fruit ripening, though polyuronide degradation may be but a single character in a large cast responsible for changes in fruit texture during ripening.

Mutant complementation and antisense gene repression data have caused renewed thought concerning the function of polygalacturonase in ripening tomato fruit and softening in particular. The approaches just described are general and can be used to analyze the function of other cell wall degrading enzymes, as well. It is hoped that in the future, these approaches will lead to a greater understanding of cell wall structure, cell wall hydrolytic enzymes, and ultimately, the growth and development of plants.

LITERATURE CITED

Abeles, F. 1973. *Ethylene in Plant Biology.* Academic Press, New York.

Ali, Z. I., and F. Brady. 1982. Purification and characterization of polygalacturonases of tomato fruits. *Austral. J. Plant Physiol.* 9:155–169

Baile, J., and R. Young. 1981. Respiration and ripening in fruits—retrospect and prospect. p. 1–39. In: J. Fiend and M. Rhodes (eds.). *Recent Advances in the Biochemistry of Fruit and Vegetables.* Academic Press, London.

Baldwin, E., and R. Pressey. 1988. Tomato polygalacturonase elicits ethylene production in tomato fruit. *J. Am. Soc. Hort. Sci.* 113:92–95.

Bennett, A., and D. DellaPenna. 1987a. Polygalacturonase: Its importance and regulation in ripening. p. 98–107. In: W. Thompson, E. Nothnagel, and R. Huffaker (eds.). *Plant Senescence: Its Biochemistry and Physiology.* American Society of Plant Physiologists, Rockville, MD.

———. 1987b. Polygalacturonase gene expression in ripening tomato fruit. p. 299–308. In: D. Nevins and R. Jones (eds.). *Tomato Biotechnology.* Alan R. Liss, New York.

Bernatzky, R., and S. Tanksley. 1986. Toward a saturated linkage map in tomato based on isozymes and random cDNA sequences. *Genetics.* 112:887–898.

Biggs, M., and A. Handa. 1989. Temporal regulation of polygalacturonase gene expression in fruits of normal, mutant, and heterozygous tomato genotypes. *Plant Physiol.* 89:117–125.

Bird, C., C. Smith, J. Ray, P. Moureau, M. Bevan, A. Bird, S. Hughes, P. Morris, D. Grierson, and W. Schuch. 1988. The tomato polygalacturonase gene and ripening-specific expression in transgenic plants. Plant Mol. Biol. 11:651–662.

Brady, C., G. MacAlpine, W. McGlasson, and Y. Ueda. 1982. Polygalacturonase in tomato fruit and the induction of ripening. Austral. J. Plant Physiol. 9:171–178.

Brady, C., S. Meldrum, W. McGlasson, and M. Ali. 1983. Differential accumulation of the molecular forms of polygalacturonase in tomato mutants. J. Food Biochem. 7:7–14.

Brady, C., W. McGlasson, J. Pearson, S. Meldrum, and E. Kopeliovitch. 1985. Interaction between the amount and molecular forms of polygalacturonase, calcium, and firmness in tomato fruit. J. Am. Soc. Hort. Sci. 110:254–258.

Brady, C., B. McGlasson, and J. Speirs. 1987. The biochemistry of fruit ripening. p. 279–288. In: D. Nevins and R. Jones (eds.). Tomato Biotechnology. Alan R. Liss, New York.

Brecht, J., D. Huber. 1986. Stimulation of tomato fruit ripening by tomato cell-wall fragments. HortScience. 21:319.

Brown, S., and M. Crouch. 1990. Characterization of a gene family abundantly expressed in Oenothera organensis pollen that shows sequence similarity to polygalacturonase. Plant Cell. 2:263–274.

Buescher, R. 1977. Factors affecting ethephon-induced red color development in harvested fruits of the rin tomato mutant. HortScience. 12:315–316

Beuscher, R., and E. Tigchelaar. 1975. Pectinesterase, polygalacturonase, Ca-cellulase activities and softening of the rin tomato mutant. HortScience. 10:624–625.

Burg, S., and E. Burg. 1967. Molecular requirements for the biological activity of ethylene. Plant Physiol. 42:144–152.

Cabrera, C., M. Alonso, P. Johnston, R. Phillips, and P. Lawrence. 1987. Phenocopies induced with antisense RNA identify the wingless gene. Cell. 50:659–663.

Colot, V., L. Robert, T. Kavanagh, M. Bevan, and R. Thompson. 1987. Localization of sequences in wheat endosperm protein genes which confer tissue-specific expression in tobacco. EMBO J. 6:3559–3564.

Cordes, S., J. Deikman, L. Margossian, and R. Fischer. 1989. Interaction of a developmentally regulated DNA-binding factor with sites flanking two different fruit-ripening genes from tomato. The Plant Cell. 1:1025–1034.

Crookes, P., and D. Grierson. 1983. Ultrastructure of tomato fruit ripening and the role of polygalacturonase isoenzymes in cell wall degradation. Plant Physiol. 72:1088–1093.

Darby, L. 1978. Isogenic lines of tomato fruit colour mutants. Hort. Res. 18:73–84.

Davies, K., G. Hobson, D. Grierson. 1988. Silver ions inhibit the ethylene stimulated production of ripening-related mRNAs in tomato. Plant Cell Environ. 11:729–738.

Davis, K., A. Darvill, P. Albersheim, and A. Dell. 1986. Host-pathogen interactions XXIX. Oligogalacturonides released from sodium polypectate by endopolygalacturonic acid lyase are elicitors of phytoalexins in soybean. Plant Physiol. 80:568–577.

Deikman, J., and R. Fischer. 1988. Interaction of a DNA-binding factor with the 5'-flanking region of an ethylene-responsive fruit ripening gene from tomato. EMBO J. 7:3315–3320.

DellaPenna, D., D. Alexander, and A. Bennett. 1986. Molecular cloning of tomato fruit polygalacturonase: Analysis of polygalacturonase mRNA levels during ripening. Proc. Natl. Acad. Sci. (USA) 83:6420–6424.

DellaPenna, D., and A. Bennett. 1988. In vitro synthesis and processing of tomato fruit polygalacturonase. Plant Physiol. 86:1057–1063.

DellaPenna, D., and J. Giovannoni. 1990. Regulation of gene expression in ripening tomatoes. In: D. Grierson (ed.). Plant Biotechnology Series, Vol. 2, Factors Governing Plant Gene Expression. Blackie and Son Limited, Glasgow, Scotland.

DellaPenna, D., D. Kates, and A. Bennett. 1987. Polygalacturonase gene expression in Rutgers, rin, nor, and Nr tomato fruits. Plant Physiol. 85:502–507.

DellaPenna, D., J. Lincoln, R. Fischer, and A. Bennett. 1989. Transcriptional analysis of polygalacturonase and other ripening associated genes in Rutgers, rin, nor, and Nr tomato fruit. Plant Physiol. 90:1372–1377.

Ecker, J., and R. Davis. 1986. Inhibition of gene expression in plant cells by expression of antisense RNA. *Proc. Natl. Acad. Sci.* (USA) 83:5372–5376.

Fillatti, J., J. Kiser, B. Rose, and L. Comai. 1987. Efficient transformation of tomato and the introduction and expression of a gene for herbicide tolerance. p. 199–210. In: D. Nevins and R. Jones (eds.). *Tomato Biotechnology.* Alan R. Liss, New York.

Frenkel, C., and S. Garrison. 1976. Initiation of lycopene synthesis in the tomato mutant *rin* as influenced by oxygen and ethylene interactions. *HortScience.* 11:20–21.

Giovannoni, J. 1990. The role of polygalacturonase in tomato fruit ripening. PdD Thesis, Univ of California, Berkeley.

Giovannoni, J., D. DellaPenna, A. Bennett, and R. Fischer. 1989. Expression of a chimeric polygalacturonase gene in transgenic *rin* (ripening inhibitor) tomato fruit results in poly-uronide degradation but not fruit softening. *Plant Cell* 1:53–63.

Giovannoni, J., D. DellaPenna, C. Lashbrook, A. Bennett, and R. Fischer. 1990. Expression of a chimeric polygalacturonase gene in transgenic *rin* (ripening inhibitor) tomato fruit. In: C. Lamb and R. Beachy (eds.). *Plant Gene Transfer.* Alan R. Liss, Inc.

Green, P., O. Pines, and M. Inouye. 1986. The role of antisense RNA in gene regulation. *Annu. Rev. Biochem.* 55:569–597.

Grierson, D. 1985. Gene expression in ripening tomato fruit. *CRC Press Crit. Rev. Plant Sci.* 3:113 132.

_____ . 1986. Molecular biology of fruit ripening. *Oxford Surveys of Plant Mol. Cell Biol.* 3:363–383.

Grierson, D., M. Purton, J. Knapp, and B. Bathgate. 1987. Tomato ripening mutants. p. 73–94. In: H. Thomas and D. Grierson (eds.). *Developmental Mutants in Higher Plants.* Cambridge University Press, Cambridge, UK.

Grierson, D., C. Smith, C. Watson, P. Morris, A. Turner, W. Schuch, C. Bird, J. Ray, and A. Hamilton. 1990. Controlling gene expression in transgenic tomatoes. In: A. Bennett and S. O'Neil (eds.). *Horticultural Biotechnology.* Alan R. Liss Inc., New York.

Grierson, D. and G. Tucker. 1983. Timing of ethylene and polygalacturonase synthesis in relation to the control of tomato fruit ripening. *Planta.* 157:174–179.

Grierson, D., G. Tucker, J. Keen, J. Ray, Bird, and W. Schuch. 1986. Sequencing and identification of a cDNA clone for tomato polygalacturonase. *Nucleic Acids Res.* 14:8595–9603.

Guilley, H., R. Dudley, G. Jonard, E. Balazs, and K. Richards. 1982. Transcription of cauliflower mosaic virus DNA: Detection of promoter sequences, and characterization of transcripts. *Cell.* 30:763–773.

Herner, R., and K. Sink. 1973. Ethylene production and respiratory behavior of the *rin* tomato mutant. *Plant Physiol.* 52:38–42.

Hobson, G. 1964. Polygalacturonase in normal and abnormal tomato fruit. *Biochem J.* 92:324–332.

_____ . 1965. The firmness of tomato fruit in relation to polygalacturonase activity. *J. Hort. Sci.* 40:66–72.

_____ . 1967. Effects of alleles at the "never-ripe" locus on ripening of tomato fruit. *Phytochemistry* 6:1337–1341.

_____ . 1968. Cellulase activity during the maturation and ripening of tomato fruit. *J. Food Sci.* 33:588–592.

Hobson, G., R. Nichols, J. Davies, and P. Atkey. 1984. The inhibition of tomato fruit ripening by silver. *J. Plant Physiol.* 116:21–29.

Holdsworth, M., W. Schuch, and D. Grierson. 1988. Organization and expression of a wound/ripening-related small multigene family from tomato. *Plant Mol. Biol.* 11:81–88.

Huber, D. 1983a. The role of cell wall hydrolases in fruit softening. *Hort. Rev.* 5:169–219.

_____ . 1983b. Polyuronide degradation and hemicellulose modifications in ripening tomato fruit. *J. Am. Soc. Hort. Sci.* 108:405–409.

Knapp, J., P. Moureau, W. Schuch, D. Grierson. 1989. Organization and expression of polygalacturonase and other ripening related genes in Ailsa Craig "Neverripe" and

"Ripening inhibitor" tomato mutants. *Plant Mol. Biol.* 12:105–116.

Knecht, D., W. Loomis. 1987. Antisense RNA inactivation of myosin heavy chain gene expression in *Dictyostelium discoideum. Science.* 236:1081–1086.

Knee, M., J. Sargent, and D. Osborne. 1977. Cell wall metabolism in developing strawberry fruits. *J. Expt. Bot.* 28:377–396.

Knegt, E., E. Vermeer, and J. Bruinsma. 1988. Conversion of polygalacturonase isoenzymes from ripening tomato fruits. *Physiol. Plant.* 72:108–114.

Kopeliovitch, E., Y. Mizrahi, H. Rabinowitch, and N. Kedar. 1980. Physiology of the tomato mutant *alcobaca. Physiol Plant.* 48:307–311.

Lincoln, J., S. Cordes, E. Read, and R. Fischer. 1987. Regulation of gene expression by ethylene during tomato fruit development. *Proc. Natl. Acad. Sci.* (USA). 84:2793–2793.

Lincoln, J., and R. Fischer. 1988a. Diverse mechanisms for the regulation of ethylene inducible gene expression. *Mol. Gen. Genet.* 212:71–75.

_____. 1988b. Regulation of gene expression by ethylene in wild-type and *rin* tomato (*Lycopersicon esculentum*) fruit. *Plant Physiol.* 88:370–374.

Lisker, N. and N. Retig. 1974. Detection of polygalacturonase and pectinlyase isoenzymes in polyacrylamid gels. *J. Chromatogr.* 96:245–249.

Luthe, D., and R. Quatrano. 1980. Transcription in isolated wheat nuclei. *Plant Physiol.* 65:305–313.

Lyons, J. and H. Pratt. 1964. Effect of stage of maturity and ethylene treatment on respiration and ripening of tomato fruits. *Proc. Am. Soc. Hortic. Sci.* 84:491–500.

Margossian, L., A. Federman, J. Giovannoni, and R. Fischer. 1988. Ethylene regulated expression of a tomato fruit ripening gene encoding a proteinase inhibitor with a glutamic residue at the reactive site. *Proc. Natl. Acad. Sci.* (USA). 85:8012–8016.

Maunders, M., M. Holdsworth, A. Slater, J. Knapp, C. Bird, W. Schuch, and D. Grierson. 1987. Ethylene stimulates the accumulation of ripening-related mRNAs in tomatoes. *Plant, Cell Environ.* 10:177–184.

McCormick, S., J. Neidermeyer, J. Fry, A. Barnson, R. Horsch, and R. Fraley. 1986. Leaf disc transformation of cultivated tomato (*L. esculentum*) using *Agrobacterium tumefaciens. Plant Cell Rep.* 5:81–84.

McGlasson, W., H. Dostol, and E. Tigchelaar. 1975. Comparison of propylene induced responses of immature fruit of normal and *rin* mutant tomatoes. *Plant Physiol.* 55:218–222.

McMurchie, E. W. McGlasson, and I. Eaks. 1972. Treatment of fruit with propylene gives information about the biogenesis of ethylene. *Nature* 237:235–236.

Melton, D. 1985. Injected anti-sense RNAs specifically block messenger RNA translation in vivo. *Proc. Natl. Acad. Sci.* (USA). 82:144–148.

Mitcham, E., C. Gross and T. Ng. 1989. Tomato fruit cell wall synthesis during development and senescence. *Plant Physiol.* 89:477–481.

Mizrahi, Y., H. Dostol, and J. Cherry. 1975. Ethylene induced ripening in attached *rin* fruit; a non-ripening mutant of tomato. *HortScience* 10:414–415.

Mizrahi, Y., R. Zohar, and S. Malis-Arad. 1982. Effect of sodium chloride on fruit ripening of the nonripening tomato mutants *nor* and *rin. Plant Physiol.* 69:497–501.

Moshrefi, M., and B. Luh. 1983. Carbohydrate composition and electrophoretic properties of tomato polygalacturonase isoenzymes. *Eur. J. Biochem.* 135:511–514.

Mutschler, M. 1984. Ripening and storage characteristics of the *alcobaca* ripening mutant in tomato. *J. Am. Soc. Hort. Sci.* 109:504 507.

Ng, T., and E. Tigchelaar. 1977. Action of the non-ripening (*nor*) mutant on fruit ripening of tomato. *J. Am. Soc. Hort. Sci.* 102:504–509.

Ogura, N., H. Nakagawa, and H. Takenhana. 1975a. Effect of high temperature-short term storage of mature green tomato fruits on changes of their chemical composition after ripening at room temperature. (Studies on the storage temperature of tomato fruits Part I). *J. Agr. Chem. Soc. Japan* 49:189–196.

_____. 1975b. Effect of storage temperature of tomato fruits on changes of their polygalacturonase and pectinesterase activities accompanied with ripening. (Studies on the storage temperature of tomato fruits Part II). *J. Agr. Chem. Soc. Japan* 49:271–274.

Pear, J., N. Ridge, R. Rasmussen, R. Rose, and C. Houck. 1989. Isolation and characterization of a fruit-specific cDNA and the corresponding genomic clone from tomato. *Plant Mol. Biol.* 13:639–651.

Picton, S., and D. Grierson. 1988. Inhibition of expression of tomato-ripening genes at high temperature. *Plant, Cell Environ.* 11:265–272.

Poovaiah, B., G. Glenn and A. Reddy. 1988. Calcium and fruit softening: Physiology and biochemistry. *Hort. Rev.* 10:107–152.

Poovaiah, B., and A. Nukaya. 1979. Polygalacturonase and cellulase enzymes in the normal Rutgers and mutant rin tomato fruits and their relationship to the respiratory climacteric. *Plant Physiol.* 64:534–537.

Pressey, R. 1984. Purification and characterization of tomato polygalacturonase converter. *Eur. J. Biochem.* 144:217–221.

_____. 1986a. Changs in polygalacturonase isoenzymes and converter in tomatoes during ripening. *HortScience* 21:1183–1185.

_____. 1986b. Extraction and assay of tomato polygalacturonases. *HortScience.* 21:490–492.

_____. 1987. Exopolygalacturonase in tomato fruit. *Phytochemistry* 26(7):1867–1870.

_____. 1988. Reevaluation of the changes in polygalacturonases in tomatoes during ripening. *Planta* 174:39–43.

Pressy, R., and J. Avants. 1973. Two forms of polygalacturonase in tomatoes. *Biochim. Biophys. Acta.* 309:363–369.

Ramirez, D., and M. Tomes. 1964. Relationship between chlorophyll and carotenoid biosynthesis in dirty-red (green flesh) mutant in tomato. *Bot. Gaz.* 125:221–226.

Rexová-Benková, L. and O. Markovic. 1986. Pectic enzymes. *Adv. Carbohydr. Chêm. Biochêm.* 33:323–385.

Rhodes, M. 1980. The maturation and ripening of fruits. p. 157–205 In: K. Thimann (ed.). *Senescence in Plants.* CRC Press, Boca Raton, Fl.

Rick, C. 1956. New mutants. *Rep. Tomato Genet. Coop.* 6:22–23.

_____. 1980. Tomato linkage survey. *Rep. Tomato Genet. Coop.* 30:2–17.

_____. 1986. Tomato mutants: Freaks anomalies, and breeders' resources. *HortScience* 21:918–919.

Robinson, R., and M. Tomes. 1968. Ripening inhibitor: A gene with multiple effects on ripening. *Rep. Tomato Genet. Coop* 18:36–37.

Rothstein, S., J. DiMaio, M. Strand, and D. Rice. 1987. Stable and heritable inhibition of the expression of nopaline synthase in tobacco expressing antisense RNA. *Proc. Natl. Acad. Sci.* (USA). 84:8439–8443.

Sato, T., S. Kusaba, H. Nakagawa, and N. Ogura. 1984. Cell-free synthesis of a putative precursor of polygalacturonase in tomato fruits. *Plant Cell Physiol.* 25:1069–1071.

Schuch, W., C. Bird, J. Ray, C. Smith, C. Watson, P. Morris, J. Gray, C. Arnold, G. Seymour, G. Tucker, and D. Grierson. 1989. Control and manipulation of gene expression during tomato fruit ripening. *Plant Mol. Biol.* 13:303–311.

Sheehy, R., M. Kramer, and W. Hiatt. 1988. Reduction of polygalacturonase activity in tomato fruit by antisense RNA. *Proc. Natl. Acad. Sci.* (USA) 85:8805–8809.

Sheehy, R., J. Pearson, C. Brady, and W. Hiatt. 1987. Molecular characterization of tomato fruit polygalacturonase. *Mol. Gen. Genet.* 208:30–36.

Sink, K., R. Herner, and L. Knowlton. 1974. Chlorophyll and carotenoids of the rin tomato mutant. *Can J. Bot.* 52:1657–1660.

Smith, C., C. Watson, J. Ray, C. Bird, P. Morris, W. Schuch, and D. Grierson. 1988. Antisense RNA inhibition of polygalacturonase gene expression in transgenic tomatoes. *Nature* 334:724–726.

Su, L., T. McKeon, D. Grierson, M. Cantwell, and S. Yang. 1984. Development of 1-aminocyclopropane-1-carboxylic acid synthase and polygalacturonase activities during the maturation of ripening tomato fruit. HortScience 19:576–578.

Tieman, D., and A. Handa. 1989. Immunocytolocalization of polygalacturonase in ripening tomato fruit. Plant Physiol. 90:17–20.

Tigchelaar, E., W. McGlasson, and M. Franklin. 1978a. Natural and ethephon-stimulated ripening of F1 hybrids of the ripening inhibitor (rin) and non-ripening (nor) mutants of tomato (Lycopersicon esculentum Mill.). Austral. J. Plant Physiol. 5:449–456.

Tigchelaar, E., W. McGlasson, and R. Buescher. 1978b Genetic regulation of tomato fruit ripening. HortScience 13:508–513.

Tigchelaar, E., M. Tomes, E. Kerr, and R. Barman. 1973. A new fruit ripening mutant, non-ripening (nor). Rep. Tomato Genet. Coop. 23:33.

Tucker, G., N. Robertson, and D. Grierson. 1980. Changes in polygalacturonase isozymes during the 'ripening' of normal and mutant tomato fruit. Eur. J. Biochem. 112:119–124.

_____. 1981. The conversion of tomato fruit polygalacturonase isoenzyme 2 into isoenzyme 1 in vitro. Eur. J. Biochem. 115:87–90.

Tucker, G., C. Schindler, and J. Roberts. 1984. Flower abscission in mutant tomato plants. Planta 160:164–167.

van der Krol, A., P. Lenting, J. Veenstra, I. van der Meer, R. Koes, A. Gerats, J. Mol, and A. Stuitje. 1988. An anti-sense chalcone synthase gene in transgenic plants inhibits flower pigmentation. Nature 333:866–869.

Walker-Simmons, M., and C. Ryan. 1986. Proteinase inhibitor I accumulation in tomato suspension cultures: Induction by plant and fungal cell wall fragments and an extracellular polysacharide secreted into the medium. Plant Physiol. 80:68–71.

Wallner, S., and J. Walker. 1975. Glycosidases in cell wall-degrading extracts of ripening tomato fruits. Plant Physiol. 55:94–98.

Walter, P., P. Gilmore, and G. Bobel. 1984. Protein translocation across the endoplasmic reticulum. Cell 38:5–8.

Yang, S. 1985. Biosynthesis and action of ethylene. HortScience 20:41–45.

Yoshida, O., H. Nakagawa, N. Ogura, and T. Sato. 1984. Effect of heat treatment on the development of polygalacturonase activity in tomato fruit during ripening. Plant Cell Physiol. 25:505–509.

Crop Physiology of Cassava

Walter E. Splittstoesser
Department of Horticulture, University of Illinois
Urbana, IL 61801

Geoffrey O. Tunya[1]
Egerton University
P.O. Njoro, Kenya

I. INTRODUCTION

Cassava (*Manihot esculenta* Crantz, Euphorbiaceae), also called mandioca (Portugese), yuca (Spanish), manioc and tapioca (French), is grown for its starchy root and is the only one of 98 species in the Euphorbiaceae widely grown as a tropical root crop. It may be the world's least expensive source of food calories and is a staple food of 400 million

[1]This review was prepared while G. O. Tunya was a graduate student at the University of Illinois.

people in the tropics (Pereira and Splittstoesser 1986b). Cassava is one of the 10 major food sources for human consumption. The practical production and utilization of cassava have been reviewed by O'Hair (1990), while genetic improvement has been reviewed by Byrne (1984).

There is a renewed interest in cassava production as a starch source (CIAT 1984b), not only for human consumption, but also for animal feed and for use in fermentation (Balagopalan et al. 1988). Major research programs are being conducted at the International Institute for Tropical Agriculture (IITA), Ibadan, Nigeria, and at Centro Infornacional de Agricultura Tropical (CIAT), Cali, Colombia. Programs in Brazil are largely aimed at utilization of cassava for ethanol production.

Cassava is produced primarily for its starchy root, and yields vary widely. Light intensity, photoperiod, location, planting date, harvest date, rainfall, fertilizer, growing season temperature, pests, and cultivar all influence yield. Frequently environmental influences exceed cultivar differences. These environmental influences affect the physiology of the crop. This chapter reviews a number of the factors which influence cassava physiology and ultimately crop growth and root formation (yield).

II. LEAF CHARACTERISTICS

Productivity of cassava in terms of calories per unit area per year is higher than that of most food crops (De Vries et al. 1967). Variation in root yield per hectare is common among cassava cultivars and has been associated with leaf area index (LAI), leaf orientation, photosynthetic efficiency, and dry matter partitioning (Ramanujam and Indira 1983).

A. Leaf Area Index

Growth models of cassava have suggested a crop growth rate (CGR) of 16 g m^{-2}d^{-1} (Keating et al. 1982a). These high CGR were associated with a high LAI (7–12) and an intense solar radiation of 18 m J m^{-2}d^{-1} (Keating et al. 1982a). However, in other plants an LAI greater than 7 did not result in an increase in CGR, even under the highest solar radiation (Black 1963; Verhagen et al. 1963). The optimum LAI needed for the maximum CGR in Sierra Leone was 4.7–6.9, depending on the cultivar (Enyi 1973), and 3.5–4.0 in Colombia (Cock et al. 1979). An LAI of more than 10.0 is rare in cassava, and the accuracy of a model at this extreme may be questionable (Keating et al. 1982a). In high latitudes, such as found in Queensland, Australia, cassava is capable of a high CGR. However, the LAI decreases as the cool season approaches. Low temperatures are largely responsible for the reduction in the CGR, although a reduction in light intensity also plays a role (Fukai et al. 1984). LAI also decreases under inadequate rainfall (Ramanujam 1985). High LAI is a result of profuse branching. Low-yielding cultivars (per plant basis) with an LAI below optimum could attain increased yields per ha by increasing their plant population density (Ramanujam 1985).

The decrease in CGR when the LAI is above 4 may be due to a short leaf life, caused by mutual shading with a consequent reduced net photosynthetic output. A large number of shaded leaves would become net consumers of photosynthates. The CGR increases with LAI until the growth rate reaches a maximum and then declines (Keating et al. 1982a). This occurs because the carbohydrates required to form and maintain the existing LAI increase approximately linearly with LAI; and as the LAI increases above optimum, less carbohydrate is available for root growth.

B. Effect of Temperature and Shade

At a temperature of 24°C or greater, it takes 2 weeks for cassava leaves to fully expand (Cock 1984a). A longer period is required at a lower temperature. Higher temperatures are also associated with a greater CGR (Keating et al. 1982a), provided solar radiation and LAI are also high.

Shade affects leaf life as shaded leaves tend to abscise. Cassava leaves have a life span of up to 125 days under ideal growing conditions (Cock 1984a), but when placed under 100% shade, leaves abscise within 10 days (Cock et al. 1979). However, 50% shade does not seem to greatly affect leaf life. Exposure to only 5–25% of full sunlight reduces leaf life by up to 20 days (Cock et al. 1979). Fukai et al. (1984) reported that a 22% reduction of solar input induced an 86% reduction in leaf area and a 47% reduction in root dry weight on young plants. A 32% reduction of solar radiation reduced cassava growth, regardless of plant age.

Under field conditions, shading typically occurs 30 days after leaf formation. In this case, leaf life is not markedly reduced although shade is 95% (Cock et al. 1979). Cassava leaves tend to become adapted to low light conditions caused by shade (Fukai et al. 1984). However, 32% or 78% shade at planting time reduces storage root growth and subsequently, yield (Aresta and Fukai 1984). This indicates that under limited photosynthesis caused by reduced solar radiation, assimilates are diverted to shoot growth, i.e., under these conditions shoots are a stronger sink than roots.

Low solar radiation also induces a large leaf area per unit weight (Fukai et al. 1984). Although the overall LAI is only slightly affected, it appears that the low light intensity substantially reduces dry matter disposition. This reduces root development and growth more than shoot growth.

C. Leaf Size and Age

The size of fully expanded leaves increases with age up to 4 months and then declines. At low temperatures or under water stress, maximum size is smaller and production of the largest leaf size is delayed (Cock 1984a). Leaves increase in size to a maximum (200–600 cm^2, depending upon cultivar), and subsequent leaves are smaller in all cultivars examined. Leaf size is greatly reduced under water stress conditions (Connor and Cock 1981; Cock 1984b) or under low N availability. Shading also limits leaf size if it occurs during leaf development (Cock et al. 1979).

Cassava cultivars with broad leaves (300–325 cm^2) are generally lower yielders than narrow-leaved (260–290 cm^2) cultivars (Ramanujam and Indira 1983; Ramanujam 1985). It appears that profusely branching narrow-leaved cultivars have the advantage of increased interception of solar radiation, while broad-leaved cultivars have mutual shading and consequently reduced dry matter accumulation with a greater-than-optimum LAI.

Leaf life span is dependent on temperature. At a temperature of 15–24°C the leaves remain on the plant for up to 200 days (Irikura et al. 1979), while at higher temperatures the life span is 120 days (Cock 1984a). During a dry period, smaller leaves are produced, but their life span is unaffected (Connor and Cock 1981). However, shading or pest damage reduces the life span of individual leaves.

Leaf age decreases when the LAI rises above 3 (Cock et al. 1979), perhaps due to mutual shading as the canopy becomes denser. However, cultivar differences exist: 'M Col 1434' has a leaf life of 99 days, while 'M Col 1513' has a life of 80 days, and both have a similar LAI (Cock et al. 1979). Although leaf life may be cultivar dependent, environmental stress and temperature also influence leaf life. Leaf longevity ranges between 37 and 54 days, from fully expanded to leaf abscission in a 10-month crop (Ramanujam 1985). The most active leaf growth occurs during the first 6 months after planting; thereafter LAI declines, sometimes reaching less than one in some cultivars as a result of prolonged water stress during the dry season in Kerala, India.

D. Leaf Formation and Branching

New leaf formation per ha depends on: (1) the rate of leaf formation at the apex; (2) the number of apices per plant; and (3) the number of plants per unit area (Cock et al. 1979). The rate of new leaf formation decreases with time, with little cultivar variation. Differences in the leaf formation rate per plant are associated with branching, where profusely branching cultivars have the highest rate of new leaf formation.

A relationship exists between the degree of branching and the number of apices per plant (Cock et al. 1979). Two types of branching are found in cassava. Lateral branches develop from auxiliary buds well below the apex. If they are removed early in a high plant population (20–40,000 plants ha^{-1} at Cali, Colombia), increased yields result. However, with a plant population of 10,000 or less, yield increases with lateral branching.

The second type of branching occurs later during the growth of the plant. The main terminal bud develops into an inflorescence, and 2–4 auxiliary buds just below it form branches of about equal size. The number of branches depends on the cultivar, as does the time at which branching occurs, varying from 2 to 12 months after planting. Reducing the number of these branches increases yield of roots (Ramanujam and Indira 1983). Profusely branching cultivars accumulate dry matter in vegetative parts (Ramanujam and Indira 1983). Non-branching types, which qenerally have a less-than-optimum LAI at a standard 90 × 90 cm spacing, could produce similar storage root yields by being planted closer together to increase plant population and, hence, LAI.

In cassava, the LAI increases to a certain level, above which the CGR remains constant or declines (Cock et al. 1979). A combination of high temperature and a high LAI under long-day conditions tends to produce a greater shoot demand at the expense of storage root dry matter accumulation (Keating et al. 1982b). This is consistent with the crop growth model (Cock et al. 1979) which assumes that storage roots of cassava receive assimilates only after the shoot requirement is satisfied. Therefore, LAI could be used to indicate the size of the shoot as a sink. Root dry matter production is predicted to increase up to an LAI of 6–11 with a decline in storage root growth rate at an LAI of 5–6.

Growth of individual leaves of cassava is slow for the first weeks after planting, and thereafter vegetative growth significantly increases (Ramanujam 1985). When planted at a spacing of 90 × 90 cm to achieve a plant population of 12,340 ha^{-1}, some cultivars cover 90% of the ground area in 45 days while others take 75 to 90 days to cover the same area. In Kerala, India, the canopy size reached a maximum 90–180 days after planting, depending on the cultivar. Cultivars with a high (4–4.9) or low (1.7–1.8) LAI are low yielders (27 tons ha^{-1}), whereas high-yielding cultivars (40 tons ha^{-1}) maintain a moderate LAI of 2.4–3 (Cock et al. 1979; Ramanujam and Indira 1983). A high LAI under favorable growing conditions, favors shoot growth at the expense of root growth.

E. Anatomy

Leaf thickness of cassava in the mesophyll areas varies from 100 to 120 μm (Pereira and Splittstoesser 1990). The adaxial epidermis of the mature leaf consists of a single layer of irregularly-shaped cells devoid of intercellular spaces. The adaxial surface is glabrous with no conspicuous surface irregularities, giving a smooth, waxy surface correlating with the typical shiny look of the cassava leaf. The epidermal cells become elongated as they associate with the vein. Of 1500 cultivars investigated, only 2% had stomata on the adaxial surface (El-Sharkawy and Cock 1987a). These stomata were found where the veins joined the midrib and were located over large substomatal cavities bordering the photosynthetic mesophyll tissue that was connected to the cells surrounding the veins.

The abaxial epidermis is a single layer 10 μm thick of 30 μm $\times$ 12 μm tabular cells (Pereira 1977; El-Sharkawy et al. 1984b). The stomatal complex is made up of two guard cells and two accessory cells (Fahn 1982) with no prominent waxy extrusions. The number of stomata per unit surface area reaches a maximum of 700 mm^{-2}. No stomata are over the veins, similar to many C$_4$ photosynthesis grasses (Laetsch 1974). Cassava, however, functions both as a C$_3$ and C$_4$ photosynthesis plant, depending upon the temperature (Pereira et al. 1986). If all stomata were completely opened, the exposed surface of the pores would occupy 1.37% of the total leaf surface.

Seven palisade parenchyma cells of cassava leaves join each adaxial epidermal cell (Pereira and Splittstoesser 1990). These palisade cells have the shape of truncated cones, with diameters for the major and minor bases of 15 μm and 7 μm, respectively, and a height of 60 to 70 μm (Table 3.1). With these dimensions, each palisade cell has a volume of 5400 μm^3 and exposes a lateral surface of 900 μm^2. An average of 25 tightly packed chloroplasts, 6 μm and 3 μm for the long and short axis, respectively, are found in each cell. Each chloroplast becomes smaller toward the tip of the cell. At the base there are three chloroplasts per cross section. Cassava contains more chlorophyll (1g m^{-2}) than many other plants (Pereira et al. 1986), probably because of the large number of chloroplasts found in the cells.

The spongy parenchyma cells are branched with a diameter of 10 μm. The spongy parenchyma area occupies 40% of the total leaf volume, with 70% being cell volume and 30% being air space (Table 3.1). For each 1 cm^2 of external surface area, there is a 10-fold increase in internal surface area, with a total palisade area of 6.22 cm^2 and a spongy area of 4.12 cm^2.

Cassava leaves have a reticular venation, with the smaller vascular bundles converging toward larger ones and ending at the midrib of each lobe. The lobe midribs fuse together before they enter the petiole. Small veins are located exclusively inside the layer of spongy parenchyma. The total length of the veins is 125 cm per cm^{-2} of leaf (Table 3.1). For a medium-sized plant with 1 m^2 of leaf area (Pereira and Splittstoesser 1986a), the veins would have a length of 12.5 km.

Each vein is surrounded by a layer of spongy parenchyma cells. In the transverse view, this tissue has the appearance of a bundle sheath (Pereira 1977), suggesting that cassava has an operational C$_4$ pathway (Pereira et al. 1986). The laticiferous characteristic of the Euphorbiaceae (Milanez 1949) is also present in the cassava leaf. The laticifers are long, branched cells without transverse divisions, located between other cells

Table 3.1. Anatomical parameters of a cassava leaf. Values expressed in terms of cm^2 were calculated considering only one surface (source: Pereira and Splittstoesser 1990).

LEAF
Thickness   1.2 × 10^{-2} cm (120 μmm)
Volume   1.2 × 10^{-2} cm^3 (1.2 × 10^6 μm^3 cm^{-2})
Number of cells 1.94 × 10^6 cell cm^{-2}

EPIDERMIS
Adaxial
  Number of cells   1.71 × 10^5 cell cm^{-2}
  Cell thickness   0.9 × 10^{-3} cm (9 μm)
  Cell length or width 2.8 × 10^{-3} cm (28 μm)
Abaxial
  Number of cells   3.8 × 10^5 cell cm^{-2}
  Cell thickness   0.9 × 10^{-3} to 2.9 × 10^{-3} cm (9 to 29 μm)
  Cell length or width 3.3 × 10^{-3} and 1.4 × 10^{-3} (33 and 14 μm)
Stomatal complex
  Maximum number   7.0 × 10^4 stomata cm^{-2}
  Average number   3.8 × 10^4 stomata cm^{-2}
  Model number   4.0 × 10^4 stomata cm^{-2}
  Open pore   1.2 × 10^{-3} × 0.4 × 10^{-3} cm (12 μm × 4 μm)
  Perimeter open pore   2.5 × 10^{-3} cm (25 μm)
  Area individual pore 3.6 × 10^{-7} cm^2 (36 μm^2)
  Total area open pores 1.4 × 10^{-2} cm^2 (1.37 × 10^6 μm^2 cm^{-2})
  % open pores   1.4

PARENCHYMA
Palisade
  Number of cells   9.6 × 10^5 cell cm^{-2}
  Volume per cell   5.4 × 10^{-9} cm^3 (5.37 × 10^3 μm^3)
  Area per cell   6.6 × 10^{-6} cm^2 (6.58 × 10^2 μm^2)
  Total cell volumee 5.2 × 10^3 cm^3 cm^{-2} (5.16 × 10^2 μm^2 cm^{-2})
  Total cell area   6.3 × cm^2 cm^{-2} (6.32 × 10^8 μm^2 cm^{-2})
  Schizogenous space 2.0 × 10^{-3} cm^3 cm^{-2} (1.04 × 10^9 μm cm^{-2})
Spongy
  Number of cells   4.7 × 10^5 cell cm^{-2}
  Volume per cell   7.2 × 10^{-2} cm^3 (7.21 × 10^3 μm^3)
  Area per cell   8.7 × 10^{-2} cm^2 (8.73 × 10^2 μm^2)
  Total cell volume 3.4 × 10^{-3} cm^2 cm^{-2} (3.4 × 10^9 μm^3 cm^{-2})
  Total cell area   4.1 cm^2 cm^{-2} (4.12 × 10^8 μm^2 cm^{-2})
  Schizogenous space 1.4 × 10^{-3} cm^3 cm^{-2}
Total number of cells   1.4 × 10^6 cell cm^{-2}
Total cell volume   8.6 × 10^{-3} cm^{-2}
% cell volume   71
Volume schizogenous space 3.4 × 10^{-3} cm^3 cm^{-2}
% schizogenous space   29
Total internal area   10.4 cm^2 cm^{-2}

VASCULAR SYSTEM
  Length 125 cm cm^{-2} (1.25 × 10^6 μm cm^{-2})

and forming an independent system inside the leaf. A 3500-μm-long laticifer is traced in a petiole. Exudate forms on these petioles (Pereira and Splittstoesser 1987), but serial sections do not reveal the presence of structural nectaries.

Cassava is used primarily for its starchy root, and its productivity in terms of calories per unit land area per unit of time appears to be significantly higher than other staple food crops. Cassava can produce 250×10^3 cal ha^{-1} day^{-1} compared to 200×10^3 for maize (Zea mays L.), 176×10^3 for rice (Oryza sativa L.), 114×10^3 for sorghum (Sorqhum halepense L.), and 110×10^3 for wheat (Triticum vulgare L.) (FAO 1984). From serial photographs made on mature leaf tissues, anatomical parameters have been calculated (Pereira and Splittstoesser 1990) (Table 3.1) which help explain the high starch production. The large number of stomata, intercellular air space, and number of chloroplasts allow the plant to obtain high photosynthetic rates, and increase photosynthetic efficiency.

III. PHOTOSYNTHESIS

Cassava is an efficient crop in converting solar energy into assimilates (Hunt et al. 1977). This is possible because cassava is efficient in CO_2 fixation, interception of solar radiation, and use of water, especially under stress.

A. C$_3$ and C$_4$ Pathways

Under a high light intensity, cassava leaves show a higher photosynthetic rate than snapbeans (Phaseolus vulgaris L.) (El-Sharkawy and Cock 1987a). Cassava fixes up to 1 mg (CO_2) m^{-1} S^{-1} (Pereira et al. 1986). The photorespiration rate of cassava was a maximum of 12% of net photosynthesis in air which was lower than that commonly found in C$_3$ plants (El-Sharkawy and Cock 1987a). The CO_2 compensation point at 25°C was 50 cm^3 m^{-3} at 20% O_2 and zero in a N atmosphere (Pereira et al. 1986). This suggests that cassava is a C$_3$ plant; cassava, however, is intermediate between C$_3$ and C$_4$ plants. Several non-food plants (species of Panicum, Flavena. and Moricandia) also exhibit photosynthetic characteristics intermediate between C$_3$ and C$_4$ plants (El-Sharkawy and Cock 1988). Earlier observations (Mahon et al. 1977a) detected high photorespiration rates under low CO_2 concentrations in some Manihot species and one cassava cultivar.

Attached cassava leaves release less CO_2 in CO_2-free air under high light intensities than typical in C$_3$ plants (El-Sharkawy and Cock 1987a). The CO_2 compensation point in attached whole leaves is 25 cm^3 CO_2 m^{-2} and 50 cm^3 CO_2 m^{-2} in detached leaves (Pereira et al. 1986). This is greater than that found for C$_3$ plants, but lower than for C$_4$ plants.

The photorespiration rate for the upper leaf surface is 6 cm^3 CO$_2$ m^{-2} when determined in cassava cultivars with stomata on both the upper and lower leaf surfaces; but the leaves release no CO$_2$ in CO$_2$-free air (El-Sharkawy and Cock 1987a). However, considerable CO$_2$ is released from the lower leaf surface under both light and dark conditions. Respiratory CO$_2$ is released through the lower leaf surface, which has numerous stomata. However, the CO$_2$ released from photorespiration and dark respiration is much less than typical C$_3$ plants. This suggests that differences in CO$_2$ exchange between cassava and a typical C$_3$ plant (i.e., less CO$_2$ released into the air and intermediate O$_2$ compensation point of whole leaves) are the result of a photosynthetic mechanism in cassava that recycles respiratory CO$_2$, possibly in the photosynthetic palisade cells.

Of 1500 cassava cultivars investigated, less than 2% had stomata on both leaf surfaces (El-Sharkawy and Cock 1987a). Most had stomata on the lower leaf surface only (El-Sharkawy et al. 1984a; El-Sharkawy and Cock 1987a). The few stomata found in the upper leaf surface were around the veins (El-Sharkawy et al. 1984a). Pereira (1977) suggested that some of the adaxial stomata may not function, as separate stomatal activity was not detected in the upper and lower leaf surfaces. When both stomatal density (from 322 to 553 mm^{-2}) and leaf area (from 3% to 6%) increased, CO$_2$ uptake also increased (Pereira 1977).

The pattern of stomatal distribution, amphistomatous or hypostomatous, is genotype dependent in cassava and influences maximum photosynthesis (El-Sharkawy et al. 1984a). A 27% reduction in photosynthesis occurred when the lower leaf surface was covered with silicone grease to prevent O$_2$ adsorption (El-Sharkawy and Hesketh 1965; El-Sharkawy and Cock 1987a).

When cassava leaves were given ^{14}CO$_2$ for short periods of time in the light, the first products of photosynthesis were C$_4$ acids (40–60% malate, aspartate, and oxalacetate), and only 30–50% was phosphoglyceric acid (Cock et al. 1987). In similar experiments with snapbeans, two-thirds of ^{14}C was found in phosphoglyceric acid; in maize, two-thirds of ^{14}C was found in C$_4$ acids. This indicates that the cassava plant is intermediate between snapbeans, a typical C$_3$ plant, and maize, a typical C$_4$ plant.

B. Response to Temperature

When attached cassava leaves were exposed to increasing temperatures under saturated light conditions, the least photorespiration was observed between 20–25°C (El-Sharkawy et al. 1984c), while maximum rates occurred at 18–20°C and 25–30°. Below 18°C and above 40°C photorespiration was 80% lower than the maximum rate observed. Similar results were found with different aged leaves of *M. esculenta* cultivars and wild species of *Manihot*. Earlier, Mahon et al. (1977b) suggested that the optimum temperature for photosynthesis in cassava was 25°C under controlled growth chamber conditions. The recent findings of El-Sharkawy et al. (1984c) are similar to results observed earlier with warm-season C_3 crops (El-Sharkawy and Hesketh 1965) and cassava (El-Sharkawy et al. 1967; Porto 1984).

The pattern of incorporation of $^{14}CO_2$ and the temperature response of photosynthesis in intact cassava leaves (El-Sharkawy et al. 1984c) suggest that two separate enzyme systems exist, each with its own temperature optimum (Cock et al. 1987). Thus, both C_3 and C_4 photosynthetic pathways appear to operate in cassava leaves. At lower temperatures, the C_3 pathway is dominant. As temperatures increase, the activity of phosphoenolpyruvate carboxylase and other C_4 enzymes become more pronounced, and the C_4 pathway dominates (Cock et al. 1987). It is known that the optimum temperature for C_3 photosynthetic species is lower (20°C) than that for C4 species (>25°C) (El-Sharawy 1965; El-Sharkawy et al. 1967; Berry and Bjorkman 1980; Bjorkman et al. 1980).

IV. MECHANISM OF DROUGHT RESISTANCE

Cassava tolerates prolonged drought coupled with high temperature (>30°C) and low atmospheric humidity (vapor pressure deficit <1.2 K Pa), conditions common in tropical savanna climates. During dry periods, cassava uses the following survival mechanisms: closure of stomata in dry air; shedding of older leaves; and production of smaller new leaves (Connor et al. 1981; El-Sharkawy and Cock 1984; Porto 1984; El-Sharkawy and Cock 1987b). Stomata rapidly close when cassava is placed in dry air, although leaf water content remains unchanged (Porto 1984; El-Sharkawy and Cock 1986). Plants stressed at 6 months of age shed more leaves and produce no new leaves compared to plants stressed at 3 months of age (Porto 1984).

Under field or controlled drought conditions, stomatal CO_2 uptake is reduced, resulting in depletion of the intercellular CO_2 and, hence, a reduction in crop growth rate. However, this allows the crop to survive under water deficits and to use the limited water more efficiently (Cock et al. 1985). Although the overall growth rate is reduced, starch continues to accumulate in storage roots (Connor et al. 1981; El-Sharkawy and Cock 1987b). The increase in root starch, which occurs during the dry period, is further enhanced by cassava's ability to recycle respiratory CO_2 when the stomata are closed. When water is no longer a limiting factor, previously stressed plants rapidly produce new leaves and develop a higher leaf area index than non-stressed plants. This results in a greater accumulation of starch in the roots during the post-recovery period (Connor et al. 1981). The previously stressed plants also show a lower leaf water potential than non-stressed plants 30–40 days after drought recovery, possibly due to an adaptation to stress acquired during the drought period (Porto 1984). The response to water stress in cassava is also partially controlled by genetic factors.

There is a relationship between water stress and leaf temperature in many desert plants (Fischer and Turner 1978). In cassava, a reduced leaf temperature can occur as a result of a change in leaf orientation with respect to the position of the sun (heliotropism) in order to assume a reflective posture (Porto 1984) or to increase leaf absorption area at midday (El-Sharkawy and Cock 1984). In the morning, higher solar radiation is observed in the top layers of a fully developed cassava canopy than in the lower layers (San Jose and Berrade 1983). However, in the afternoon, the heliotropic response of the cassava canopy results in an increased depth of light penetration. Dissipation of the excess radiant energy assumes a life saving significance when the stomata are closed in response to a low vapor pressure density and excess solar radiant energy (Osmond et al. 1980; Cock 1984b). Stomata rapidly close under a low vapor pressure density (El-Sharkawy and Cock 1984, 1986). It appears that in cassava, the problem of excess energy dissipation is regulated by an internal recycling of CO_2, via photosynthesis or photorespiration. Furthermore, amphistomatal leaves with the same conductance value (El-Sharkawy et al. 1985) or stomatal density (El-Sharkawy and Cock 1987b) are less sensitive to variations in vapor pressure density. These leaves maintain open stomata for longer periods during the day, leading to an increased photosynthetic rate. Many F_1 cassava seedlings from Cali, Colombia (Cock 1984b; El-Sharkawy et al. 1984a; El-Sharkawy and Cock 1987b) had a higher frequency of amphistomatal leaves. These plants should be less sensitive to variations in vapor pressure density. Heritability of the amphistomatal trait needs to be studied. Amphistomatal cultivars would maintain high photosynthetic rates in dry air and would be suited to areas where soil moisture stress can be minimized (El-Sharkawy et al. 1984d; El-Sharkawy and Cock 1984, 1986). However, the more common hypostomatic cultivars, with stomata sensitive to humidity, are advantageous in areas with limited soil moisture and low relative humidity. Most cassava cultivars are hypostomatic.

V. STARCH AND DRY MATTER ACCUMULATION IN ROOTS

Starch yield is directly correlated to dry matter accumulation in cassava roots (Kayode 1983) and is dependent upon various environmental factors (Keating et al. 1985). The roots are not a strong sink, and under cool temperatures or drought conditions starch is recycled between the roots and shoots. Thus, starch content in the storage roots ranges from 20–36%, depending upon time of harvest (Ternes et al. 1978).

During the first 60 days of growth, storage root dry matter increases slowly, but then increases rapidly during the next 60 days (Ramanujam 1985; Manrique 1987). Dry matter accumulation may be ranked as stem > leaves > fibrous roots > storage roots (Aresta and Fukai 1984; Manrique 1987). After 60 days of growth, starch accumulation in fibrous and storage roots is about equal (Aresta and Fukai 1984). Kawano et al. (1987) stressed the existence of an ideal age to harvest cassava roots, when starch accumulation was greatest. Ramanujam (1985) showed that storage root growth reached a maximum value of 400g plant^{-1} 4–6 months after planting, and this was independent of the cultivar used (Ramanujam 1985; Manrique 1987).

In older cassava plants (Wholey and Booth 1979) root starch decreased from week 48 to 56 as shoot regrowth occurred, and starch then increased again. When cassava is grown for several years, and passes through a period of little or no growth, annual growth rings occur in the storage root (O'Hair 1989). These rings occur because about 12% of the starch is removed from the root for regrowth. The highest starch content is found in the youngest tissue as the starch removed from the older tissue is not replaced. The root peel contains about 20% starch, the parenchyma and secondary xylem cells about 40%, and the central pith about 30% (O'Hair 1989). However, most of the root is composed of parenchyma and secondary xylem cells, and thus, most of the starch is found in this tissue. More starch accumulates in the root tissue closest to the plant than in parts further away (O'Hair et al. 1983). The enzymes responsible for starch degradation and sucrose utilization are high in young leaves, and 300–900 mM sucrose is found in the phloem (Pereira and Splittstoesser 1986b). Degradative enzymes decrease in older leaves, and sucrose is then again transported into the storage roots.

Although photoperiod affects the dry matter balance between shoots and storage roots (Keating et al. 1985), it is not a requirement for storage root initiation. However, storage root development is greater under a 10h photoperiod than under 16h. Root elongation was reduced 50% (Aresta and Fukai 1984) when sunlight was reduced 32% by shading, but even 78% shade had no effect on storage root initiation. Shading greatly affects plant and root growth at an early stage, but not at later stages (Fukai et al. 1984). Manrique (1987) showed that dry matter production at an early stage is related to photosynthesis. When cassava is intercropped, it recovers when the other crop is removed, producing as much root yield as non-intercropped plants. Root dry matter is related to shoot size once plants reaches optimum growth. Leaf area has a direct effect on root yield. Under conditions of limited carbohydrate production, the shoot is a stronger sink than roots (Aresta and Fukai 1984), and little dry matter accumulates in roots. Root dry matter is also influenced by water stress. Moisture-stressed cassava accumulates more starch in the roots during the recovery period when watering is resumed. Non-stressed plants develop their source (leaves) and sink (roots) simultaneously (Connor et al. 1981), but the roots do not form a strong sink. Under optimum growing conditions, roots do not compete with the shoots, and the assimilates are used to develop and sustain canopy growth (Manrique 1987).

Soil nitrogen also influences starch accumulation in roots. Initially nitrogen stimulated leaf and shoot growth (Holmes and Wilson 1979) and nitrate reductase activity is high (Pereira and Splittstoesser 1986b). When nitrogen is depleted, nitrate reductase activity declines in mature leaves (Pereira and Splittstoesser 1986b). Storage roots are initiated, and the plant switches to a carbohydrate metabolism favoring starch accumulation in the roots (Holmes and Wilson 1979). This change may occur in response to an increased carbohydrate demand or a change in the plant hormone balance which occurs upon storage root initiation (Holmes and Wilson 1979).

Peeled cassava roots contain from 1.5–5.1% protein on a dry weight basis as estimated by the Kjeldahl method (Splittstoesser 1977). Cassava also contain a considerable amount of non-protein water-soluble amino acids (Martin and Splittstoesser 1975); thus the actual protein content ranges from 1.2–2.7%. For human diets, these proteins are deficient in methionine, cysteine, tyrosine, isoleucine (Splittstoesser and Rhodes 1975), and tryptophan (Splittstoesser and Martin 1975). Protein content in the roots varies because total starch content in the roots varies with time.

VI. CYANIDE CONTENT

Poisoning from eating cassava was probably recognized when it was first domesticated. Various processing methods are used to reduce the toxicity of cassava roots (Cooke and Coursey 1981).

Cassava is often classified as bitter or sweet, according to the amount of cyanide present. Bitter cultivars are generally those with high amounts of cyanide (Sinha and Nair 1968; Pereira et al. 1981; Dufour 1988), but sweet cultivars are inconsistent in their levels of cyanide (Sinha and Nair 1968). Some sweet or non-bitter cultivars contain very high amounts of cyanide, suggesting that factors other than the absence of cyanide are responsible for sweetness (Sinha and Nair 1968; Pereira et al. 1981). The presence or absence of glucose might be the most important factor in determining the organoleptic property of "bitterness" or "sweetness", and not cyanide alone (Sinha and Nair 1968).

Bitter cultivars contain from 0.2–400 mg kg^{-1} (FW) cyanide (Nestel and MacIntyre 1973; Dufour 1988). The high cyanide concentration is, in part, caused by the inclusion of the root epidermis. The total cyanide content in the cassava root prepared for eating ranges from 1–100 mg per 100 g of fresh weight (Bradbury 1988).

A large amount of bitter cassava is grown throughout the producing regions, and this suggests that when properly prepared, they are not very dangerous as a food staple. Bitter cultivars are probably grown for their cultural and traditional values (Pereira et al. 1981; Dufour 1988). Sadasivan (1970) suggested that the presence of cyanide in the leaf exudate may protect the plant from insect attack. However, the exudate does not contain cyanide, but rather contains mostly reducing sugars and fructofuranosides and is not toxic to insects (Pereira and Splittstoesser 1987). Glycosides have been detected in the exudate of some cultivars (Fahn 1979), but these are not toxic to insects.

A. Distribution

Cyanide is found in all tissues of cassava. Young leaves usually contain higher amounts of cyanide than older leaves (Pereira et al. 1981; Cooke and DeLaCruz 1982). Usually, roots contain lesser amounts than leaves during early stages of growth, and this trend was reversed as the plant matures (Cooke et al. 1978; Gomez and Valdivieso 1984), but this does not occur with all cultivars (Pereira et al. 1981). Large variations in HCN content occur among cultivars, their ages, and different plant parts (DeBruijn 1973; Zitnak 1973). However, the root epidermis of wet grown cassava contains relatively lower amounts of HCN than cassava grown in drier areas. The cyanide content of the root is also influenced by fertilizer application (Jones 1959; DeBruijn 1973). Nitrogen applications result in decreased cyanide. In most cultivars, under most cultural conditions, the root epidermis contains 5–10 times more cyanide (Nestel and MacIntyre 1973) than the root parenchyma tissue (Sinha and Nair 1968; Gomez and Valdivieso 1984; Dufour 1988). Often more than 90% of the root cyanide is concentrated in the root epidermis.

B. Properties of Cyanide

Cyanide is found in cassava as free and bound cyanogenic glycosides, mostly linamarin (hydroxy-isobutyro-nitrile-β-D-glucose) and, in much lesser amounts, lotanstralin (L-hydroxyl-L-methyl-butyronitrile-β-D-glucose). When cyanogenic glycosides are hydrolyzed, they release HCN (Coursey 1973; DeBruijn 1973; Zitnak 1973; Oke 1983; Gomez and Valdivieso 1984), which is highly volatile (bp 26°C) (Zitnak 1973) and a potent mammalian poison (Cooke and Coursey 1981; Oke 1983). Only small amounts of cyanide are found as free HCN (Cooke and Coursey 1981; Gomez and Valdivieso 1984; Dufour 1988). Perhaps this explains the controversy concerning Koch's classification (cited in Bolhuis 1954) which was widely used in older literature. Based on cyanide content, he classified cassava into three groups: (1) innocuous varieties which contained $<$ 50 mg cyanide kg^{-1} peeled root (FW); (2) moderate to poisonous varieties which contained 50–100 mg cyanide kg^{-1} peeled root (FW); and (3) dangerously poisonous varieties which contained $>$ 100 mg cyanide kg^{-1} peeled root (FW). A more appropriate classification is that of Rogers and Fleming (1973) which uses taximetric methods instead of cyanide content. The accurate determination of free and bound cyanide has only recently been resolved (Cooke 1978). Although the procedure is time consuming, its accuracy is extremely sensitive. Bound cyanide in cassava tissues tends to be quite stable and requires a specific enzyme system (β-glycosidase or Linamarase) to cleave the sugar from the molecule and release free HCN from the resulting cyanohydrin (Coursey 1973; Zitnak 1973; Pereira et al. 1981). All "bitter" or "sweet" cassava cultivars contain cyanogenic glycosides capable of releasing HCN, depending on the cultivar and environmental conditions (Coursey 1973; Zitnak 1973).

Toxic levels of HCN can be found in improperly prepared cassava food and feed products (Cooke and Maduagwu 1978; Oke 1984. If linamarin is itself non-toxic, presumably all that is required would be heating sufficient to denature the hydrolytic enzyme and prevent the release of free HCN. Traditional methods of processing are aimed at bringing the enzyme and substrate together by cell rupture, followed by elimination of the liberated HCN (Nestel and MacIntyre 1973; Cooke and Coursey 1981; Oke 1983). HCN concentration is reduced to a safe level by juice extraction; leaching in cold water to remove HCN present, followed by drying or boiling to deactivate enzymatic hydrolysis and volatilize the HCN; or fermentation combined with one or more of the above which removes most of the free and bound HCN. Traditional processing cannot completely remove all cyanogenic glycosides without rendering the product unusable (Oke 1984). Therefore, people who consume traditional cassava products invariably ingest some cyanide.

Cassava roots are also processed commercially to extract the starch which is used as cattle feed, in industry, and as tapioca. By-products of the extraction process are also used as cattle feed. HCN is released during processing and removed along with the effluent, leaving pure white cassava starch (Balagopalan et al. 1988).

The physiological effect of ingesting cyanide from cassava is only poorly understood (Dufour 1988), but a toxicity may occur when large quantities of cassava are eaten over a period of time. About 60 mg of HCN are considered lethal (Cooke and Coursey 1981), but some cyanide can be detoxified by the human body (Davis 1981; Oke 1983). Cassava toxicity is most common in parts of Africa where consumption is highest, and is associated with goiter, neuropathy (a nervous system disorder), and thiamine deficiency (HCN inactivates thiamine) (Nestal and MacIntyre 1973). These toxic effects can be reduced by supplements of iodine, methionine and cysteine, and thiamine.

VII. PROPAGATION

Commercial cassava is exclusively established from planting stem cuttings from mature cassava plants. The use of botanical seed is currently limited to breeding programs (Byrne 1984). In some cases, the availability of plantable stems has retarded the rate of expansion of cassava production (Akoroda et al. 1987). Dependence upon a vegetative means for commercial propagation is a deterrent to rapid multiplication, improvement, and dissemination of cassava germplasm to farmers.

A. Stem Planting Material

Traditional propagation methods are very slow. It takes three years for a newly introduced cultivar to reach commercial production (Akoroda et al. 1987). Recently CIAT (Roca et al. 1980; Cock 1985) has developed a rapid propagation technique with the potential to produce 300,000 to 450,000 plantable, 30-cm-long cassava stem cuttings from a single mature plant. A technique developed by IITA (1988) involves taking a single or double node cutting (mini-stem cuttings) from mature stems and inducing them to sprout in a clear plastic bag. The sprouted cuttings, with or without roots, are then distributed to growers for planting. This technique increases planting material, reduces handling cost per unit cutting, and increases rapid distribution of elite clones.

Cuttings from different sections of the stem have a varying influence on subsequent growth and yield of cassava (Garcia and Rodriques 1983; CIAT 1984a; Guritno 1985; Lazo-Casteleanos and Betancourt 1985). Secondary and tertiary shoots are the poorest, and whole mid-section stem cuttings give the best growth and yield response. However, Guritno (1985) reported better growth and higher yields from basal stem cuttings. Large cuttings produce more vigorous plants in the initial stages of growth (Bariars 1984; CAIT 1984a). The ideal cutting size is 25–30 cm long and 2–2.5 cm thick. Stem cuttings shorter than 20 cm produce fewer primary stems and a lower overall shoot weight (CIAT 1984a). Planting techniques also affect yield. Planting cuttings vertically result in plants with heavier tops than planting horizontally (CIAT 1984a; Godo 1984); and plants produce higher (CIAT 1984a) or equivalent (Godo 1984) root yields.

Some growers cut back the cassava plants prior to harvesting (Udealor and Ezedenma 1987). This provides easy access at harvesting and assists in recovery of plantable stems. Cutting back stems at an early stage of growth reduces root yield, possibly as a result of a reversed flow of carbohydrates to regenerating stems (O'Hair 1989). However, when plants are cut back after 9 or 12 months of growth, root yield is unaffected. Usually top growth stops or slows down earlier than root enlargement (CIAT 1984a). Cultivars with an optimum dry matter accumulation in the roots at 10 months had an optimum stem maturity 2 months earlier.

Obtaining healthy planting material is an integral part of cassava production (Leihner 1984). To produce stem cuttings on a large scale requires different cultural practices than are required for root production. The use of older stem cuttings, which are strongly vegetative, results in vigorous growth (CIAT 1984a), probably because the old stems contain higher levels of carbohydrate reserves. A greater amount of stem cuttings was obtained by spacing closer than 90×90 cm, which increased competition between plants and resulted in thinner stems, but greater numbers (Aresta and Fukai 1984; CIAT 1984a). Nitrogen (Holmes and Wilson 1979) and irrigation (Wholey 1977) also increase stem yield. These practices are not normally used for cassava root production.

Stem cuttings should be free of cassava diseases commonly spread by planting materials. Cassava mosaic virus (CMV), found in Africa and India, can reduce root yield up to 70% (Muimba-Kankolongo and Phuti 1987). CMV is spread by a whitefly vector (Bemisia tabaci Genn.) (Hahn et al. 1980). The degree of infection in a cassava plantation remains stable when the vector is not present, and the infection level increases when it is present. Cultivars resistant to CMV have been reported (Acland 1971; Hahn et al. 1980), but none of these are widely used (Acland 1971) as breeding material. Bock and Guthrie (1982) controlled CMV by growing cassava from mosaic-free cuttings. Robertson (1985) observed a slow spread of CMV in a cassava plantation and attributed this to the inefficiency of the whitefly as a vector. This suggests that under normal growing conditions, CMV could be kept at a minimum even when the disease is generally present, provided that planting was established from virus-free material. Under these conditions, a program to provide disease-free planting material to growers on a 2- or 3-year cycle could keep the level of CMV at a minimum.

B. Tissue Culture

In vitro culture (micropropagation) provides a higher multiplication rate for cassava than traditional propagation methods. Pest and disease elimination, ease of germplasm maintenance, and safe exchange of germplasm are other reasons for increased interest in cassava tissue culture (Kartha and Gamborg 1975; Kartha 1982, 1986; Smith et al. 1986a; Szabados et al. 1987).

1. Virus elimination. Virus could be completely eliminated in cassava through meristem culture in combination with thermotherapy (Kartha and Gamborg 1975; Adejore and Coults 1981; Mabanza 1987; Szabados et al. 1987). This was first achieved in culture with 0.4-mm meristem tips which were found to contain low levels of virus (Kartha et al. 1974). However, when diseased cassava plants were grown at 35°C for 30 days, and 0.8-mm meristem tips were removed and cultured, they produced virus-free plants. It has been suggested (Kartha 1986) that heat may have either inactivated the virus, interfered with their reproduction mechanism, or reduced virus movement into the rapidly growing meristems. Once a virus-free plant is obtained, it can be used for micropropagation to produce large numbers of plants for distribution:

2. Microcuttings. Cassava grown in vitro exhibits strong apical dominance and tends not to branch. For this reason, microcuttings have not been used for large-scale propagation. Smith et al. (1986a) excised axillary buds from node explants, and with certain exogenous hormones induced these buds to produce multiple shoots. These shoots were used to initiate additional cultures which then regenerated complete plants.

3. Organogenesis. Regeneration of plantlets via adventitious systems from callus tissue has proved very difficult (Kartha et al. 1974; Nair et al. 1979; Stamp and Henshaw 1982; Smith et al. 1986a, 1986b). Tilquin (1979) has reported shoot formation from stem-derived callus segments. Kartha et al. (1974) observed root formation, but not shoot formation, on callus grown on standard Murashige-Skoog (Murashige and Skoog 1962) medium supplemented with gibberellic acid (GA_3) and naphthaleneacetic acid (NAA). Root formation was suppressed by 6-benzylamino purine (BA). Many different media and exogenous hormones have been evaluated with limited success for plantlet production (Nair et al. 1979; Stamp and Henshaw 1982, 1986; Smith et al. 1986a; Szabados et al. 1987). Anther culture has been unsuccessful for plantlet formation (Liu and Chen 1979). Hidajat (1987) used the enzymes peroxidase and phosphatase as markers to identify shoot and root formation in callus tissue. Both enzymes were found in cassava callus, but were not found in young meristematic root or shoot initials; rather, they were found in mature parenchyma cells. Considerably more research is needed to identify the relationship between callus formation and root and shoot regeneration, as influenced by exogenous and endogenous hormones and media.

4. Somatic embryogenesis. Somatic embryogenesis in cassava was first reported by Stamp and Henshaw (1982). Somatic embryos have been induced from mature zygote embryos, axillary and terminal buds, cotyledons, and immature leaves from plants grown in vivo and in vitro (Stamp and Henshaw 1982, 1986, 1987a; Szabados et al. 1987). Somatic embryos can be transformed into normal seedlings (Williams and Meheswaran 1986).

Somatic embryos can be readily induced from leaf tissue (Stamp and Henshaw 1982, 1987a; Szabados et al. 1987). The age of the leaf greatly influences the response. As the cassava leaf matures, fewer embryos are induced (Stamp and Henshaw 1986, 1987a), as observed with other plants (Stamp and Henshaw 1987a).

Media and growth regulator composition and concentration are critical factors influencing somatic embryogenesis. For example, somatic embryos were initiated in a "parental" medium with a relatively high growth regulator concentration (Stamp and Henshaw 1981, 1986). However, further development of the embryos required a transfer to a medium with a lower growth regulator ("proliferation" medium) concentration. Somatic embryos developed into normal plantlets when they were returned to the parental medium. Embryos sub-cultured on parental medium developed secondary embryos from the cotyledons of the primary embryos (Stamp and Henshaw 1982, 1986, 1987b). All of these embryos resembling zygotic embryos (Stamp and Henshaw 1987b) arose adaxially near the shoot tip of the primary embryonic axes. This resulted in a continuous production system of secondary embryos in vitro which produced over 50 plantlets every 30 days per explant (Stamp and Henshaw 1986; Szabados et al. 1987). Maintaining these cultures for 20 months did not affect their regenerative ability. These secondary embryos also produced plantlets when returned to the parental medium. These plantlets grew into normal plants in the greenhouse (Stamp and Henshaw 1986) and field (Szabados et al. 1987).

In the above studies, secondary embryos did not arise from unorganized callus, but rather directly from organized tissue structures in the leaves and cotyledons (Stamp and Henshaw 1982, 1986; Stamp 1984; Szabados et al. 1987). The ability to produce large numbers of somatic embryos from callus could be more efficient in terms of plant production (Sharp et al. 1980; Wang and Vasil 1982; Hackman et al. 1985; Williams and Meheswaran 1986), but somatic embryos have not been produced directly from cell suspension culture.

Embryogenesis can be very efficient in terms of plant production, but can result in somaclonal variation (Skirvin 1978). Such variation can be a problem for propagators, but could be of major importance to plant breeders who wish to select improved forms of cassava cultivars.

VIII. CONCLUDING REMARKS AND FUTURE RESEARCH

Cassava is produced in more than 80 countries, but its production in North America and Europe is limited to regions that have a long frost-free growing season, such as areas in southern Florida, Texas, and Italy. Nevertheless, it is one of the major food crops of the world. Although a wide variation in root yield exists, only limited changes in yield have been reported. Thus, the potential for rapid yield improvements appears to be considerably greater than for crops intensively studied for many years such as maize, rice, wheat, or potatoes.

Cassava is often grown as a subsistence crop. With increased yields, marketable surpluses can be expected to enter the market economy and assume increasing importance. For this to occur in any magnitude, present information needs to be extended to tropical producers. Future research is needed to: (a) reduce labor inputs in all aspects of produotion, including mechanization of root harvest which accounts for 30% of total labor costs; (b) improve handling and marketability of the crop, particularly in urban centers; (c) determine the optimum LAI, leaf size, age, and anatomy for optimum root yields; (d) improve cropping systems, particularly intercropping; (e) characterize how temperature influences the photosynthetic pathways and shoot-root partitioning of carbon; (f) evaluate the importance of drought tolerance on photosynthesis and starch yield; (g) improve production and handling of planting material, particularly tissue culture methodology; (h) genetically reduce cyanogenic glucosides in the root to reduce nutritional problems; and (i) breed and select plants for crop efficiency as well as insect and disease resistance or tolerance.

LITERATURE CITED

Acland, J. D. 1971. East Africa Crops. FAO, Rome, Italy.

Adejore, G. O., and R. H. A. Coutts. 1981. Eradication of cassava mosaic disease from Nigerian cassava clones by meristem culture. Plant Cell Tissue Organ Cult. 1:25–32.

Akoroda, M. O., E. O. Oyinlola, and T. Gebremeskel. 1987. Plantable stem supply system for Institute of International Tropical Agriculture cassava varieties in Oyo, State of Nigeria. Agric. Systems 24:305–317.

Aresta, R. B., and S. Fukai. 1984. Effects of solar radiation on growth of cassava (Manihot esculenta Crantz). II. Fibrous root length. Field Crops Res. 9:361–371.

Balagopalan, C., G. Padmaya, S. K. Nanda, and S. N. Moorthy. 1988. Cassava in Food, Feed and Industry. CRC Press, Boca Raton, FL.

Berry, J., and O. Bjorkaman. 1980. Photosynthetic response and adaptation to temperature in higher plants. Annu. Rev. Plant Physiol. 31:491–543.

Bjorkaman, O., M. R. Badger, and B. A. Armond. 1980. Response and adaptation of photosynthesis to high temperature. p. 233–349. In: N. C. Turner and P. J. Kramer (eds.). Adaptation of Plants to Water and High Temperature Stress. Wiley, New York.

Black, J. N. 1963. The interrelationship of solar radiation and leaf area index in determining the rate of dry matter production of swards of subterranean clover (Trifolium subterraneum L.). Austral. J. Agric. Res. 14:20–28.

Bock, K. R., and J. Guthrie. 1982. Control of CMD in Kenya. Tropical Pest Manag. 28:219–222.

Bolhuis, G. G. 1954. Toxicity of cassava roots. Neth. Agric. J. 2:176–185.

Bradbury, J. H. 1988. The chemical composition of tropical root crops. ASEAN Food J. 4(1):3–13.

Briars, J. R. 1984. Contribution to the vigor and initial development of cassava cuttings. Revista de la Facultad de Agronomica de la universidad Central de Venezuela. Alcance 33:15–54.

Byrne, D. 1984. Breeding cassava. *Plant Breed. Rev.* 2:73–134.

CIAT. 1984a. Cassava planting materials: Conditions for its production. *Cassava Newsl.* 8:11–1.

_____. 1984b. Cassava in Asia: Its potential and research development needs. CIAT, Cali, Colombia.

Cock, J. H. 1973. Cyanide toxicity in relation to the cassava research program of CIAT in Colombia. p. 37–40. In: B. Nestel and R. MacIntyre (eds.). *Chronic Cyanide Toxicity.* International Development Research Centre, Ottawa, Canada.

_____. 1984a. Cassava. p. 529–549. In: P. R. Goldworth and N. M. Fisher (eds.). *The Physiology of Tropical Field Crops.* Wiley, New York.

_____. 1984b. Strategies of cassava plant for resisting drought. *Cassava Newsl.* 8:4–10.

_____. 1985. *Cassava: New Potential for a Neglected Crop.* Westview Press. Boulder, CO.

Cock, J. H., D. Franklin, G. Sandoval, and P. Juri. 1979. The ideal cassava plant for maximum yield. *Crop Sci.* 19:271–279.

Cock, J. H., M. C. P. Porto, and M. A. El-Sharkawy. 1985. Water use efficiency of cassava. III. Influence of air humidity and water stress on gas exchange of field grown cassava. *Crop Sci.* 25:265–272.

Cock, J. H., N. M. Riano, M. A. El-Sharkawy, Y. E. Lopez, and G. Bastides. 1987. C_3–C_4 intermediate photosynthetic characteristics of cassava (Manihot esculenta Crantz). II. Initial products of $^{14}CO_2$ fixation. *Photosynthesis Res.* 12:237–241.

Connor, D. J., and J. H. Cock. 1981. Response of cassava to water shortage. II. Canopy dynamics. *Field Crops Res.* 4:285–296.

Connor, D. J., J. H. Cock, and G. H. Parra. 1981. Response of cassava to water shortage. I. Growth and yield. *Field Crops Res.* 4:181–200.

Cooke, R. D. 1978. An enzymic assay for the total cyanide content of cassava (Manihot esculenta Crantz). *J. Sci. Food Agric.* 29:345–352.

Cooke, R.D., and Coursey, D. G. 1981. Cassava: A major cyanide containing food crop. p. 93–114. In: B. Vennesland, E. E. Conn, C. J. Knowles, J. Wisely, and F. Wissing (eds.). *Cyanide in Biology.* Academic Press, New York.

Cooke, R. D., and E. M. DeLaCruz. 1982. Changes in cyanide content of cassava (Manihot esculenta Crantz) tissue during plant development. *J. Sci. Food Agric.* 33:269–275.

Cooke, R. D., A. K. Howland, and S. K. Khan. 1978. Screening cassava for low cyanide using an enzymatic assay. *Exptl. Agric.* 14:367–372.

Cooke, R. D., and E. N. Maduagwu. 1978. The effect of simple processing on cyanide content of cassava chips. *J. Food Technol.* 13:299–306.

Coursey, D. G. 1973. Cassava as food: Toxicity and technology. p. 27–36. In: B. Nestel and R. MacIntyre (eds.). *Chronic Cassava Toxicity.* International Development Research Center, Ottawa, Canada.

Davis, R. H. 1981. Cyanide detoxification in domestic fowl. p. 51–60. In: B. Vennesland, E. E. Conn, C. K. Knowles, J. Westley, and F. Wissing (eds.). *Cyanide in Biology.* Academic Press, New York.

DeBruijn, G. H. 1973. The cyanogenic character of cassava. p. 43–48. In. B. Nestel and R. MacIntyre (eds.). *Chronic Cassava Toxicity.* International Development Research Center, Ottawa, Canada.

DeVries, C. A., J. D. Ferwerda, and M. Flach. 1967. Choice of food crops in relation to actual potential production in the tropics. *Neth. J. Agr. Sci.* 15:241–248.

Dufour, D. L. 1988. Cyanide content of cassava (Manihot esculenta Euphorbiaceae) cultivars used by Tukanoan Indians in Northwest Amazon. *Econ. Bot.* 42:255–266.

El-Sharkawy, M. A. 1965. Factors limiting photosynthesis rates of different plant species. Ph. D. Dissertation. Univ. Arizona, Tucson, AZ.

El-Sharkawy, M. A., and J. H. Cock. 1984. Water use efficiency of cassava. I. Effects of air humidity and water stress on stomatal conductance and gas exchange. *Crop Sci.* 24:497–502.

———. 1986. Humidity factors in stomatal control and its effect on crop productivity. p. 187–198. In: R. Marcelle, H. Clijstes and M. Van Poucke (eds.). Biological Control of Photosynthesis. Martinus Nijhoff Dordrecht, Boston, MA.

———. 1987a. C_3–C_4 intermediate photosynthetic characteristics of cassava (Manihot esculenta Crantz). I. Gas exchange. Photosynthesis Res. 12:219–235.

———. 1987b. Response of cassava to water stress. Plant Soil. 100:345–360.

———. 1988. Photosynthesis in cassava. Cassava Newsl. 12:14–15.

El-Sharkawy, M. A., J. H. Cock, and G. DeCadena. 1984a. Stomatal characteristics among cassava cultivars and their relation to gas exchange. Exptl. Agric. 20:67–76.

———. 1984b. Influence of differences in leaf anatomy on net photosynthetic rates of some cultivars of cassava. Photosynthesis Res. 5:235–242.

El-Sharkawy, M. A., J. H. Cock, and A. A. Held. 1984c. Photosynthetic responses of cassava cultivars (Manihot esculenta Crantz) from different habitats to temperature. Photosynthesis Res. 5:243–250.

———. 1984d. Water use efficiency of cassava. II. Differing sensitivity of stomata to air humidity in cassava and other warm climate species. Crop Sci. 24:503–507.

El-Sharkawy, M. A., J. H. Cock, and A. P. Hernandez. 1985. Stomatal response to air humidity and its relation to stomatal density in a wide range of warm climate species. Photosynthesis Res. 7:137–149.

El-Sharkawy, M. A., and J. D. Heskett. 1965. Photosynthesis among species in relation to characteristics of leaf anatomy and CO_2 diffusion resistances. Crop Sci. 5:517–521.

El-Sharkawy, M. A., R. S. Loomis, and W. A. Williams. 1967. Apparent assimilation of carbon dioxide by different plant species. Physiol. Plant. 20:171–186.

Enyi, B. A. C. 1973. Growth rate of 3 cassava varieties (Manihot esculenta Crantz) under varying population densities. J. Agric. Sci. 81:15–28.

Fahn, A. 1979. Secretory Tissue in Plants. Academic Press, New York.

———. 1982. Plant Anatomy. 3rd ed. Pergamon Press, New York. FAO. 1984. Production yearbook. Food and Agriculture Organization. Vol. 38. Rome.

Fischer, R. A. and N. C. Turner. 1978. Plant productivity in the arid and semi-arid zones. Annu. Rev. Plant Physiol. 29:277–317.

Fukai, S., A. B. Alcoy, A. B. Llamelo, and R. D. Patterson. 1984. Effects of solar radiation on growth of cassava (Manihot esculenta Crantz). I. Canopy development and dry matter growth. Field Crops Res. 9:347–360.

Garcia, M., and M. Rodrigues. 1983. Comparative study on cassava cuttings from different parts of the plant. Viandas Tropicales. 6:39–49.

Godo, G. H. 1984. Yield components as influenced by methods of planting cassava cuttings. p. 219–224. In: Proc. Sixth Symp. of the International Society for Tropical Root Crops, International Potato Center. Lima, Peru.

Gomez, G., and M. Valdivieso. 1984. Changes in cyanide content of cassava tissues as affected by plant age and variety. p. 323–328. In. Proc. Sixth Symp. of the International Society for Tropical Root Crops, International Potato Center, Lima, Peru.

Guritno, B. 1985. Influence of Planting Material on Plant Performance in Cassava. Univ. of Branijaya, Malang, Indonesia.

Hackman, I., L. C. Fowke, S. Von Arnold, and T. Erikson. 1985. The development of somatic embryo tissue cultures initiated from immature embryos in tissue cultures. Plant Sci. 38:53–59.

Hahn, S. K., E. R. Terry, and K. Leuschner. 1980. Breeding cassava for resistance to cassava mosaic disease. Euphytica 29:673–683.

Hidajat, E. B. 1987. Histochemical localization of some enzymes in callus of cassava (Manihot esculenta Crantz). Cytologia 52:671–678.

Holmes, E. B., and L. A. Wilson. 1979. Effect of nitrogen supply on early growth and development and nitrate reductase activity in two cassava cultivars. p. 487–495. In: Proc. 5th Symp. of the International Society for Tropical Root Crops, Manila, The Philippines.

Hunt, L. A., D. M. Wholey, and J. H. Cock. 1977. Growth physiology of cassava. *Field Crops Abst.* 30:77–89.

IITA. 1988. Research Highlights Reports. Ibadan, Nigeria.

Irikura, Y., J. H. Cock, and K. Kawano. 1979. The physiological basis of genotype temperature interaction in cassava. *Field Crops Res.* 2:227–239.

Jones, W. O. 1959. *Manioc in Africa.* Stanford Univ. Press, Stanford, CA.

Kartha, K. K. 1982. In vitro growth responses and plant regeneration from cryopreserved meristems of cassava (*Manihot esculenta* Crantz). *Z. Pflanzenphysiol.* 107:133–140.

_____. 1986. Production and indexing of disease-free plants. p. 319–329. In: L. A. Withers and P. G. Alderson (eds.). *Plant Tissue Culture and Its Agricultural Applications.* Butterworths, London, U.K.

Kartha, K. K., and O. L. Gamborg. 1975. Elimination of cassava mosaic disease by meristem culture. *Phytopathol.* 65:826–828.

_____. 1979. Cassava tissue culture: Principles and applications. p. 711–725. In: W. R. Sharp, P. O. Larsen, E. F. Paddock, and V. Raghaven (eds.). *Plant Cell and Tissue Culture: Principles and Applications.* Ohio State Univ. Press, Columbus, OH.

Kartha, K. K., O. L. Gamborg, F. Coustebel, and J. P. Shyluk. 1974. Regeneration of cassava plants from apical meristems. *Plant Sci. Let.* 2:102–103.

Kawano, K., W. M. G. Fuduka, and U. Cenpukdee. 1987. Genetic and environmental effects on dry matter content of cassava root. *Crop Sci.* 27:71–74.

Kayode, G. O. 1983. Effects of various planting and harvesting times on the yield of four cassava varieties in the tropical rainforest region. *J. Agric. Sci.* 101:633–636.

Keating, B. A., J. P. Evenson, and S. Fukai. 1982a. Environmental effects on growth and development of cassava (*Manihot esculenta* Crantz). II. Crop growth rate and biomass yield. *Field Crops Res.* 5:283–292.

_____. 1982b. Environmental effects on growth and development of cassava (*Manihot esculenta* Crantz). I. Crop development. *Field Crops Res.* 5:271–281.

Keating, B. A., G. L. Wilson, and J. P. Evenson. 1985. Effects of photoperiod on growth and development of cassava (*Manihot esculenta* Crantz). *Austral. J. Plant Physiol.* 12:261–30.

Laetsch, W. M. 1974. The C_4 Syndrome: A structural analysis. *Annu. Rev. Plant Physiol.* 25:27–52.

Lazo-Casteleanos, R., and F. F. Betancourt. 1985. Relationship between origin of cassava cuttings and yields in Camaguey-Tunas Penplane. Ciencia y Technica en la Agricultura. *Viandas Tropicales* 8:57–67.

Leinher, D. E. 1984. Production of planting material in cassava: Some agronomic implications. In: Proc. Sixth Symp. of the International Society for Tropical Root Crops. International Potato Center, Lima, Peru.

Liu, M. C. and W. H. Chen. 1979. Organogenesis and chromosome number in callus derived from cassava anthers. *Can. J. Bot.* 56:1287–1290.

Mabanza, J. 1987. Improvement of cassava (*Manihot esculenta* Crantz) by in vitro culture. p. 35. In: E. R. Terry, M. O. Akoroda and O. B. Arene (eds.). Proc. Triennial Symp. of the International Society for Tropical Root Crops, African Branch. Owen, Nigeria. 1986. Tropical Root Crops and the African Food Crisis, Proc. Ottawa, Canada.

Mahon, J. D., S. B. Lowe, and L. A. Hunt. 1977a. Variation in the rate of photosynthetic CO_2 uptake in cassava cultivars and related species of *Manihot. Photosynthetica* 11:131–138.

Mahon, J. D., S. B. Lowe, L. A. Hunt, and M. Thiagarajah. 1977b. Environmental effects on photosynthesis and transpiration in attached leaves of cassava (*Manihot esculenta* Crantz). *Photosynthetica* 11:121–130.

Manrique, L. A. 1987. Leaf area development and growth performance of cassava germplasm on a strongly acid utisol of Panama. *J. Plant Nutrition* 10:677–698.

Martin, F. M., and W. E. Splittstoesser. 1975. A comparison of total protein and amino acids of tropical roots and tubers. *Tropical Root Tuber Crops Newsl.* 8:7–15.

Milanez, F. R. 1949. Segunda nota sobre os laticiferos. *Lilloa* 16:193–211.

Muimba-Kankolongo, A., and J. Phuti. 1987. Relationship of cassava mosaic severity in planting material to mosaic development, growth and yield of cassava in Zaire. *J. Exptl. Agric.* 23:221–225.

Murashige, T., and F. Skoog. 1962. A revised medium for rapid growth and bioassays with tobacco tissue culture. *Physiol. Plant.* 15:474–497.

Nair, N. G., K. K. Kartha, and O. L. Gamborg. 1979. Effect of growth regulators on plant regeneration from shoot apical meristems of cassava (*Manihot esculenta* Crantz) and on the cutting of internodes in vitro. *Z. Pflanzenphysiol.* 95:51–56.

Nestel, B., and R. MacIntyre. 1973. Chronic cassava toxicity. International Development Centre. Ottawa, Canada.

O'Hair, S. K. 1989. Cassava root starch content and distribution varies with tissue age. *HortScience* 24:505–506.

――――. 1990. Tropical root and tuber crops. *Hort. Rev.* 12:(in press).

O'Hair, S. K., R. B. Forbes, S. J. Locascio, J. R. Rich, and R. L. Stanley. 1983. Starch and glucose distribution within cassava roots as affected by cultivar and location. *HortScience* 18:735–737.

Oke, O. L. 1983. The mode of cyanide detoxification. p. 97–104. In: B. Nestel and R. MacIntyre (eds.). *Chronic Cassava Toxicity.* International Development Research Center, Ottawa, Canada.

――――. 1984. Processing and detoxification of cassava. p. 329–336. In: Proc. of the Sixth Annual Symp. International Society of Tropical Root Crops, International Potato Center, Lima, Peru.

Osmond, C. B., K. Winter, and S. B. Powles. 1980. Adaptive significance of carbon dioxide cycling during photosynthesis in water stressed plants. p. 139–154. In: N. C. Turner and P. K. Kramer (eds.). *Adaptation of Plants to Water and High Temperature Stress.* Wiley, New York.

Pereira, J. F. 1977. Fisiologia de la yuca. Escuela de Ingenieria Agronomica, Universidad deOriente, Monagas, Venezuela.

Pereira, J. F., D. S. Seigler, and W. E. Splittstoesser. 1981. Cyanogenesis in sweet and bitter cultivars of cassava. *HortScience* 16:776–777.

Pereira, J. F., and W. E. Splittstoesser. 1986a. A rapid method to estimate leaf area of cassava plants. *HortScience* 21:1218–1219.

――――. 1986b. Nitrate reduction by cassava. *Plant Cell Physiol.* 27:925–927.

――――. 1987. Exudate from cassava leaves. *Agric. Ecosystems Environ.* 18:191–194.

――――. 1990. Anatomy of the cassava leaf. *Interamer. Soc. Trop. Hort.* 34:73–78.

Pereira, J. F., W. E. Splittstoesser, and W. L. Ogren. 1986. Photosynthesis in detached leaves of cassava. *Photosynthetica* 20:286–292.

Porto, M. C. M. 1984. Physiological mechanisms of drought tolerance in cassava (*Manihot esculenta* Crantz. Ph. D. Dissertation. Univ. Arizona, Tucson.

Ramanujam, T. 1985. Leaf density profile and efficiency in partitioning dry matter among high and low yielding cultivars of cassava (*Manihot esculenta* Crantz). *Field Crops Res.* 10:291–303.

Ramanujam, T. and P. Indira. 1983. Canopy structure on growth and development of cassava (*Manihot esculenta* Crantz). *Turrialba* 33:321–327.

Robertson, I. M. 1985. Whitefly, a vector of cassava mosaic disease in Kenya. *Insect Sci. Appl.* 8:797–801.

Roca, W. M., A. Rodriguez, L. F. Pitena, R. C. Barba, and J. C. Toro. 1980. Improvement of propagation techniques for cassava using single leaf bud cutting: preliminary report. *Cassava Newsl.* 8:4–5.

Rogers, D. J., and H. S. Fleming. 1973. A monograph of *Manihot esculenta* with an explanation of the taximetric methods used. *Econ. Bot.* 27:1–113.

Sadasivan, K. V. 1970. The composition of leaf exudate and leachate of tropical foliage. *Sci. Cult.* 36:608–609.

San Jose, J. J., and F. Berrade. 1983. Transfer of mass and energy in a cassava (Manihot esculenta Crantz cv. Cubana) community. I. Microclimate and water vapor exchange in a Savanna climate. Ann. Bot. 52:507–520.

Sharp, W. R., M. R. Sondahl, L. S. Caldes, and S. B. Maraffa. 1980. The physiology of in vitro asexual embryogenesis. Hort. Rev. 2:268–310.

Sinha, S. K., and T. V. R. Nair. 1968. Studies on the variability of cyanomic glucoside content in cassava tubers. Indian J. Agric. Sci. 38:958–963.

Skirvin, R. M. 1978. Natural and induced variation in tissue culture. Euphytica 27:241–266.

Smith, M. K., B. J. Biggs, and K. J. Scott. 1986a. On in vitro propagation of cassava (Manihot esculenta Crantz). Plant Cell Tissue Organ Cult. 6:221–228.

_____ . 1986b. Cassava tissue culture. Cassava Newsl. 10:6–7.

Splittstoesser, W. E. 1977. Protein quality and quantity of tropical roots and tubers. HortScience 12:294–298.

Splittstoesser, W. E., and F. W. Martin. 1975. The tryptophan content of tropical roots and tubers. HortScience 10:23–24.

Splittstoesser, W. E., and A. M. Rhodes. 1975. Protein and amino acid values of some tropical root crops. Ill. Res. 15(4):6–7.

Stamp, J. A. 1984. In vitro plant regeneration studies with cassava (Manihot esculenta Crantz). Ph. D. Dissertation. Univ. Birmingham, U.K.

Stamp, J. A., and G. G. Henshaw. 1982. Somatic embryogenesis in cassava. Z. Pflanzenphysiol. 105:183–187.

_____ . 1986. Adventitious regeneration in cassava. p. 149–157. In: L. A. Withers and P. G. Alderson (eds.). Plant Tissue Culture and Its Agricultural Application. Butterworths, London, U.K.

_____ . 1987a. Somatic embryogenesis from clonal leaf tissues of cassava. Ann. Bot. 59:445–450.

_____ . 1987b. Secondary somatic embryo-genesis and plant regeneration in cassava. Ann. Bot. 59:227–333.

Szabados, L., R. Hoyos, and W. Roca. 1987. In vitro embryogenesis and plant regeneration in cassava. Plant Cell Rpts. 6:248–251.

Ternes, N., E. Mondardo, and V. J. Vizzotto. 1978. Variacao do teor deamido na cultura da mandioca em Santa Catarina. Empresa Catarinense de Pesquisa Agropecuaria. Florianoplis, SC, Brazil. Indicacao de Pesquisa 23.

Tilguin, J. P. 1979. Plant regeneration from stem callus of cassava. Can. J. Bot. 57:1761–1763.

Udealor, A., and F. O. Ezedenma. 1987. Effects of cutting back cassava stems on the yield and quality of the roots. p. 35. In: E. R. Terry, M. O. Akoroda, and O. B. Arene (eds.). Proc. Triennial Symp. of International Society for Tropical Root Crops. African Branch. Oweh, Nigeria. Tropical Root Crops and the African Food Crisis, Proc. Ottawa, Canada.

Verhagan, A. M. W., W. A. Williams, and E. H. Hall. 1963. Plant production in relationship to foliage illumination. Ann. Bot. (N. S.) 27:627–639.

Wang, D. M., and I. K. Vasil. 1982. Somatic embryogenesis and plant regeneration from inflorescence segments of Pennisetum purpureum Schum. (Napier or Elephant grass). Plant Sci. Lett. 25:147–154.

Wholey, D. M. 1977. Changes during storage of cassava planting materials and their effect on regeneration. Tropical Sci. 19:205–216.

Wholey, D. W., and R. H. Booth. 1979. Influence of variety and planting density on starch accumulation in cassava roots. J. Sci. Fd. Agric. 30:165–170.

Williams, E. G., and G. Meheswaran. 1986. Somatic embryogensis: Factors influencing coordinated behavior of cells as an embryogenic group. Ann. Bot. 57:443–462.

Zitnak, A. 1973. Assay method for hydrocyanic acid in plant tissue and their application in studies of cyanogenic glucosides in Manihot esculenta. p. 89–96. In: B. Nestel and R. MacIntyre (eds.). Chronic Cassava Toxicity. International Development Research Centre, Ottawa, Canada.

Preplant Physiological Seed Conditioning

Anwar A. Khan
Department of Horticultural Sciences
New York State Agricultural Experiment Station, Cornell University
Geneva, NY 14456

I. INTRODUCTION

The rationale for preplant seed treatment is to mobilize the seed's own resources and to augment them with external resources to maximize improvement in stand establishment and yield. Mobilization of these resources is particularly critical when using seeds of modern cultivars which are selected for high yields often under optimal growing conditions. Physiological and non-physiological treatments are needed to remove several soil, climatic and hydrological constraints. Physiological seed treatments that improve or enhance seed performance and that are based primarily on seed hydration include presoaking, wetting and drying, humidification, osmoconditioning, matriconditioning, and pregermination (Table 4.1). Other physiological treatments that promote germination or remove blocks to germination require physical or chemical stimuli in addition to seed hydration. These include moist chilling, thermal shock, irradiation, aeration, and hormone permeation.

There are several non-physiological seed treatments that can remove mechanical, soil, and environmental constraints and that may indirectly improve seed germination and stand establishment. These include seed scarification, coating or pelleting, treatment with bioactive chemicals, and inoculation or coating of seeds with beneficial microbes (Table 4.1).

This review is not intended to be all-encompassing. The subjects of wetting and drying (Henckel 1964; Heydecker and Coolbear 1977), humidification (Hegarty 1978; Justice and Bass 1978), osmoconditioning or priming (Heydecker 1977; Heydecker and Coolbear 1977; Khan et al. 1978, 1980/81; Bradford 1986), and pregermination (Currah et. al. 1974;

Table 4.1.   Preplant seed treatments to improve germination and performance of seeds.

Physiological treatments based on seed hydration

    Pregermination soak at low water potential (Physiological seed conditioning)

        Conditioning

        Matriconditioning

        Osmoconditioning (priming, osmopriming, osmotic priming)

    Pregermination soak at zero water potential

        Presoaking

        Wetting and drying (hardening, drought hardening)

    Other hydration treatment

        Humidification (humidifying)

        Pregermination (chitting)

Physiological treatments based on seed hydration plus other stimuli

        Aeration (oxygenation)

        Irradiation

        Permeation with hormones

        Stratification (moist chilling)

        Thermal shock

Non-physiological treatments

        Chemical or mechanical seed scarification

        Seed coating or seed pelleting

        Treatment with pesticides and nutrients

        Inoculation or coating with beneficial microbes

Gray 1981) have been extensively covered in previous articles. Recently, conditioning of seeds on solid carriers with high osmotic and/or matric potential has attracted attention (Bennett and Waters, 1984, 1987a, 1987b; Khan and Taylor 1986; Kubik et al. 1988; Taylor et al. 1988; Zuo et al. 1988a, 1988b; Khan et al. 1990) and forms the subject matter of a detailed study by this reviewer (Khan et al. 1990). Numerous articles have appeared also on seed coating and pelleting (Millier and Sooter 1967; Johnson 1975; Halmer 1988), treatments of seeds with growth regulators and protectants (Callan 1975; Khan 1978; Maude 1986), and treatments with beneficial microbes (Weller 1988; Digat 1989; Callan et al. 1990; Harman and Taylor 1990).

In the late 1970's, an attempt was made to integrate seed priming or osmoconditioning with several other beneficial treatments, such as irradiation, moist-chilling, and permeation with bioactive chemicals to improve seed performance (Khan 1977; Khan et al. 1979). This review focuses on the advances in the preplant low water potential seed hydration (physiological seed conditioning) treatments and their integration with other useful seed treatments to improve seedling emergence and stand establishment. A glossary of terms describing the various physiological seed treatments is presented in Table 4.2; species and binomials cited in the text are presented in Table 4.3.

Table 4.2. Glossary of terms used in the text to describe physiological preplant seed treatments to improve germination and stand establishment.

Aeration, oxygenation: A germination promoting treatment applied during seed hydration by bubbling air or oxygen.

Bio-priming: A combined treatment involving seed bacterization and seed moisturizing. The term refers to hydration of previously inoculated sweet corn seeds with moistened vermiculite (see "matriconditioning").

Conditioning: Enhancement of physiological and biochemical events (those associated with rapidity and synchrony of germination, repair and improved germination potential) in seeds during controlled hydration and suspension of germination by low water potential of the imbibing medium (liquid or moist solid). Seed is hydrated until water potential of the seed and the imbibing medium become equal (reach equilibrium water potential). Conditioning is usually conducted at 15 to 20°C.

Hardening, wetting and drying, drought hardening: Soaking of seeds with a limited amount of water followed by drying and, sometimes, repeating the wetting-drying cycle.

Humidification, humidifying: Exposure of seeds to a water vapor saturated (100% RH) or nearly saturated atmosphere. Equilibrium moisture content of seeds at specific RH varies with crop species.

Irradiation: Exposure of seeds in a hydrated state to light in order to break seed dormancy or to improve the rate of germination.

Matriconditioning: Enhancement of physiological and biochemical events (see "conditioning") in seeds during suspension of germination by low matric potential and negligible osmotic potential of the imbibing medium. Materials such as diatomaceous earth and vermiculite, with high adsorptive and capillary forces, are mixed with known

Table 4.2. Continued.

---

amounts of water and seeds to establish a moisture equilibrium (when water potential is equal on both sides) between seeds and the matrix needed for conditioning.

Moisturizing: A treatment to partially hydrate seeds on a material with hydrophilic surfaces in order to prevent germination and to invigorate the seed. The term refers to hydration of seeds on moistened vermiculite (see "matriconditioning").

Osmoconditioning: Enhancement of physiological and biochemical events (see "conditioning") in seeds during suspension of germination by low osmotic potential and negligible matric potential of the imbibing medium. Salts or non-penetrating organic solutes in liquid medium or matrix solution are used to establish equilibrium water potential between seed and the osmotic medium needed for conditioning.

Permeation with hormones: Soaking of seeds for a few minutes to several hours in aqueous or non-aqueous (organic solvents) hormone solutions. Response of seed to permeated hormones is elicited during subsequent hydration.

Pregermination, chitting: Seed soak under optimal hydration and temperature conditions to a point when the radicle is just visible, usually 1 to 2 millimeter in length. The term is used in connection with gel seeding or fluid drilling of seeds.

Presoaking: Pregermination water soak of seeds in unlimited amount of water at low to moderate temperatures.

Priming, osmopriming, osmotic priming: These terms are used interchangeably with osmoconditioning.

Solid matrix priming: Similar to "conditioning" (see above). The term has been used to refer to seed conditioning by osmotic and/or matric component of the solid matrix water potential. Materials such as Agro-Lig with high osmotic solute content and those with hydrophilic surfaces such as expanded calcined clay and vermiculite have been used to condition seeds.

Stratification, moist-chilling: A dormancy breaking seed treatment generally conducted at 1–10°C during prolonged seed soak in water.

Thermal shock: A dormancy breaking seed treatment achieved in a hydrated state by a brief shift to high temperature (e.g. 1 h at 40°C) followed by a return to ambient temperature (20–25°C) condition.

Table 4.3. List of crops, weeds, and binomials cited in the text.

Common name	Binomial
Barley	Hordeum vulgare L.
Beet	Beta vulgaris L.
Brussels sprout	Brassica oleracea (Gemmifera Group)
Cabbage	Brassica oleracea (Capitata Group)
Cantaloupe	Cucumis melo L.
Carrot	Daucus carota var. sativus (Hoffn.)
Corn cockle	Agrostemma githago L.
Celery	Apium graveolens L.
Common ragweed	Ambrosia artemisiifolia L.
Cotton	Gossypium hirsutum L.
Cucumber	Cucumis sativus L.
Curly dock	Rumex crispus L.
Dusty miller	Senecio cineraria D.C.
Eggplant	Solanum melongena L.
Giant foxtail	Setaria faberi Herrm.
Impatiens (orange green leaved)	Impatiens novette
Kale	Brassica oleracea (Acephala Group)
Kohlrabi	Brassica oleracea (Gongylodes Group)
Leek	Allium porrum L.
Maize	Zea mays L.
Lettuce	Lactuca sativa L.
Lima bean	Phaseolus vulgaris L.
Loblolly pine	Pinus taeda L.
Melon	Cucumis melo L.
Muskmelon	Cucumis melo L.
Onion	Allium cepa L.
Parsnip	Pastinaca sativa L.
Pea	Pisum sativum L. var. sativum
Peanut	Arachis hypogea L.
Pepper	Capsicum anuum L.
Perennial goosefoot, Good-King-Henry	Chenopodium bonus-henricus L.
Primrose (crown scarlet)	Primula acaulis (L.) J. Hill = vulgaris Huds.
Radish	Raphanus sativus L.
Redroot pigweed	Amaranthus retroflexus L.
Rice	Oryza sativa L.
Ryegrass	Lolium perenne L.
Salvia	Salvia splendens, F. Sellow ex Roem. & Schult.
Shortleaf pine	Pinus echinata Mill.
Snap bean	Phaseolus vulgaris L.
Sorghum	Sorghum bicolor (L.) Moench
Spinach	Spinacia oleracea L.
Soybean	Glycine max (L.) Merr.
Sugar beet	Beta vulgaris L.
Sweet corn	Zea mays L.
Tomato	Lycopersicon esculentum Mill.
Turnip	Brassica rapa (Rapa Group)
Watermelon	Citrullus lanatus (Thumb.) Matsumi & Nakai
Wheat	Triticum aestivum L.
Witchweed	Striga asiatica (L.) Kuntze

II. ON TERMINOLOGY OF LOW WATER POTENTIAL
SEED HYDRATION TREATMENTS

The term "priming of seed" was coined by Malnassy (1971) as reported by Heydecker (1973/74) to describe a treatment to improve germination at low temperature. Heydecker and his co-workers have extensively used the term "priming" to describe an osmotic seed treatment to enhance seed processes (Heydecker 1973/74, 1977; Heydecker and Coolbear 1977). The term "osmotic conditioning" or "osmoconditioning" (Khan et al. 1978, 1979) was proposed as an alternative to priming because of two concerns: (1) priming did not tell what it meant; and (2) the term had been previously used to describe "priming" DNA fragments required for the enzymatic synthesis of DNA (Kornberg 1957; Bollum 1963). Heydecker (1973/1974) had the same concern and preferred the terms "halopriming" and "osmopriming" over priming to describe seed hydration in salt and polyethylene glycol (PEG) solution, respectively. Since they were first proposed, all three terms—priming, osmopriming and osmoconditioning—continue to be used extensively at the present time to describe preplant osmotic seed treatments in salts, PEG (usually PEG 8000), and other osmotic solutes (see Literature Cited).

The terms osmoconditioning, priming, and osmopriming historically have been used in the context of osmotic seed treatment. Consistent with such usage, seed hydration by moistened solid carriers with large amounts of organic or inorganic solutes have been referred to as "osmoconditioning", "priming," "osmotic priming," or "solid matrix priming" (Peterson 1976; Khan and Taylor 1986; Taylor et al. 1988, Zuo et al. 1988a, 1988b). Other workers have used the terms "moisturizing" (Bennett et al. 1987a, 1987b, Callan et al. 1990) and "solid matrix priming" (Kubik et al. 1988, Berg et al. 1989) to describe seed hydration by surface-active, non-osmotic carriers. The author has proposed retention of the term osmoconditioning for controlled seed hydration by osmotic solutes in the carrier (liquid or solid) and has proposed the term "matriconditioning" for hydration by moistened carriers with high water adsorptive matric forces (Khan et al. 1990). As matriconditioning does not involve osmotic solutes it should be distinguished from osmoconditioning, priming, or solid matrix priming. The term "conditioning" is appropriate when seed hydration is controlled by either osmotic or matric forces of the carrier or when the magnitude or the relative contributions of such forces are not known.

Despite its usage in the seed industry the term "seed enhancement" implies improvement or activation of seed processes as a result of a treatment, such as osmoconditioning, and is, therefore, inappropriate as a seed treatment term. In this review, priming and osmoconditioning are used interchangeably, while matriconditioning is used to describe seed conditioning by non-osmotic, matric forces of the carrier substance (see sections IV. A and IV. B and Table 4.2 for further explanations of the terms).

III. SEED HYDRATION IN LIQUID MEDIUM OR HUMID AIR

When a seed is hydrated, physiological and biochemical changes occur. A prolonged seed hydration, particularly at low water potential, profoundly influences the rapidity, synchrony, and percentage of seeds that germinate. Several seed hydration procedures have been developed to improve the rate and uniformity of seedling emergence. One approach has been to soak seeds with limited or unlimited amounts of water at low to moderate temperatures (presoaking, wetting and drying). Another approach has been to shorten the time between sowing and emergence by pregerminating (chitting) seeds under optimal hydration and temperature conditions prior to planting. The third approach has been to hydrate the seed in moist air (humidification). The fourth and more popular approach has been to hydrate seeds in low water potential solution of inorganic and organic solutes for extended periods of time (osmoconditioning).

Seed hydration at low water potential (seed conditioning) while preventing the radicle from protruding allows the seed to achieve many of the biochemical and physiological changes that are consistent with a rapid and synchronous seedling emergence upon planting. Seeds may also respond positively during conditioning and other hydration treatments to various other physical, chemical, and biological factors such as aeration, moist chilling, thermal shock, irradiation, seed coating, hormones, and beneficial microbes, but these may be characterized as germination promotive, dormancy breaking, or protective treatments that could be integrated with seed hydration treatments (Table 4.1) to maximize seed performance.

A. Presoaking

Improving germination by presoaking seeds in water has been known for many years (Lang 1965). Kidd and West (1918, 1919) first demonstrated that short periods of presoaking had a favorable effect on subsequent percentage germination and seeding growth. The favorable effects were maintained after redrying. Recent studies confirmed these results. Seeds of leek, pepper, and lettuce subjected to a presoak treatment in water at 3–20°C germinated at temperatures formerly too high for germination (Gelmond 1965; Heydecker 1973/74; Guedes and Cantliffe 1980). Hydration of lettuce seeds by short duration presoaking in water appears to bring about changes that are similar but not necessarily of the same magnitude to changes obtained during osmoconditioning (Guedes and Cantliffe 1980; Karssen and Weges 1987) (see section III. E. 4).

B. Pregermination

The rationale behind pregermination (synonym: chitting) is to eliminate the variable effects of weather and soil conditions on germination and to obtain a rapid and uniform emergence of seedlings from the soil. Seeds are first germinated under ideal conditions, usually in aerated water at temperatures of 18–20°C, followed immediately by sowing (Currah et al. 1974; Darby and Salter 1976). Technology has been developed for fluid drilling of pregerminated seeds using alginate and other types of gels (Currah et al. 1974; Likorish and Darby 1976; Gray 1981). Fluid drilling of pregerminated seeds has proved successful for a number of small, slow germinating vegetable seeds, such as celery, parsnip, and carrot (Bussel and Gray 1976; Darby and Salter 1976; Gray and Steckel 1977). The major advantages of fluid drilling pregerminated seeds have been rapid emergence, greater synchronization of emergence, ability to plant in cooler soils, and the use of gel as a delivery system for nutrients and pesticides (Finch-Savage 1984; Giammichele and Pill 1984; Pill and Mucha 1984; Furutani et al. 1985). Among the disadvantages are the drying of gel in the field, particularly at high temperatures, forming a hard film around the germinated seeds; susceptibility of germinated seeds to damage during handling; and the rapid loss of viability during storage of germinated seeds (Gray 1981; Kahn and Motes 1988).

C. Hardening

Hardening (synonyms: wetting and drying, drought hardening) appears to impart drought tolerance and is particularly effective in improving percentage germination of old seeds (Henckel 1964; Gray and Steckel 1977). Seeds are hydrated at 15–25°C for various times with a limited amount of water with final water content of seeds usually ranging from 30–70% of the air-dry weight (May et al. 1962; Henckel 1964). The treatment effect resembles osmoconditioning in that it reduces the time to 50% (T_{50}) germination (Wickham and Nichols 1976; Gray and Steckel 1977). However, a considerable caution is needed as the period of hydration before drying may be critical and may adversely affect seed germination (Berrie and Drennen 1971; Sen and Osborne 1974). Even the uptake of a limited amount of water could be injurious to such seeds as bean, lima bean, sweet corn, and soybean. These seeds are susceptible to imbibitional injury (Obendorf and Hobbs 1970; Roos and Pollock 1971), and care must be taken to slow the entry of water into the seed if the advantages of seed conditioning by partial seed hydration is to be realized (see sections III. E. 1. and VI. A. 5).

D. Humidification

Humidification is achieved by exposure of seeds to a water vapor saturated (100% R.H.) or nearly saturated atmosphere. The procedure is familiar to those working with seed storage and seed deterioration (Justice and Bass 1978). A relationship between the relative humidity and seed moisture content has been developed for various species (Justice and Bass 1978). In most species, moisture uptake is rather rapid for the first 2–3 days and then decreases. Hegarty (1978) has summarized some of the disadvantages and difficulties in humidifying seeds at high R.H. These include (1) difficulty in maintaining constant temperature conditions and preventing water condensation on the seeds; (2) impossibility of keeping a uniform moisture content during the treatment; (3) extreme sensitivity of the seed hydration process to differences in carbon dioxide and oxygen concentrations in confined atmospheres; (4) the danger of seed deterioration during protracted period of seed hydration; and (5) substantially less hydration of seeds in water vapor equilibrated atmosphere than in seeds imbibed in contact with liquid phase, especially the part of the seed that can retain "free" water which can be utilized by the seed during germination.

In spite of the difficulties mentioned above, seed activation and germination can be achieved in vapor phase over a range of water potentials (Owen 1952). Seeds equilibrated with moisture in a saturated atmosphere show rapid germination (Heydecker and Coolbear 1977). Similar physiological and biochemical changes occur whether hydration is achieved by hardening, humidification, or osmoconditioning (Perl 1978; Knypl et al. 1980; Taylorson 1986; Rao et al. 1987).

E. Osmotic Conditioning

The purpose of osmotic conditioning (synonyms: osmoconditioning, priming, osmopriming, osmotic priming) is to reduce the germination time, synchronize germination, and to improve the percentage germination and seedling stand. Osmoconditioning begins when a seed is hydrated in a low water potential osmotic solution. After the moisture equilibrium is achieved, further uptake of water by the seed is prevented. Seed osmoconditioning can be effectively achieved by adjusting the concentration of the osmoticum in the conditioning solution to a level just preventing germination. A number of osmotica, including PEG, KNO_3, K_3PO_4, KH_2PO_4, $MgSO_4$, NaCl, glycerol, and mannitol, have been used to osmocondition seeds of many plant species. The amount of water entering the seed during osmoconditioning may vary depending upon whether the osmoticum is a salt or a non-penetrating organic molecule. For example, the amounts of water taken up by 'Ruby Queen' beet seeds after a 24-h soak at 15°C in water, −1.2 MPa PEG, −1.2 MPa KNO_3 and −1.2MPa NaCl solutions were 94, 57, 84, and 87% (on dry weight basis), respectively; after 96h the corresponding values were 119, 61, 88, and 102% (Khan, unpublished). Seed osmoconditioning generally requires 2–21 days depending on the species. The osmotic potential of conditioning solutions is generally kept at −0.8 to −1.6 MPa, and the temperature during conditioning is generally maintained at 15–20°C, although there are some exceptions. Seed conditioning is best achieved in the light for seeds in which germination is under phytochrome control. Research has advanced on several fronts on seed osmoconditioning or seed priming in the last several years.

1. Biomass or seed yield. The basic methodology for seed priming or osmoconditioning was developed by Heydecker and his coworkers. Two excellent reviews (Heydecker 1977; Heydecker and Coolbear 1977) summarize the development of the methodology, the effectiveness of this procedure in laboratory studies in comparison with other preplant seed treatments, and its potential usefulness in improving seed performance and crop productivity. It has become increasingly apparent that osmoconditioning is particularly suited to improving the performance of slow emerging small vegetable seeds under cold and/or wet conditions and where the paramount consideration is increasing the biomass of the vegetative portion of the plant rather than seed or fruit yield.

Osmoconditioning has proved effective in improving emergence or yield in carrot (Szafirowska et al. 1981; Brocklehurst and Dearman 1983b; Haigh et al. 1986), beet (Khan et al. 1983), sugar beet (Durant et al. 1983), celery (Rennick and Tiernan 1978; Brocklehurst and Dearman 1983b), parsley (Heydecker and Coolbear 1977; Pill 1986; Rabin et al. 1988), parsnip (Gray et al. 1984), leek (Brocklehurst et al. 1984; Dearman et al. 1987), onion (Brocklehurst and Dearman 1983b, 1984; Haigh et al 1987), pepper (Rivas et al. 1984; Passam et al. 1989; Bradford et al. 1990), tomato (Wolfe and Sims 1982; Haigh et al. 1986; Alvarado et al. 1987; Barlow and Haigh 1987; Argerich et al. 1990), Brussels sprouts (Khan et al. 1980/81), cabbage (Khan et al. 1981), kale (Rao et al. 1987), turnip (Rao et al. 1987), lettuce (Cantliffe et al. 1981; Valdes et al. 1985), cantaloupe (Bradford 1985), cucumber (Passam et al. 1989), melon (Passam et al. 1989), muskmelon (Bradford et al. 1988), and watermelon (Sachs 1977).

Seed osmoconditioning in liquid medium has proved less successful in improving the performance of large seeds. A 4–8-day osmoconditioning of soybean seeds in −0.86 to −1.19 MPa PEG 8000 solution, amended with 0.2% thiram, at 15°C improved the rate of germination and emergence at suboptimal temperatures in laboratory and growth chamber studies (Knypl and Khan 1981), but had little effect on seedling emergence or yield in several early spring field plantings under cold, wet conditions (Khan et al. 1980/81). In a field study with several soybean cultivars, Helsel et al. (1986) found that osmoconditioning in PEG promoted rapid and uniform emergence of seedlings in early plantings. The conditioning had little effect on yield, however. Bennett and Waters (1987a, 1987b) compared the effectiveness of seed osmoconditioning with two other partial seed hydration treatments, seed hardening and moisturizing on vermiculite (see matriconditioning in section IV. B), in improving sweet corn seed performance in cool soil. Seed hardening and moisturizing improved early emergence by as much as 20% while osmoconditioning lowered emergence.

2. Effect of drying. Drying reduces to some extent the advantages gained by osmoconditioning. Only a slight reduction in germination time occurred by air-drying osmoconditioned soybean seeds for up to 72 h at 25°C (Knypl and Khan 1981). The treatment, however, did not improve emergence or yield in field trials (Khan et al. 1980/81). Conditioning of parsley seeds in −1.2 MPa PEG solution reduced the T_{50} for radicle protrusion at 15°C from about 1 week to 1 or 2 days for surface dried, 2 or 3 days for air-dried and 1 week stored, and 3 or 4 days for 10-week stored seeds (Ely and Heydecker 1981). Air drying of wheat, barley, and sorghum seeds reduced the advantages gained by conditioning in −1.0 MPa PEG solution at 10°C (Bodsworth and Bewley 1981). Drying back PEG osmoconditioned leek seeds slightly delayed germination and emergence compared with seeds which had been conditioned but not dried back (Brocklehurst et al. 1984). When carrot, celery, and onion seeds after conditioning in PEG for 2 weeks were dried back at 30°C, germination was 0.2–0.7 day later than when they were dried back at 15°C (Brocklehurst and Dearman 1983a). Percentage emergence was unaffected by drying back osmoconditioned seed in onion but reduced in carrot and celery (Brocklehurst and Dearman 1983b). The conditioned and dried-back seeds emerged later than those seeds which were only conditioned by up to 1.5, 2.6, and 2.6 days in carrot, celery, and onion, respectively. Several other workers have found that drying back following conditioning of onion seeds reduces germination and emergence time and increases the time spread of emergence when compared to conditioned but undried seeds (Furutani et al. 1986; Haigh et al. 1986). Lettuce seeds conditioned for 6–9 h and air dried at 21°C had higher germination percentage and rate compared to seeds oven dried at 32°C (Guedes and Cantliffe 1980). These results indicate that drying (an essential prerequisite for storage and handling) after conditioning not only delays subsequent germination but might also be injurious to some seeds and should be conducted at relatively low temperatures in order to retain most of the advantages of seed conditioning.

3. Storage and aging. Seeds stored for various periods following osmoconditioning and drying retain most of the advantages of the treatment. In addition, osmoconditioning reverses some of the adverse effects of aging. Air-drying and storage for up to 28 days of tomato and carrot seeds following osmoconditioning in $K_3PO_4 + KNO_3$ at −1.6 MPa at 15°C

had no effect on subsequent emergence in the field (Haigh et al. 1986).
Air-drying of conditioned onion seeds, however, reduced the percentage
emergence. Conditioning onion seeds in PEG before subjecting them to
accelerated aging (40°C, 18% moisture content) delayed the loss of
viability due to aging (Dearmann et al. 1986). Osmoconditioning (7 days)
and humidyfying in moist air (3 days) of lettuce seeds were more effective than a 2-h soaking in cystein, KI, or thiosulfate solution in reducing
the frequency of chromosomal aberration, increasing the rate of root
growth, and decreasing morphological abnormalities due to seed
deterioration (Rao et al. 1987). Pepper seeds conditioned in −0.99 MPa
mannitol prior to storage at 5°, or 35°C for up to 6 months, retained the
capacity for high rate and final percentage germination (Georghiou et al.
1987, Thanos et al. 1989). Carrot and leek seeds conditioned in PEG for 10
and 14 days, and stored at 10°C for 12 months, retained the ability for
early germination; a 17-day conditioning had some adverse effect in
carrot after 12 months storage (Dearman et al. 1987).

Bradford and his co-workers have conducted a detailed study on the
relationship of osmoconditioning with seed vigor and performance.
Conditioning of tomato seeds in −1.25 MPa KNO_3 or PEG for 7 days
followed by drying retained the high viability during the 18-month
storage at 10° and 20°C in a sealed container at 6% moisture content
(Alvarado and Bradford 1988a). Conditioned seeds lost vigor and viability
more rapidly than the untreated seeds when stored at 30°C. In general,
osmoconditioned tomato seeds showed reduced storability despite
improved germination rate (Alvarado and Bradford 1988a; Argerich et al.
1989). Alvarado and Bradford (1988b) found that the effectiveness of
repriming to restore the germination performance of aged primed seed
depended on the degree of deterioration, on the initial advantage gained
due to first priming and on priming solution. Bradford et al. (1989)
recently reported that the conditioning of tomato seeds in salt solution (5
days) and of lettuce seeds in water (0.5 to 10 h) at 20°C followed by drying
did not extend their mean viability period. Their data indicated that the
ability to withstand thermal deterioration in the dry state declined during
imbibition.

Pill and Frett (1989) have developed a rapid and practicable way of
sowing seeds at predetermined spacing in hydroxyethylcellulose (HEC,
Natrosol 250 LR) plasticised with triacetin. After a 4-week storage in
HEC sheets, conditioned (in −1.0 MPa PEG, 15°C for 1 week) tomato
seeds performed better (earlier seedling emergence and increase in
seedling shoot weight) than the untreated seeds similarly stored in HEC
sheets. Taken together, these studies indicate that the basic advantages of
seed osmoconditioning remains intact in most seeds following drying and
short-term storage. Long-term storage, particularly under adverse conditions, could be more detrimental to the conditioned than to the untreated
seeds and should be avoided, if possible.

4. Prevention of thermoinhibition and thermodormancy. Thermo-dormancy in lettuce is a well-known phenomenon. When lettuce seeds are imbibed at supraoptimal temperatures germination is inhibited (thermoinhibition), and if kept at these temperatures for an extended period of time an induction of a secondary dormancy or thermodormancy occurs (Khan 1980/81). Presoaking of lettuce and a number of other seeds in water or PEG solution at low to moderate temperatures improves their ability to germinate at supraoptimal temperatures. Gelmond (1965) reported that presoaking leek seeds in water at 3°C for an extended period of time allowed the seeds to germinate better at supraoptimal tempera-tures of 25–30°C. Similarly, Heydecker (1973/74) reported that soaking pepper seeds in water at 10°C or treating lettuce seeds with PEG solution at 10°C allowed them to germinate at 30°C, a temperature that would have been too high for their germination. A prolonged osmotic conditioning of 'Grand Rapids' lettuce seeds in −0.84 MPa at 15°C for 14 days only slightly improved the seedling emergence at high temperatures (Khan 1977, 1978).

Short duration osmoconditioning at 15–18°C or presoaking in water at low temperatures could be highly effective in promoting germination of lettuce and other seeds at supraoptimal temperatures. A 1% K_3PO_4 or water soak for 6–9 h at 15°C greatly reduced the thermoinhibition of seeds of several lettuce cultivars, the salt soak being more effective (Guedes and Cantliffe 1980). Up to 86% germination at 35°C occurred when 'Minetto' lettuce seeds were soaked for 20 h in 1% K_3PO_4 or water at 15°C (Cantliffe et al. 1984). Conditioning with 1% K_3PO_4 of previously aged (3 or 5 days at 41°C and 100% R.H.) lettuce seeds was unsuccessful in over-coming thermoinhibition, indicating the necessity of using vigorous seeds if thermodormancy was to be avoided (Perkins-Veazie and Cantliffe 1984). Conditioning of 'Saladin' lettuce seed in 1% K_3PO_4 in water for 6–12 h had no effect on germination at 20°C but significantly increased the germination at 35°C; the conditioning in salt was more effective (Wurr and Fellows 1984).

The alleviation of high temperature stress by osmotic treatment has been effective in other seeds besides lettuce. In celery a 14-day seed conditioning at 15–20°C in −1.1ヌ MPa PEG solution improved the per-centage germination at 25° and 30°C, and in spinach seeds conditioned in −1.03 to −1.18 MPa PEG solution germination improved at 30°C (Nakamura et al. 1982). Similarly, Atherton and Farooque (1983) found that conditioning of spinach seeds at 10°C for 14 days in −1.25 MPa PEG solution improved the subsequent germination at 30°C. They also found that the treated seeds could be dried back and stored for at least 30 days at 5°C without the loss of the high temperature germination capacity. A conditioning of corn cockle seeds in PEG 4000 solution allowed them to germinate at 30°C (De Klerk 1986). The germination percentage of salvia seeds at 20° and 25°C was unaffected by conditioning in −0.8 MPa PEG

8000 solution at 15°C, but the percentage was greater at 10°, 15°, and 30°C. At 35°C, conditioned seeds had 44% and 65% germination, while the untreated seeds failed to germinate. Storage of conditioned seeds at 5°C, however, adversely affected the capacity for total or early germination (Carpenter 1989). Osmoconditioned (in −1.0 to −1.2 MPa PEG solution) and untreated dusty miller seeds showed no differences in germination percentages at 20° and 25°C, but conditioned seeds germinated 42% and 81% higher at 30° and 35°C, respectively (Carpenter 1990).

IV. SEED HYDRATION WITH MOISTENED SOLID CARRIERS

It has long been known that seeds restricted from germinating by high to moderate salt concentrations in the soil (resembling a solid carrier with osmotic properties) not only retain their full viability for extended periods of time but may actually show higher percentage germination (Ayers and Hayward 1948; Bernstein et al. 1955). Solid carriers such as sand, vermiculite, and peat moss have been extensively used in combination with water to after-ripen or stratify seeds to improve germination or to break dormancy (Barton 1965; Stokes 1965). However, the use of solid carriers in conjunction with seed and water to hydrate the seed in order to shorten the time and to improve the rate and the synchrony of germination is a relatively new area of research and has considerable merits (Bennett and Waters 1984, 1987a, 1987b; Kubik et al 1988; Taylor et al. 1988; Zuo et al. 1988a, 1988b; Berg et al. 1989; Khan et al. 1990; Parera and Cantliffe 1990). Research on solid carriers for conditioning seeds is motivated by and seems desirable for several reasons. Conditioning of seeds in polyethylene glycol solution, with low oxygen solubility and high viscosity, has not proved suitable for large seed, such as soybean, bean, and sweet corn. Poor performance of the treated seeds may be due to injury sustained during conditioning or during post-conditioning operations. Cost considerations and logistics of handling large quantities of seeds also favor solid carriers over PEG solution for seed conditioning. Inorganic salt solutions, too, are not ideally suited for seed conditioning as they are known to cause injury to seeds, particularly during long exposures.

Although several studies in the past have used solid carriers for hydrating seeds, only recently has it been shown that osmotic and matric component of the water potential can separately bring about seed conditioning (see sections IV. A and IV. B). Depending upon the physical and/or the chemical characteristics of the carrier substance, the water entering the seed can be regulated by the osmotic potential of the dissolved salts or other osmotically active solutes in the carrier matrix, by the matric or the physical forces of the carrier material itself, or by a combination of the two processes.

A. Osmotic Conditioning

Several approaches have been used to condition seeds on moistened solid carriers in order to improve their performance. Both osmotic and matric components of the water potential were likely involved in the conditioning of onion seeds in a slurry mixture of PEG 6000, vermiculite, and water, although the author did not make the distinction (Peterson 1976). The conditioning of red beet seeds directly in the moist soil after planting was achieved by a preplant incorporation of 1.1–3.95 mg PEG 8000 into a proprietary coating material (Moran) as indicated by an improvement in the emergence rate, final stand, and yield (Khan and Taylor 1986). Again, the relative contribution of the osmotic and the matric forces in the conditioning process was not determined.

Recently, a number of organic and inorganic carriers (e.g., vermiculite, sodium polypropionate gel, Agro-Lig [Leonardite shale], bituminous soft coal, expanded calcined clay, GrowSorb, Celite, Micro-Cel), in combination with water, have been used to condition seeds (Bennett and Waters 1984, 1987a, 1987b; Kubik et al. 1988; Taylor et al. 1988; Zuo et al. 1988a, 1988b; Berg et al. 1989; Khan et al. 1990; Parera and Cantliffe 1990). Both osmotic and matric forces may be involved in the conditioning of seeds on semi-solid carriers, such as sodium polypropionate gel (Zuo et al. 1988a, 1988b). When seed conditioning (termed solid matrix priming) is achieved largely by the solute potential or the osmotic component of the water potential, as was clearly the case with Agro-Lig (Taylor et al. 1988), the term osmoconditioning is appropriate. Seed conditioning in Agro-Lig was conducted at 60–95% of its own weight in water, and the water content of the carrier at the end of conditioning ranged from 35% to 50%. The amount of water needed to saturate Agro-Lig is 30–40%. As matric potential of a saturated matrix is zero (Hadas 1982), the conditioning was evidently achieved by the solute content of the matrix. This is consistent with the high osmotic potential of Agro-Lig (>97% of the total water potential) needed for conditioning (Taylor et al. 1988). The situation is analogous to osmoconditioning with PEG or salts on saturated solid matrix, such as filter paper and sand, with relatively low water holding capacity.

The term matriconditioning seems appropriate to describe controlled hydration in the presence of moistened carriers, with high matric component (high water retentive physical forces of the carrier matrix) and negligible solute component of the water potential, as was clearly the case with Zonolite vermiculite and Micro-Cel (Khan et al. 1990).

B. Matric Conditioning

Osmoconditioning or priming (liquid or solid), employs the osmotic properties of the solute(s) to condition seeds while matric conditioning or matriconditioning is dependent upon the matric properties of the carrier material. The matric potential component of the carrier matrix depends upon the carrier texture, structure, and water content. In a carrier matrix (or soil), the matric water potential component is derived from adsorptive, interfacial tension, attraction and adhesion between carrier matrix, matrix-air and matrix-water interfaces while osmotic water potential component is derived from solute concentration and composition in the matrix solution (Hadas 1982).

1. Early observations. Matriconditioning, like osmoconditioning, may have had its origins long before its use in preplant seed conditioning. Field trials were conducted in Enkhuizen, the Netherlands, in 1978 with untreated and coated lettuce seeds (Royal Sluis's Splitkote procedure) under relatively dry (only 4 mm rainfall during 9 days after first planting when emergence counts were made) and moderate temperature conditions; seed coating resulted in marked improvement in seedling emergence (see Table 4.6). The coating material, designed for outdoor planting, may have served as a solid matrix easily blending into the soil matrix, thus bringing about seed matriconditioning and the consequent improvement in seedling emergence. Previous studies with beet seeds indicated that when a period of relative dryness of the field soil after planting was followed by a wet period, there was a resurgence of seedling emergence (Khan et al. 1983). This indicated that the dry or water unsaturated soil matrix might create a low water potential microenvironment favoring partial hydration of seeds in a manner similar to that achieved by osmotically active substances during preplant osmoconditioning. As the water potential (matric potential + osmotic potential) of most soils is predominantly a function of the matric properties of the solid phase, the possibility of developing an efficient preplant seed conditioning treatment which would utilize the matric potential component of the solid substances has been explored.

2. Characteristics of solid substances needed for matriconditioning. The substances for matriconditioning should ideally: (a) have a high matric potential and a negligible solute or osmotic potential; (b) have negligible water solubility; (c) have low chemical reactivity; (d) have high water holding capacity, high flowability, capacity to remain dry, free flowing powder; (e) have variable particle size, structure, and porosity; (f) have high surface area; (g) have high bulk value and low bulk density, providing results at very low levels of addition; and (h) have the ability to adhere to seed surface. Among the substances which possessed these characteristics were the various grades of Celite (diatomaceous silica) and Micro-Cel (produced by hydrothermal reaction of diatomaceous silica, hydrated lime and water), the proprietary products of Manville; and the finer grades of Zonolite vermiculite (vermiculite is hydrated magnesium-aluminum-silicate; vermiculite concentrate is exfoliated to produce Zonolite vermiculite), the proprietary products of W. R. Grace and Company (Khan et al. 1990; Khan, unpublished) (see section IV. B. 5 for further discussion of their properties). Other solid carriers, including other grades of vermiculite (Terra-Lite, vermiculite #3), expanded calcined clay, and GrowSorb (the origin or characteristic of this material is not fully described) (Bennett and Waters 1984, 1987a, 1987b; Kubik et al. 1988; Berg et al. 1989; Callan et al. 1990; Parera and Cantliffe 1990) have also been used to condition vegetable seeds; however, the matric potential component of these substances needed for seed conditioning has not been reported.

3. Development of the procedure for matriconditioning. Micro-Cel E and Zonolite vermiculite (grades #4 and #5) were used for matriconditioning of several vegetable and flower seeds (Khan et al. 1990). These substances have a high matric potential and little or no solute or osmotic potential. Seeds were matriconditioned at 15°C for 7–14 days in the light in loosely capped glass jars by mixing seeds with water (or aqueous solution or suspension of the bioactive chemicals) and the carrier (Micro-Cel E or zonolite vermiculite). The seeds and the carrier rapidly reached a moisture equilibrium; thereafter little or no water was taken up by the seed. The water lost by evaporation during the entire conditioning period was less than 6% of the water added to the conditioning mixture. The type of seeds and the carriers used, the ratio of seed to carrier to water, and the percentage of water remaining in the seeds and the carrier matrix at the termination of matriconditioning are shown for a number of seeds in Table 4.4. The relationship between matric potential and the water holding capacity of Micro-Cel E and Zonolite vermiculite #5 has been determined by a pressure plate extractor. When this relationship was extrapolated to the data in Table 4.4, the water contents of Micro-Cel E, in equilibrium with the water content of various seeds at the end of matriconditioning, corresponded to the matric potentials ranging from −0.4 to −1.5 MPa (Khan et al. 1990), well within the range needed for seed conditioning. The matric potential needed for matriconditioning was similarly estimated for Zonolite vermiculite. Matriconditioned seeds were planted, with or without washing and with or without drying by forced air at 25°C. Emergence was determined in Cornell Peat-Lite Mix at 12 h, 20°C day/10°C night temperature regime.

The matriconditioning procedure (Khan et al 1990) resembles the procedure described by Kubik et al. (1988) (termed solid matrix priming) but differs significantly in several ways: (A) The ratio of seed to carrier was 1:1 in the procedure of Kubik et al. compared to 1 part of seed to 0.2–0.4 part of carrier (except celery seeds in which 1 part seed to 0.8 part carrier was used) by weight in the matriconditioning procedure. (B) Kubik et al. used nearly the same weight of carrier and water (in level 3 SMP), while the weight of water was 3–6 times greater than the weight of Micro-

Table 4.4. The type of seeds and the carriers used, the ratio of seed to carrier to water by weight, the duration of conditioning and the equilibrium water content of the seed and the carrier at the end of conditioning (Khan et al. 1990).

Seed cv.	Carrier	Seed: Carrier: Water ratio	Duration of conditioning (days)	Water content (%) seed[z]	carrier[y]
Red Beet 'Cardenal'	Micro-Cel E	16: 3.2: 18.4	7	46	323
Red beet 'Cardenal'	Vermiculite #5	16: 4.8: 16.0	7	42	177
Sugarbeet 'E-4'	Micro-Cel E	16: 3.2: 12.0	7	41	145
Sugarbeet 'E-4'	Vermiculite #5	16: 4.0: 10.0	7	36	87
Onion 'Texas early grano'	Micro-Cel E	16: 4.8: 20.0	7	71	163
Tomato 'Jackpot'	Micro-Cel E	16: 4.8: 24.0	7	84	197
Pepper 'Rino'	Micro-Cel E	16: 4.8: 24.0	7	81	210
Carrot 'Nantes'	Micro-Cel E	16: 8.0: 32.0	7	88	287
Celery 'FM 1218'	Micro-Cel E	16: 12.8: 64.0	14	86	368
Lettuce 'Mesa 659'	Micro-Cel E	16: 4.0: 24.0	0.83	88	242
Snap bean 'BBL 47'	Micro-Cel E	20: 8.0: 26.0	2	63	162
Snap bean 'BBL 47'	Vermiculite #5	20: 8.0: 20.0	2	49	123
Soybean 'Hodgson'	Micro-Cel E	20: 8.0: 24.0	2	76	106
Soybean 'Hodgson'	Vermiculite #5	20: 8.0: 20.0	2	62	92

[z]Initial moisture content of seeds, 5–7%.
[y]Initial moisture content of Micro-Cel E, 4.3% and of Zonolite vermiculite #5, 0.1%.

Cel E and 2.5–3.3 times greater than the weight of Zonolite vermiculite #5 in the matriconditioning procedure. (C) In the matriconditioning procedure, the water content of the seed remained high after reaching the moisture equilibrium. In the procedure of Kubik et al. the moisture content of the seed continued to decline during conditioning with most of the water lost by 7–10 days. (D) The conditioning temperature of 25°/20°C used by Kubik et al. was higher than the 15°C used for matriconditioning of seeds.

Drying of the seeds or carrier during conditioning might reduce the moisture equilibrium water content of the seed and carrier that would adversely affect the conditioning process. Decreasing the water potential of PEG solution to a value lower than approximately -1.2 MPa reduced its conditioning effect presumably by limiting the water necessary for conditioning of soybeans (Knypl and Khan 1981). It would be interesting to compare how drying during and following conditioning influences the improvements by matriconditioning.

4. Improvements in seedling emergence. The effect of matriconditioning of a number of vegetable seeds in Micro-Cel E on seedling emergence and fresh weight of seedling tops are shown in Table 4.5. Untreated seeds and the osmoconditioned seeds (conditioned in -1.2 MPa PEG 8000 solution for the same duration as those matriconditioned in Micro-Cel E) were included for comparison. In all cases, matriconditioning was superior to osmoconditioning in PEG solution. Drying (air-drying for 2 h at 25°C) after matriconditioning had little effect on total emergence in red beet, sugarbeet, and onion, but significantly reduced the total emergence, the T_{50} for emergence, and the fresh weight of seedlings in carrot, celery, and pepper. Drying had no effect on total emergence in tomato, but the T_{50} for emergence and the fresh weight of seedlings were reduced. The pattern of performance of matriconditioned seeds in the Peat-Lite Mix did not change after one month of storage at 7°C and 28% R.H. (Khan, unpublished).

5. Advantages of matriconditioning and differences from osmoconditioning.
   a. *High matric potential.* In matriconditioning, seed hydration is controlled by the water retentive physical forces of the carrier matrix. In contrast, seed hydration during osmoconditioning is controlled by the osmotic forces of the dissolved substances in the carrier.
   b. *Good water holding capacity.* The carriers used for matriconditioning, such as Micro-Cel E and Zonolite vermiculite, have high water holding capacity, thus requiring a low carrier:water ratio by weight (1 part Micro-Cel E to 3–6 part water; 1 part Zonolite vermiculite #5 to 2.5–3.3 part water) for conditioning seeds (Khan et al. 1990). Much higher carrier:water ratio by weight (1 part carrier to 0.6–0.95 part water) is needed for a solid carrier, like Agro-Lig, that relies predominantly on high solute content for seed conditioning (Taylor et al. 1988). The water status of the seed is not likely to change much due to a small loss of water during seed conditioning when using a carrier such as Micro-Cel E. The water holding capacities of Agro-Lig, Zonolite vermiculite #5, and Micro-Cel E at 0.01 MPa were estimated to be 32%, 450%, and 560%, respectively. The relative water holding capacities (or the amounts of water needed to achieve liquid consistency) of some of the carriers used for seed conditioning are shown in Figure 4.1.
   c. *Insoluble in water.* Materials found successful for matricondi-

Table 4.5. The effect of matriconditioning in Micro-Cel E on the performance of vegetable seeds at 20°/10°C (Khan et al. 1990).

Seed cv.	Treatment[z]	Total emergence (%)	T_{50}^y (d)	Top fresh wt.[x]
Red beet	Matriconditioned	155a[w]	3.8c	1.29a (13)
'Cardenal'	Matriconditioned + Dried	156a	3.9c	1.25a (13)
	−1.2 MPa PEG 8000	140b	5.5b	1.03b (13)
	Untreated	131b	7.5a	0.81c (13)
Sugarbeet	Matriconditioned	88ab	2.3c	1.46a (13)
'E-4'	Matriconditioned + Dried	95a	3.4b	1.38a (13)
	−1.2 MPa PEG 8000	88ab	3.6b	0.94c (13)
	Untreated	82b	4.9a	0.72d (13)
Onion	Matriconditioned	98a	3.9c	0.62a (15)
'Texas early grano'	Matriconditioned + Dried	97a	4.0c	0.61a (15)
	−1.2 MPa PEG 8000	92b	6.8b	0.48b (15)
	Untreated	93b	7.9a	0.36c (15)
Tomato	Matriconditioned	95a	4.3d	1.32a (15)
'FM jackpot'	Matriconditioned + Dried	94a	5.3c	1.12b (15)
	−1.2 MPa PEG 8000	88a	8.2b	0.80c (15)
	Untreated	89a	11.8a	0.69d (15)
Pepper	Matriconditioned	96a	7.4b	2.26a (21)
'Rino'	Untreated	82b	14.1a	1.18b (21)
Carrot	Matriconditioned	88a	5.0b	0.58a (18)
'Nantes'	Matriconditioned + Dried	74b	8.5a	0.42b (18)
	−1.2 MPa PEG 8000	78b	8.4a	0.36b (18)
	Untreated	89a	9.3a	0.31b (18)
Celery	Matriconditioned	92a	6.9c	0.19a (17)
'FM 1218'	Matriconditioned + Dried	68c	9.4b	0.10b (17)
	−1.2 MPa PEG 8000	86a	9.3b	0.06bc(17)
	Untreated	78b	13.8a	0.03c (17)

[z]Matriconditioning was conducted at 15°C in light in a mixture of seed:carrier:water as shown in Table 4.4. After matriconditioning seeds were planted with or without air-drying in Cornell Peat-Lite Mix. Seeds conditioned in PEG 8000 were washed and wipe-dried before planting. Emergence recorded at 12-h day, 20°/10°C, night temperature regime.
[y]Time to 50% emergence.
[x]Fresh weight of 15 tops. Data in parentheses are days after planting.
[w]Means in columns for each seed type separated by DMRT at 5% level.

tioning, such as the various grades of Micro-Cel, Celite, and Zonolite vermiculite, are water insoluble. In contrast osmoconditioning depends upon ionizing and nonionizing solutes, such as salts, alcohols, sugars, and nonpenetrating organic molecules.

   d. *Good inoculation and delivery system.* Micro-Cel and Zonolite

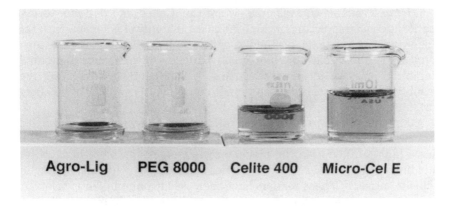

Figure 4.1.   Relative water holding capacities (or water needed to achieve liquid consistency) of the same weight of some of the carriers used for preplant physiological seed conditioning (Khan, unpublished).

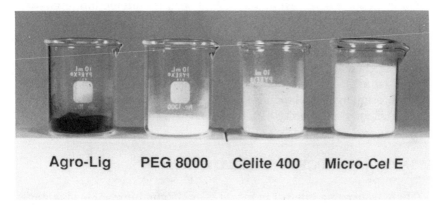

Figure 4.2.   The relative volumes occupied by the same weight of some of the carriers used for physiological seed conditioning (Khan, unpublished).

vermiculite have characteristics[1] that make them ideal carriers for liquid inoculants and for multiplying beneficial microorganisms (bacteria and fungi) around the seed and for delivering them to target areas in the rhizosphere for the control of destructive soilborne insects and diseases during seedling establishment.

e. *High bulk value.* Because of the high bulk value (e.g. bulk densities of Micro-Cel E and Zonolite vermiculite #5 are 88 and 120 kg/m^3, respectively) of the carriers used for matriconditioning, a small amount of carrier relative to seed can be used (Khan et al. 1990), facilitating treatment, handling, and transport of large amounts of seeds. The relative volumes occupied by the same weight of some of the carriers used in seed conditioning are shown in Figure 4.2.

f. Large available surface area. Micro-Cel, Celite, and Zonolite vermiculite have characteristics[1] which make them highly efficient carriers for
incorporating nutrients, pesticides, and hormones. The high surface area
appears to allow ready exchange and diffusion of toxic substances and
inhibitors present in some seeds (e.g., sugar beet), thus diluting the
inhibitor content of the seed and consequently improve seedling
establishment.

g. Enzyme and microbe immobilization. Use of carriers as bio-catalysts
and for seed conditioning requires that they be relatively inert chemically,
have high surface area, be mechanically stable, and have high water
holding capacity. These characteristics are found in celite and synthetic
silicate carriers[1]. These carriers have been used to immobilize a variety of
enzymes (amylases, proteases, and lipases have been immobilized, personal communication from Dr. Mark Jones, Manville Corporation), cells,
and microbes[1]. Further development and judicious use of such
biocatalyst carriers may improve both seed conditioning (e.g., by
mobilizing seed reserves) and seed and seedling protection (e.g., with
beneficial microbes).

h. Blends with soil. There is no need to remove the carrier adhering to
the seed surface following matriconditioning. The carrier particles surrounding the seed easily blend into the soil, which has similar matric
properties as the carrier substance. Upon planting, the excess amount of
water present in the soil reduces the matric forces of the carrier, restoring
the water uptake capacity of the seed needed for germination.

V. INTEGRATION OF SEED TREATMENTS

There is growing interest in recent years in combining the advantages
of various physiological and non-physiological treatments in a variety of
ways to maximize improvement in germination and seedling establishment. An integrative approach satisfies the most pressing demands of a
growing seedling.

A. Osmotic or Matric Conditioning with Pregermination

Studies with parsnip seeds showed that osmoconditioning or
osmoconditioning followed by pregermination improved the percentage
seedling emergence when compared to the untreated and pregerminated

---

[1]Micro-Cel synthetic calcium silicate functional fillers, Bulletin FF-427, 1985; Celite
catalyst carriers for biomolecule, enzyme, and cell immobilization, Bulletin FF-410, 1990;
Functional fillers for industrial applications, Bulletin FF 396, 1986; Manville Sales
Corporation, Wayne NJ 07470. Vermiculite the mineral for the '80's, Bulletin V 102, 1983;
Zonolite vermiculite, Bulletin V 104, 1985; W. R. Grace & Co., Cambridge, MA 02140.

seeds (Gray et al. 1984). Although both osmoconditioning and pregermination reduced the mean emergence time, the effect of the combined treatment, in general, was greater than either treatment alone. Germinating (for 4 days at 15°C in aerated water) the osmoconditioned (in −1.2MPa PEG 8000 for 3 weeks at 15°C) parsley seeds before fluid drilling decreased the time to 50% emergence by 52% and increased the shoot weight by 192% relative to raw seed performance (Pill 1986). Although the percentage emergence was unaffected by the various treatments, the T_{50} for germination was affected in the order: raw seed>osmoconditioned air-dried seed>osmoconditioned fluid drilled seed>osmoconditioned, germinated fluid drilled seed. A similar trend was found in shoot growth improvement. In a recent study, increased grain yields were observed when fluid drilled soybeans were previously conditioned (solid matrix primed) on GrowSorb (Berg et al. 1989).

B. Stratification with Osmotic Conditioning

Stratification or moist-chilling of seeds is a dormancy breaking treatment that restores full germinability and is generally conducted at 1–10°C for prolonged periods of time (Stokes 1965; Lewak and Rudnicki 1977). In contrast, osmoconditioning reduces germination time and invigorates the seed. However, the two physiological treatments can be combined to further improve germination. Moist-chilling treatment of perennial goosefoot seeds at 4°C for 28–30 days allowed the seeds to germinate nearly 100% (Khan and Karssen 1980). A 7-day osmoconditioning of moist-chilled seeds in the light in −0.86 MPa PEG solution reduced the time to germination, while the osmotic treatment in the dark increased the time to germination and reduced the rate of germination and eventually led to the induction of a secondary dormancy. Stratification of loblolly and short leaf pine seeds followed by osmoconditioning in −0.8 to −1.0 MPa PEG solution reduced the germination time when compared to seed given only a stratification treatment (Hallgren 1989).

Other dormancy breaking treatments can also be combined with osmoconditioning. In true potato seed, nondormancy by prolonged storage was a requirement for effective osmoconditioning (in −1.0 MPa KNO_3 + K_3PO_4 in light) (Pallais and Fong 1988).

C. Growth Regulator with Seed Coating

Seed coating and pelleting technology have been developed to suit different needs: for seed singulation; for adding bioactive chemicals, nutrients, and beneficial microbes; and for sowing seeds under dry and wet conditions (Millier and Sooter 1967; Johnson 1975; Halmer 1988). Several studies have examined the advantages of growth regulator permeation into seeds (via aqueous soak or via organic solvents, such as

Table 4.6.   Effect of fusicoccin (FC) and cotylenin E (CN), with or without seed coating (Splitkote), on seedling emergence in 'Grand Rapids' lettuce seeds in two field plantings (Khan, unpublished).

	Emergence (%)	
Treatments	Planting 1[z]	Planting 2[y]
Untreated	8c[x]	—
FC	43b	—
CN	71a	—
Untreated + Splitkote	53b	35b
FC + Splitkote	66ab	50a
CN + Splitkote	66ab	50a

[z]300 seeds/row, 3 replicates planted July 18, harvested July 27, 1978.
[y]260 seeds/row, 6 replicates planted August 14, harvested August 29, 1978.
[x]Means in columns separated by DMRT at 5% level. Seeds were planted in 5 m rows, 2 cm deep. The average soil temperature was approximately 20°C and the soil was drier in the 2nd trial than in the first.

acetone and dichloromethane), with or without seed coating (Palevitch and Thomas 1974; Khan et al. 1976, 1979; Khan 1977; Thomas et al. 1978; Sachs et al. 1981, 1982). In several instances, seed coatings have been shown to delay germination by interfering with oxygen and gas exchange; this effect can be alleviated to varying degrees by incorporating growth regulators into seeds prior to coating them (Khan et al. 1977, 1979; Thomas et al. 1978; Sachs et al. 1981, 1982).

Certain types of coatings could also have a promotive effect on seedling emergence. In 1978, replicated field trials with 'Grand Rapids' lettuce seeds compared the effects of seed coating (Royal Sluis Splitkote); permeation of seeds with fungal toxins fusicoccin (0.5 mM) and cotylenin E (0.5 mM) via acetone; and the permeation of growth regulators followed by seed coating. The results of two field plantings are shown in Table 4.6 (Khan, unpublished). Both the seed coating and the application of growth regulators improved the performance when compared to untreated raw seeds (planting #1). An improvement over that by coating alone was obtained when growth regulators were combined with coating in both plantings. In a recent study, an improvement in germination of rice seed was accomplished by encapsulating the seed in a gel coating that included an air bubble (Kouno 1989).

D. Osmotic Conditioning with Seed Coating

Several attempts have been made to combine the advantages of osmoconditioning with those of seed coating. 'Empire' lettuce seeds coated with a clay based material (Royal Sluis Splitkote D) subsequent to osmoconditioning (24 h in −1.5 MPa PEG 8000 solution at 15°C) emerged

46–69% compared to 18–21% for the untreated seeds under field conditions where the soil temperature exceeded 35°C (Valdes et al. 1985). The effect of osmoconditioning in producing germination in lettuce seeds at high temperatures (up to 37°C) was retained following coating and prolonged storage (Valdes and Bradford 1987). In tomato seeds, osmoconditioning plus coating (Splitkote Special) improved the seedling emergence over coating alone, but there was no improvement over osmoconditioned seeds (Bennett 1988).

E. Presoaking, Osmotic Conditioning, or Pregermination with Bioactive Chemicals

Improved seed performance has been achieved by incorporating plant growth regulators and pesticides during presoaking, osmoconditioning, and other presowing treatments (Sharples 1973; Palevitch and Thomas 1974; Braun and Khan 1976; Khan 1977; Thomas et al. 1978; Khan et al. 1979). Inclusion of growth regulators, such as gibberellin (GA), during prolonged dark osmotic treatment in PEG solution reduced the time to germination, improved the rate of germination, and prevented the induction of secondary dormancy in lettuce, perennial goosefoot, and curly dock seeds (Khan and Karssen 1980; Khan and Samimy 1982; Hemmat et al. 1985). Seeds of several lettuce cultivars were osmoconditioned for 24 h in −1.2 MPa PEG 8000 solution amended with 10 mM ethephon or 10 mM ethephon plus 0.01 mM kinetin followed by drying in forced air at 25°C. The combined treatment proved highly effective in improving the emergence of lettuce seeds at high temperature regimes of 35°/25°C and 37°/30°C (Prusinski and Khan, unpublished). In a study with carrot seeds, an inclusion of 200 mg.liter^{-1} N-substituted phthlimide during a 14-day soak in −1.0 MPa PEG solution at 15°C further enhanced the improvements by osmoconditioning (Pill and Finch-Savage 1988).

Pesticides and nutrients have been effectively combined with the seed hydration treatments. Osmoconditioning or pregermination alone has been found to improve seedling emergence in disease infested soil (Taylor et al. 1985; Bradford et al. 1988; Osburn and Schroth 1988). Improvements may result from decreased leakage of exudates in osmoconditioned seeds (Osburn and Schroth 1988). Several fungicides, including metalaxyl, etridiazole, and captan, incorporated into magnesium silicate gel controlled damping-off in tomato seedlings caused by Pythium aphanidermatum (Giammichele and Pill 1984). The addition of metalxyl during osmoconditioning of sugarbeet seeds in NaCl or PEG resulted in an additive increase in the control of preemergence damping-off by Pythium ultimum, greater than in any of the treatments alone (Osburn and Schroth 1989). Improvements in the growth of carrot seedlings have resulted by the addition of nutrients, such as monosodium phosphate and ammonium polyphosphate, to gels used for fluid drilling (Finch-Savage and Cox 1982).

F. Matric Conditioning with Growth Regulators

Matriconditioning in Micro-Cel E and Zonolite vermiculite #5 combined with growth regulators improved seed performance of a number of
species (Khan et al. 1990). The matriconditioning of 'BBL 47' snap bean
seeds in Micro-Cel E, with or without the addition of GA_3 (matriconditioning mixture by weight:20 part seed:8 part carrier:26 part water or
0.001 mM GA_3), at 15°C for 2 days, improved performance (Fig. 4.3). The
T_{50} for emergence of the treated seeds planted in Cornell Peat-Lite Mix at
20°, 12 h day/10°C, night regime was 5.5 days compared to 6.8 and 8.6
days for matriconditioned and untreated seeds, respectively. The fresh
weight of seedling tops cut at the soil surface 15 days after planting differed significantly and were 1.49, 1.40, and 1.30 g per plant in matriconditioned plus GA_3, matriconditioned, and untreated seeds, respectively.
Similar results were obtained when Zonolite vermiculite #5 was used for
matriconditioning (matriconditioning mixture by weight:20 parts seed:8
part carrier:20 part water or 0.001 mM GA_3). In a field trial at Geneva, NY,
conducted in the spring of 1990, matriconditioning of 'BBL 47' snap bean
seeds in the presence of 0.001 mM GA_3 reduced the time of emergence but
not the final percentage of seedlings that emerged and had no adverse
effect on seedling growth (Khan, unpublished).

Figure 4.3.   Improvement in emergence of 'BBL 47' snap bean seeds at 20°/10°C by
preplant matriconditioning in Micro-Cel E (Khan, unpublished).

The T_{50} for emergence of 'Hodgson' soybean matriconditioned for 2 days at 15°C in Micro-Cel E (matriconditioning mixture by weight:20 part seed:8 part carrier:24 part water or 0.001 mM GA_3) or Zonolite vermiculite (matriconditioning mixture by weight:20 part seed:8 part carrier:20 part water or 0.002 mM GA_3) was also reduced, again the combined effect of matriconditioning and GA_3 was greater than by matriconditioning alone (Khan et al. 1990).

Matriconditioning in Micro-Cel E with the addition of ethephon has proved effective in preventing thermoinhibition at supraoptimal temperatures in lettuce and celery seeds (Khan et al. 1990). In a detailed study with 'Mesa 659' lettuce seeds, a 20-h matriconditioning in Micro-Cel E with the addition of ethephon (matriconditioning mixture by weight:4g seed:1g carrier:6g water or 10 mM ethephon) at 15°C allowed 94% emergence from a peat-lite mix at 12 h, 37°C day/30°C night temperature regime. Emergence from seeds that received no treatment, matriconditioning alone, a 24-h osmoconditioning (in −1.12 MPa PEG solution), and osmoconditioning plus 10 mM ethephon were 0%, 84%, 34% and 71%, respectively. The effectiveness of matriconditioning plus ethephon in preventing thermoinhibition did not diminish even after 4 months of storage at 7°C and 28% R.H.

The influence of matriconditioning in Micro-Cel E (matriconditioning mixture by weight:1 part seed:0.8 part carrier:4 part water or 10 mM ethephon or 10 mM ethephon + 1 mM GA_3) and osmoconditioning on Agro-Lig (osmoconditioning mixture:1 part seed:1.5 part Agro-Lig:1.35 part water or 10 mM ethephon or 10 mM ethephon + 1 mM GA_3) in the presence or absence of hormones was studied in 'FM 1218' celery seeds (Khan et al. 1990; Khan, unpublished). After conditioning at 15°C for 7 days, the seeds were rinsed, blotted dry, and sown in Cornell Peat-Lite Mix at 26°C in continuous light. Seedling emergence, 12 days after planting, from seeds conditioned in Micro-Cel, Agro-Lig, Micro-Cel + ethephon, Agro-Lig + ethephon, Micro-Cel + GA, Agro-Lig + GA, Micro-Cel + ethephon + GA, and Agro-Lig + ethephon + GA were 76%, 53%, 72%, 56%, 78%, 77%, 79%, and 82% and the T_{50} for emergence were 5.7, 6.9, 5.7, 6.8, 5.6, 6.8, 5.7, and 6.7 days, respectively. Only 1% emergence occurred in untreated seeds. These results show that osmoconditioning or matriconditioning of celery seeds is an effective way to prevent thermoinhibition. While growth regulators combined with Micro-Cel E failed to improve emergence, the addition of GA to Agro-Lig or Agro-Lig + ethephon significantly enhanced the emergence (77–82%) over that achieved by Agro-Lig (53%) or Agro-Lig + ethephon (56%) alone. This may be related to improved activity of GA at the low pH of Agro-Lig (about pH 4.0).

A 9-day matriconditioning in Micro-Cel E (matriconditioning mixture by weight:1 part seed:0.8 part carrier:2 part water or 10 mM ethephon + 0.05 mM GA_3) of primrose seeds (Zaadunie Research, NL) reduced the T_{50}

for emergence from 17.2 days to 11 days and the addition of ethephon plus GA_3 further reduced the T_{50} for emergence to 9.8 days (Khan et al. 1990). Matriconditioning (matriconditioning mixture same as in primrose) of impatiens seeds in Micro-Cel E reduced the T_{50} for emergence from 12.2 days to 8.8 days and increased the shoot weight by about 50%. Addition of 0.05 mM GA_3 during conditioning had little effect on the T_{50} for emergence but doubled the shoot weight when compared to the untreated seeds (Khan et al. 1990).

G. Osmotic Conditioning, Matric Conditioning, or Pregermination with Beneficial Microbes

There is growing interest in manipulating the rhizosphere microflora to enhance the activity of certain beneficial microorganisms in order to control plant pathogens affecting seedling establishment. There are numerous reports on the application of beneficial microbes as seed coating to improve the health and the nutritional status of the growing seedlings. Among the more popular fungi that have been used to control damping-off and other plant diseases are Trichoderma harzianum (Elad et al. 1986; Mihuta-Grimm and Rowe 1986; Sivan et al. 1987; Smith and Wehner 1987), Pythium oligandrum (Al-Hamdani et al. 1983; Lutchmeah and Cooke 1985; Martin and Hancock 1987), and Chaetomium globosum (Walther and Gindrat 1988). Several beneficial bacteria, such as Bacillus subtilis (Gaikwad et al. 1987; Tschen 1987) and pseudomonads (e.g., Pseudomonas fluorescens and Pseudomonas putida) (Suslow and Schroth 1982; Iswandi et al. 1987; Digat 1988, 1989), have been applied as seed coating or as inoculants for the purpose of protecting the growing seedlings against soilborne fungi and insects; others, such as Rhizobia (Jawson et al. 1989) and Azospspirillum brasliense (Bashan 1986; Bashan et al. 1988), have been used to improve the nutritional status, growth, and yield of cereal and noncereal crops.

In several instances the application of beneficial microbes has been integrated with gel seedling and physiological seed conditioning to maximize seed performance, particularly under conditions of high disease incidence or to improve the nutritional status of the growing seedlings. Fluid-drilling a gel matrix containing ungerminated radish seeds and a 5-day-old culture of Trichoderma (antagonist and seeds added to gel just before gel seeding) was found superior to other delivery systems for the biocontrol of Rhizoctonia solani on radish in field studies (Mihuta-Grimm and Rowe 1986). Application of Trichoderma during conditioning (termed "solid matrix priming") of seeds in the presence of Agro-Lig, a solid carrier with high solute potential, was effective in controlling the preemergence damping-off caused by Pythium ultimum (Harman and Taylor 1988). The conditioning (termed "bio-priming") of sweet corn seeds on vermiculite in the presence of a naturally occurring strain of *Pseudomonas fluorescens* protected the seeds from preemergence damping-off caused by *Pythium ultimum* (Callan et al. 1990). In a field study, the inoculation of soybean seeds with *Bradyrhizobium japonicum* during conditioning (termed "solid matrix priming") on a solid carrier, GrowSorb, followed by gel seeding did not increase nodulation (Berg et al. 1989). This study, however, demonstrated that the delivery of *Bradyrhizobia* in gel was possible for fluid-drilled soybean.

VI. PHYSIOLOGICAL AND BIOCHEMICAL ASPECTS

Various physiological and biochemical changes occur in seeds during or as a result of seed hydration. Most studies have been made with seed osmoconditioning, and it is likely that similar (although not identical) changes might occur during other hydration treatments, such as matriconditioning, short duration water soak, humidification, and hardening. In giant foxtail seeds, for example, either a brief exposure to 0.3 MPa PEG solution, a slow hydration in humid air, or drying after partial hydration improved the germination in a similar fashion (Taylorson 1986).

A. Physiological Changes and Practical Implications

Factors such as hormones, quality and quantity of light, low and high temperature, accessibility to oxygen, and the level and length of hydration profoundly influence the degree to which a seed can be conditioned. Depending upon the interaction with these factors, a seed may become dormant, develop high germination potential, show rapidity and improved synchrony in germination, acquire tolerance for cold temperature, or develop a capacity to alleviate high temperature.

1. Enhancement in germination potential. Physiological and biochemical changes occur during osmoconditioning that allow seeds to develop a high germination potential (radicle thrust) or the ability to remove the seed coat restraint. This might be the basis for an improvement in the rate of germination and emergence often seen in osmoconditioned seeds. The improved germination potential can be demonstrated readily by germinating the conditioned and untreated seeds in solutions of PEG with a progressively lower osmotic potential to counter the radicle thrust. Embryos of lettuce seeds excised after osmoconditioning in −1.2 MPa PEG 8000 solution for 10 days at 15°C showed an increase in the radicle thrust at 25°C which was 0.4 to 0.6 MPa greater than the thrust generated by the embryos from seeds soaked in water for 6 h (nonconditioned seeds) (Khan and Samimy 1982). Similarly, osmoconditioning of parsley seeds in PEG solution at 25°C improved the germination potential as indicated by their ability to more effectively counter the water stress (Akers et al. 1987).

'Emperor' lettuce seeds osmoconditioned for 24 h in −1.2 MPa PEG 8000 solution at 15°C developed a germination potential (equivalent to water potential of PEG solution allowing 50% germination) of 0.26 MPa during soaking at 35°C. The addition of 20 mM ethephon during osmoconditioning further enhanced the germination potential to 0.53 MPa (Fig. 4.4) (Prusinski and Khan, unpublished). Similarly, the addition of 0.1 mM kinetin during a 24-h osmoconditioning enhanced the germination potential to 0.48 MPa. Only 20% to 30% of the untreated seeds germinated in water at 35°C (zero germination potential). The ability of osmoconditioned seeds, particularly those conditioned in the presence of ethephon and/or kinetin, to generate high germination potential appears to be the basis for the alleviation of a wide variety of stresses imposed by supraoptimal temperatures, salinity, and drought in lettuce and perhaps

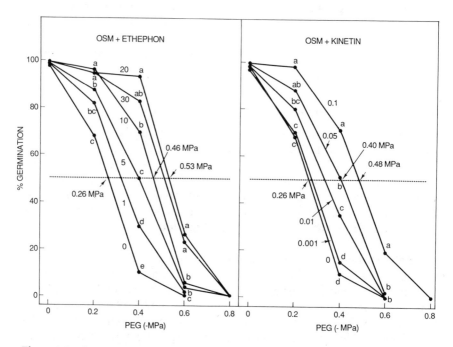

Figure 4.4. Improvement in germination potential at 35°C of 'Emperor' lettuce seeds by a 24-h osmoconditioning (OSM) in −1.2 MPa PEG solution and a further improvement upon the addition of various concentrations of ethephon and kinetin during osmoconditioning. After conditioning seeds were rinsed and germinated for 48 h in PEG 8000 solutions of progressively decreasing water potential of 0 to −0.8 MPa at 35°C. Germination potential at 35°C was equivalent to water potential of PEG solution allowing 50% germination. Untreated seeds germinated less than 50% in water (zero germination potential) (Prusinski and Khan, unpublished).

other seeds. The development of a high radicle thrust during osmoconditioning may be related to activation and/or synthesis of numerous enzymes, increased metabolic activity, and the production of osmotically active substances.

2. Role of gibberellin, ethylene and cytokinin. The advent of polyethylene glycol for germination enhancement studies provided a great opportunity for investigating the role of plant hormones in controlling germination under stressful conditions. The observations that germination is similarly influenced by the osmotic potentials of PEG and the matric potentials of soil (Hadas 1976, 1977) greatly aided such studies. Early studies with seeds of lettuce, celery, perennial goosefoot, and curly dock indicated that GA added to PEG solution replaced the light effect in reducing germination time and in preventing dormancy induction (Khan and Karssen 1980; Khan et al. 1980/81; Hemmat et al. 1985). The promotive effect of GA during osmoconditioning of lettuce seeds was not influenced by the addition of abscisic acid (ABA) in the PEG solution (Khan and Samimy 1982). This lent credence to the hypothesis that hormones have at least two sites of action in the control of germination, a primary site (GA site), and a preventive-permissive site (secondary site) at which cytokinin-ABA interaction might occur (Khan 1971).

Aminoethoxyvinyl glycine (AVG), a specific inhibitor of the conversion of S-adenosyl methionine to 1-aminocyclopropane-1-carboxylic acid (ACC) in the ethylene biosynthetic pathway (Yang 1984), failed to inhibit the germination advancement effect of osmoconditioning in the light or in the presence of GA in the dark (Table 4.7), indicating that ethylene biosynthesis was not involved in the germination advancement effect of seed osmoconditioning. These results support the suggestion that the action of GA is directly related to germination while the role of ethylene, cytokinin, and ABA is secondary, involving the removal or imposition of stress on germination.

There is evidence for the participation of ACC during the short duration presoaking of lettuce seeds in water (for 4, 6, or 9 h at 15–20°C) in the alleviation of thermoinhibition. The addition of 1 mM AVG during the presoak period prevented the relief of thermoinhibition; the effect was reversed by adding 1 mM ACC or 0.1 mM ethephon (Table 4.8). It has been reported previously that the utilization of ACC for ethylene production by lettuce seeds is greatly reduced under conditions of stress such as high temperature, salinity, or low water potentials (Khan and Huang 1988; Khan and Prusinski 1989). The same may be true for ACC biosynthesis in seeds kept continuously at supraoptimal temperatures. It seems that the activation of ACC synthase and/or the production of ACC occurs during the presoaking at 15–20°C, and that the ACC produced participates in the alleviation of thermoinhibition. Studies indicate that ACC synthesized during or shortly after the short duration presoaking at moderate

ANWAR A. KHAN

Table 4.7. Effect of aminoethoxyvinyl glycine (AVG) and/or 1-aminocyclopropane-1-carboxylic acid (ACC) applied during osmoconditioning in light or darkness plus gibberelin A_{4+7} on subsequent germination in water of 'Grand Rapids' lettuce seeds (Khan, unpublished).

Treatment[z]	Ethylene[y] $(nl.g^{-1})$	Germination[x] (%)	
		3h	7h
Light			
Untreated[w]	0c	0b	0b
Osmoconditioned	14b	95a	98a
Osmoconditioned + ACC	410a	95a	99a
Osmoconditioned + AVG	0c	91a	98a
Osmoconditioned + AVG + ACC	444a	93a	99a
Darkness			
Untreated + GA[w]	0c	0b	0b
Osmoconditioned + GA	19b	97a	99a
Osmoconditioned + GA + ACC	407a	97a	98a
Osmoconditioned + GA + AVG	0c	93a	99a
Osmoconditioned + GA + AVG + ACC	390a	97a	98a

[z] Seeds were osmoconditioned in −1.2 MPa PEG 8000 solution for 4 d at 15°C in the presence of 1 mM ACC, 1 mM AVG, 1 mM GA_{4+7} (in darkness), or combinations thereof.
[y] For ethylene determination 0.5g seeds (original wt.) were rinsed in water and incubated in light or darkness for 6 h at 25°C in tubes sealed with rubber septa. Ethylene was determined in 1 ml gas samples.
[x] Seeds were washed and germinated in water at 25°C in light and in darkness.
[w] Untreated dry seeds were soaked in water or GA_{4+7} for ethylene and germination determination.

Table 4.8. The relief of thermoinhibition of 'Emperor' lettuce seeds by a 4-h water presoak at 15°C, its inhibition by aminoethoxyvinyl glycine (AVG), and the reversal of the AVG inhibition by 1-aminocyclopropane-1-carboxylic acid (ACC) and ethephon (Khan, unpublished).

4 h Presoak at 15°C[z]	48 h Soak at 35°C	Seeds germinated (%)
None	Water	20
Water	Water	93
Ethephon[y]	Ethephon	94
AVG	AVG	0
Water	AVG	54
AVG	Water	1
AVG	AVG + ACC	95
AVG	AVG + Ethephon	96

[z] After 4 h presoak at 15°C in various solutions seeds were rinsed with water and transferred to the same or other solutions at 35°C for additional 48 h soak.
[y] Concentrations: ethephon, 0.1 mM; AVG, 1 mM; ACC, 1 mM.

Table 4.9. Effect of no presoak and a 4-h presoak in water and ethephon at 15°C on seedling emergence of 'Mesa 659' and 'Emperor' lettuce seeds at a supraoptimal temperature regime (Khan, unpublished).

		Emergence (%)[z]			
		4-h presoak in ethephon			
Cultivar	No presoak	0mM	0.1mM	1.0mM	10mM
Emperor	9	49	59	63	98
Mesa 659	0	0	0	1	81

[z]Dry seeds and seeds presoaked for 4 h in various concentrations of ethephon were planted in Cornell Peat-Lite Mix and emergence recorded after 6 days at 12-h day, 37°/30°C, night temperature regime.

temperatures participates in the alleviation of thermoinhibition. This process is distinctly different from the long-term advancement effect of osmoconditioning in which ethylene biosynthesis does not participate.

The effectiveness of short duration presoaking in water, or in solutions of ACC or ethephon, to prevent thermoinhibition in lettuce seeds on filter paper, where most of the studies have been conducted, is greatly reduced when the seeds are planted in a peat-lite mix. 'Mesa 659' lettuce seeds presoaked in 10 mM ACC or 10 mM ethephon for 1 h failed to emerge when planted in a peat-lite mix at 25°/35°C (Khan 1990). A 4-h presoak of 'Emperor' and 'Mesa 659' lettuce seeds in water or low concentrations of ethephon at 15°C was largely ineffective in preventing thermoinhibition (Table 4.9). High percentages of emergence occurred only in seeds soaked for 4 h in 10 mM ethephon. The responsiveness of 'Mesa 659' and 'Emperor' lettuce seeds to presoaking treatments was highly variable. A large variability in seeds of lettuce cultivars with respect to ethylene production has been found, and a good relationship exists among the cultivars between the ethylene producing capacity of the seeds and the ability to alleviate high temperature and other stresses (Prusinski and Khan 1990). Thus, it seems that both the production of ACC and its utilization is greatly reduced at high temperatures. Seeds of lettuce cultivars, such as 'Emperor,' with a greater ability to produce ethylene, appear to be better equipped to deal with high temperatures.

Aging of seeds decreases the ability to produce ethylene (Takayanagi and Harrington 1971; Samimy and Taylor 1983) and to convert ACC into ethylene (Khan and Seshu 1987; Jilani et al. 1989), and this is correlated with the decreased ability for germination and seedling growth. Aged lettuce and snap bean seeds showed reduced capacity to convert ACC to ethylene and to remove the seed coat restraint (Khan, unpublished). A short duration presoaking treatment prevented thermodormancy in unaged but not in aged lettuce seeds (Perkins-Veazie and Cantliffe 1984). This may be related to the reduced ability of aged seeds to produce ethylene and to develop a germination potential that was strong enough to remove the seed coat restraint.

Short duration hydration of seeds in water or osmotic solution had little effect on germination percentage at moderate temperatures of 20° and 25°C but was effective in alleviating high temperature effect on germination (Wurr and Fellows 1984; Carpenter 1989, 1990). This is consistent with the suggestion that ethylene does not participate in germination under normal conditions, and the requirement of ethylene biosynthesis for germination is induced only under stressful conditions, such as high temperature, salinity, osmotic stress, and seed coat restraint (Khan and Huang 1988; Khan and Prusinski 1989).

The germination patterns of weed seeds in soils of the temperate regions have been extensively studied. The induction of a secondary dormancy in such soils during spring and early summer is a common occurrence (Taylorson 1972; Karssen 1980/81). Hormones are able to break or prevent the induction of dormancy in seeds of witchweed, common ragweed, curly dock, perennial goosefoot, redroot pigweed, and others (Eplee 1975; Schoenbeck and Egley 1981; Samimy and Khan 1983a, 1983b; Hemmat et al. 1985). In some cases the seasonal germination and dormancy patterns of weed seeds in the field have been simulated by soaking them in PEG solution in light, hormones (in darkness), or darkness (Khan and Karssen 1980; Hemmat et al. 1985). The use of PEG solution has provided a convenient and rapid way to study the dormancy and germination pattern of weed seeds in both light and in darkness in the presence of various bioactive chemicals. These studies suggest that a judicious deployment of plant hormones such as GA, cytokinin, ethylene, or their combinations could be used to prevent the induction of seed dormancy and thus control weed seed populations in soil.

3. Light, darkness, and high temperature. Light does not appear to be necessary during short duration osmoconditioning (about 24 h) or humidification (1–5 days) in order to promote subsequent dark germination of lettuce seeds at inhibitory (20–25°C in 'Grand Rapids' lettuce seeds) or non-inhibitory (25°C in 'Mesa 659' lettuce seeds) temperatures (Hsiao and Vidaver 1973; Khan et al. 1980/81; Khan and Samimy 1982). Short duration presoaking in water or osmotic solution induces changes that allow seeds of several lettuce cultivars to germinate at 35°C in light or darkness (Guedes and Cantliffe 1980). Karssen and Weges (1987) working with 'Musette' lettuce seeds showed that a short duration presoaking in water or osmotic soak at 15°C in the dark improved the subsequent germination at 28° and 30°C in the dark.

Irradiation is essential during prolonged osmoconditioning of seeds with phytochrome control for reducing germination time (Khan et al. 1979, 1980/81). This is clearly shown in seeds of lettuce, celery, perennial goosefoot, and curly dock (Khan et al. 1979; Khan and Karssen 1980;

Khan et al. 1980/81; Hemmat et al. 1985). Keeping these seeds in osmotic solution in the dark for extended periods either induced a secondary dormancy or decreased the rate of germination to a level below that in the untreated seeds. The secondary dormancy can be induced during osmotic treatment in the light as well as in the dark in seeds such a curly dock, but the rate of induction of dormancy in the dark is more rapid than in the light (Samimy and Khan 1983a; Hemmat et al. 1985). In seeds of modern cultivars of carrot, beet, onion, bean, and maize, irradiation is not necessary during prolonged dark soaking; these seeds evidently lack a mechanism for dormancy induction (Khan and Zeng 1985).

4. Hydration level, dormancy, and performance. At the same water potential, the induction of dormancy as shown by suspension of germination or reduction in the rate of germination is more rapid in darkness than in light; at high than at low temperature; and after a long than a short imbibition period (Hsiao and Vidaver 1973; Khan and Samimy 1982; Samimy and Khan 1983a; Hemmat et al. 1985; Karssen and Weges 1987). The rate of germination decreased or dormancy intensified as the soaking time of lettuce seeds in moist air, water, or −1.2 MPa PEG 8000 increased (Hsiao and Vidaver 1973; Khan and Samimy 1982; Karssen and Weges 1987). In curly dock seeds soaked in water at 30°C, the rate of dormancy induction in the dark was more rapid than in the light, as indicated by subsequent germination at 20°C (Samimy and Khan 1983a). Curly dock seeds soaked in −1.57 MPa PEG 8000 solution at 25°C acquired a deeper dormancy than seeds soaked in the osmoticum at 15°C.

Changes in the water potential of PEG 8000 solution has a profound effect on dormancy induction and performance of seeds. In curly dock, the dormancy intensified with an increase in the concentration of PEG 8000 from 0 to 45g/100 ml water (−2.1MPa) in the imbibing medium at 25°C in the light as determined by subsequent germination at 20°C in the light (Samimy and Khan 1983a). Further increase in PEG concentration led to a decrease in the intensity of dormancy. There was a gradual decrease in the intensity of dormancy induction in 'Grand Rapids' lettuce seeds with an increase in PEG 8000 concentration from 0 to −1.23 MPa (Karssen and Weges 1987). These studies indicate that seed performance is profoundly influenced by the seed characteristics, the hydration level, and the interaction of hydrated seeds with a variety of environmental factors.

5. Resistance to imbibitional and chilling injury. Imbibitional injury to seeds is generally associated with a rapid entry of water and is greatly influenced by seed moisture content, temperature and water potential of the imbibing medium (Pollock 1969; Hobbs and Obendorf 1972; Hadas 1976). The physiological basis for the imbibitional injury has been extensively studied in soybeans by Vertucci and Leopold (1983, 1984). At moisture content below 8%, water was strongly bound to the tissues which rendered the seed susceptible to imbibitional injury. They hypothesized that the initial wetting reaction rather than the long-term imbibitional rate was linked to chilling injury.

The phenomenon of imbibitional injury is a well-known characteristic of chilling-sensitive seeds, such as beans, maize, soybean, and cotton. Reduction in the hydration rates due to seed coat characteristics, by soaking in soil or soilless media with low matric or osmotic potential, or by coating with hydrophobic materials, often improves seed performance (El-Sharkawi and Springuel 1977; Priestley and Leopold 1986; Taylor and Dickson 1987). Aged seeds or seeds with damaged seed coats have been shown to produce relatively healthier seedlings when subjected to osmotic stress (Woodstock and Tao 1981; Woodstock and Taylorson 1981; Tilden and West 1985; Dell'Aquila and Taranto 1986).

Seed osmoconditioning, matriconditioning, or other hydration treatment has proved effective in improving seed performance or stand establishment under cold, wet conditions. In soybeans, osmoconditioning in PEG solution promoted rapid, uniform emergence in early plantings (Helsel et al. 1986). Hardening and moisturizing treatments, the latter resembling matriconditioning, were more effective than osmoconditioning in improving early emergence in sweet corn (Bennett and Waters 1987a, 1987b). Matriconditioning of snap bean seeds for 2 days in Micro-Cel E, followed by planting in Cornell Peat-Lite Mix, greatly improved the emergence and the shoot fresh weight at 20°/10°C. Addition of GA_3 further improved the emergence rate and seedling fresh weight (see Figure 4.3).

B. Biochemical Changes

Several molecular and biochemical changes are initiated as a result of osmoconditioning or presoaking of seeds. These include increases in the synthesis of macromolecules, activities of several enzymes, and metabolic rate.

1. DNA synthesis. Early studies with tomato and lettuce seeds indicated that little DNA synthesis was associated with osmoconditioning; large increases in synthetic activity occurred only after germination (Coolbear and Grierson 1979; Khan 1980/81). In lettuce seeds, the timing of DNA synthesis varied a great deal depending on the presowing treatment. The incorporation of ^{3}H-thymidine into DNA occurred just prior to radicle protrusion but was delayed in relation to radicle protrusion in osmoconditioned seeds or seeds treated with cotylenin E (Khan 1980/81). As fungal toxins, fusicoccin and cotylenin E, preferentially influence cell elongation and not cell division or DNA synthesis (Galli et al. 1975; Marre 1979), it is quite possible that osmoconditioning might preferentially influence events related to cell elongation. Osmoconditioning decreased the mean germination time in aged wheat seeds, and this was correlated with the resumption of cell division, seedling length, incorporation of ^{3}H-thymidine into DNA, and incorporation of ^{3}H-leucine into proteins (Dell'Aquila and Taranto 1986; Dell'Aquila 1987; Dell'Aquila and Tritto 1989).

Bray et al. (1989a) found detectable levels of ^{3}H-thymidine incorporation into DNA of leek seeds during osmoconditioning in absence of cell division. As in the case of osmoconditioned lettuce seeds, large increases in DNA synthesis occurred only after a lag period of relatively low synthetic activity, 6–9 h in leek, 6 h in lettuce (Khan 1980/81; Bray et al. 1989a). In a further study, Bray et al. (1989b) reported that the low level of DNA synthesis during osmoconditioning is mainly a replicative cytoplasmic type rather than a repair synthesis. It appears that the initial effect of osmoconditioning may largely be on cell elongation rather than on cell division and DNA synthesis.

2. RNA and protein synthesis. During or immediately following osmoconditioning in lettuce, tomato, and leek seeds, much greater amounts of RNA than DNA were synthsized (Khan et al. 1978; Coolbear and Grierson 1979; Fu et al. 1988; Bray et al. 1989a). It does not appear, however, that RNA synthesized during osmoconditioning or at the time of radicle protrusion participates in the advancement of germination time. Cordycepin, an inhibitor of RNA synthesis, strongly inhibited the synthesis of RNA, including that of poly A (+) RNA, in perennial goosefoot seeds conditioned in −0.86 MPa PEG solution for 7 days in the light. However, this inhibitor had no effect on germination when present either during osmoconditioning, germination, or both (Khan and Karssen 1981). When these seeds were osmotically treated for 7 days in the dark, the RNA synthesizing ability and the germinability of the seeds decreased greatly.

Cordycepin also failed to inhibit the germination of lettuce seeds previously given a 4-day light osmotic treatment or a 4-day dark osmotic treatment in the presence of GA_{4+7}, in spite of the fact that it strongly inhibited the amount and the rate of RNA synthesized (Khan 1980/81; Khan and Samimy 1982). The inhibitor, however, strongly inhibited the germination of untreated seeds. The inability of RNA synthesis inhibitor to influence germination but to strongly inhibit RNA synthesized by the seeds was also shown during soaking of lettuce seeds in the presence of the fungal toxin, cotylenin E (Khan 1980/81; Khan and Samimy 1982). Thus it appears that changes occurring during osmotic treatment independently modulate germination and RNA synthesis and that RNA synthesis is not necessary for germination occurring by cell elongation. These studies point out that a mere correlation between RNA synthesis and germination could be misleading.

There is good evidence that protein synthesized during or following osmoconditioning participates in the improvements associated with osmoconditioning. Several workers have shown an enhancement of protein synthesis during osmoconditioning and/or during subsequent germination in seeds of lettuce, tomato, pepper, leek, and wheat (Koehler 1967; Khan et al. 1978, 1980/81; Mazor et al. 1984; Cobb et al. 1988; Bray et al. 1989a; Dell'Aquila and Tritto 1989). Unlike cordycepin, cycloheximide, an inhibitor of protein synthesis, has been shown to inhibit both germination and protein synthesis in osmoconditioned lettuce and perennial goosefoot seeds (Khan et al. 1980/81; Khan and Karssen 1981). Cycloheximide (0.2 mM) applied during osmoconditioning or during water soak inhibited the rate of protein synthesis and early germination in lettuce seeds (Table 4.10). The effect of this inhibitor was, however, short lived and all osmoconditioned seeds germinated by 24 h. Although cycloheximide inhibited protein synthesis in all cases, its effect on germination was strongest in seeds germinated in the presence of cotylenin E, a growth regulator known to promote cell elongation (Marre 1979) (Table 4.10).

Table 4.10. Effect of 0.2 mM cycloheximide (CH) on protein synthesis and germination of untreated (UNT) and variously treated 'Grand Rapids' lettuce seeds. Some data are from Khan et al. (1980/81).

Treatment[z]	Protein synthesis[y] (CPM $\times 10^{-3}$)				Germination[x] (%)		
	1.3h	2.3h	4.3h	10.3h	7h	24h	48h
UNT							
−CH	—	3.7	3.4	2.3	0	96	98
+CH	—	2.2	2.2	0.9	0	36	77
OSM							
−CH	4.9	4.0	5.0	—	96	97	—
+CH	3.5	1.5	1.4	—	47	100	—
OSM (+ CH)							
−CH	3.8	4.2	4.8	—	87	98	—
+CH	3.1	1.8	1.2	—	47	100	—
CN							
−CH	2.9	4.1	6.3	—	5	98	99
+CH	2.2	3.6	3.7	—	0	0	36
LSD 5%	0.3	0.4	0.8	0.3	5	6	5

[z]Seeds were osmoconditioned (OSM) in −0.9 MPa PEG solution at 15°C for 4 day in the presence or absence of 0.2 mM CH. Osmoconditioned seeds were rinsed with water.
[y]After soaking for different times at 25°C, embryos were excised and, at indicated times, incubated in 1 ml solution containing 50 ug chloramphenicol and 2 μCi ^{3}H-leucine (62 Ci/mmol) for 30 min at 25°C, then washed and ground in 5% TCA. Radioactivity in TCA-insoluble fraction determined after centrifugation at 20,000g.
[x]Seeds were germinated at 25°C in water or 0.05 mM CN, with or without the addition of 0.2 mM CH.

In pepper seeds nearly twice as much amino acid was incorporated into protein during the first 24 h when imbibed in PEG solution versus water (Mazor et al. 1984). In another study with these seeds, protein synthesis increased during all osmoconditioning treatments (Cobb et al. 1988).

Pea seed pretreated with −0.3 MPa PEG solution at 25°C for 36 h failed to show an increase in protein synthesis over that achieved by the untreated seeds (Dell'Aquila and Bewley 1989). However, in wheat seeds pretreated with −0.37 MPa PEG solution for 36 h at 20°C, the rate of protein synthesis increased greatly over that of untreated seeds, and the enhanced protein synthesizing capacity persisted even after drying (Dell'Aquila and Tritto 1989). The soluble protein content and germination increased in pea, tomato, and spinach seeds osmoconditioned (osmoprimed) in 10% sodium polypropionate for 4 days (Zuo et al. 1988a, 1988b). Recently quantitative and qualitative differences were shown in proteins synthesized in osmoconditioned versus untreated seeds and during osmoconditioning versus germination in leek seeds using two-dimensional polyacrylamide gel electrophoresis (Bray, personal communication). A vast majority of proteins synthesized at the end of the osmoconditioning period also appeared to be synthesized during germination. Taken together, these studies indicate that improvements by osmoconditioning, possibly related to cell elongation, may be influenced by an enhancement in protein synthesizing activities.

3. Enzymatic and metabolic activities. Enhancements in enzyme and metabolic activities are common features during seed osmoconditioning and seem to be related to seed invigoration during germination. Increases in the respiratory activities and the formation of adenosine triphosphate (ATP), needed for synthesis of macromolecules, membranes, and cell wall materials, have been reported during or following osmoconditioning. Increasing amounts of ATP accumulated during 24-h osmoconditioning in PEG in seeds of kohlrabi, spinach, eggplant, and pepper were similar to that accumulated during a 4–5-h imbibition in water (Mazor et al. 1984). In another study with pepper, respiration rates during osmoconditioning did not differ from the controls; after osmoconditioning, however, the treated seeds had significantly higher rates of respiration (Smith and Cobb, 1989). Drying of the seeds after osmoconditioning greatly reduced the ATP content (Mazor et al. 1984; Fu et al. 1988). This might be related to the loss in the ability to synthesize active metabolites that participate in germination improvement. Increased respiratory activities in osmoconditioned (in 10% sodium polypropionate for 4 days at 20°C) seeds were also reported in pea, tomato, and spinach by Zuo et al. (1988b). Only traces of guanosine triphosphate (GTP) and cytidine triphosphate (CTP), and low levels of ATP and uridine triphosphate (UTP) were found during osmoconditioning of leek seeds; their levels increased sharply after a lag period of 6 h on transfer to water (Bray et al. 1989a)

Increased activity of several acid phosphatases and esterases were found in osmoconditioned lettuce seeds as a result of activation or de novo synthesis (Khan et al. 1978). Increases in the activities of malate dehydrogenase, glucose-6-phosphate dehydrogenase, isocitrate lyase, and aldolase were reported in pepper seed as a result of osmoconditioning (Smith and Cobb 1988). In peanut, osmoconditioning increased the activities of isocitrate lyase, ATPase, and acid phosphatase (Fu et al. 1988). Osmoconditioning in 10% sodium polypropionate enhanced the activities of amylase in pea, tomato, and spinach seeds and of peroxidase in tomato and spinach (Zuo et al. 1988b). In summary, it seems that a large-scale mobilization of storage reserves occurs during osmoconditioning that provides substrates which are readily utilized for germination. Many of these small molecules evidently have osmotic properties, which result in the lowering of the water potential of the seed, and that in turn accounts for the ability of the conditioned seeds to readily absorb water, to germinate rapidly, and to tolerate stressful environments.

VII. CONCLUDING REMARKS

Preplant seed hydration treatments, particularly at low water potential (physiological seed conditioning), mobilize seed's resources that are utilized for rapid and synchronous germination, improved emergence rate, and larger stand size and yield. Seed conditioning is ideally suited to combat or lessen the impact of a number of soil and climatic constraints, including cold and wet conditions, thermoinhibition, drought, and incidence of diseases. The conditioning treatment is versatile enough to be integrated with other preplant physiological and non-physiological treatments to combat other internal and external constraints. These include chilling to break the primary dormancy, permeation with hormones to prevent the induction of secondary dormancy, and treatment with pesticides and beneficial microbes to combat seed and seedling diseases during stand establishment. Conditioning of seeds on moistened solid carriers with high matric or osmotic component is an effective alternative to conditioning seeds in liquid media. The physiology and biochemistry of seed invigoration as a result of seed conditioning are largely unknown, but may involve changes in membrane integrity, breakdown of seed reserves, and the production of osmotically active substances and other metabolites needed for rapid water uptake and germination and for improved seed performance.

Nature is a wonderful teacher, and a seed is one of the simplest teaching aids. Several examples can be cited. The age-old observation that certain leguminous seeds at low moisture content are susceptible to injury if planted in cold, wet soil led to the development of strategies to reduce wet soil injuries (increasing the moisture content, using seeds with semi-hard seed coat, using hydrophobic coatings) (Pollock 1969; Hobbs and Obendorf 1972; Taylor and Dickson 1985; Priestley and Leopold 1986). It is not surprising that matriconditioning in Micro-Cel E or vermiculite, which simulates the controlled hydration in a water unsaturated soil microenvironment (Khan et al. 1983), also improved the subsequent performance of bean and soybean seeds in a cold, wet environment (see Figure 4.3). Lettuce seeds exposed to high temperatures and osmotic stress develop a secondary dormancy. The ACC biosynthesis and the conversion of ACC to ethylene are strongly inhibited during an exposure to high temperature, salinity, or low water potential PEG solution (Khan and Huang 1988; Khan and Prusinski 1989; Khan 1990; Table 4.8). Permeation of ACC, an ethylene precursor, into lettuce seeds during the presoak or during osmotic or matric conditioning, appears to compliment the advantages of seed hydration treatments in the alleviation of stress and the prevention of secondary dormancy, presumably by improving the conversion of ACC into ethylene. In deepwater or floating rice, plant establishment is dependent upon the ability of young seedlings to elongate rapidly, a typical GA response, to avoid injury. Deepwater and floating rices respond to a GA spray by elongating, while several of the non-elongating types, highly tolerant of submergence, do not (Khan et al. 1988). In perennial ryegrass and winter wheat, calcium peroxide seed coating improved the seedling establishment in wet soil and in the presence of straw residues, respectively. In dry soil or in the absence of straw residues no benefit was obtained by the peroxide coating (Christian and Miller 1984; Naylor and Prentice 1986). These examples are instructive and illustrate the differing capabilities and needs of seeds and seedlings for plant establishment, requiring different strategies for improving plant performance.

In the future, it is likely that seed conditioning will be conducted on solid carriers with greater frequency. Natural and synthetic carriers with distinctive physical and/or chemical properties will be sought. An effective integration of seed conditioning with other promotive treatments will depend upon a clearer understanding of the invigoration process and its interplay with the dynamics of the environment. In recent years public concern about large-scale pesticide use and degradation of the environment has stimulated research on alternative ways to control pests. There is growing evidence that improved membrane integrity and other changes during seed conditioning decrease the leakage of exudates from the seeds and thereby reduce the incidence of damping-off, a major cause of seed and seedling rot. Insects might similarly be attracted to exudates, and the damage to seeds and seedlings might similarly be reduced by preplant seed conditioning. Thus, seed conditioning offers a way to decrease the amount of pesticides used as seed dressing or as soil amendment. The use of beneficial fungi and bacteria in combination with seed conditioning and other hydration treatments to control insect pests and plant

Bernstein, L., A. J. Mackenzie, and B. A. Krantz. 1955. The interaction of salinity and planting practice on the germination of irrigated row crops. *Proc. Soil Sci. Soc. Amer.* 19:240–243.

Berrie, A. M. M., and D. S. H. Drennan. 1971. The effect of hydration-dehydration on seed germination. *New Phytologist* 70:135–142.

Bodsworth, S., and J. D. Bewley. 1981. Osmotic priming of seeds of crop species with polyethylene gloycol as a means of enhancing early and synchronous germination at cool temperatures. *Can. J. Bot.* 59:672–676.

Bollum, F. J. 1963. 'Primer' in DNA polymerase reaction. p. 1–26. In: *Progr. Nucleic Acid Res.* Academic Press, New York.

Bradford, K. J. 1985. Seed priming improves germination and emergence of cantaloupe at low temperatures. *HortScience* 20:598.

———. 1986. Manipulation of seed water relations via osmotic priming to improve germination under stress conditions. *HortScience* 21:1105–1112.

Bradford, K. J., C. A. Argerich, D. Peetambar, O. Somasco, A. Tarquis, and G. E. Welbaum. 1988. Seed enhancement and seed vigor. Proc. Intern. Conf. Stand Estab. Hort. Crops, Lancaster, PA. p. 1–35.

Bradford, K. J., C. A. Argerich, A. M. Tarquis. 1989. Influence of priming on seed deterioration characteristics. 3rd International Workshop on Seeds, NATO Adv. Res. Workshop, p. 8. Williamsburg, VA. (Abstr.).

Bradford, K. J., D. M. May, B. J. Hoyle, Z. S. Skibinski, S. J. Scott, and K. B. Tyler. 1988. Seed and soil treatments to improve emergence of muskmelon from cold or crusted soils. *Crop Sci.* 28:1001–1005.

Bradford, K. J., J. J. Steiner, and S. E. Trawatha. 1990. Seed priming influence on germination and emergence of pepper seed lots. *Crop Sci.* 30:718–721.

Braun, J. W., and A. A. Khan. 1976. Alleviation of salinity and high temperature stress by plant growth regulators permeated into lettuce seeds via aceton. *J. Am. Soc. Hort. Sci.* 101:716–721.

Bray, C. M., P. A. Davison, M. Ashraf, and R. M. Taylor. 1989a. Biochemical changes during osmopriming of leek seeds. *Ann. Bot.* 63:185–193.

———. 1989b. Biochemical changes during seed osmopriming. Third Intl. Workshop on seeds. A NATO Adv. Res. Workshop., Williamsburg, VA. p. 11 (Abstr.).

Brocklehurst, P. A., and J. Dearman. 1983a. Interactions between seed priming treatments and 9 seed lots of carrot, celery and onion: I. Laboratory germination. *Ann. Appl. Biol.* 102:577–584.

———. 1983b. Interactions between seed priming treatments and 9 seed lots of carrot, celery and onion: II. Seedling emergence and plant growth.*Ann. Appl. Biol.* 102:585–593.

———. 1984. A Comparison of different chemicals for osmotic treatment of vegetable seed. *Ann. Appl. Biol.* 105:391–398.

Brocklehurst, P. A., J. Dearman, and R. K. L. Drew. 1984. Effects of osmotic priming on seed germination and seedling growth in leek. *Scientia Hort.* 24:201–210.

Bussel, W. T., and D. Gray. 1976. Effects of pre-sowing seed treatments and temperatures on tomato (*Lycopersicon esculentum* Mill) seed germination and sedling emergence. *Scientia Hort.* 5:101–109.

Callan, I. W. 1975. Achievements and limitations of seed treatments. p. 271–174. In: *Outlook on Agriculture*, vol. 8. Imperial Chemical Industries Limited, Haslemere.

Callan, N. W., D. E. Mathre, and J. B. Miller. 1990. Bio-priming seed treatment for biological control of *Pythium ultimum* preemergence damping-off in sh2 sweet corn. *Plant Disease* 74:368–372.

Cantliffe, D. J., J. M. Fischer, and T. A. Nell. 1984. Mechanism of seed-priming in circumventing thermodormancy in lettuce. *Plant Physiol.* 75:290–294.

Cantliffe, D. J., K. D. Schuler, and A. C. Guedes. 1981. Overcoming seed dormancy in heat sensitive romaine lettuce by seed priming. *HortScience* 16:196–198.

Carpenter, W. J. 1989. *Salvia splendens* seed germination and priming for rapid and uniform plant emergence. *J. Am. Soc. Hort. Sci.* 114:247–250.

_____. 1990. Priming dusty miller seeds: role of aeration, temperature and relative humidity. *HortScience* 25:299–302.

Christian, D. G., and D. .P. Miller. 1984. Effect of calcium peroxide seed coating on the establishment and yield of winter wheat sown by direct drilling in the presence of straw residues. *J. Sci. Food Agr.* 35:606–608.

Cobb, B. G., P. T. Smith, and W. Matthew. 1988. Protein and amino acid metabolism during priming and subsequent germination of *Capsicum annuum. HortScience* 23:795 (Abstr.).

Coolbear, P. and D. Grierson. 1979. Studies on changes in the major nucleic acid components of tomato seeds (*Lycopersicon esculentum* Mill.) resulting from osmotic presowing treatment. *J. Expt. Bot.* 30:1153–1162.

Currah, I. E., D. Gray, and T. H. Thomas. 1974. The sowing of germinating vegetable seeds using a fluid drill. *Ann. Appl. Biol.* 76:311–318.

Darby, R. J., and P. J. Salter. 1976. A technique for osmotically pre-treating and germinating quantities of small seeds. *Ann. Appl. Biol.* 83:313–315.

Dearman, J., P. A. Brocklehurst, and R. L. K. Drew. 1986. Effects of osmotic priming and aging on onion seed germination. *Ann. Appl. Biol.* 108:639–648.

_____. 1987. Effects of osmotic priming and ageing on the germination and emergence of carrot and leek seeds. *Ann. Appl. Biol.* 111:717–722.

De Klerk, G. L. 1986. Advantageous and detrimental effects of osmotic pre-sowing treatment on the germination performance of *Agrostemma githago* seeds. *J. Expt. Bot.* 37:765–774.

Dell'Aquila, A. 1987. Mean germination time as a monitor of the seed aging. *Plant Physiol. Biochem.* (Paris) 25:761–768.

Dell'Aquila, A., and J. D. Bewley. 1989. Protein synthesis in the axes of polyethylene glycol-treated pea seed and during subsequent germination. *J. Expt. Bot.* 40:1001–1007.

Dell'Aquila, A., and G. Taranto. 1986. Cell division and DNA synthesis during osmopriming treatment and following germination in aged wheat embryos. *Seed Sci. Technol.* 14:333–342.

Dell'Aquila, A., and V. Tritto. 1989. Ageing and osmotic priming in wheat seeds: Effects upon certain components of seed quality. *Ann. Bot.* (in press).

Digat, B. 1988. Strategies for bacterization with *Rhizobacteria. Bull. Organ. Eur. Mediterr. Prot. Plant* (OEPP) 18:29–35.

_____. 1989. Strategies for seed bacterization. *Acta Hort.* 253:121–130.

Durrant, M. J., P. A. Payne, and J. M. Maclaren. 1983. The use of water and some inorganic salt solutions to advance sugar beet seed: II. Experiments under controlled and field conditions. *Ann. Appl. Biol.* 103:517–526.

Elad, Y., Y. Zvieli, and I. Chet. 1986. Biological control of *Macrophomina phaseolina* Tassi Goid by *Trichoderma harzianum. Crop Prot.* 5:288–292.

El-Sharkawi, H. M., and I. Springuel. 1977. Germination of some crop plant seeds under reduced water potential. *Seed Sci. Technol.* 5:677–688.

Ely, P. R., and W. Heydecker. 1981. Fast germination of parsley (*Petroselinum crispum*) seeds. *Scientia Hort.* 15:127–136.

Eplee, R. E. 1975. Ethylene: A witchweed seed germination stimulant. *Weed Sci.* 23:433–436.

Finch-Savage, W. E. 1984. A comparison of seedling emergence from dry-sown and fluid-drilled carrot seeds. *J. Hortic. Sci.* 59:403–410.

Finch-Savage, W. E., and C. J. Cox. 1982. Effect of adding plant nutrients to the gel used for fluid drilling early carrots. *J. Agric. Sci.* (Camb.) 99:295–303.

Fu, J. R., X. H. Lu, R. Z. Chen, B. Z. Zhang, Z. S. Liu, Z. S. Li, and D. Y. Cay. 1988. Osmoconditioning of peanut (*Arachis hypogea* L.) seeds with PEG to improve vigor and some biochemical activities. *Seed Sci. Technol.* 16:197–212.

Furutani, C. S., B. H. Zandstra, and H. C. Price. 1985. Low temperature germination of celery seeds for fluid drilling. J. Am. Soc. Hort. Sci. 110:153–156.

_____. 1986. The effects of osmotic solute composition and duration and temperature of priming on onion seed germination. Seed Sci. Technol. 14:545–552.

Gaikwad, S. J., B. Sen, and S. U. Meshram. 1987. Effect of bottlegourd seed coating with antagonists on seedling quantum of the pathogen inside the seedlings and populations of the soil against Fusarium oxysporum. Plant Soil 101:205–210.

Galli, M. G., E. Sparvoli, and M. Caroi. 1975. Comparative effects of fusicoccin and gibberellic acid on the promotion of germination and DNA synthesis initiation in Haplopappus gracilis. Plant Sci. Lett. 5:351–357.

Gelmond, H. 1965. Pretreatment of leek seeds as a means of overcoming superoptimal temperatures of germination. Proc. Int. Seed Test Assoc. 30:737–742.

Georghiou, K., C. A. Thanos, and H. C. Passam. 1987. Osmoconditioning as a means of counteracting the ageing of pepper seed during high temperature storage. Ann. Bot. 60:279–286.

Giammichele, L. A., and W. G. Pill. 1984. Protection of fluid-drilled tomato seedlings against damping-off by fungicide incorporation in a gel carrier. Hortic. Sci. 19:877–879.

Gray, D. 1981. Fluid drilling of vegetable seeds. Hort. Rev. 3:1–27.

Gray, D., P. A. Brocklehurst, J. R. A. Steckel, and J. Dearman. 1984. Priming and pre-germination of parsnip (Pastinaca sativa L.) seed. J. Hort Sci. 59:101–108.

Gray, D., and J. R. A. Steckel. 1977. Effects of presowing treatments on the germination and establishment of parsnips. J. Hort. Sci. 52:525–534.

Guedes, A. C., and D. J. Cantliffe. 1980. Germination of lettuce (Lactuca sativa) at high temperature after seed priming. J. Am. Soc. Hort. Sci. 105:777–781.

Hadas, A. 1976. Water uptake and germination of leguminous seeds under changing external water potential in osmotic solutions. J. Expt. Bot. 27:480–489.

_____. 1977. A simple laboratory approach to test and estimate seed germination performance under field conditions. Agron. J. 69:582–588.

_____. 1981. Seed-soil contact and germination. p. 507–527. In: A. A. Khan (ed.), The Physiology and Biochemistry of Seed Development, Dormancy and Germination. Elsevier, Amsterdam.

Haigh, A. M., E. W. R. Barlow, F. L. Milthrope, and P. J. Sinclair. 1986. Field emergence of tomato (Lycopersicon esculentum), carrot (Daucus carota) and onion (Allium cepa) seeds primed in an aerated salt solution. J. Am. Soc. Hort. Sci. 111:660–665.

Hallgren, S. W. 1989. Effects of osmotic priming using aerated solution of polyethylene glycol on germination of pine seeds. Ann. Sci. For. (Paris) 46:31–38.

Halmer, P. 1988. Technical and commercial aspects of seed pelleting and film coating. p. 191–204. British Crop Protection Council, Thornton Heath.

Harman, G. E., and A. G. Taylor. 1988. Improved seedling performance by integration of biological control agents at favorable pH levels with solid matrix priming. Phytopathology 77:520–525.

_____. 1990. Development of an effective biological seed treatment system. p. 415–426. In: D. Hornby (ed.), New Strategies of Control of Soilborne Plant Pathogens by Nonconventional Means. C. A. B. International, Wallingford.

Hegarty, T. W. 1978. The physiology of seed hydration and dehydration, and the relation between water stress and the control of germination: A review. Plant Cell Environ. 1:101–119.

Helsel, D. G., D. R. Helsel, and H. C. Minor. 1986. Field studies on osmoconditioning soybeans, Glycine max. Field Crops Res. 14:291–298.

Hemmat, M., G.-W. Zeng, and A. A. Khan. 1985. Responses of intact and scarified curly dock (Rumex crispus) seeds to physical and chemical stimuli. Weed Sci. 33:658–654.

Henckel, P. A. 1964. Physiology of plants under drought. Annu. Rev. Plant Physiol. 15:363–386.

Heydecker, W. 1973/74. Germination of an idea: The priming of seeds. University of Nottingham School of Agriculture Report. p. 50–67.

_____. 1975. "Seed priming"—the treatment of the future? Grower 554–555. (Sep. 27).

_____. 1977. Stress and seed germination. p. 240–282. In: A. A. Khan (ed.), The Physiology and Biochemistry of Seed Dormancy and Germination. Elsevier/North-Holland, Amsterdam.

Heydecker, W., and P. Coolbear. 1977. Seed treatments for improved performance-survey and attempted prognosis. Seed Sci. Technol. 5:353–425.

Hobbs, P. R., and R. L. Obendorf. 1972. Interaction of initial seed moisture and imbibitional temperature on germination and productivity of soybean. Crop Sci. 12:664–667.

Hsiao, A.-I., and W. Vidaver. 1973. Dark reversion of phytochrome in lettuce seeds stored in a water-saturated atmosphere. Plant Physiol. 51:459–463.

Iswandi, A., P. Bossier, J. Vandenabeele, and W. Verstraete. 1987. Effect of seed inoculation with the rhizopseudomonad strain 7NSK2 on the root microbiota of maize (Zea mays) and barley (Hordeum vulgare). Biol. Fertil. Soils 3:153–158.

Jawson, M. D., A. J. Franzluebbers, and R. K. Berg. 1989. Bradyrhizobium japonicum survival in and soybean inoculation with fluid gels. Appl. Environ. Microbiol. 55:617–622.

Jilani, G., R. C. Saxena, and A. A. Khan. 1989. Ethylene production as an indicator of germination and vigor loss in stored rice seed infested by Rhizopertha dominica (F.) (Coleoptera: Bostrychidae). J. Stored Prod. Res. 25:175–178.

Johnson, I. J. 1975. New developments in seed pelleting and seed coating with special reference to rangeland improvement. p. 281–283. In: Outlook on Agriculture, vol. 8. Imperial Chemical Industries Limited, Haslemere.

Justice, O. L., and L. N. Bass. 1978. Principles and practices of seed storage. U.S. Dept. Agr., Agric. Handb. 506. U.S. Govt. Printing Office, Washington, D.C.

Kahn, B., and J. E. Motes. 1988. Comparison of fluid drilling with conventional planting methods for stand establishment and yield of spring and fall broccoli crops. J. Am. Soc. Hort. Sci. 113:670–674.

Karssen, C. M. 1980/81. Patterns of change in dormancy during burial of seeds in soil. Isr. J. Bot. 29:65–73.

Karssen, C. M., and R. Weges. 1987. Osmoconditioning of lettuce seeds and induction of secondary dormancy. Acta Hort. 215:165–171.

Khan. A. A. 1971. Cytokinins: Permissive role in seed germination. Science 171:353–359.

_____. 1977. Preconditioning, germination and performance of seeds, p. 283–316. In: A. A. Khan (ed.), The Physiology and Biochemistry of Seed Dormancy and Germination. Elsevier, Amsterdam.

_____. 1978. Incorporation of bioactive chemicals into seeds to alleviate environmental stress. Acta Hort. 83:225–234.

_____. 1980/81. Hormonal regulation of primary and secondary seed dormancy. Isr. J. Bot. 29:207–224.

_____. 1990. Enhanced sensitivity of germination and growth processes to ethylene under stress, p. 1258–1270. In: S. K. Sinha, P. V. Sane, S. C. Bhargava and P. K. Agarwal (eds.). Proc. Int. Congr. Plant Physiol., New Delhi, India.

Khan, A. A., J. W. Braun, K.-L. Tao, W. F. Millier, and R. F. Bensin. 1976. New methods for maintaining seed vigor and improving performance. J. Seed Technol. 1:33–57.

Khan, A. A., and X.-L. Huang. 1988. Synergistic enhancement of ethylene production and germination with kinetin and 1-aminocyclopropane-1-carboxylic acid in lettuce seeds exposed to salinity stress. Plant Physiol. 87:847–852.

Khan, A. A., and C. M. Karssen. 1980. Induction of secondary dormancy in Chenopodium bonus-henricus L. seeds by osmotic and high temperature treatments and its prevention by light and growth regulators. Plant Physiol. 66:175–181.

_____. 1981. Changes during light and dark osmotic treatment independently modulating germination and ribonucleic acid synthesis in Chenopodium bonus-henricus seeds. Physiol. Plant. 51:269–276.

Khan, A. A., C. M. Karssen, E. F. Leue, and C. H. Roe. 1979. Preconditioning of seeds to improve performance. p. 395–413. In: T. K. Scott (ed.), Plant Regulation and World Agriculture. Plenum, New York.

Khan, A. A., H. Miura, J. Prusinski, and S. Ilyas. 1990. Matriconditioning of seeds to improve seedling emergence. Proc. National Symp. Stand Estab. Hort. Crops. Minneapolis, MN. p. 19–40.

Khan, A. A., N. H. Peck, and C. Samimy. 1980/81. Seed osmoconditioning: Physiological and biochemical changes. Isr. J. Bot. 29:133–144.

Khan, A. A., N. H. Peck, A. G. Taylor, and C. Samimy. 1983. Osmoconditioning of beet seeds to improve emergence and yield in cold soil. Agron. J. 75:788–794.

Khan, A. A., and J. Prusinski. 1989. Kinetin enhanced 1-aminocyclopropane-1-carboxylic acid utilization during alleviation of high temperature stress in lettuce seeds. Plant Physiol. 91:733–737.

Khan, A. A., and C. Samimy. 1982. Hormones in relation to primary and secondary seed dormancy. p. 203–241. In: A. A. Khan (ed.), The Physiology and Biochemistry of Seed Development, Dormancy and Germination. Elsevier, Amsterdam.

Khan, A. A., and D. V. Seshu. 1987. Using ethylene to monitor the influence of adverse climatic factors and to predict plant performance. p. 103–122. In: Weather and rice. Proc. Internat. Workshop on the Impact of Weather Parameters on Growth and Yield of Rice, 7–10 April, 1986. International Rice Res. Inst., Los Banos, Manila.

Khan, A. A., A. Szafirowska, and N. H. Peck. 1981. Osmoconditioning of seed. New York's Food Life Sci. Quart. 13:9–13.

Khan, A. A., K.-L. Tao, J. S. Knypl, B. Borkowska, and L. E. Powell. 1978. Osmotic conditioning of seeds: Physiological and biochemical changes. Acta Hort. 83:267–278.

Khan, A. A., and A. G. Taylor. 1986. Polyethylene glycol incorporation in table beet seed pellets to improve emergence and yield in wet soil. HortScience 21:987–989.

Khan, A. A., R. Thakur, M. Akbar, D. HilleRisLambers, and D. V. Seshu. 1988. Hormonal regulation of elongation in floating rice during submergence. Crop Sci. 28:121–128.

Khan, A. A., and G.-W. Zeng. 1985. Dual action of respiratory inhibitors. Inhibition of germination and prevention of dormancy induction in lettuce seeds. Plant Physiol. 77:817–823.

Kidd, F., and C. West. 1918. Physiological predetermination: The influence of the physiological condition of the seed upon the course of subsequent growth and upon the yield. I. The effect of soaking seeds in water. Ann. Appl. Biol. 5:1–10.

———. 1919. The influence of temperature on the soaking of seeds. New Phytologist 18:35–39.

Knypl, J. S., K. M. Jans, and A. Radziwonowska-Jozwiak. 1980. Is enhanced vigor in soybean (Glycine max) dependent on activation of protein turnover during controlled hydration of seeds? Physiol. Veg. 18:157–161.

Knypl, J. S., and A. A. Khan. 1981. Osmoconditioning of soybean seeds to improve performance at suboptimal temperatures. Agron. J. 73:112–116.

Koehler, D. E. 1967. Studies on a treatment hastening germination of tomato seeds. M.S. thesis, Purdue University West Lafayette, IN.

Kornburg, A. 1957. Pyrophosphorylases and phosphorylases in biosynthetic reactions. p. 191–240. In: F. F. Nord (ed.), Advances in Enzymology, vol. XVIII. Interscience, New York.

Kouno, Y. 1989. Method of applying gel coating to plant seeds. U. S. Patent-4808430, February 28, 1989.

Kubik, K. K., J. A. Eastin, J. D. Eastin, and K. M. Eskridge. 1988. Solid matrix priming of tomato and pepper. Proc. Int. Conf. Stand Est. Hortic. Crops, Lancaster, PA. p. 86–96.

Lang, A. 1965. Effects of some internal and external conditions on seed germination. Encyclop. Plant Physiol. 15/2:848–893.

Lewak, S., and R. M. Rudnicki. 1977. After-ripening in cold-requiring seeds, p. 193–217. In: A. A. Khan (ed.), The Physiology and Biochemistry of Seed Dormancy and Germination. Elsevier, Amsterdam.

Lickorish, G. R., and R. J. Darby. 1976. A hand-operated fluid drill for small plot experiments. Exptl. Agric. 12:299–303.

Lutchmeah, R. S., and R. C. Cooke. 1985. Pelleting of seed with antagonist Pythium oligandrum for biological control of damping-off. Plant Pathol. (London) 34:528–531.

Malnassy, T. G. 1971. Physiological and biochemical studies on a treatment hastening the germination of seeds at low temperature. Ph.D. Thesis, Rutgers University, New Jersey.

Marre, E. 1979. Fusicoccin: A tool in plant physiology. Annu. Rev. Plant Physiol. 30:273–288.

Martin, F. N., and J. G. Hancock. 1987. The use of Pythium oligandrum for biological control of preemergence damping-off caused by Pythium ultimum. Phytopathology 77:1013–1020.

Maude, R. B. 1986. Treatment of vegetable seeds. p. 239–261. In: Seed Treatment. British Crop Protection Council, Thornton Heath.

May, L. M., E. J. Milthrope, and F. L. Milthrope. 1962. Pre-sowing hardening of plants to drought. Field Crop Abstracts. 15:93–98.

Mazor, L., M. Perl, and M. Negbi. 1984. Changes in some ATP-dependent activities in seeds during treatment with polyethylene glycol and during the redrying process. J. Expt. Bot. 35:1119–1127.

Mihuta-Grimm, L., and R. C. Rowe. 1986. Trichoderma sp. as biocontrol agents of Rhizoctonia damping-off of radish Raphanus sativus in organic soil and comparison of 4 delivery systems. Phytopathology 76:306–312.

Millier, W. F., and C. Sooter. 1967. Improving emergence of pelleted vegetable seed. Trans. Amer. Soc. Agr. Eng. 10:658–666.

Nakamura, S., T. Teranishi, and M. Aoki. 1982. Promoting effect of polyethylene glycol on the germination of celery and spinach seeds. J. Japan. Soc. Hort. Sci. 50:461–467.

Naylor, R. E. L., and C. G. Prentice. 1986. Effect of calcium peroxide seed coating on germination of perennial ryegrass seeds. Ann. Appl. Biol. 108:610–618.

Obendorf, R. L., and P. R. Hobbs. 1970. Effect of seed moisture on temperature sensitivity during imbibition of soybean. Crop Sci. 10:563–566.

Osburn, R. M., and M. N. Schroth. 1988. Effect of osmopriming on exudation and subsequent damping-off caused by Pythium ultimum. Amer. Phytopath. Soc. 78:1246–1250.

⸺. 1989. Effect of osmopriming sugar beet seed on germination rate and incidence of Pythium ultimum damping-off. Plant Dis. 73:21–24.

Owen, P. C. 1952. The relation of germination of wheat to water potential. J. Expt. Bot. 3:188–203.

Palevitch, D., and T. H. Thomas. 1974. Thermodormancy release of celery seed by gibberellin, 6-benzylaminopurine and ethephon applied in organic solvent. J. Expt. Bot. 25:981–986.

Pallais, N., and N. Fong. 1988. Influence of dormancy on the effectiveness of priming true potato seed. HortScience 23:796 (Abstr.).

Parera, C. A., and D. J. Cantliffe. 1990. Improved stand establishment of sh2 sweet corn by solid matrix priming. Proc. National Symp. Stand Estab. Hort. Crops, Minneapolis, MN. p. 91–96.

Passam, H. C., P. I. Karavites, A. A. Papandreou, C. A. Thanos, and K. Georghiou. 1989. Osmoconditioning of seeds in relation to growth and fruit yield of aubergine pepper, cucumber and melon in unheated greenhouse cultivation Scientia Hort. 38:207–216.

Perkins-Veazie, P., and D. J. Cantliffe. 1984. Need for high quality seed for effective priming to overcome thermodormancy. J. Am. Soc. Hort. Sci. 109:368–372.

Perl, M. 1978. Invigoration of cotton seedlings by treatment of seeds for pregermination activities. J. Expt. Bot. 30:183–192.

Peterson, J R. 1976. Osmotic priming of onion seeds—the possibility of a commercial scale treatment. *Scientia Hort.* 5:207–214.

Pill, W. G. 1986. Parsley emergence and seedling growth from raw, osmoconditioned and pregerminated seeds. *HortScience* 21:1134–1136.

Pill, W. G., and W. E. Finch-Savage. 1988. Effect of combining priming and plant growth regulator treatments on the synchronization of carrot seed germination. *Ann. Appl. Biol.* 113:383–390.

Pill, W. G., and J. J. Frett. 1989. Performance of seeds embedded in hydroxyethyl cellulose sheets. *Scientia Hort.* 38:193–200.

Pill, W. G., and C. F. Mucha. 1984. Performance of germinated, imbibed and dry petunia seed fluid-drilled in two gels with nutrient additives. *Scientia Hort.* 22:181–188.

Pollock, B. M. 1969. Imbibition temperature sensitivity of lima beans controlled by initial seed moisture. *Plant Physiol.* 44:907–911.

Priestley, D. A., and A. C. Leopold. 1986. Alleviation of imbibitional chilling injury by use of lanolin. *Crop Sci.* 26:1252–1254.

Prusinski, J., and A. A. Khan. 1990. Relationship of ethylene production to stress alleviation in seeds of lettuce cultivars. *J. Am. Soc. Hort. Sci.* 115:294–298.

Rabin, J., G. A. Berkowitz, and S. W. Akers. 1988. Field performance of osmotically primed parsley seed. *HortScience* 23:554–555.

Rao, S. C., S. W. Akers, and R. M. Ahring. 1987. Priming *Brassica* seed to improve emergence under different temperatures and soil moisture conditions. *Crop Sci.* 27:1050–1053.

Rao, N. K., E. H. Roberts, and R. H. Ellis. 1987. The influence of pre-storage and post-storage hydration treatment on chromosomal aberrations, seedling abnormalities and viability of lettuce seeds. *Ann. Bot.* 60:97–108.

Rennick, G. A., and P. I. Tiernan. 1978. Some effects of osmopriming on germination, growth and yield of celery. *Seed Sci. Technol.* 6:695–700.

Rivas, M., F. J. Sundstrom, and R. L. Edwards. 1984. Germination and crop development of hot pepper after seed priming. *HortScience* 19:279–281.

Roos, E. E., and B. M. Pollock. 1971. Soaking injury in lima beans. *Crop Sci.* 11:78–81.

Sachs, M. 1977. Priming of watermelon seeds for low-temperature germination. *J. Am. Soc. Hort. Sci.* 102:175–178.

Sachs, M., D. J. Cantliffe, and T. A. Nell. 1981. Germination studies of clay-coated sweet pepper seeds. *J. Am. Soc. Hort. Sci.* 106:385–389.

_____. 1982. Germination behavior cf sand coated sweet pepper seeds. *J. Am. Soc. Hort. Sci.* 107:412–416.

Samimy, C., and A. A. Khan. 1983a. Secondary dormancy, growth regulator effects, and embryo growth potential in curly dock (*Rumex crispus*) seeds. *Weed Sci.* 31:153–158.

_____. 1983b. Effect of field application of growth regulators on secondary dormancy of common ragweed (*Ambrosia Artemisiifolia*) seeds. *Weed Sci.* 31:299–303.

Samimy, C., and A. G. Taylor. 1983. Influence of seed quality on ethylene production of germinating snap beans. *J. Am. Soc. Hort. Sci.* 108:767–769.

Schoenbeck, M. W., and G. H. Egley. 1981. Phase sequence of redroot pigweed seed germination response to ethylene and other stimuli. *Plant Physiol.* 68:175–179.

Sen, S., and D. J. Osborne. 1974. Germination of rye embryos following hydration-dehydration treatments: enhancement of protein and RNA synthesis and earlier induction on DNA replication. *J. Expt. Bot.* 25:1010–1019.

Sharples, G. C. 1973. Stimulation of lettuce seed germination at high temperature by ethephon and kinetin. *J. Am. Soc. Hort. Sci.* 98:209–212.

Sivan, A., O. Ucko, and I. Chet. 1987. Biological control of crown rot of tomato by *Trichoderma harzianum* under field conditions. *Plant Dis.* 71:587–582.

Smith, E. M., and F. C. Wehner. 1987. Biological and chemical measure integrated with deep soil cultivation against crator disease of wheat. *Phytopathology* 19:87–90.

Smith, P. T., and B. G. Cobb. 1988. Retained enzymatic activity of dry seeds after osmoconditioning. *HortScience* 23:795 (Abstr.).

_____. 1989. Respiration of Capsicum annuum seed during and after osmoconditioning. Plant Physiol. 89:173 (Abstr.).

Stokes, P. 1965. Temperature and dormancy. Encyclop. Plant Physiol. 15/2:746–803.

Suslow, T. V., and M. N. Schroth. 1982. Rhizobacteria of sugar beets. Effects of seed application and root colonization on yield. Phytopathology 72:199–206.

Szafirowska, A., A. A. Khan, and N. H. Peck. 1981. Osmoconditioning of carrot seeds to improve seedling establishment and yield in cold soil. Agron J. 73:845–848.

Takayanagi, K., and J. F. Harrington. 1971. Enhancement of germination rate of aged seeds by ethylene. Plant Physiol. 47:521–524.

Taylor, A. G., and M. H. Dickson. 1987. Seed coat permeability in semi-hard snap bean seeds: Its influence on imbibitional chilling injury. J. Hortic. Sci. 62:183–189.

Taylor, A. G., Y. Hadar, J. M. Norton, A. A. Khan, and G. E. Harman. 1985. Influence of presowing seed treatments of table beets on the susceptibility to damping-off caused by Pythium. J. Am. Soc. Hort. Sci. 110:516–519.

Taylor, A. G., D. E. Klein, and T. H. Whitlow. 1988. SMP: Solid matrix priming of seeds. Scientia Hort. 37:1–11.

Taylorson, R. B. 1972. Phytochrome controlled changes in dormancy and germination of buried weed seeds. Weed Sci. 20:417–422.

_____. 1986. Water stress-induced germination of giant foxtail Setaria faberi seeds. Weed Sci. 34:871–875.

Thanos, C. A., K. Georghiou, and H. C. Passam. 1989. Osmoconditioning and ageing of pepper seeds during storage. Ann. Bot. 63:65–69.

Thomas, T. H., N. L. Biddington, and D. Palevitch. 1978. Improving the performance of pelleted celery seeds with growth regulator treatments. Acta Hort. 83:235–243.

Tilden, R. L., and S. H. West. 1985. Reversal of the effects of aging in soybean seeds. Plant Physiol 77:584–586.

Tschen, J. S. M. 1987. Control of Rhizoctonia solani by Bacillus subtilis. Trans. Mycol. Soc. Japan 28:483–494.

Valdes, V. M., and K. J. Bradford. 1987. Effects of seed coating and osmotic priming on the germination of lettuce seeds. J. Am. Soc. Hort. Sci. 112:153–156.

Valdes, V. M., K. J. Bradford, and K. S. Mayberry. 1985. Alleviation of thermodormancy in coated lettuce Lactuca sativa cultivar 'Empire' by seed priming. HortScience 20:1112–1114.

Vertucci, C. W., and A. C. Leopold. 1983. Dynamics of imbibition in soybean embryos. Plant Physiol. 72:190–193.

_____. 1984. Bound water in soybean seed and its relation to respiration and imbibitional damage. Plant Physiol. 75:114–117.

Walther, D., and D. Gingrat. 1988. Biological control of damping-off of sugar beet and cotton with Chaetomium globosum or a fluorescent Pseudomonas sp. Can. J. Microbiol. 34:631–637.

Weller, D. M. 1988. Biological control of soilborne plant pathogens in the rhizosphere with bacteria. Ann. Rev. Phytopathol. 26:379–407.

Wickham, B. D., and M. A. Nichols. 1976. Germination studies with "hardened" vegetable seed. New Zeal. J. Exptl. Agr. 4:457–461.

Wolfe, D. W., and W. L. Sims. 1982. Effects of osmoconditioning and fluid drilling of tomato seed on emergence rate and final yield. HortScience 17:936–937.

Woodstock, L. W., and K.-L. Tao. 1981. Prevention of imbibition injury in low vigor soybean embryonic axies by osmotic control of water uptake. Physiol. Plant. 51:133–139.

Woodstock, L. W., and R. B. Taylorson. 1981. Soaking injury and its reversal with poly-ethylene glycol in relation to respiratory metabolism in high and low vigor soybean seeds. Physiol. Plant. 53:263–268.

Wurr, D. C. E., and J. R. Fellows. 1984. The effect of grading and priming of crisp lettuce cultivar 'Saladin' on germination at high temperature, seed vigor and crop uniformity. *Ann. Appl. Biol.* 105:345–352.

Yang, S. F., and N. E. Hoffman. 1984. Ethylene biosynthesis and its regulation in higher plants. *Annu. Rev. Plant Physiol.* 33:155–189.

Zuo, W., C. H. Hang, and G. Zheng. 1988a. Effects of osmotic priming with sodium polypropionate (SPP) on seed germination. Proc. Intern. Conf. Stand Estab. Hort. Crops, Lancaster, PA. p. 114–123.

———. 1988b. Physiological effects of priming with SPP on seeds of pea, tomato and spinach. Proc. Intern. Conf. Stand Estab. Hort. Crops, Lancaster, PA. p. 124–133.

5

Environmental Influences on Seed Size and Composition

M.Fenner
Department of Biology
University Southampton S09 5NH England

I. INTRODUCTION

Size and chemical composition are important aspects of seed quality. Both can be influenced by the conditions under which seeds develop. Although these characteristics are relatively constant in certain species such as the weed *Senecio vulgaris* L. (Fenner 1986), many crop plants show considerable plasticity in both seed size and content. Even a modest increase in seed size can have important consequences for successful seedling establishment and competitive ability, and many seeds have increased commercial value with improved protein or oil content.

Different environmental factors affect seed size and quality in various ways. Published examples have proliferated in recent years, but the data (involving food, fiber, and ornamental crops as well as weeds and non-economic species) are very thinly distributed amongst horticultural, agricultural, physiological, and ecological publications. There is a need for a synthesis of the scattered literature to see if any general trends can be found, so that broad conclusions can be drawn. This review aims firstly to gather together the evidence on environmental effects on seeds during development; secondly, to point out general patterns in the way that specific factors affect seed size and quality; and thirdly, to suggest possible physiological explanations and practical consequences of these effects.

This review does not consider those effects which are due to the genetic contribution of the parent to the offspring, or to "maternal" effects transmitted via the cytoplasm or endosperm during development. These distinctions are covered in a review by Roach and Wulff (1987). Nor does it deal with environmental effects on the genetic constitution of the embryo (e.g., mutations due to ultraviolet light). This review is only concerned with strictly phenotypic effects: how the growing conditions of the plant can affect the development of the seed and determine its size and chemical content. It begins with some general observations about these two seed characteristics, and then deals with environment-induced parental effects. The influence of the major abiotic factors are dealt with separately. The concluding section summarizes the data and suggests direction for future research. This review concentrates mainly on papers published in the last 15 years (1975–90). Parental effects on *germination* are dealt with in Gutterman (1980/81, 1982, 1991) and in Fenner (1991). Roach and Wulff (1987) provide a wide-ranging review of maternal (and paternal) effects on seed size, composition, and germinability.

The common names for familiar crops are used in the text of this paper. The Latin binomials for these are given in Table 5.1.

II. SEED SIZE AND CHEMICAL COMPOSITION

A. Variability

Within virtually all species which have been investigated there exists considerable variation in mean seed size between populations. These differences may be the result of natural selection or genetic drift in different geographical regions, or they may be due to a combination of genetic and environmental factors. Only experiments in which the different populations are grown in the same environment can distinguish between the genetic and environmental components (e.g., in *Chenopodium bonus-henricus* L. [Dorne 1981], *Prunella vulgaris* L. [Winn and Werner 1987]). Even within individual plants, seed size variation can have a genetic basis since each embryo will normally be genetically distinct, at least in cross-pollinated plants. The use of mean seed weight for making comparisons

Table 5.1   Latin names and authorities for common crops mentioned in the text.

Adzuki bean	Vigna angularis (Willd.) Ohwi & Ohashi
Barley	Hordeum vulgare L.
Broad bean	Vicia faba L.
Carrot	Daucus carota L.
Chickpea	Cicer arietinum L.
Cocoa	Theobroma cacao L.
Common bean	Phaseolus vulgaris L.
Cotton	Gossypium hirsutum L.
Cowpea	Vigna unguiculata (L.) Walp. ssp. unguiculata
Flax	Linum usitatissimum L.
Lettuce	Lactuca sativa L.
Lupin	Lupinus angustifolius L.
Maize	Zea mays L.
Oat	Avena sativa L.
Oilseed rape	Brassica napus L.
Pea	Pisum sativum L.
Peanut	Arachis hypogaea L.
Perennial rye grass	Lolium perenne L.
Pigeon pea	Cajanus cajan (L.) Millsp.
Rice	Oryza sativa L.
Sorghum	Sorghum bicolor (L.) Moench.
Soybean	Glycine max (L.) Merr.
Sugar beet	Beta vulgaris L.
Sunflower	Helianthus annuus L.
Tomato	Lycopersicon esculentum Mill.
Turnip	Brassica rapa L.

often masks the considerable size differences among seeds within individual plants. Hendrix (1984) reports a two-fold variation in seed weight within plants of *Pastinaca sativa* L. These differences are thought to be due mainly to variations in resources available to individual seeds because of differences in the micro-environment of each position on the parent plant. Within the general large-scale parental environment each seed exists in its own local micro-environment which determines the resources available to it. In spring wheat, distal kernels have slower growth rates and shorter seed-filling periods than proximal ones (Simmons and C· Jokston 1979). The growth of maize kernels at the tips of cobs appears to be limited by inadequate assimilation supply (Hanft et al. 1986). In sorghum, grains at the base of the panicles are smaller and have a slower growth rate than those at the distal end (Muchow 1990). These position effects on size can have long-term consequences for the individual seeds. For example, in one experiment, oats grown from primary kernels yielded 14.5% more grain and 13.1% more straw than plants grown from secondary kernels (Brinkman 1979).

The extent of internal competition among seeds on the same plant can be readily quantified by removing seeds or fruits at an early stage of

development and measuring the effect on the remaining seeds. Removal
of pods from soybeans results in larger seeds in the remaining pods (Egli
et al. 1985, 1987a), and similar effects of surgical removal of fruits are
known for wheat (Pinthus and Millet 1978; Radley 1978), carrot (Gray et
al. 1986), maize (Woronecki et al. 1980), Sesbania species (Marshall et al.
1985), and the lily, Clintonia borealis Ait. (Raf.) (Galen et al. 1985). Kiniry
et al. (1990) reduced kernel numbers per ear in maize by bagging ears to
control pollination. Seed number was reduced by 15–45%, and individual seed weight increased by 19–25%. Individual seed weight can be
readily reduced by defoliation of the branch bearing the fruit, as for
example in soybean (Fehr et al. 1985; Egli et al. 1987a), Rubus
chamaemorus L. (Ågren 1989), Trifolium pratense L. (Rincker et al. 1977),
Catalpa speciosa Warder (Stephenson 1980), and Desmodium paniculatum L. DC (Wulff 1986a). Predation of leaves by insects can have the
same effect, as has been demonstrated in Rumex obtusifolius L. and R.
crispus L. (Bentley et al. 1980), Senecio jacobaea L. (Crawley and
Nachapong 1985), and wheat (Wratten 1975).

The chemical composition of a seed can also be influenced by its position on the plant. In Abutilon theophrasti Medic. the amounts of nitrogen,
phosphorus, and potassium in the seeds vary with fruit position (Benner
and Bazzaz 1985). In oilseed rape the relative proportions of seven fatty
acids also vary with the position of the pod (Diepenbrock and Geisler
1979). This variation among seeds from the same plant is undoubtedly
due to competition for nutrients among fruits or among seeds within the
same fruit, with some seeds being more favorably placed in this respect.
First and second kernels in a wheat spikelet have higher nitrogen concentrations than third and fourth kernels (Simmons and Moss 1978). When
Ries et al. (1976) supplied a nitrogen fertilizer to wheat, the resulting
protein increase was greatest in the grains of the proximal florets within
each spikelet. In soybeans, depodding also increases the concentration of
protein in the seeds of the remaining pods (Openshaw et al.1979).

B. Heteromorphism

The maternal control of seed size is very clearly seen in those species
which produce two or more distinct seed types which may differ in both
size and shape. This somatic polymorphism (usually dimorphism) is seen
in many members of the Asteraceae, in which seeds produced by ray
florets on the capitulum margins tend to be larger than seeds from the disk
florets. They also tend to differ in their germination requirements (e.g.,
Bidens pilosa L. [Forsyth and Brown 1982] and their dispersability. Much
of the variation among seeds from the same batch may be hidden by being
physiological rather than morphological. Thus, within any batch of seeds
set to germinate, a proportion will usually germinate while the others
remain dormant, indicating a physiological dimorphism within the

population. The parental environment can affect the relative proportions of the two morphs which the plant produces. For example,
increased nutrient availability causes a shift between the production of
the two morphs seen in the annual grass *Amphicarpum purshii* Kunth.
(Cheplick 1989). The seasonal change in weather conditions may account
for the fact that the ratio of ray to disk seeds increases in *Heterotheca latifolia* Buckl. (Venable and Levin 1985) and *Tragopogon dubious* L.
(McGinley 1989).

C. Advantages of Greater Seed Size

Although a large seed may be a handicap for dispersal in the wild, many
studies indicate that larger individual seeds in a population often exhibit
marked advantages over smaller ones. The likelihood of emergence in the
field is greatly enhanced because of the ability to establish at greater
depths. This has been well demonstrated in perennial rye grass (Naylor
1980), *Raphanus raphanistrum* L. (Stanton 1984), *Pennisetum americanum* (L.) Leeke (Lawan et al. 1985), *Lithospermum carolinense* (Walt.)
MacM. (Weller 1985), *Prunella vulgaris* (Winn 1985), and *Desmodium
paniculatum* (Wulff 1986b). In some cases, the rate of germination itself is
greater in the larger seeds (Weis 1982; Verkaar and Schenkeveld 1984;
Morse and Schmitt 1985), although this is by no means a universal feature of larger seeds. Hendrix (1984), Lafond and Baker (1986), and Roach
(1986) all report species in which the smaller seeds germinated more
rapidly, so that the advantage of early germination may, in these cases at
least, partly compensate for their smaller initial size.

The main advantage of increased seed size is the larger seedlings which
they produce. In the many trials which have compared plants resulting
from small and large seeds of the same species, the initial size advantage
is usually maintained into maturity (Howe and Schupp 1985; Wurr et al.
1986; Ellison 1987). In many cases the advantage is exhibited right
through to seed production, with plants derived from large seeds
producing significantly more seeds (Brinkman 1979; Kumar et al. 1979;
Dharmalingam and Ramakrishnan 1981; Murray et al. 1984; Stanton
1985). For many horticultural purposes this increased reproductive
output may be as important as the increased plant size. Although large
seeds often show these advantages over small ones, a number of studies
have found no long-term benefit (e.g., Dhillon and Kler 1978; Gill and
Singh 1979; Singh and Gill 1981: Cideciyan and Malloch 1982; Lahiri et
al. 1982; Choudhari et al. 1984).

The advantage of greater size in the seed is perhaps best expressed in
those cases where the performance of the progeny has been tested in competitive situations. Wulff (1986c) grew small and large seeds of
Desmodium paniculatum singly, in pure stands of the same-sized seed,
and in mixtures of small and large seeds. Only in the last case did plants

derived from large seeds produce significantly more seeds than plants derived from small seeds. The special benefit of large seededness in competitive situations is confirmed by Gross (1984) for six monocarpic perennials, by Dolan (1984) for *Ludwigia leptocarpa* (Nutt.) Harao, and by Crawley and Nachapong (1985) for *Senecio jacobaea* L. However, in bean, the largest 10% are often damaged mechanically, and the middle 80% are the most productive.

D. Chemical Composition

In addition to influencing the size of a seed, the parental environment can also determine its chemical composition. This can have important consequences for the growth of the resulting seedlings. Seedling vigor in wheat is related to higher levels of nitrogen in the grain (Bulisani and Warner 1980). Seedling competitive ability in *Abutilon theophrasti* has also been attributed to the same cause (Parrish and Bazzaz 1985). Low levels of phosphorus in pea can adversely affect the growth of the plants right through to adulthood, especially in phosphorus deficient environments (Austin 1966b). The concentrations of certain trace elements in seeds may also be significant for growth. Singh and Bharti (1985) found that the manganese content of wheat grain showed a direct linear relationship with plant height in different cultivars.

III. FIELD EVIDENCE

A. Location Effects

Some of the clearest evidence for phenotypic variation in seed size and chemical content comes from experiments in which the same cultivar is grown simultaneously in various locations, resulting in the production of seeds with different characteristics. Kahn and Stoffella (1985) grew nine cultivars of cowpea in Florida and Oklahoma and obtained marked differences in most of the components of yield, including mean seed size (133 mg in Florida vs. 165 mg in Oklahoma). Identical strains of the annual grass *Taeniatherum asperum* (Simonkai) Nevski grown in contrasting sites in Washington state also showed this effect, with mean seed weight of 5.4 mg in cool, moist Pullman and 4.6 mg in warm, dry Hooper (Nelson et al. 1970). Both these studies suggest that low temperatures in the field are associated with greater seed weight. Location effects on seed size are reported by Ellis and Brown (1986) for barley, McGraw et al. (1986) for *Lotus corniculatus* L., and Winn and Werner (1987) for *Prunella vulgaris*.

The chemical content of seeds from the same cultivars can vary with location. Baenziger et al. (1985) report considerable variation in the protein content in the grain of winter wheat grown simultaneously on several sites. This may be due to any number of soil or climatic factors, and not just soil nutrient levels. Soil water content at different sites can affect the grain protein content in wheat (Karathanasis 1980). The concentration of manganese in barley grains also varies markedly among sites (Longnecker and Uren 1990).

B. Year Effects

Another field indicator of the environmental influence on seed quality is the annual variation which occurs in the same location. Mean seed size and mineral content change on the same plant from year to year, presumably reflecting the annual variations in the weather conditions. Soybean has been shown to have a marked annual fluctuation in seed size when grown under the same cultivation regime at the same site (Collins 1981; Gubbels 1981; Bharati et al. 1986; Egli et al. 1987; Elmore 1987). This appears to be due to differences in weather conditions, e.g., rainfall in the case of soybean monitored by Wright et al. (1984), and temperature in the case of carrot investigated by Gray et al. (1983). Year effects on seed size have also been found in sugar beet (Wood et al. 1980) and *Lotus corniculatus* (McGraw et al. 1986). In the case of lettuce, Wurr et al. (1986) found that *seedling* size was significantly affected by year of production. Weather conditions are also thought to be responsible for year-to-year variation in seed chemical quality in peanut (Ketring et al. 1978), pea (Gubbels 1981), and *Limnanthes alba* Hartweg ex Benth. (Crane et al. 1981). Such environmental effects may interact with genetic factors.

C. Season Effects

In addition to year-to-year variation, many studies report marked differences in seeds produced at different times within any one season. The most common pattern is a decline in size as the season progresses. Cavers and Steele (1984) monitored seed weight from individual plants in eight short-lived weed species throughout one growing season. In all cases late seeds were smaller than early ones. A reduction of 25% was recorded in the case of *Melilotus alba* Medic. with lesser, although significant, reductions in the other species. Smaller seeds from late-formed fruits are reported for *Rumex crispus* L. (Maun and Cavers 1971), cotton (Leffler et al. 1977), sugar beet (Malik 1977), oilseed rape (Clarke 1979), soybean (Gbikpi and Crookson 1981; Egli et al. 1987b), *Leontodon hispidus* L. (Fuller et al. 1983), *Daucus carota* L. and *Scabiosa columbaria* L. (Verkaar and Schenkeveld 1984), *Sesbania* spp. (Marshall et al. 1985), *Geranium carolinianum* L. (Roach 1986), *Clarkia unguiculata* Lindl. (Smith-Huerta and Vasek 1987), and *Prunella vulgaris* (Winn and Werner 1987).

Seasonal decline in seed size appears to be a very general effect. It may be caused by a decline in resources (most probably nutrients and water) with time. In addition, especially in annual species, the physiological condition of the plant will alter as the plant ages. The "season effect" is often difficult to distinguish from the "position effect" on the plant. Sheldrake and Saxena (1979) note that in chickpea, later nodes have pods with smaller seeds. They attribute this to a progressive reduction in assimilate supplies because of declining leaf area as the plants senesce, but the changing seasonal conditions may also be a factor. This problem can be investigated by sowing seeds of annual plants over a period of time in the spring and early summer and noting the effect of ripening time on the size of seeds from equivalent positions on the plants. In many cases where the effect of sowing (and hence ripening) times on seed size has been investigated, later ripening often results in lower seed size, e.g., lupin (Perry 1975) and sunflower (Unger 1986). For some species this may be due to higher mean temperatures in the late season, as in the case of wheat in New South Wales (McDonald et al. 1983) and winter field pea (Murray et al. 1984).

Seeds matured at different times in the season can also vary in their chemical quality, such as oil content in rapeseed (Auld et al. 1984). Late season soybeans have higher levels of protein (Gbikpi and Crookston 1981). The oil quality of sunflower seeds tends to improve gradually over the season, with the best oil being produced under the cooler conditions of late summer (Unger 1986). This change in oil quality with time may not, however, be due entirely to the late season temperatures, as daylength is also decreasing at this time.

In most of these examples from the field, the change in seed size and quality can be attributed to some environmental difference among sites, or to variations in weather from year to year, or to changes within a season. However, because of the complexity of the field situation, it is difficult to attribute causes to any specific environmental factor. Wet years also tend to be cooler. Warm days in mid-summer are accompanied by a long photoperiod in temperate climates. A great deal of experimental work has been carried out in recent years on the influence of individual factors (and combinations of them) on seeds, and from this research it is possible to derive certain generalizations about the effects of the major factors.

IV. ENVIRONMENTAL FACTORS

A. Temperature

Numerous studies on a wide range of species show that high temperatures during development result in smaller seeds, and vice versa. Examples include *Dactylis glomerata* L. and perennial rye grass (Shimzu et al. 1979), common bean (Siddique and Goodwin 1980), wheat (Campbell et al. 1981; Wardlaw et al. 1989), *Pennisetum typhoides* S. & H. (Ong 1983), various Apiaceae (Putievsky 1983), lupin (Downes and Gladstones 1984a), flax (Green 1986), *Desmodium paniculatum* (Wulff 1986a), and sorghum (Kiniry and Musser 1988). A notable exception is sugar beet (Wood et al. 1980). In most cases the reduction in mean seed weight is within the range 15–25%. The 38% decrease seen in one treatment in Huxley and Summerfield's (1976) study on cowpea is exceptionally high. The reduction in seed size is due to the differential effect of temperature on the seed ripening and seed filling processes. High temperatures increase the rate of ripening and so reduce the period of time available for the assimilation of photosynthesis by the seed. At higher temperatures the proportional reduction in the duration of seed growth is usually much greater than any increase in seed growth rate (resulting in smaller seeds), as seen in wheat (Ford 1976; Sofield et al. 1977; Chowdhury and Wardlaw 1978); broad bean (Dekhuijzen and Verkerke 1986); and in rice, barley, and sorghum (Chowdhury and Wardlaw 1978). For *Pennisetum americanum*, Fussell et al. (1980) found that seed filling rates were constant at all temperatures tested (in the range 21/16°C to 33/28°C), so that seed weight was directly related to the period during which the grain was able to assimilate dry matter. Similar results have been obtained for rice (Negato and Ebata 1965), *Paspalum dilatum* Poiret (Pearson and Shah 1981), and jojoba *Simmondsia chinensis* (Link) C. Schneider (Wardlaw and Dunstone 1984). Wiegland and Cuellar (1981) showed that in the case of wheat, each increase of 1°C in mean daily air temperature during grain filling shortened ripening time by 3.1 days, and reduced each kernel by a mean of 2.8 mg. In carrot, Gray et al. (1988) found that the apparent decrease in seed size at higher temperatures is due to reduction in the pericarp tissue only, while the embryo and endosperm were not significantly affected.

   A crucial factor in determining the effect of temperature on seed weight is the exact stage in the plant's life cycle at which it experiences the relevant temperatures. Largest seeds of lupins were obtained when temperatures were high before flowering and low after flowering (Downes and Gladstones 1984b). In wheat, Ford and Thorne (1975) found no pre-anthesis effect of temperature on seed weight, but cool post-anthesis temperatures gave larger grains. Johnson and Kanemasu (1983) applied three temperature treatments at five growth stages to winter wheat. Grain weights were significantly increased by high temperatures applied early in development, but were reduced by late applications of high temperatures. Essentially similar results were obtained for wheat by Wardlaw (1970) and by Warrington et al. (1977). Egli and Wardlaw (1980) narrowed the period of maximum sensitivity to temperature in soybean to the period of flowering and pod set. High temperatures (33/28°C) applied during this time reduced seed growth rate by 36% regardless of the temperature during the subsequent seed filling period. In wheat the greatest grain size reductions with high temperatures occurred during a 10-day period commencing 7–10 days after anthesis (Tashiro and Wardlaw 1990). Grains from the upper florets of a spikelet were more sensitive to high temperatures than those at the base.

A particularly interesting phenomenon is the differential effect that night vs. day temperatures have on seed size. The results of three investigations differ in this respect. Doto and Whittington (1981) found that in cowpea both night and day temperatures act in the same direction (warm conditions giving smaller seeds). However, Huxley and Summerfield (1976) working on the same species found that mean seed weight was decreased by 18.2% in warm nights, but *increased* by 14.8% in warm days. Experiments by Seddigh and Jolliff (1984b) on soybean suggest that warm nights *increase* seed size. It may be that each species differs in this respect, but clearly the differential effects of day and night temperatures on seed size need further investigation .

Most of the experiments carried out on the effect of temperature on seed chemical composition have been done with the aim of determining how a particular seed product is affected. For example, in sunflower the quality of the oil is markedly influenced by temperature during seed development. The major fatty acid components of the oil are the diunsaturated linoleic acid (18:2) and the mono-unsaturated oleic acid (18:1). At low temperatures (ca. 12°C) linoleic predominates in the ratio of about (6:1). At higher temperatures (ca. 28°C) the proportions are almost equal (Harris et al. 1980). Goyne et al. (1979) obtained a strong negative correlation between linoleic acid and the mean temperature after the start of flowering. Since the linoleic acid component is the more valuable, cool temperatures during seed development are essential for a high quality product. The effect of temperature on the ratio of linoleic:oleic acid has also been confirmed for *wild* sunflowers (Seiler 1983). The ratio can be readily manipulated by delaying the sowing date so that the seeds develop under cooler conditions late in the season (Keefer et al. 1976; Unger 1980; Owen 1983; Jones 1984). Knowing the relationship between temperature and the linoleic:oleic acid ratio, it is possible to predict the likelihood of an area with a known climatic regime producing a crop of the required quality (Harris et al. 1980). Similar alterations in oil quality with temperature are seen in seeds of flax (Green 1986), cocoa (Wright et al. 1983), and in the wax quality in jojoba seeds (Dunstone et al. 1984).

Temperature also affects the protein concentration in the seeds of a number of species. In general, high temperatures are associated with higher levels of protein, as in spring wheat (Campbell and Davidson 1979; Campbell et al. 1981). Late season cool temperatures can reduce the protein content of soybean (Radford et al. 1977). The higher protein concentration at high temperatures may simply be the result of a reduction in carbohydrate accumulation. In the case of rapeseed, protein content may be inversely related to oil content (Hodgson 1979).

B. Nutrients

Evidence for the effect of nutrient supply on seed size and chemical quality comes mostly from fertilizer experiments. In most cases specific nutrients (usually nitrogen, phosphorus, or potassium, singly or in combination) are applied, and all components of yield measured. Experiments in which the levels of several nutrients are added simultaneously have nearly all resulted in an increase in seed size. Parrish and Bazzaz (1985) grew *Abutilon theophrasti* in dilutions of a standard nutrient solution (×1.0, 0.5, 0.25, and 0.125) and obtained a consistent increase in seed size with increasing nutrients supplied. The seeds from the poorest-supplied plants were 25% lighter than the controls. Broadly similar results have been found in soybean (Boswell and Anderson 1976), tomato (Varis and George 1985), *Asclepias syriaca* L. (Willson and Price 1980), and *Desmodium paniculatum* (Wulff 1986a). This positive relationship between general nutrient supply and seed weight may simply be a reflection of a higher growth rate of the seed during the filling period. However, some species may be much less plastic in this respect. No significant effect on mean seed size in *Senecio vulgaris* was obtained by Fenner (1986) over a five-fold dilution of a standard nutrient solution, even though the lowest nutrient level was markedly limiting for parental growth.

Where the level of a *single* nutrient is varied, the results for seed size are much less consistent. For example, taking nitrogen fertilizer additions alone, seed size increases have been recorded for perennial rye grass (Ene and Bean 1975), soybean (Ham et al. 1975), turnip (Chakrabarti 1983), and maize (Eck 1984). In contrast, seed size *reductions* with added nitrogen have been found in wheat (Gardener and Jackson 1976; Dexter et al. 1982; Read and Warder 1982; Bauer et al. 1985; Frederick and Marshall 1985), soybean (Garcia and Hanway 1976), and tomato (George et al. 1980). In other experiments no significant effect of added nitrogen could be detected, e.g., in wheat (Miezan et al. 1977; Johnson et al. 1984). It is possible that these apparently contradictory results could be explained in terms of a bell-shaped response curve. Experiments conducted using nitrogen levels in the suboptimal range would record a positive response to increased fertilizer levels, while those conducted in the supraoptimal range would record negative responses. An interesting confirmation that both high and low levels of a mineral nutrient may reduce seed weight comes from the work of Heenan and Campbell (1980) on the effects of a range of manganese levels on soybean.

Both the size and number of the seeds produced by a plant are probably determined by the nutrient status of the parent plant at the time of flower bud initiation, since much of the nutrient content of the seeds may be translocated from the vegetative tissues. The effect, therefore, of the timing of nutrient applications on seed size is crucial. In general, the earlier the added nutrient is supplied, the more effective it will be. Langer and Liew (1973) added nitrogen at two levels at pre-flowering, ear emergence, and post-ear emergence phases of development in wheat. Early application of high nitrogen markedly increased mean grain weight (by up to 21%); late applications had virtually no effect. The timing of nitrogen fertilizer (as urea) has also been shown to affect the size of the grain in rice (Humphreys et al. 1987).

The mineral nutrient supply to the parent plant is the most important single determinant of seed chemical composition. As a general rule, addition of a particular mineral nutrient results in increased concentrations of that mineral in the seeds. Added phosphate increased the phosphorus concentration in pea (Peck et al. 1980), soybean (Cassman et al. 1981), wheat (Porter and Paulsen 1983), and common bean (Vieira 1986). Added nitrogen almost invariably results in a higher seed protein content, e. g., in wheat (Dexter et al. 1982; Altman et al. 1983; Glenn et al. 1985), rice (Allen and Terman 1978), maize (Warrent et al. 1980; Wolfson and Shearer 1981), and cottonseed (Elmore et al. 1979). Calcium fertilizer added to peanut increased the calcium concentration in the seeds (Coffelt and Hallock 1986).

In legumes, the nitrogen (and hence protein) levels of the seeds can be markedly influenced by the particular strain of *Rhizobium* supplied. Certain strains are apparently more efficient at nitrogen fixation than others, at least in combination with a given legume. For example, in soybean the use of the *Rhizobium* strain USDA 110 increased seed protein by 13–16% compared with the strain USDA 31 (Israel 1981). Similar protein enhancement in soybean by inoculating with particular strains of *Rhizobium* have been reported by Hanus et al. (1981) and Rennie and Debetz (1984). The effect is clearly similar to that of nitrogen additions in non-nitrogen fixing species.

The addition of trace elements to crops also usually results in increased concentrations in the seed. This has been found for manganese in soybean (Boswell et al. 1981; Parker et al. 1981), copper in wheat (Hill et al. 1979; Loneragan et al. 1980), zinc in soybean (Raboy and Dickinson 1984), cadmium and selenium in lettuce and wheat (Cary 1981), boron in soybean (Touchton and Boswell 1975), and cobalt in lupin (Robson et al. 1980). It is possible that the normal seed complement of trace elements may be well in excess of that required for the growth of the plant, at least for the first season. Meagher et al. (1952) calculated that seeds of certain legumes contained ten times the amount of molybdenum needed for the adult plant. Woodbridge (1969) grew pea plants in boron-free conditions for four generations before the seed boron content was sufficiently reduced to cause deficiency symptoms. Seeds may even preferentially accumulate certain trace elements. When additional copper was supplied to wheat, the increased level in the grain was disproportionately high in comparison with the small amount supplied (Loneragan et al. 1980).

An interesting aspect of the effects of parental nutrient supply is the change seen in the amino acid composition of the seed proteins. Nitrate applied to wheat, for instance, increased the proportions of glutamate, proline, and phenylalanine, while it reduced the proportions of threonine, serine, glycine, alanine, and valine (Dubetz et al. 1979). Similar shifts in amino acid composition with added nitrogen are seen in cottonseed (Leffler 1977; Elmore et al. 1979) and maize (Rendzig and Broadbent 1979; Wolfson and Shearer 1981). The addition of sulphur can also alter the amino acid balance of the seed proteins. Hojjati (1976) induced a marked increase in the amino acid methionine in common bean by the addition of sulphate fertilizer. The endosperm of sulphur-deficient wheat contains less methionine and cysteine, and more non-sulphur amino acids such as arginine and aspartic acid than endosperm of control plants (Wrigley et al. 1980). When Gayler and Sykes (1985) subjected soybean to sulphur-deficiency, the level of total protein was almost unaffected, but there was a 40% decrease in the level of glycinins, storage proteins which contain a high proportion of methionine and cysteine. There was a simultaneous increase in the level of beta-conglycinins which contain less of the sulphur amino acids. Essentially similar results have been found in lupin (Gillespie et al. 1978) and pea (Randall et al. 1979; Chandler et al. 1983, 1984). Differences in the availability of nitrogen can also alter the balance of proteins in barley (Turley and Ching 1986). A remarkable flexibility in producing alternative types of storage proteins is seen in the experiments of Randall et al. (1979) in which a switch to the synthesis of specific proteins was induced by restoring sulphur levels more than half way through seed development. All these studies suggest that there are probably many genes which code for seed reserve proteins, and that their expression may be determined by the balance of nitrogen and sulphur available.

In species with oily seeds, a frequent result of changes in nutrient supply is a shift in the balance between the oil and the protein contents. In general, and increase in nutrients (especially nitrogen) results in more protein but less oil. Boswell and Anderson (1976) applied various fertilizer treatments to soybean and obtained a clear reciprocal relationship between oil and protein contents. This relationship, one of the most consistent effects of parental nutrition on seed chemical composition, has been noted in soybean (Ham et al. 1975; Sesay and Shibles 1980; Poole et al. 1983), in sunflower (Mathers and Stewart 1982), and in meadowfoam (*Limnanthus alba*) (Crane et al. 1981).

C. Drought

For most species, the effect of a general prolonged drought is a reduction in seed size. Soybean seed size declines in dry years (Wright et al. 1984), and drought experiments confirm this observation (Meckel et al. 1984; Hill et al. 1986). The reduction in seed dry weight may be considerable. The mean weight of *Desmodium paniculatum* seeds was reduced from 7.1 mg in well-watered controls to 4.8 mg by restricting watering from the time of flowering (Wulff 1986a). Similar effects have been reported for *Avena fatua* L. (Sawhney and Naylor 1982), *Clarkia unguiculata* (Smith-Huerta and Vasek 1987), maize (Eck 1986), *Sorghum bicolor* (Benech Arnold et al. 1991), *Trifolium subterraneum* L. (Andrews et al. 1977), and wheat (Sionit et al. 1980). In addition to a general reduction in seed size, there may be a thickening of the seed coat (Hill et al. 1986). Overall, the effect of drought is to reduce total weight of seed produced, but there is often a partial trade off between number and size (Woodward and Begg 1976; Izzeldin et al. 1980; Gales and Wilson 1981).

The effect of reduced water supply is highly dependent upon the timing of the stress relative to anthesis. Stress applied before or at flowering usually reduces seed number rather than size (Eck 1986). If adequate supplies of water subsequently become available, the seeds may be larger (though fewer) than in unstressed plants (Sionit and Kramer 1977). This may be due to reduced competition between the seeds on the same plant. Drought suffered late in the post-anthesis phase is much more effective at reducing seed size than earlier applied stress, e.g., in maize (Harder et al. 1982) and cowpea (Turk et al. 1980). Precisely timed irrigation can also affect seed size and number. Soybean irrigated at flowering produced more (but smaller) seeds than non-irrigated controls. When irrigated at the seed enlargement stage, seed weight was increased considerably with little effect on numbers (Korte et al. 1983). An interesting complication is that, at least in one case, drought applied early in the reproductive period may alter the dry matter partitioning of the plant to increase harvest index (Snyder et al. 1982).

Drought can also result in an increase in the concentration of storage proteins in the seeds. A highly significant negative relationship between percent protein and water supply in winter wheat was obtained by Karathanasis et al. (1980), and the phenomenon is widely reported in other species such as perennial rye grass (Ene and Bean 1975), maize (Harder et al. 1982), spring wheat (Nutall et al. 1979), soybean (Pikaard and Cherry 1984), common bean and adzuki bean (Robinson 1983). Increased salinity (which results in physiologically induced drought) can also induce higher seed protein concentrations, e.g., in wheat (Francois et al. 1986). The increase may be quite considerable. Jurgens et al. (1978) obtained a 33% increase in protein concentration in maize subjected to drought during grain fill, though this was accompanied by an 18% drop in oil concentration. As with other environmental effects on seed quality, the timing of the drought period determines whether the protein level is affected. High moisture stress applied to spring wheat during the boot stage increased protein content, whereas this was unaffected by stress applied later in the growth cycle (Campbell et al. 1981).

D. Light

In general, reduced illuminance during seed development results in a reduction in size. This has been shown in wheat (Wardlaw 1970; Spiertz 1974; Ford and Thorne 1975; Jenner 1979), carrot (Gray et al. 1986), pea (Gubbels 1981), soybean (Peet and Kramer 1980; Egli et al. 1987a), maize (Kiniry and Ritchie 1985), and *Trifolium subterraneum* (Collins et al. 1978). The reduced light intensity no doubt reduces seed growth rate, while having little effect on the duration of the seed filling period. For example, in wheat, duration as such was unaffected by a two-fold range in mean daily radiation after flowering (Sofield et al. 1974). The shorter seed filling period which usually occurs with high radiation levels may be due to the increase in temperature which normally accompanies the higher radiation.

It may be for the same reason that short days tend to result in smaller seeds. Examples are seen in *Aegilops ovata* L. (Datta et al. 1972), pea (Reid 1979), perennial rye grass (Bean 1980), carraway (*Carum carvi* and dill (*Anethum graveolens*) (Putievsky 1983), and *Desmodium paniculatum* (Wulff 1986a). As with reduced illuminance, short days reduce the overall rate of assimilation while leaving the seed filling period relatively constant. However, there are a small number of anomalous cases in the literature which defy these generalizations. Seed size increased in shade in *Trifolium subterraneum* (Taylor 1976) and with short days in *Coriandrum sativum* (Putievsky 1983).

As with other environmental factors, the *timing* of any light treatment has a crucial effect on the outcome. Seed number (and hence size) is often determined at the very earliest stages of allocation. In *Chenopodium rubrum* L. the effect of a longer photoperiod (15 vs. 12h) is most strongly exerted during the development of the floral primordia (Cook 1975). In this case, longer photoperiods resulted in smaller (but more numerous) seeds. Seed size in *Avena fatua* can even be influenced by the daylength experienced by the plant before anthesis (Klinck and Sim 1977).

Little work has been done on the effects of light conditions on the chemical quality of seeds. Gubbels (1981) found an increase in the protein content of pea with shading, and Seiler (1983) suggests that the oil content of wild sunflower is influenced by daylength. Short days are also known to increase the endogenous gibberellins in pea (Ingram and Browning 1979) and in *Lactuca scariola* L. (Gutterman et al. 1975).

E. Competition

Competition from neighboring plants can impose multiple stresses upon the individual by depriving it of light, nutrients, and water, as well as reducing the temperature of its environment. Except in very harsh environments, where individuals may be widely spaced, most plants grow in close association with other plants. Evidence for the effect of competition from other plants on seed size comes mainly from trials in which plants are grown at various densities, or from experiments on the effects of weeds on crops.

The majority of cases reported show that seed size decreases with severity of competition (based on increased density of the test plant or of accompanying weeds). A striking example of an effect of inter-specific competition on individual seed weight is provided by Bhaskar and Vyas (1988) in their experiments on wheat grown with the weed *Chenopodium album* L. Mean grain weight of the wheat declined from 45.9 mg when grown weed-free, to 6.4 mg when the weeds outnumbered the crop by a factor of three. However, this case is quite exceptional. Most competition-induced reductions are in the range 15–25%. One of the few studies that have sought to clarify the mechanism involved is that of Poneleit and Egli (1979) in maize. In their experiments the rate of seed filling was similar over the whole range of density levels investigated, but the duration of the filling period was reduced by 2–5 days.

Within a species, the effect of competition on seed weight can vary considerably from one cultivar to another, e.g., in common bean (Westermann and Crothers 1977). It can also depend upon the particular competitor as some species have much more effect than others (Wahua and Miller 1978; Elmore and Jackobs 1984). The effect may even be confined to certain seeds within the plant; e.g., seeds in the upper canopy in soybean (which normally tend to be smaller) were increased in size by 24% when plant density was reduced, while seeds in the lowest part of the canopy were virtually unaffected (Weil and Ohlrogge 1976).

There is evidence that there is a wide plateau in the response curve for most species, with size change only evident at the extremes of competition. Thus in trials with *Lotus corniculatus*, mean seed size increased only at densities which were too low for optimum yield per unit area (McGraw et al. 1986). At the other extreme, seed size in *Oenothera biennis* L. decreased only at very high parent plant densities (Kromer and Gross 1987).

Because of the complexity of the effects of one plant on another (e.g., the temperature reduction might be expected to have effects opposite to those of nutrient, light, and water deprivation), it is not surprising that in some cases competition *increases* seed size. This was found in pigeon pea (Akinola and Whiteman 1975) and in winter wheat (Frederick and Marshall 1985). Where competition reduces the number of seeds produced, the remaining ones may be larger as a result of weaker *internal* competition for resources (Reeves 1976). The timing of competition may determine the balance between size and number (Fischer and Wilson 1975).

Seed chemical content can also be affected by the competition experienced by the parent plants. Experiments investigating the effects of crop density on seed quality generally indicate that, as competition increases, protein concentration declines while oil concentration increases. In sunflower, the increase in oil concentration may be small (Jones 1984; Majid and Schneiter 1987) or only occur in some cases (Prunty 1981), but the general trend is clear. A reduction in protein (or nitrogen) content of the seeds with increased plant density has been found in rice (Nandisha and Mahadevappa 1984), wheat, and barley (Read and Warder 1982). Weeds can have the same effect (Nakoneshny and Friesen 1961). The reciprocal effect of competition on oil and protein is well illustrated in the experiments of Robinson et al. (1980) on sunflower, where oil increased by 13% and protein decreased by 32% with parent plant density. The reduced level of protein is presumably due to reduced availability of nitrogen under highly competitive conditions, and the lower level of nitrogen may result in surplus fixed carbon being diverted to oil synthesis. This suggests that for oil crops, dense planting may be of some advantage, providing that total productivity is not impaired.

F. Interactions Between Factors

In many of the experiments investigating the environmental effects on seeds, two or more factors may be altered simultaneously. In such cases there are often interactions (either positive or negative) between the factors. For example, in soybean, there is a marked effect of temperature on the response of the seed to daylength. Raper and Thomas (1978) grew soybean under short and long day conditions under three temperature regimes (day/night temperatures of 22/18°, 26/22°, and 30/26°C). In line with most other studies, the seed weight declined with increasing temperature, but under long days the temperature effect was much more pronounced. However, at low temperatures the short day seeds were smaller, whereas, at higher temperatures the long day seeds were smaller. Thus a long photoperiod may have the effect of increasing or decreasing seed weight depending on the temperature at which the test is conducted. The effect of fertilizers can also be temperature dependent. In spring wheat, the level of grain protein was increased by nitrogen additions at day/night temperatures of 22/12°C, but was unaffected by nitrogen at 27/12°C (Campbell and Davidson 1979; Campbell et al. 1981).

Soil type can determine the seed response to drought. Miller and Burke (1983) found that seed weight in common bean either increased or decreased with irrigation depending on which of two contrasting soils the plants were grown on.

In many experiments, nutrients have been applied to the plants in various combinations. In such cases interactions are often seen between two or more elements in the seed. The addition of phosphorus may be necessary to obtain the maximum response to added nitrate, while the application of phosphate alone may even reduce the nitrogen content of the seeds, as in sunflower (Blamey and Chapman 1981). Jones et al. (1977) obtained a phosphorus × potassium interaction in soybean. Phosphorus concentration was highest when phosphate was applied without potassium, and lowest when potassium was applied without phosphate. Similar cases (in which one element influences the level of another in the seed) have been reported for nitrogen and sulphur (Rasmussen et al. 1975; Randall et al. 1981), phosphorus and zinc (Peck et al. 1980; Raboy and Dickinson 1984), cobalt and selenium (Cary 1981), and calcium and iron (Coffelt and Hallock 1986).

V. CONCLUDING REMARKS

Environmental factors clearly affect seed quality, and an understanding of this phenomenon may help improve commercial seed production. In practice, certain factors are more readily controlled than others, at least under field conditions. Nutrient levels and moisture are perhaps the easiest to manipulate by means of added fertilizers and irrigation. There is little that can be done in the field to alter temperature and daylength, except to choose field locations or sowing dates to tailor the crop's growth period to the seasonal change in weather. Even where conditions can be effectively controlled to produce the desired effects on the seeds, the circumstances may be uneconomical in other respects. For example, increased seed protein in pea was found to occur only at a level of added nitrogen which was above that needed for maximum yield (Trevino and Murray 1975).

It may well be that when the physiological mechanisms which control seed quality are understood, then both size and chemical content will be more readily manipulated by chemical applications than by altering the physical environment. There is little doubt that the environmentally induced effects are mediated by changes in the amount, translocation, or activity of endogenous growth substances. Seed size, for example, may be at least partly regulated by abscisic acid (ABA). This is indicated by several lines of evidence. Large-seeded genotypes of soybean maintain an endogenous ABA concentration more than 50% greater than that of small-seeded genotypes (Schussler et al. 1984). ABA injected into individual wheat grains during development resulted in increased absorption of photosynthate (Dewdney and McWha 1979), and this hormone also causes an increase in assimilate of sucrose in barley (Tietz and Dingkuhn 1981). Applied ABA increased final seed size in common bean (Clifford et al. 1987). During seed development ABA content of most seeds increases at first, comes to a maximum (often at about the same time as dry weight peaks), and then declines rapidly (King 1982). A possible explanation for the increase in seed size associated with low temperatures is that, under such conditions, the process of development is slowed down, thereby prolonging the period of high ABA content (and hence assimilation), resulting in larger seeds (Goldbach and Michael 1976).

Seed size can be manipulated by the application of a wide range of natural growth substances and synthetic compounds. Increases in seed weight have been obtained by the application of dalapon in sorghum (Santakumari and Reddy 1980), by butanedioic 2, 2-dimethylhydrazide (daminozide) in buckwheat (*Fagopyrum esculentum*) (Gubbels 1979), and by naphthaleneacetic acid (NAA) on sugar beet (Malik and Shakara 1977). Decreased seed weights have been induced by applications of morphactins in various field crops (Dybing and Lay 1981), and by NAA in *Litchi chinensis* (Sharma and Dhillon 1986). Decreased seed size is, of course, a desirable feature in the production of certain fruit crops such as citrus and table grape.

Seed chemical content can also be manipulated by hormonal treatments. Protein levels have been increased by applications of gibberellic acid in rice (Mukherjee and Prabhakar 1980), Phosfon-D in cowpea (Jain and Yadava 1980), and by dalapon in sorghum (Santakumari and Reddy 1980). Morphactins can increase oil content in seeds of flax, soybean, wheat, and oat (Dybing and Lay 1982). The balance of oleic to linoleic acid in peanut can be changed by applications of the growth regulator butanedioic 2, 2-dimethylhydrazide (Mozingo et al. 1986). Mineral uptake in sorghum grains is markedly altered by spraying with methiocarb (Duncan and Boswell 1981). The mechanism of action of these compounds is largely unknown.

The effects of the various environmental factors on seed characteristics have now been well established, at least in broad outline. High temperatures reduce seed size, increase protein content, and alter the balance of fatty acids. Added mineral nutrients increase seed size and, in the case of nitrogen, can also modify the relative proportions of the various storage proteins present. Drought generally reduces seed size (especially if it occurs after seed number has been determined), but increases protein concentration. Shade and short days result in smaller seeds. Future research on the effects of the parental environment on seeds during development should concentrate on (1) determining the physiological mechanisms whereby the environment controls the action of the various endogenous growth substances; (2) the practical effect (both in artificial and natural circumstances) of phenotypic changes in seed quality; (3) the question as to why the seeds of some species are more phenotypically plastic than others; and (4) the cost-effectiveness of altering the environment for seed production for particular cultivars.

LITERATURE CITED

Ågren J. 1989. Seed size and number in Rubus chamaemorus: Between-habitat variation, and effects of defoliation and supplemental pollination. J. Ecol. 77:1080–1092.

Akinola, J. O, and P. C. Whiteman. 1975. Agronomic studies on pigeon pea (Cajanus cajan (L.) Millsp.). II. Responses to sowing density. Austral. J. Agr. Res. 26:57–66.

Allen, S. E., and G. L. Terman. 1978. Yield and protein content of rice as affected by rate, source, method, and time of applied N. Agron. J. 70:238–242.

Altman, D. W., W. L. McCuistion, and W. E. Kronstad. 1983. Grain protein percentage, kernel hardness, and grain yield of winter wheat with foliar applied urea. Agron. J. 75:87–91.

Andrews, P., W. J. Collins, and W. R. Stern. 1977. The effect of withholding water during flowering on seed production in Trifolium subterraneum L. Austral. J. Agr. Res. 28:301–307.

Auld, D. L., B. L. Bettis, and M. J. Dial. 1984. Planting date and cultivar effect on winter rape production. Agron. J. 76:197–200.

Austin, R. B. 1966. The influence of the phosphorus and nitrogen nutrition of pea plants on the growth of their progeny. Plant & Soil 24:359–368.

Baenziger, P. S., R. L. Clements, M. S. McIntosh, W. T. Yamazaki, T. M. Starling, D. J. Sammons, and J. W. Johnson. 1985. Effect of cultivar, environment, and their interaction and stability analyses on milling and baking quality of soft red winter wheat. Crop Sci. 25:5–8.

Bauer, A., A. B. Frank, and A. L. Black. 1985. Estimation of spring wheat grain dry matter assimilation from air temperature. Agron. J. 77:743–752.

Bean, E. W. 1980. Factors affecting the quality of herbage seeds. p. 593–604. In: P. D. Hebblethwaite (ed.), Seed Production. Butterworth, London.

Benech Arnold, R. L., M. Fenner, and P. J. Edwards. 1991. Changes in germinability, ABA levels and ABA embryonic sensitivity in developing seeds of Sorghum bicolor (L.) Moench. induced by water stress during grain filling. New Phytol. 118 (in press).

Benner, B. L., and F. A. Bazzaz. 1985. Response of the annual Abutilon theophrasti Medic. (Malvaceae) to timing of nutrient availability. Am. J. Bot. 72:320–323.

Bentley, S., J. B. Whittaker, and A. J. C. Malloch. 1980. Field experiments on the effects of grazing by a chrysomelid beetle (Gastrophysa viridula) on seed production and quality in Rumex obtusifolius and Rumex crispus. J. Ecol. 68:671–674.

Bharati, M. P., D. K. Whigham, and R. D. Voss. 1986. Soybean response to tillage and nitrogen, phosphorus and potassium fertilization. Agron. J. 78:947–950.

Bhaskar, A., and K. G. Vyas. 1988. Studies on competition between wheat and Chenopodium album L. Weed Res. 28:53–58.

Blamey, F. P. C., and J. Chapman. 1981. Protein, oil and energy yields of sunflower as affected by N and P fertilization. Agron. J. 73:583–587.

Boswell, F. C., and D. E. Anderson. 1976. Long-term residual fertility and current N-P-K application effects on soybeans. Agron. J. 68:315–318.

Boswell, F. C., K. Ohki, M. B. Parker, L. M. Shuman, and D. O. Wilson. 1981. Methods and rates of applied manganese for soybeans. Agron. J. 73:909–912.

Brinkman, M. A. 1979. Performance of oat plants grown from primary and secondary kernels. Can. J. Plant Sci. 59:931–937.

Bulisani, E. A., and R. L. Warner. 1980. Seed protein and nitrogen effects upon seedling vigor in wheat. Agron. J. 72:657–662.

Campbell, C. A., and H. R. Davidson. 1979. Effect of temperature, nitrogen fertilization and moisture stress on yield, yield components, protein content and moisture use efficiency of Manitou spring wheat. Can. J. Plant Sci. 59:963–974.

Campbell, C. A., H. R. Davidson, and G. E. Winkleman. 1981. Effect of nitrogen, temperature, growth stage and duration of moisture stress on yield components and protein content of Manitou spring wheat. Can. J. Plant Sci. 61:549–563.

Cary, E. E. 1981. Effect of selenium and cadmium additions to soil on their concentrations in lettuce and wheat. *Agron. J.* 73:703–706.

Cassman, K. G., A. S. Whitney, and R. L. Fox. 1981. Phosphorus requirements of soybean and cowpea as affected by mode of N nutrition. *Agron. J.* 73:17–22.

Cavers, P. B., and M. G. Steele. 1984. Patterns of change in seed weight over time on individual plants. *Am. Nat.* 124:324–335.

Chakrabarti, A. K. 1983. Effect of nitrogen and phosphorus on turnip seed crop. *Seed Res.* 11:87–90.

Chandler, P. M., T. J. V. Higgins, P. J. Randall, and D. Spencer. 1983. Regulation of legumin levels in developing pea seeds under conditions of sulphur deficiency. Rates of legumin synthesis and levels of legumin mRNA. *Plant Physiol.* 71:47–54.

Chandler, P. M., D. Spencer, P. J. Randall, and T. J. V. Higgins. 1984. Influence of sulphur nutrition on developmental patterns of some major pea seed proteins and their mRNAs. *Plant Physiol.* 75:651–657.

Cheplick, G. P. 1989. Nutrient availability, dimorphic seed production, and reproductive allocation in the annual grass *Amphicarpum purshii. Can. J. Bot.* 67:2514–2521.

Choudhari, S. D., V. K. Shinde, and N. L. Bhale. 1984. Effect of seed size on the grain yield of sorghum. *Seed Res.* 12:48–52.

Chowdhury, S. I., and I. F. Wardlaw. 1978. The effect of temperature on kernel development in cereals. *Austral. J. Agr. Res.* 29:205–223.

Cideciyan, M. A., and A. J. C. Malloch. 1982. Effect of seed size on germination, growth and competitive ability of *Rumex crispus* and *Rumex obtusifolius. J. Ecol.* 70:227–232.

Clarke, J. M. 1979. Intra plant variation in number of seeds per pod and seed weight in *Brassica napus* 'Tower'. *Can. J. Plant Sci.* 59:959–962.

Clifford, P. E., C. E. Offler, and J. W. Patrick. 1987. Injection of growth regulators into seeds growing in situ on plants of *Phaseolus vulgaris* with a double fruit sink system. *Can. J. Bot.* 65:612–615.

Coffelt, T. A., and D. L. Hallock. 1986. Soil fertility responses of virginia-type peanut cultivars. *Agron. J.* 78:131–137.

Collins, W. J. 1981. The effect of length of growing season, with and without defoliation, on seed yield and hard-seededness in swards of subterranean clover. *Austral. J. Agr. Res.* 32:783–792.

Collins, W. J., R. C. Rossiter, and A. R. Monreal. 1978. The influence of shading on seed yield in subterranean clover. *Austral. J. Agr. Res.* 29:1167–1175.

Cook, R. E. 1975. The photoinductive control of seed weight in *Chenopodium rubrum* L. *Am. J. Bot.* 62:427–431.

Crane, J. M., W. Calhoun, and T. A. Ayres. 1981. Seed and oil characteristics of fertilized meadowfoam. *Agron. J.* 73:255–256.

Crawley, M. J., and M. Nachapong. 1985. The establishment of seedlings from primary and regrowth seeds of ragwort (*Senecio jacobaea*). *J. Ecol.* 73:255–261.

Datta, S. C., Y. Gutterman, and M. Evenari. 1972. The influence of the origin of the mother plant on yield and germination of their caryopses in *Aegilops ovata. Planta* 105:155–164.

Dekhuijzen, H. M., and D. R. Verkerke. 1986. The effect of temperature on development and dry matter accumulation of *Vicia faba* seeds. *Ann. Bot.* 58:869–885.

Dewdney, S. J., and J. A. McWha. 1979. Abscisic acid and the movement of photosynthetic assimilates towards developing wheat (*Triticum aestivum* L.) grains. *Z. Pflanzenphysiol.* 92:183–186.

Dexter, J. E., W. L. Crowle, R. R. Matsuo, and F. G. Kosmolak. 1982. Effect of nitrogen fertilization on the quality characteristics of five north American amber durum wheat cultivars. *Can. J. Plant Sci.* 62:901–912.

Dharmalingam, C., and V. Ramakrishnan. 1981. Studies on the relative performance of sized seed in peanut (*Arachis hypogaea*) cv. Pol. 2. *Seed Res.* 9:57–66.

Dhillon, G. S., and D. S. Kler. 1978. Influence of seed size on the growth and yield of triticale. *Seed Res.* 6:110–117.

204   M. FENNER

Diepenbrock, W., and G. Geisler. 1979. Compositional changes in developing pods and seeds of oilseed rape (Brassica napus L.) as affected by pod position on the plant. Can. J. Plant Sci. 59:819–830.

Dolan, R. W. 1984. The effect of seed size and maternal source on individual size in a population of Ludwigia leptocarpa (Onagraceae). Am. J. Bot. 71:1302–1307.

Dorne, A. J. 1981. Variation in seed germination inhibition of Chenopodium bonus-henricus in relation to altitude of plant growth. Can. J. Bot. 59:1893–1901.

Doto, A. L., and W. J. Whittington. 1981. Responses of cow pea (Vigna unguiculata) varieties and their hybrids to variation in day and night temperature regimes. Ann. Appl. Biol. 97:213–219.

Downes, R. W., and J. S. Gladstones. 1984a. Physiology of growth and seed production in Lupinus angustifolius L. II. Effect of temperature before and after flowering. Austral. J. Agr. Res. 35:501–509.

_____.1984b. Physiology of growth and seed production in Lupinus angustifolius L. III. Effects of defoliation and lateral branch excision on dry matter and seed production at different growth temperatures. Austral. J. Agr. Res. 35:511–520.

Dubetz, S., E. E. Gardiner, D. Flynn, and A. I. De La Roche. 1979. Effect of nitrogen fertilizer on nitrogen fractions and amino acid composition of spring wheat. Can. J. Plant Sci. 59:299–305.

Duncan, R. R., and F. C. Boswell. 1981. Seed element uptake, grain yield, and bird damage of methiocarb-treated sorghum hybrids. Agron. J. 73:290–292.

Dunstone, R. L., M. L. Tonnet, I. F. Wardlaw, and A. Shani. 1984. Effect of temperature on seed development in jojoba (Simmondsia chinensis (Link) Schneider). II. Wax content and composition. Austral. J. Agr. Res. 35:693–700.

Dybing, C. D., and C. Lay. 1982. Oil and protein in field crops treated with morphactins and other growth regulators for senescence delay. Crop Sci. 22:1054–1058.

Eck, H. V. 1984. Irrigated corn yield responses to nitrogen and water. Agron. J. 76:421–428.

_____.1986. Effects of water deficits on yield, yield components, and water use efficiency of irrigated corn. Agron. J. 78:1035–1040.

Egli, D. B., R. D. Guffy, L. W. Meckel, and J. E. Leggett. 1985. The effect on source-sink alterations on soybean seed growth. Ann. Bot. 55:395–402.

Egli, D. B., and I. F. Wardlaw. 1980. Temperature response of seed growth characteristics of soybeans. Agron. J. 72:560–564.

Egli, D. B., R. A. Wiralaga, and E. L. Ramseur. 1987a. Variation in seed size in soybean. Agron. J. 79:463–467.

Egli, D. B., R. A. Wiralaga, T. Bustamam, Yu Zhen-Wen, and D. M. TeKrony. 1987b. Time of flower opening and seed mass in soybean. Agron J. 79:697–700.

Ellis, R. P., and J. Brown. 1986. Yield in spring barley at contrasting sites in England and Scotland. Ann. Appl. Biol. 109:613–617.

Ellison, A. M. 1987. Effect of seed dimorphism on the density-dependent dynamics of experimental populations of Atriplex triangularis (Chenopodiaceae). Am. J. Bot. 74:1280–1288.

Elmore, C. D., W. I. Spurgeon, and W. O. Thom. 1979. Nitrogen fertilization increases N and alters amino acid concentration of cottonseed. Agron. J. 71:713–716.

Elmore, R. W. 1987. Soybean cultivar response to tillage systems. Agron. J. 79:114–119.

Elmore, R. W., and J. A. Jackobs. 1984. Yield and yield components of sorghum and soybeans of varying plant heights when intercropped. Agron. J. 76:561–564.

Ene, B. N., and E. W. Bean. 1975. Variations in seed quality between certified seed lots of perennial rye grass and their relationships to nitrogen supply and moisture status during seed development. J. Brit. Grassland Soc. 30:195–199.

Fehr, W. R., B. D. Lynk, and G. E. Carlson. 1985. Performance of semideterminate and indeterminate soybean genotypes subjected to defoliation. Crop Sci. 25:24–26.

Fenner, M. 1986. The allocation of minerals to seeds in Senecio vulgaris plants subjected to nutrient shortage. J. Ecol. 74:385–392.

Fenner, M. 1991. The effects of the parent environment on seed germinability. Seed Sci. Res. 1 (in press).

Fischer, K. S., and G. L. Wilson. 1975. Studies of grain production in Sorghum bicolor (L. Moench). IV. Some effects of increasing and decreasing photosynthesis at different stages of the plant's development on the storage capacity of the inflorescence. Austral. J. Agr. Res. 26:25–30.

Ford, M. A., I. Pearman, and G. N. Thorne. 1976. Effects of variation in ear temperature on growth and yield in spring wheat. Ann. Appl. Biol. 82:317–333.

Ford, M. A., and G. N. Thorne. 1975. Effects of variation in temperature and light intensity at different times on growth and yield of spring wheat. Ann. Appl. Biol. 80:283–299.

Forsyth, C., and N. A. C. Brown. 1982. Germination of the dimorphic fruits of Bidens pilosa L. New Phytol. 90:151–164.

Francois, L. E., E. V. Maas, T. J. Donovan, and V. L. Youngs. 1986. Effect of salinity on grain yield and quality, vegetative growth, and germination of semi-dwarf and durum wheat. Agron. J. 78:1053–1058.

Frederick, J. R., and H. G. Marshall. 1985. Grain yield and yield components of soft red winter wheat as affected by management practices. Agron. J. 77:495–499.

Fuller, W., C. E. Hance, and M. J. Hutchings. 1983. Within-season fluctuations in mean fruit weight in Leontodon hispidus L. Ann. Bot. 51:545–549.

Fussell, L. K., C. J. Pearson, and M. J. T. Norman. 1980. Effect of temperature during various growth stages on grain development and yield of Pennisetum americanum. J. Expt. Bot. 31:621–633.

Galen, C., R. C. Plowright, and J. D. Thomson. 1985. Floral biology and regulation of seed set and seed size in the lily, Clintonia borealis. Am. J. Bot. 72:1544–1552.

Gales, K., and N. J. Wilson. 1981. Effects of water shortage on the yield of winter wheat. Ann. Appl. Biol. 99:323–334.

Garcia, R. L., and J. J. Hanway. 1976. Foliar fertilization of soybeans during the seed-filling period. Agron. J. 68:653–657.

Gardener, B. R., and E. B. Jackson. 1976. Fertilization, nutrient composition, and yield relationships in irrigated spring wheat. Agron. J. 68:75–78.

Gayler, K. R., and G. E. Sykes. 1985. Effects of nutritional stress on the storage proteins of soybeans. Plant Physiol. 78:582–585.

Gbikpi, P. J., and R. K. Crookson. 1981. Effect of flowering date on accumulation of dry matter and protein in soybean seeds. Crop Sci. 21:652–655.

George, R. A. T., R. J. Stephens, and S. Varis. 1980. The effect of mineral nutrients on the yield and quality of seeds in tomato. p. 561–567. In: P. D. Hebblethwaite (ed.), Seed Production. Butterworth, London.

Gill, S. S., and H. Singh. 1979. Effect of seed size and sowing dates on germination and yield of radish (Raphanus sativus L.) roots. Seed Res. 7:58–62.

Gillespie, J. M., R. J. Blagrove, and P. J. Randall. 1978. Effect of sulphur supply on the seed globulin composition of various species of lupin. Austral. J. Plant Physiol. 5:641–650.

Glenn, D. M., A. Carey, F. E. Bolton, and M. Vavra. 1985. Effect of N fertilizer on protein content of grain, straw and chaff tissues in soft white winter wheat. Agron. J. 77:229–232.

Goldbach, H., and G. Michael. 1976. Abscisic acid content of barley grains during ripening as affected by temperature and variety. Crop Sci. 16:797–799.

Goyne, P. J., B. W. Simpson, D. R. Woodruff, and J. D. Churchett. 1979. Environmental influence on sunflower achene growth, oil content and oil quality. Austral. J. Expt. Agr. Anim. Husb. 19:82–88.

Gray, D., J. R. A. Steckel, J. Dearman, and P. A. Brocklehurst. 1988. Some effects of temperature during seed development on carrot (Daucus carota) seed growth and quality. Ann. Appl. Biol. 112:367–376.

Gray, D., J. R. A. Steckel, and J. A. Ward. 1983. Studies on carrot seed production: Effects of plant density on yield and components of yield. J. Hort. Sci. 58:83–90.

———. 1986. The effect of cultivar and cultural factors on embryo-sac volume and seed weight in carrot (Daucus carota L.). Ann. Bot. 58:737–744.

Green, A. G. 1986. Effect of temperature during seed maturation on the oil composition of low-linolenic genotypes of flax. Crop Sci. 26:961–965.

Gross, K. L. 1984. Effects of seed size and growth form on seedling establishment of six monocarpic perennial plants. J. Ecol. 72:369–387.

Gubbels, G. H. 1979. Yield and weight per seed in buckwheat after applications of growth regulators and anti-transpirants. Can. J. Plant Sci. 59:857–859.

———. 1981. Quality, yield and seed weight of green field peas under conditions of applied shade. Can. J. Plant Sci. 61:213–217.

Gutterman, Y. 1980/81. Influences on seed germinability: Phenotypic maternal effects during seed maturation. Isr. J. Bot. 29:105–117.

———. 1982. Phenotypic maternal effect of photoperiod on seed germination. p. 67–79. In: A. A. Khan (ed.), The Physiology and Biochemistry of Seed Development, Dormancy and Germinability. Elsevier Biomedical Press, Amsterdam.

———. 1991. Maternal effects on seeds during development. In: M. Fenner (ed.), Seeds— The Ecology of Regeneration in Plant Communities. CAB International, Oxford.

Gutterman, Y., T. H. Thomas, and W. Heydecker. 1975. Effects on the progeny of applying different day length and hormone treatments to parent plants of Lactuca scariola L. Physiol. Plant. 34:30–38.

Ham, G. E., I. E. Liener, S. D. Evans, R. D. Frazier, and W. W. Nelson. 1975. Yield and composition of soybean seed as affected by N & S fertilization. Agron. J. 67:293–297.

Hanft, J. M., R. J. Jones, and A. B. Stumme. 1986. Dry matter accumulation and carbo-hydrate concentration patterns of field-grown and in vitro cultured maize kernels from the tip and middle ear positions. Crop Sci. 26:568–572.

Hanus, F. J., S. L. Albrecht, R. M. Zablotowicz, D. W. Emerich, A. R. Sterling, and H. J. Evans. 1981. Yield and N content of soybean seed as influenced by Rhizobium japonicum inoculants possessing the hydrogenase characteristic. Agron. J. 73:368–372.

Harder, H. J., R. E. Carlson, and R. H. Shaw. 1982. Yield, yield components, and nutrient content of corn grain as influenced by post-silking moisture stress. Agron. J. 74:275–278.

Harris, H. C., J. R. McWilliam, and V. J. Bofinger. 1980. Prediction of oil quality of sun-flower from temperature probabilities in eastern Australia. Austral. J. Agr. Res. 31:477–488.

Heenan, D. P., and L. C. Campbell. 1980. Growth, yield components and seed composition of two soybean cultivars as affected by manganese supply. Austral. J. Agr. Res. 31:471–476.

Hendrix. S. D. 1984. Variation in seed weight and its effects on germination in Pastinaca sativa (Umbelliferae). Am. J. Bot. 71:795–802.

Hill, J., A. D. Robson, and J. F. Loneragan. 1979. The effects of copper and nitrogen supply on the distribution of copper in dissected wheat grains. Austral. J. Agr. Res. 30:233–237.

Hill, H. J., S. H. West and K. Hinson. 1986. Effect of water stress during seedfill on imper-meable seed expression in soybean. Crop Sci. 26:807–812.

Hodgson, A. S. 1979. Rapeseed adaptation in northern New South Wales. III. Yield, yield components and grain quality of Brassica campestris and Brassica napus in relation to planting date. Austral. J. Agr. Res. 30:19–27.

Hojjati, S. M. 1976. Amino acid patterns of kidney beans grown under different S and K regimes. Agron. J. 68:668–671.

Howe, H. F., and E. W. Schupp. 1985. Early consequences of seed dispersal for a neotropical tree (Virola surinamensis). Ecology 66:781–791.

Humphreys, E., W. A. Muirhead, F. M. Melhuish, and R. J. G. White. 1987. Effects of time of urea application on combine-sown calrose rice in south-east Australia. 1. Crop response and N-uptake. Austral. J. Agr. Res. 38:101–112.

Huxley, P. A. and R. J. Summerfield. 1976. Effects of daylength and day/night tempera-tures on growth and seed yield of cowpea cv. K2809 grown in controlled environments.

Ann. Appl. Biol. 83:259–271.

Ingram, T. J., and G. Browning. 1979. Influence of photoperiod on seed development in the genetic line of peas G2 and its relation to changes in endogenous gibberellins measured by combined gas chromatography-mass spectrometry. *Planta* 146:423–432.

Israel, D. W. 1981. Cultivar and *Rhizobium* strain effects on nitrogen fixation and remobilization by soybeans. *Agron. J.* 73:509–516.

Izzeldin, H., L. F. Lippert, and F. H. Takatori. 1980. An influence of water stress at different growth stages on yield and quality of lettuce seed. *J. Am. Soc. Hort. Sci.* 105:68–71.

Jain, N., and R. B. R. Yadava. 1980. Effect of Phosfon-D on seed yield and quality of cowpea (*Vigna unguiculata* (L.) Walp) plants. *Seed Res.* 8:42–45.

Jenner, C. F. 1979. Grain-filling in wheat plants shaded for brief periods after anthesis. *Austral. J. Plant Physiol.* 6:629–641.

Johnson, J. W., W. L. Hargrove, J. T. Touchton, and W. T. Yamazaki. 1984. Influence of N fertilization on wheat milling and baking quality. *Crop Sci.* 24:904–906.

Johnson, R. C., and E. T. Kanemasu. 1983. Yield and development of winter wheat at elevated temperatures. *Agron. J.* 75:561–565.

Jones, G. D., J. A. Lutz, and T. J. Smith. 1977. Effects of phosphorus and potassium on soybean nodules and seed yield. *Agron. J.* 69:1003–1006.

Jones, O. R. 1984. Yield, water use efficiency, and oil concentration and quality of dryland sunflower grown in the southern high plains. *Agron. J.* 76:229–235.

Jurgens, S. K., R. R. Johnson, and J. S. Boyer. 1978. Dry matter production and translocation in maize subjected to drought during grain fill. *Agron. J.* 70:678–682.

Kahn, B. A., and P. J. Stoffella. 1985. Yield components of cowpeas grown in two environments. *Crop Sci.* 25:179–182.

Karathanasis, A. D., V. A. Johnson, G. A. Peterson, D. H. Sander and R. A. Olsen. 1980. Relation of soil properties and other environmental factors to grain yield and quality of winter wheat grown at international sites. *Agron. J.* 72:329–336.

Keefer, G. D., J. E. McAllister, E. S. Uridge, and B. W. Simpson. 1976. Time of planting effects on development, yield and oil quality of irrigated sunflower. *Aust. J. Expt. Anim. Husb.* 16:417–422.

Ketring, D. L., C. E. Simpson, and O. D. Smith. 1978. Physiology of oil seeds. VII. Growing season and location effects on seedling vigor and ethylene production by seeds of three peanut cultivars. *Crop Sci.* 18:409–413.

King, R. W. 1982. Abscisic acid in seed development. p. 157–184 In: A. A. Khan (ed.), *The Physiology and Biochemistry of Seed Development, Dormancy and Germination.* Elsevier Biomedical Press, Amsterdam.

Kiniry, J. R. and R. L. Musser. 1988. Response of kernel weight of sorghum to environment early and late in grain filling. *Agron. J.* 80:606–610.

Kiniry, J. R., and J. T. Richie. 1985. Shade-sensitive interval of kernel number of maize. *Agron. J.* 77:711–715.

Kiniry, J. R., C. A. Wood, D. A. Spanel, and A. J. Bockholt. 1990. Seed weight response to decreased seed number in maize. *Agron. J.* 54:98–102.

Klinck, H. R., and S. L. Sim. 1977. Influence of temperature and photoperiod on growth and yield components in oats (*Avena sativa*). *Can. J. Bot.* 55:96–106.

Korte, L. L., J. E. Specht, J. H. Williams, and R. C. Sorensen. 1983. Irrigation of soybean genotypes during reproductive ontogeny. II. Yield component responses. *Crop Sci.* 23:528–533.

Kromer, M., and K. L. Gross. 1987. Seed mass, genotype and density effects on growth and yield of *Oenothera biennis* L. *Oecol.* 73:207–212.

Kumar, T. N. A., K. Giriraj, T. G. Prasad, and T. S. Vidyashankar. 1979. Influence of seed test weight on yield and growth parameters in sunflower. *Seed Res.* 7:141–144.

Lafond, G. P., and R. J. Baker. 1986. Effects of genotype and seed size on speed of emergence and seedling vigor in nine spring wheat cultivars. *Crop Sci.* 26:341–346.

Lahiri, A. N., S. Kathju, and K. A. Shankarnarayan. 1982. Comparative performance of

Cenchrus ciliaris pastures raised from large and small seeds. Seed Sci. and Technol. 10:207–215.

Langer, R. H. M., and F. K. Y. Liew. 1973. Effects of varying nitrogen supply at different stages of the reproductive phase on spikelet and grain production and on grain nitrogen in wheat. Austral. J. Agr. Res. 24:647–656.

Lawan, M., F. L. Barnett, B. Khaleeq, and R. L. Vanderlip. 1985. Seed density and seed size of pearl millet as related to field emergence and several seed and seedling traits. Agron. J. 77:567–571.

Leffler, H. R., C. D. Elmore, and J. D. Hesketh. 1977. Seasonal and fertility-related changes in cottonseed protein quantity and quality. Crop Sci. 17:953–956.

Loneragan, J. F., K. Snowball, and A. D. Robson. 1980. Copper supply in relation to content and redistribution of copper among organs of the wheat plant. Ann. Bot. 45:621–632.

Longnecker, N. E., and N. C. Uren. 1990. Factors influencing variability in manganese content of seeds, with emphasis on barley (Hordeum vulgare) and white lupins (Lupinus albus). Austral. J. Agr. Res. 41:29–37.

Majid, H. R., A. A. Schneiter. 1987. Yield and quality of semidwarf and standard-height sunflower hybrids grown at five plant populations. Agron. J. 79:681–684.

Malik, K. B., and S. A. Shakara. 1977. Effect of growth regulators on seed development and indeterminate type of growth in sugar beet. Agr. Pakistan 28:65–75.

Marshall, D. L., D. A. Levin, and N. L. Fowler. 1985. Plasticity in yield components in response to fruit predation and date of fruit initiation in three species of Sesbania (Leguminosae). J. Ecol. 73:71–81.

Mathers, A. C., and B. A. Stewart. 1982. Sunflower nutrient uptake, growth, and yield as affected by nitrogen or manure, and plant population. Agron. J. 74:911–915.

Maun, M. A., and P. B. Cavers. 1971. Seed production and dormancy in Rumex crispus. II. The effects of removal of various proportions of flowers at anthesis. Can. J. Bot. 49:1841–1848.

McDonald, G. K., B. G. Sutton, and F. W. Ellison. 1983. The effect of time of sowing on the grain yield of irrigated wheat in the Namoi Valley, New South Wales. Austral. J. Agr. Res. 34:229–240.

McGinley, M. A. 1989. Within and among plant variation in seed mass and pappus size in Tragopogon dubious. Can. J. Bot. 67:1298–1304.

McGraw, R. L., P. R. Beuselinck, and R. R. Smith. 1986. Effect of latitude on genotype × environment interactions for seed yield in birdsfoot trefoil. Crop Sci. 26:603–605.

Meagher, W. R., C. M. Johnson, and P. R. Stout. 1952. Molybdenum requirements of leguminous plants supplied with fixed nitrogen. Plant Physiol. 27:223–230.

Meckel, L., D. B. Egli, R. E. Phillips, D. Radcliffe, and J. E. Leggett. 1984. Effect of moisture stress on seed growth in soybeans. Agron. J. 76:647–650.

Miezan, K., E. G. Heyne, and K. F. Finney. 1977. Genetic and environmental effects on the grain protein content in wheat. Crop Sci. 17:591–593.

Miller, D. E., and D. W. Burke. 1983. Response of dry beans to daily deficit sprinkler irrigation. Agron. J. 75:775–778.

Morse, D. H., and T. Schmitt. 1985. Propagule size, dispersal ability, and seedling performance in Asclepias syriaca. Oecologia 67:372–379.

Mozingo, R. W., J. L. Steele, and C. T. Young. 1986. Growth regulator effects on the composition of seed of five peanut cultivars. Agron. J. 78:645–648.

Muchow, R. C. 1990. Effect of high temperature on the rate and duration of grain growth in field-grown Sorghum bicolor (L.) Moench. Austral. J. Agr. Res. 41:329–337.

Mukherjee, R. K., and Prabhakar, B. J. 1980. Effect of gibberellin on rice yield response to nitrogen applied at heading, and quality of seeds. Plant Soil 55:153–156.

Murray, G. A., J. B. Swensen, and D. L. Auld. 1984. Influence of seed size and planting date on the performance of Austrian winter field peas. Agron. J. 76:595–598.

Nagato, K., and Ebata, M. 1965. Effect of high temperature during ripening period on the development and quality of rice kernels. Proc. Crop Sci. Soc. Jap. 34:59–66.

Nakoneshny, W., and Friesen, G. 1961. The influence of a commercial fertilizer treatment on weed competition in spring-sown wheat. *Can. J. Plant Sci.* 41:231–237.

Nandisha, B. S., and Mahadevappa, M. 1984. Influence of mother-plant nutrition and spacing on planting value of rice seeds (*Oryza sativa* L.). *Seed Res.* 12:25–32.

Naylor, R. E. L. 1980. Effects of seed size and emergence time on subsequent growth of perennial rye grass. *New Phytol* 84:313–318.

Nelson, J. R., G. A. Harris, and C. J. Goebel. 1970. Genetic vs. environmentally induced variation in medusahead (*Taeniatherum asperum* (Simonkai) Neuski). *Ecology* 51:526–529.

Nutall, W. F., H. G. Zandstra, and K. E. Bowren. 1979. Yield and N percentage of spring wheat as affected by phosphate fertilizer, moisture use, and available soil P and N. *Agron. J.* 71:385–391.

Ong, C. K. 1983. Response to temperature in a stand of pearl millet (*Pennisetum typhoides* S. & H.). *J. Expt. Bot.* 34:337–348.

Openshaw, S. J., H. H. Hadley, and C. E. Brokoski. 1979. Effects of pod removal upon seeds of nodulating and non-nodulating soybean lines. *Crop Sci.* 19:289–290.

Owen, D. F. 1983. Differential response of sunflower hybrids to planting date. *Agron. J.* 75:259–262.

Parker, M. B., F. C. Boswell, K. Ohki, L. M. Shuman, and D. O. Wilson. 1981. Manganese effects on yield and nutrient concentration in leaves and seed of soybean cultivars. *Agron. J.* 73:643–646.

Parrish, J. A. D., and F. A. Bazzaz. 1985. Nutrient content of *Abutilon theophrasti* seeds and the competitive ability of the resulting plants. *Oecologia* 65:247–251.

Pearson, C. J., and S. G. Shah. 1981. Effects of temperature on seed production, seed quality and growth of *Paspalum dilatatum*. *J. Appl. Ecol.* 18:897–905.

Peck, N. H., D. L. Grunes, R. M. Welch, and G. E. Macdonald. 1980. Nutritional quality of vegetable crops as affected by phosphorus and zinc fertilizers. *Agron. J.* 72:528–534.

Peet, M. M., and P. J. Kramer. 1980. Effects of decreasing source/sink ratio in soybeans on photosynthesis, photorespiration, transpiration and yield. *Plant Cell & Environment.* 3:201–206.

Perry, M. W. 1975. Field environment studies on lupins. II. The effects of time of planting on dry matter partition and yield components of *Lupinus angustifolius* L. *Austral. J. Agr. Res.* 26:809–818.

Pikaard, C. S., and J. H. Cherry. 1984. Maintenance of normal or supranormal protein accumulation in developing ovules of *Glycine max* L. Merr. during PEG-induced water stress. *Plant Physiol.* 75:176–180.

Pinthus, M. J., and E. Millet. 1978. Interactions among number of spikelets, number of grains and grain weight in the spikes of wheat (*Triticum aestivum* L.). *Ann. Bot.* 42:839–848.

Poneleit, C. G. and D. B. Egli. 1979. Kernel growth rate and duration in maize as affected by plant density and genotype. *Crop Sci.* 19:385–388.

Poole, W. D., G. W Randall, and G. E. Ham. 1983. Foliar fertilization of soybeans. I. Effect of fertilizer sources, rates and frequency of application. *Agron. J.* 75:195–200.

Porter, M. A., and G. M. Paulsen. 1983. Grain protein response to phosphorus nutrition of wheat. *Agron. J.* 75:303–305.

Prunty, L. 1981. Sunflower cultivar performance as influenced by soil water and plant population. *Agron. J.* 73:257–260.

Putievski, E. 1983. Effects of daylength and temperature on growth and yield components of three seed spices. *J. Hort. Sci.* 58:271–275.

Raboy, V., and D. B. Dickinson. 1984. Effect of phosphorus and zinc nutrition on soybean seed phytic acid and zinc. *Plant Physiol.* 75:1094–1098.

Radford, R. L., C. Chavengsaksongkram, and T. Hymowitz. 1977. Utilization of nitrogen to sulphur ratio for evaluating sulphur-containing amino acid concentration in seed of *Glycine max* and *G. soya*. *Crop Sci.* 17:273–277.

Radley, M. 1978. Factors affecting grain enlargement in wheat. *J. Expt. Bot.* 29:919–934.

Randall, P. J., K. Spencer, and J. R. Freney. 1981. Sulphur and nitrogen fertilizer effects on wheat. I. Concentrations of sulphur and nitrogen and the nitrogen to sulphur ratio in grain, in relation to the yield response. *Austral. J. Agr. Res.* 32:203–212.

Randall, P. J., J. A. Thompson, and H. E. Schroeder. 1979. Cotyledonary storage proteins in *Pisum sativum*. IV. Effects of sulphur, phosphorus, potassium and magnesium deficiencies. *Austral. J. Plant Physiol.* 6:11–24.

Raper, C. D., and J. F. Thomas. 1978. Photoperiodic alteration of dry matter patitioning and seed yield in soybeans. *Crop Sci.* 18:654–656.

Rasmussen, P. E., R. E. Ramig, R. R. Allmaras, and C. M. Smith. 1975. Nitrogen-sulphur relations in soft white winter wheat. II. Initial and residual effects of sulphur application on nutrient concentration, uptake, and N/S ratio. *Agron. J.* 67:224–228.

Read, D. W. L., and F. G. Warder. 1982. Wheat and barley responses to rates of seeding and fertilizer in Southwestern Saskatchewan. *Agron. J.* 74:33–36.

Reeves, T. G. 1976. Effect of annual rye grass (*Lolium rigidum* Gand.) on yield of wheat. *Weed Res.* 16:57–63.

Reid, J. B. 1979. Flowering in *Pisum*: Effect of the parental environment. *Ann. Bot.* 44:461–467.

Rendig, V. V., and F. E. Broadbent. 1979. Proteins and amino acids in grains of maize grown with various levels of applied N. *Agron. J.* 71:509–512.

Rennie, R. J., and S. Dubetz. 1984. Multistrain vs. single strain *Rhizobium japonicum* inoculants for early maturing (00 and 000) soybean cultivars: N2 fixation quantified by 15N isotope dilution. *Agron. J.* 76:498–502.

Ries, S. K., G. Ayers, V. Wert, and E. H. Everson. 1976. Variation in protein, size and seedling vigor with position of seed in heads of winter wheat cultivars. *Can. J. Plant Sci.* 56:823–827.

Rincker, C. M., J. G. Dean, C. S. Garrison, and R. G. May. 1977. Influence of environment and clipping on the seed-yield potential of three red clover cultivars. *Crop Sci.* 17:58–60.

Roach, D. A. 1986. Timing of seed production and dispersal in *Geranium carolinianum*: Effects on fitness. *Ecology* 67:572–576.

Roach, D. A., and R. D. Wulff. 1987. Maternal effects in plants. *Ann. Rev. Ecol. Syst.* 18:209–335.

Robinson, R. G. 1983. Yield and composition of field bean and adzuki bean in response to irrigation, compost, and nitrogen. *Agron. J.* 75:31–35.

Robinson, R. G., J. H. Ford, W. E. Lueschen, D. L. Rabas, L. J. Smith, D. D. Warnes and J. V. Wiersma. 1980. Response of sunflower to plant population. *Agron. J.* 72:869–871.

Santakumari, M., and C. R. Reddy. 1980. The use of dalopon to improve the nutritive value of sorghum (*Sorghum bicolor* L.) seeds. *Seed Res.* 8:33–37.

Sawhney, R., and J. M. Naylor. 1982. Dormancy studies in seed of *Avena fatua*. 13. Influence of drought stress during seed development on duration of seed dormancy. *Can. J. Bot.* 60:1016–1020.

Schussler, J. R., M. L. Brenner, and W. A. Brun. 1984. Abscisic acid and its relationship to seed filling in soybeans. *Plant Physiol.* 76:301–306.

Seddigh, M., and G. D. Jolliff. 1984. Night temperature effects on morphology, phenology, yield and yield components of indeterminate field-grown soybean. *Agron. J.* 76:824–828.

Seiler, G. J. 1983. Effect of genotype, flowering date, and environment on oil content and oil quality of wild sunflower seed. *Crop Sci.* 23:1063–1068.

Sesay, A., and R. Shibles. 1980. Mineral depletion and leaf senescence in soya bean as influenced by foliar nutrient application during seed filling. *Ann. Bot.* 45:47–55.

Sharma, S. B. and B. S. Dhillon. 1986. Effect of zinc sulphate and growth regulators on fruit and seed size of litchi (*Litchi chinensis* Sonn). *J. Res. Punjab Agr. Univ.* 23:233–236.

Sheldrake, A. R., and N. P. Saxena. 1979. Comparisons of earlier-and later-formed pods of chickpeas (*Cicer arietinum* L.). *Ann. Bot.* 43:467–473.

Shimzu, N., T. Komatsu, and F. Ikegaya. 1979. Studies on seed development and ripening in temperate grasses. II. Effects of temperature on seed development and ripening and germination behavior in orchard grass (*Dactylis glomerata*) and Italian rye grass (*Lolium multiflorum*). *Bull. Nat. Grassld. Inst.* 15:70–87.

Siddique, M. A., and P. B. Goodwin. 1980. Seed vigor in bean (*Phaseolus vulgaris* L. cv. Apollo) as influenced by temperature and water regime during development and maturation. *J. Expt. Bot.* 31:313–323.

Simmons, S. R., and R. K. Crookston. 1979. Rate and duration of growth of kernels formed at specific florets in spikelets of spring wheat. *Crop Sci.* 19:690–693.

Simmons, S. R., and D. N. Moss. 1978. Nitrogen and dry matter accumulation by kernels formed at specific florets in spikelets of spring wheat. *Crop Sci.* 18:139–143.

Singh, D. K., and S. Bharti. 1985. Seed manganese content and its relationship with the growth characteristics of wheat cultivars. *New Phytol.* 101:387–391.

Singh, H., and S. S. Gill. 1981. Effect of seed size and sowing dates on seed crop of pea. *Seed Res.* 9:122–125.

Sionit, N., H. Hellmers, and B. R. Strain. 1980. Growth and yield of wheat under CO_2 enrichment and water stress. *Crop Sci.* 20:687–690.

Sionit, N., and Kramer, P. J. 1977. Effect of water stress during different stages of growth of soybean. *Agron. J.* 69:274–278.

Smith-Huerta, N. L., and F. C. Vasek. 1987. Effects of environmental stress on components of reproduction in *Clarkia unguiculata*. *Am. J. Bot.* 74:1–8.

Snyder, R. L., R. E. Carlson, and R. H. Shaw. 1982. Yield of indeterminate soybeans in response to multiple periods of soil-water stress during reproduction. *Agron. J.* 74:855–859.

Sofield, I., L. T. Evans, M. G. Cook, and I. F. Wardlaw. 1977. Factors influencing the rate and duration of grain filling in wheat. *Austral. J. Plant Physiol.* 4:785–797.

Sofield, I., L. T. Evans, and I. F. Wardlaw. 1974. The effects of temperature and light on grain filling in wheat. In: R. L. Bieleski, A. R. Ferguson, and M. M. Cresswell (eds.), Mechanism of regulation of plant growth. *Bull. Roy. Soc. New. Zeal.* 12:909–915.

Spiertz, J. H. J. 1974 Grain growth and distribution of dry matter in the wheat plant as influenced by temperature, light energy and ear size. *Neth. J. Agr. Sci.* 22:207–220.

Stanton, M. L. 1984. Seed variation in wild radish: Effect of seed size on components of seedling and adult fitness. *Ecology* 65:1105–1112.

———. 1985. Seed size and emergence time within a stand of wild radish (*Raphanus raphanistrum* L.): The establishment of a fitness hierarchy. *Oecologia* 67:524–531.

Stephenson, A. G. 1980. Fruit set, herbivory, fruit reduction and the fruiting strategy of *Catalpa speciosa* (Bignoniaceae). *Ecology* 61:57–64.

Tashiro, T., and I. F. Wardlaw. 1990. The effect of high temperature at different stages of ripening on grain set, grain weight and grain dimensions in the semi-dwarf wheat 'Banks'. *Ann. Bot.* 65:51–62.

Taylor, G. B. 1976. The inhibitory effect of light on seed development in subterranean clover (*Trifolium subterraneum* L.). *Austral. J. Agr. Res.* 27:207–216.

Tietz, A., and M. Dingkuhn. 1981. Regulation of assimilate transport in barley by the abscisic acid content of young caryopses. *Z. Pflanzenphysiol.* 104:475–479.

Touchton, J. T. and F. C. Boswell. 1975. Effects of B application on soybean yield, chemical composition and related characteristics. *Agron. J.* 67:417–420.

Trevino, I. C., and G. A. Murray. 1975. Nitrogen effects on growth, seed yield, and protein of seven pea cultivars. *Crop Sci.* 15:500–502.

Turk, K. J., A. E. Hall, and C. W. Asbell. 1980. Drought adaptation of cowpea. I. Influence of drought on seed yield. *Agron. J.* 72:413–420.

Turley, R. H., and T. M. Ching. 1986. Storage proteins accumulation in 'Scio' barley seed as affected by late foliar applications of nitrogen. *Crop Sci.* 26:778–782.

Unger, P. W. 1980. Planting date effects on growth, yield and oil of irrigated sunflower. *Agron. J.* 72:914–916.

———. 1986. Growth and development of irrigated sunflower in the Texas high plains.

Agron. J. 78:508–515.

Varis, S., and R. A. T. George. 1985. The influence of mineral nutrition on fruit yield, seed yield and quality in tomato. *J. Hort. Sci.* 60:373–376.

Venable, D. L. and D. A. Levin. 1985. Ecology of achene dimorphism in *Heterotheca latifolia.* I. Achene structure, germination and disperal. *J. Ecol.* 73:133–145.

Verkaar, H. J., and A. J. Schenkeveld. 1984. On the ecology of short-lived forbs in chalk grasslands: Semelparity and seed output of some species in relation to various levels of nutrient supply. *New Phytol.* 98:673–682.

Vieira, R. F. 1986. The influence of soil phosphorus fertilizer levels on the chemical composition, physiological quality and field performance of *Phaseolus vulgaris* seeds [in Portugese]. *Revista Ceres* (Brazil) 33:173–188.

Wahua, T. A. T., and D. A. Miller. 1978. Relative yield totals and yield components of intercropped sorghum and soybeans. *Agron. J.* 70:287–291.

Wardlaw, I. F. 1970. The early stages of grain development in wheat: Response to light and temperature in a single variety. *Austral. J. Biol. Sci.* 23:765–774.

Wardlaw, I. F., I. A. Dawson, and P. Munibi. 1989. The tolerance of wheat to high temperatures during reproductive growth. II. Grain development. *Austral. J. Agr. Res.* 40:15–24.

Wardlaw, I. F., and R. L. Dunstone. 1984. Effect of temperature on seed development in jojoba *(Simmondsia chinensis* (Link) Schneider). I. Dry matter changes. *Austral. J. Agr. Res.* 35:685–691.

Warren, H. L., D. M. Huber, C. Y. Tsai, and D. W. Nelson. 1980. Effect of nitrapyrin and N fertilizer on yield and mineral composition of corn. *Agron. J.* 72:729–732.

Warrington, I. J., R. L. Dunstone, and L. M. Green. 1977. Temperature effect at three development stages on the yield of the wheat ear. *Austral. J. Agr. Res.* 28:11–27.

Weil, R. R., and A. J. Ohlrogge. 1976. Components of soybean seed yield as influenced by canopy level and interplant competition. *Agron. J.* 68:583–587.

Weis, I. M. 1982. The effects of propagule size on germination and seedling growth in *Mirabilis hirsuta. Can. J. Bot.* 60:1868–1874.

Weller, S. G. 1985. Establishment of *Lithospermum carolinense* on sand dunes: The role of nutlet mass. *Ecol.* 66:1893–1901.

Westermann, D. T., and S. E. Crothers. 1977. Plant population effects on the seed yield components of beans. *Crop Sci.* 17:493–496.

Wiegland, C. L., and J. A. Cuellar. 1981. Duration of grain filling and kernel weight of wheat as affected by temperature. *Crop Sci.* 21:95–101.

Willson, M. F., and P. W. Price. 1980. Resource limitation of fruit and seed production in some *Asclepias* species. *Can. J. Bot.* 58:2229–2233.

Winn, A. A. 1985. Effects of seed size and microsite on seedling emergence of *Prunella vulgaris* in four habitats. *J. Ecol.* 73:831–840.

Winn, A. A., and P. A. Werner. 1987. Regulation of seed yield within and among populations of *Prunella vulgaris. Ecol.* 68:1224–1233.

Wolfson, J. l., and G. Shearer. 1981. Amino acid composition of grain protein of maize grown with and without pesticides and standard commercial fertilizers. *Agron. J.* 73:611–613.

Wood, D. W., R. K. Scott, and P. C. Longden. 1980. The effects of mother-plant temperature on seed quality in *Beta vulgaris* L. (Sugar Beet). p. 257–270. In: P. D. Hebblethwaite (ed.), Seed Production. Butterworth, London.

Woodbridge, C. G. 1969. Boron deficiency in pea *Pisum sativum* cv. 'Alaska'. *J. Am. Soc. Hort. Sci.* 94:542–544.

Woodward, R. G., and J. E. Begg. 1976. The effect of atmospheric humidity on the yield and quantity of soya bean. *Austral. J. Agr. Res.* 27:501–508.

Woronecki, P. P., R. A. Stehn, and R. A. Dolbeer. 1980. Compensatory responses of maturing corn kernels following simulated damage by birds. *J. Appl. Ecol.* 17:737–746.

Wratten, S. D. 1975. The nature of the effects of the aphids *Sitobion avenae* and

Metopolophium dirhodum on the growth of wheat. *Ann. Appl. Biol.* 79:27–34.

Wright, D. C., J. Janick, and P. M. Hasegawa. 1983. Temperature effects on in vitro lipid accumulation in asexual embryos of *Theobroma cacao* L. *Lipids* 18:863:867.

Wright, D. L., F. M. Shokes, and R. K. Sprenkel. 1984. Planting method and plant population influence on soybeans. *Agron. J.* 76:921–924.

Wrigley, C. W., D. L. Du Cros, M. J. Archer, P. G. Downie, and C. M. Roxburgh. 1980. The sulphur content of wheat endosperm and its relevance to grain quality. *Aust. J. Plant Physiol.* 7:755–766.

Wulff, R. D. 1986a. Seed size variation in *Desmodium paniculatum*. I. Factors affecting seed size. *J. Ecol.* 74:87–97.

_____ . 1986b. Seed size variation in *Desmodium paniculatum*. II. Effects on seedling growth and physiological performance. *J. Ecol.* 74:99–114.

_____ . 1986c. Seed size variation in *Desmodium paniculatum*. III. Effects on reproductive yield and competitive ability. *J. Ecol.* 74:115–121.

Wurr, D. C. E., J. R. Fellows, D. Gray, and J. R. A. Steckel. 1986. The effects of seed production techniques on seed characteristics, seedling growth and crop performance of crisp lettuce. *Ann. Appl. Biol.* 108:135–144.

6

Formation and Spread of Ice in Plant Tissues*

Edward N. Ashworth
Department of Horticulture, Purdue University
West Lafayette, IN 47907

I. INTRODUCTION

The freezing of water within plant tissues is an important physical process which can limit both the production and the distribution of horticultural crops. For example, the growing season of annual crops is often established by the length of the frost-free period. Whereas in perennial crops, low midwinter temperatures or untimely cold temperatures prior to acclimation in the fall or after deacclimation in the spring can damage overwintering buds, developing blossoms, or vegetative structures. Although low temperatures are a major constraint to horticultural production, it is generally not low temperatures *per se* which limit survival. Instead, the principal hurdle which is confronted by plant cells as the temperature drops below 0°C is the crystallization of ice within

*Purdue Agricultural Experiment Station Article No. 12649. I would like to thank Colleen Thomas and Vicki Stirm for their assistance in preparing this review, and Drs. M. Burke, T. Chen, B. Joly, E. Proebsting, and C. Weiser for their helpful suggestions.

tissues. With the notable exception of those crops sensitive to chilling injury, ice formation is a prerequisite for injury, and avoiding crystallization will avoid injury (Levitt 1980). This is particularly crucial in tissues which lack tolerance to ice formation. In these tissues, the point at which ice formation is initiated marks the lethal temperature (Burke et al. 1976). Avoiding ice formation is also a strategy noted in some plant tissues which exhibit considerable winter hardiness. In these plants, ice will form in one portion of the plant but fails to spread into adjacent areas such that water in these tissues will remain liquid well below the melting point (Burke et al. 1976). In contrast, many plants are able to survive freezing and tolerate the formation and spread of ice within their tissues (Burke et al. 1976; Levitt 1980). Species exhibit a remarkable range of tolerance from temperatures slightly below 0°C to temperatures below that of liquid nitrogen (−196°C).

Rather than review the means by which plants are injured by freezing or discuss the possible adaptations involved in winter hardiness, this review focuses on the formation and spread of ice within plant tissues. This aspect of plant cold hardiness is often overlooked. Yet understanding the freezing processes is basic to understanding plant response to freezing temperatures. For coverage of other aspects of plant cold hardiness, the reader is referred to several books (Li and Sakai 1978, 1982; Levitt 1980; Li 1987, 1989; Sakai and Larcher 1987) and reviews on the freezing process in plants (Mazur 1969; Burke et al. 1976), the role of the plasma membrane (Steponkus 1984), and the role of protein metabolism (Guy 1990). In addition, several appropriate review articles have appeared in this series covering nutritional aspects of cold hardiness (Pellett and Carter 1981), freezing injury in herbaceous plants (Li 1984), cryopreservation (Sakai 1984), freezing injury in citrus (Yelonosky 1985), and frost protection (Reiger 1989).

II. ICE NUCLEATION

Although the melting point of ice is 0°C, the freezing temperature is not as clearly defined. Water and dilute aqueous solutions often remain as metastable liquids below their melting points. This phenomena is called supercooling or undercooling (Bigg 1953; Fletcher 1970; Hallett 1968; Franks 1982), and aqueous solutions within plant tissues have been observed to supercool as well (Burke et al. 1976). The growth of ice crystals within a supercooled liquid must be preceded by a process known as nucleation. When water is supercooled, conditions are energetically favorable for ice formation. However, to form ice, a small volume of liquid must first crystallize, and this crystal must in turn grow until all the liquid is frozen (Hallett 1968; Fletcher 1970; Franks 1982). This initial small crystal embryo is in an energetically unfavorable state due to its very large surface:volume ratio and the positive free energy associated with its liquid-crystalline interface. Therefore, a free energy barrier must be overcome before these small embryo crystals will continue to grow (Fletcher 1970; Franks 1981, 1982). Within a supercooled aqueous solution, water molecules are rapidly and continually associating and disassociating into small "ice-like clusters." The water has been described to consist of a mixture of "flickering clusters" (Nemethy 1968) of hydrogen bonded molecules along with less ordered water molecules. Water molecules alternate rapidly between these two forms. Small clusters are unstable and tend to disappear rather than grow. However, if a cluster reaches a critical size, the inclusion of additional water molecules will lower the total free energy and the cluster will continue to grow. To become an active nucleus and trigger further crystallization, a cluster must reach a critical size and have a lifetime sufficient to permit additional water molecules to diffuse to it. Both the lifetime of a cluster and the critical size needed to overcome the free energy barrier are inversely proportional to temperature (Nemethy 1968; Franks 1981) At temperatures just below 0°C, the number of water molecules needed to form a cluster large enough to initiate crystallization is very large, and the lifetime of a cluster would be short. As a result, the probability of forming an active ice nucleus would be very low. However, as the temperature declines, the critical size required to form a nucleus becomes smaller and the cluster's lifetime longer. Therefore, the probability of ice nucleation increases with decreasing temperature (Bigg 1953; Fletcher 1970; Franks 1981 1982, 1987). The rate of nucleation is very dependent upon temperature and exhibits a sharp temperature threshold (Fletcher 1970; Franks 1981. 1982) At temperatures approaching −40°C, the critical size of a cluster necessary to form a nucleus is sufficiently small, and the probability of its existence sufficiently high, that water crystallizes spontaneously. This point is defined as the *homogeneous ice nucleation* point, and it delineates the extent to which water can be supercooled.

A. Heterogeneous Ice Nucleation

Although water has the capacity to supercool to temperatures approaching −40°C, freezing is generally initiated at much warmer temperatures. Heterogeneous ice nucleation describes the situation where nucleation of the ice phase is catalyzed on foreign surfaces or suspended particles (Fletcher 1970; Franks 1981, 1982, 1987). In heterogeneous ice nucleation as in homogeneous ice nucleation, a cluster of water molecules must grow beyond a critical size to initiate crystallization The difference between the two is that in heterogeneous ice nucleation, foreign surfaces apparently help organize water molecules either to reduce the number of molecules necessary to form a critical nucleus, or increase the probability of a critical nucleus being formed (Fletcher 1970;

Franks 1981, 1982, 1987). Foreign particles and surfaces vary in effectiveness at catalyzing ice crystallization and these differences are thought to reflect a range of effectiveness in aligning water molecules on their surfaces into "ice-like" clusters.

B. Factors Affecting Ice Nucleation

The subdivision of bulk water into numerous small droplets has often been used as a means to promote supercooling and study ice nucleation (Vonnegut 1948; Bigg 1953; Vali and Stansbury 1966; Fletcher 1970; Rasmussen and Mackenzie 1972; Michelmore and Franks 1982), and a linear relationship between the logarithm of drop diameter and the mean freezing temperature was recognized (Levine 1950; Bigg 1953). Bigg (1953) also demonstrated that the extent of supercooling depended on the rate of cooling. He concluded that the effects of temperature, droplet volume, and time on the nucleation of supercooled water could be accounted using a probability model. This proposal has been termed the "stochastic hypothesis" and implies that when a group of water droplets were cooled below 0°C, all droplets have the same probability of nucleation. This model predicts that the probability of ice nucleation would increase exponentially with decreasing temperature. Larger droplets, or samples cooled at a slower rate, would have a greater probability of nucleation and therefore supercool less and freeze at warmer temperatures.

An alternate explanation to account for the relationship between droplet volume and the extent of supercooling is that heterogeneous ice nucleation depends on the presence of impurities suspended within the water. The suspended impurities or "motes" have a characteristic temperature range over which they effectively catalyze ice formation. Therefore, the temperatures at which a population of water droplets froze would be directly related to the distribution of motes within the droplets (Levine 1950; Vali and Stansbury 1966; Vali 1971). These motes or impurities are randomly distributed among droplets. Each individual droplet freezes at a particular temperature characteristic of the single most effective impurity within that drop. This model of heterogeneous ice nucleation has been termed the "singular hypothesis" (Vali and Stansbury 1966), since each droplet exhibits a single characteristic freezing temperature. The singular hypothesis also accounts for the effect of droplet volume on supercooling, since the likelihood of a water droplet containing a mote would increase as droplet volume increased. Likewise, the observation that ice nucleation increases with decreasing temperature could be accounted for by the increased probability of droplets containing motes effective at lower temperatures.

Attempts to distinguish the stochastic and singular hypotheses have examined the effect of cooling rate on ice nucleation. The stochastic hypothesis predicts that the probability of freezing is constant over time, such that slower cooling rates would lead to less supercooling and that if water droplets were held at a constant temperature, additional freezing events would be observed over time. In contrast, the singular hypothesis does not predict any influence of cooling rate on ice nucleation. Instead, freezing would depend entirely upon sample temperature. As soon as the temperature characteristic of the most effective ice nucleating site is reached, freezing is initiated. Therefore, neither slow cooling rates nor prolonged exposure to a fixed subzero temperature would increase the number of nucleation events. Several reports have provided evidence that ice nucleation is dependent upon cooling rate (Vonnegut 1948; Biggs 1953; Gokhale 1965; Vali and Stansbury 1966; Hallett 1968). Droplets cooled at slower rates froze at warmer temperatures. Likewise, when water was held at constant subzero temperatures, additional freezing events were observed with time (Vali and Stansbury 1966; Ashworth et al. 1985a). These observations are inconsistent with the singular hypothesis of ice nucleation, but on the other hand are not completely consistent with the stochastic hypothesis either. For example, when the cooling of water droplets was interrupted and samples held at a constant temperature, additional freezing events were noted. However, instead of the probability of nucleation being constant over time as predicted by the stochastic hypothesis, the occurrence of additional nucleation events diminished with time (Vali and Stansbury 1966). When cooling was resumed in these experiments, the frequency of ice nucleation events was diminished over the next 1°C, but then returned to rates observed during the uninterrupted cooling experiments (Vali and Stansbury 1966). Again, these observations are not entirely consistent with a purely stochastic nature of freezing and instead suggest that heterogeneous ice nucleating substances are active over a characteristic temperature range.

Another method to distinguish whether ice nucleation occurs via the stochastic or singular hypothesis is to expose samples to repeated freezing and thawing. Since the singular hypothesis is based on the premise that a droplet would freeze at a temperature characteristic of the most effective mote present within a sample, repeat experiments should produce consistent results. In contrast, if the stochastic hypothesis is correct, each droplet would have an equal probability of freezing such that upon thawing and refreezing a set of samples the results would be completely independent of those noted earlier. When Vali (1969) conducted such an experiment, he noted that the freezing of most samples was reproducible, suggesting a singular nature of ice nucleation. However, in his experiments some inconsistencies were noted, and 5% of the specimens initiated freezing at temperatures 2°C different from those initially observed. Vali (1969) suggested that the inconsistencies in his results could have risen from three different sources: fluctuations inherent in the process of embryo cluster formation (stochastic effect);

nonuniformities of temperatures within the sample, such that the site catalyzing ice nucleation was exposed to a temperature different from that measured; or alterations in the catalytic site of an ice nucleating substances (effect on singular characteristic). The second potential source of variation suggested by Vali (1969) would be primarily a function of the experimental system employed and could be measured for a specific set of conditions. However, the first and third mentioned sources of variation would be far more difficult to establish.

Since neither the singular or stochastic hypothesis for ice nucleation could account for the experimental observations, Vali and Stansbury (1966) proposed a modified singular hypothesis to explain heterogeneous ice nucleation. They noted that while cooling rates affected ice nucleation, it represented a small effect compared to temperature. Instead, components within the solution had a characteristic temperature range at which they were effective at catalyzing ice nucleation. Properties of these nucleating sites determined the mean size of an embryo (as in the singular hypothesis), yet fluctuations about the mean size were the result of a stochastic element (Vali and Stansbury 1966). The net result would be that a cluster of water molecules could reach a critical size for subsequent crystallization even if the average size of embryos associated with a given catalytic site was considerably smaller. The likelihood of such an embryo forming would increase with time. This model accounted for the detectable stochastic element and time dependence noted in droplet freezing experiments. However, it also noted that the nature of the ice nucleation agent itself largely determined the temperature at which ice formation was initiated.

The concentration of substances effective in catalyzing ice nucleation within an aqueous solution can have a pronounced effect on the temperature at which freezing is initiated. This was illustrated in the work of Zachariassen and Hammel (1988). In their study, ice nucleation active substances extracted from the afro-alpine plant, *Lobelea telekii* Schweinf., and the freeze-tolerant beetle, *Eleodes blanchardi* Blaisd., were prepared in such a manner that either the concentration of extract was serially diluted but the sample volume kept uniform, or the quantity of the extract was held constant and the sample volume altered tenfold. Upon cooling the samples prepared from either source, they observed that the quantity of ice nucleation active extract determined the temperature at which freezing was initiated, not the sample volume or concentration. In the undiluted form, each extract exhibited a characteristic temperature range at which freezing was initiated. If the quantity of extract in the sample was sufficiently reduced, the samples would supercool to a lower temperature before freezing. Presumably, the original extract had been sufficiently diluted such that there was now a low probability that a sample would contain an effective nucleating agent. At intermediate dilutions, samples froze over a broader temperature range, suggesting that some samples contained effective nucleating agents whereas others did not (Zachanassen and Hammel 1988).

The presence of dissolved solutes also influences the extent to which aqueous solutions supercool. Theoretically, if a dissolved ion or molecule acts to stabilize small ice-like clusters, it should aid ice nucleation. Whereas if the solute disrupts the cluster, further supercooling would be expected (Fletcher 1970). The effects of a number of solutes on both heterogenous and homogeneous ice nucleation have been tested using both droplet freezing assays (Reid et al. 1985) and water in oil emulsions (Rasmussen and MacKenzie 1972; Franks 1982; Mathias et al. 1984). Solutes depressed both the heterogeneous and homogeneous ice nucleation temperature as a linear function of solute concentration. The depression of nucleation temperature was greater than the depression of melting point by an equivalent concentration of each of the solutes tested. In general, a concentration of solute which lowered the melting point 1°C depressed the temperature of ice nucleation by 2°C (Franks 1982).

III. ICE NUCLEATION IN PLANT TISSUES

Although aqueous solutions readily supercool, water within plant tissues generally supercool very little under field conditions. Hoar frost, or frozen dew on plant surfaces, forms at air temperatures just below 0°C (Chandler 1954; Single and Marcellos 1981), and numerous observations have noted it to be widely distributed on vegetative tissues and windshields alike. Direct measurements of ice formation in both herbaceous (Anderson and Ashworth 1985; Ashworth et al. 1985c) and in woody species (Ashworth et al. 1985b; Ashworth and Davis 1984, 1986) growing in the field also showed that most of the water within plant tissues supercooled little before freezing. These measurements used remote data loggers to monitor the temperatures of plants and an inert reference during numerous natural frost episodes. When freezing was initiated, the release of the heat of fusion raised tissue temperature relative to the temperature of the inert reference. On calm evenings when temperatures gradually declined, freezing was readily detected and noted as a distinct exotherm (Fig. 6.1). However, on many evenings air temperatures fluctuated rapidly, and the only way to distinguish a freezing exotherm from a response to changing air temperature was to compare tissue and reference temperatures. The warming of tissue temperature relative to reference temperature was indicative of a freezing event (Fig. 6.2). Studies monitoring freezing in a range of woody plant species noted that none supercooled more than 2–3°C before freezing (Table 6.1) (Ashworth et al. 1985b; Ashworth and Davis 1986). Freezing was monitored on numerous nights under a variety of environmental conditions (advective vs. radiation frosts). Since a range of species and plant types were examined (conifers, deciduous hardwoods, shrubs, and brambles), it appeared that the lack of appreciable supercooling was a common feature of woody plants. Similar results were obtained by monitoring the freezing of five herbaceous species in the field. The median freezing temperatures of tomato, bean, maize, cotton and soybean were −3.2, −2.7. −2.5, −2.5, and −2.7, respectively (Anderson and Ashworth 1985; Ashworth et al. 1985c).

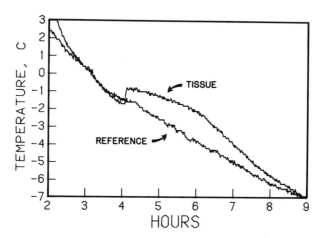

Fig. 6.1. Thermal analysis of an intact nectarine tree growing in the orchard. The plot is a comparison of reference and scaffold limb temperature. A distinct exotherm was observed near −2°C due to the freezing of water within the plant tissue. Measurements began at dusk with the elapsed time indicated on the X axis (Ashworth et al. 1985b).

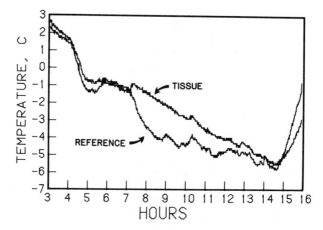

Fig. 6.2. Thermal analysis of an intact nectarine tree growing in the orchard. The plot is a comparison of reference and branch temperature. The freezing of water within the tissue was detected near −1.5°C as the divergence of tissue and reference temperature plots. Measurements began at dusk with the elapsed time indicated on the X axis (Ashworth et al. 1985b).

Table 6.1. Temperatures at which freezing was initiated in woody plants under field conditions. Data compiled from Ashworth et al. (1985b) and Ashworth and Davis (1986).

Species	Mean ice initiation temperature (°C)
Peach	−1.6
Apple	−1.3
Pear	−2.1
Beech	−1.6
Dogwood	−1.8
Holly	−1.6
Juniper	−1.3
Thornless Blackberry	−1.6
White Pine	−1.2
Yew	−2.0

A. Factors Affecting Ice Nucleation in Plant Tissues

Although it appears well established that ice formation is initiated in plants after only limited supercooling, the components or agents responsible for initiating ice formation, and the environmental factors which influence this process, have not been resolved. Investigators have used a variety of methods to study ice nucleation in plants under controlled conditions. A noninclusive list included: using suspensions of homogenized tissues (Rajashekar et al. 1983; Ashworth and Davis 1984; Ashworth et al. 1985); excised plant parts (Salt and Kaku 1967; Marcellos and Single 1979; Lindow et al. 1982; Proebsting et al. 1982; Rajashekar et al. 1983; Ashworth and Davis 1984); and potted plants (Cary and Mayland 1971; Rajashekar et al. 1983; Yelenosky 1983; Ashworth et al. 1985). Tissues have been assayed submerged in water (Anderson et al. 1982; Lindow et al. 1982; Ashworth and Davis 1984; Hirano and Upper 1986) or cooled in air using a variety of controlled temperature devices (Salt and Kaku 1967; Cary and Mayland 1970; Arny et al. 1976; Marcellos and Single 1979). Cooling rates have likewise varied from 1° to 3°C/min (Salt and Kaku 1967; Marcellos and Single 1979) to 24-h periods at a constant temperature (Ashworth et al. 1985a, 1985c).

Not surprisingly such differences in assay techniques have often resulted in conflicting observations and interpretations. For example, results from some laboratories have led investigators to conclude that plants generally lack intrinsic ice nuclei active at temperatures above −5° to −10°C (Marcellos and Single 1976, 1979; Arny et al. 1976; Lindow et al. 1982; Rajashekar et al. 1983). In other studies, intrinsic components responsible for initiating ice formation at −2°C have been reported (Ashworth and Davis 1984; Gross et al. 1984, 1988; Ashworth et al. 1985; Andrews et al. 1986; Anderson et al. 1987). Some of these differences could be due to differences among the plant species examined. However, in describing several factors which affect ice nucleation in plants, we believe that differences in the experimental protocols used to examine ice nucleation could account for some of the varied results and interpretations (Ashworth and Davis 1984; Anderson and Ashworth 1985; Ashworth et al. 1985a, 1985c).

1. Specimen size. Several experiments have demonstrated that small specimens of plant tissue will supercool to lower temperatures than either larger specimens or whole plants. Experiments with peach and cherry shoots demonstrated a logarithmic relationship between specimen size and freezing temperatures (Ashworth and Davis 1984; Andrews et al. 1986) (Fig. 6.3). In large potted trees frozen in a refrigerated chamber or outdoors (Ashworth and Davis 1984), and in mature trees growing in the orchard (Ashworth et al. 1985b), ice formed at approximately −2°C. However, when ice nucleation was assayed using samples less than one gram fresh weight, supercooling to temperatures below −6°C was noted (Ashworth and Davis 1984). Therefore, larger specimens (20 g) were subsequently used when assaying ice nucleation, so that results would be representative of the freezing of intact trees in the field (Ashworth and Davis 1984; Ashworth et al. 1985b). A similar effect of specimen size on supercooling was reported in *Buxus* leaves (Kaku 1971), conifer needles (Kaku and Salt 1968), tomato plants (Anderson and Ashworth 1985), and in maize, cotton, bean, and soybean seedlings (Ashworth et al. 1985c). If the ice nucleation activity of a tissue was assayed using either small specimens or homogenates containing a limited sample of tissue, unrepresentative results were obtained. This is illustrated in the data presented in Table 6.2. Ice nucleation activity of bean tissue was

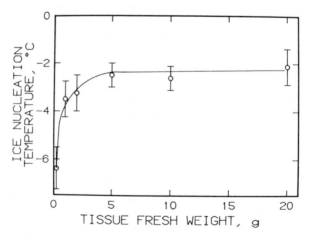

Fig. 6.3. Effect of specimen size on the temperature at which ice formation was initiated within peach shoots. Specimens were cooled at about 1.5°C/h. Error bars indicate SD (Ashworth and Davis 1984).

Table 6.2. Effect of assay technique and tissue mass on the temperature at which ice nucleation occurred in bean. Data from Ashworth et al. (1985c).

Specimen	Mean tissue mass	Temperature of nucleation (°C)	
		1st	Median
2% homogenate, 10μl	0.2mg	−9.8	−10.5
leaf discs, 6.5 mm OD	4.5mg	−7.5	−11.0
2% homogenate, lml	20mg	−7.0	−8.0
Seedlings in lab	2.7g	−1.3	−3.9
Seedlings in field	2.7g	−1.3	−2.7

monitored using techniques representative of those reported in the literature. In one case, bean tissues were homogenized in distilled water, and one hundred 10μl samples of a 2% tissue suspension were assayed for ice nucleation activity. Among the 100 drops examined, the first drop to freeze had supercooled to −9.8°C, and the median freezing temperature was −10.5°C. Based on such data, one might conclude that bean seedlings did not contain components which could serve as heterogeneous ice nucleating agents at temperatures above −9°C. Similarly, experiments with leaf discs or larger volumes of tissue homogenate also led to the conclusion that bean seedlings lack ice nucleating agents active at temperatures above −7°C. In both cases, these conclusions would be incorrect. Bean seedlings exposed to a frost stress in either the laboratory or under natural conditions in the field froze at much warmer temperatures (Table 6.2). Therefore, predictions based on the freezing of small tissue specimens are inappropriate. Parallel experiments using maize. cotton, soybean, and soil suspensions yielded similar results (Ashworth et al. 1985c), with these latter experiments using soil, demonstrating that this effect was not unique to plant tissues.

There are three possible explanations for the observed effect of specimen size on the temperature of ice nucleation. One is that ice nucleation active components are infrequent and unevenly distributed within the tissue. Therefore, the probability of a large sample containing an ice nucleating agent active at warm temperatures is greater than that of a small sample. An effect of specimen size on freezing temperature would not be observed if ice nucleation active agents were abundant and evenly distributed throughout the tissue. This situation was described by Anderson and Ashworth (1985). They observed that tomato plants inoculated with populations of ice nucleation active (INA) bacteria froze between −2° to −3°C regardless of the sample size. While the inoculated plants contained numerous ice nucleation active agents, control plants did not, and a logarithmic relationship between specimen size and the temperature of ice nucleation was observed (Anderson and Ashworth 1985).

The second explanation as to how specimen size could influence the extent of supercooling is that ice nucleation is a function of probability, and the larger the sample mass, the greater the probability that a nucleation event would occur. A third possibility incorporates both the likelihood that an effective ice nucleating agent would be present within a small specimen, and the probability that a nucleation event would be catalyzed. This possibility is most consistent with the modified singular hypothesis of ice nucleation (Vali and Stansbury 1966), that heterogeneous ice nucleating substances have a characteristic temperature range at which they are effective at catalyzing ice formation, but that a stochastic element influences whether in fact an ice nuclei of critical size is formed.

Regardless of which explanation accounts for the effect of specimen size on nucleation temperature, it is clear that predictions of plant freezing temperature based upon small tissue samples are not valid and tend to overestimate the extent of supercooling. Therefore, reports that some plant species lack intrinsic ice nucleating substances active above $-5°$ to $-10°C$ which were based on small tissue samples may need to be reevaluated. It should be stressed that subdividing a large specimen into smaller components does not alter the fundamental properties of ice nucleating substances within the tissue. Rather, subdividing the tissue leads to a sequestering of nucleation sites. Therefore, instead of a single ice nucleus initiating ice formation within the entire specimen, only one of the subsamples would freeze, and the remainder would continue to supercool. The temperature range over which the entire population of subsamples freezes would depend on the nature of the associated ice nucleating agents and their distribution among the original specimen. Problems arise when a limited subsample of the original tissue is assayed, and the results used to make subsequent predictions on the freezing behavior of the entire specimen. For example, when assaying the freezing of 10 μl samples from a 2% bean tissue homogenate (Table 6.2), each sample contains approximately 0.2 mg of tissue. Therefore, assaying 100 of these samples would only be assessing less than 1% of the original specimen. This situation would be even further exaggerated if such small samples were used to predict the freezing temperature of large objects such as intact trees or the soil profile.

The use of small tissue specimens to predict the freezing temperatures of intact plants will also be inappropriate if ice nucleating agents are not uniformly distributed among the various tissues of the plant. For example, Modlibowska (1962) observed that spurs or flowers detached from apple trees would supercool more than comparable tissue left attached. She noted that the smaller the detached portion, the more the tissue would supercool. Similar observations were reported for floral buds and developing fruit (Table 6.3) of *Pyrus* and *Prunus* species (Proebsting et al. 1982; Andrews et al. 1986). Again, specimens detached from

Table 6.3. Mean temperatures of ice nucleation for developing fruits of peach, cherry, and pear as influenced by attachment to stem tissue 30 cm long. Data from Proebsting et al. (1982).

	Ice nucleation temperature (°C)	
Species	Fruit attached	Fruit detached
Sweet cherry	−2.2	−8.0
Peach	−3.8	−8.3
Pear	−3.0	−6.2

the stem supercooled significantly more than comparable tissues left attached. Since the woody stem tissues of deciduous fruit trees were more effective at heterogeneous ice nucleation than the corresponding floral tissues and developing fruits, Proebsting and coworkers (1982; Andrews et al. 1986) concluded that ice formation was first initiated within the woody stem and subsequently spread into the attached blossoms or developing fruits. Therefore, freezing excised flower blossoms or similar tissues in order to predict the freezing characteristics of intact plants would not be appropriate.

2. Cooling rate. During natural freeze stresses, temperatures generally decline gradually at rates less than 2°C/h (Levitt 1980; Ashworth et al. 1985; Steffens et al. 1989). Comparable cooling rates have not always been used when assaying ice nucleation. The primary reasons for using faster cooling rates are that experiments could be conducted faster, equipment used more efficiently, and research conveniently scheduled. Using faster cooling rates has often been justified on the basis that the singular hypothesis of ice nucleation predicts that ice nucleation is strictly a function of temperature (Lindow et al. 1982; Hirano et al. 1985). Experiments to determine whether cooling rates influence ice nucleation in plant tissues have indicated otherwise. Studies with both herbaceous species such as bean, maize, cotton, soybean, and tomato (Cary and Mayland 1970; Anderson and Ashworth 1985; Ashworth et al. 1985c), and with woody species such as citrus (Yelenosky 1983) and peach (Figure 6.4) (Ashworth et al. 1985a), demonstrated an increased proportion of specimens freezing with time at constant temperatures. For example, in the experiments using peach shoots, tissues were maintained at a constant temperature for 24 h and temperature monitored at one-minute intervals. Numerous freezing events occurred even after 12 h at a constant temperature. The rate of ice nucleation was a function of temperature. Tissues held at −2°C nucleated faster than comparable specimens held at either −1° or −1.5°C (Ashworth et al. 1985a). The observed increase in the number of samples frozen over time did not appear to result from slow temperature equilibration, fluctuations in sample

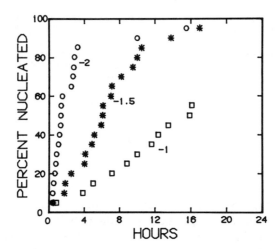

Fig. 6.4. The effect of temperature on the rate of ice nucleation in peach shoots. Twenty-gram segments of peach shoots were maintained at a constant temperature of either −1°, −1.5°, or −2°C for 24 h. The proportion of shoots which froze over time was detected using thermal analysis (Ashworth et al. 1985a).

temperature, or the slow propagation of ice. Tissue temperatures were monitored at one-minute intervals throughout the 24-h period. Specimens rapidly equilibrated to the test temperature and deviated little during the 24-h exposure. In addition, once ice formation was initiated it propagated rapidly throughout the tissue. An increased rate of nucleation at lower temperatures was also observed in bean. maize, cotton, and soybean seedlings (Ashworth et al. 1985c).

Time dependent ice nucleation was not only a property of plant tissues, but has also been observed during the freezing of insects (Salt 1961) and sterilized suspensions of kaolin (Ashworth et al. 1985c). In addition, time dependent ice nucleation has been observed under a variety of different experimental conditions, suggesting that the effect was not an artifact of the experimental protocol.

The combined evidence of these studies indicates that although temperature has a greater effect on ice nucleation than the length of exposure, the effect of time is important. Considering that cooling rates in nature are generally slow, experiments which use rapid cooling rates have the potential to overestimate the extent of supercooling. However, there are instances when rapid cooling rates are appropriate. White and Weiser (1964) noted that arborvitae foliage exposed to full sun in winter would cool rapidly (−7°C/min) at sunset. Therefore, both cooling rate and the length of exposure to subzero temperatures should be adjusted to mimic appropriate field conditions when designing experiments to determine the temperatures at which ice formation is initiated within plant tissues.

3. Surface moisture. On clear, calm nights, plant tissue temperatures often drop below the dew point and water begins to condense on plant surfaces. As temperatures drop below 0°C, hoar frost forms on the plant surfaces. This external ice appears to inoculate the freezing of internal tissue moisture, and ice subsequently spreads throughout the tissue (Cary and Mayland 1970; Single and Marcellos 1981). In contrast, if plant temperatures drop below the freezing point of the tissue water, yet do not reach the dew point; ice nucleation presumably occurs internally (Cary and Mayland 1970). While both of the above conditions have occurred in the field, the latter was generally observed within growth chambers during simulated freezing stresses. In an effort to simulate hoar frost episodes in the laboratory, many researchers have applied moisture to plant surfaces prior to conducting laboratory freezing experiments. The presence of surface moisture raised the temperature of ice nucleation compared to tissues with dry surfaces in a variety of species (Modlibowska 1962; Cary and Mayland 1970; Thomas and Barber 1974; Andrews and Proebsting 1983; Yelenosky 1983; Ashworth et al. 1985c). Experiments in which a fine mist was applied to maize, bean, cotton, and soybean seedlings noted 3°C less supercooling in the latter three species. Less of an effect was noted on maize, and this may have reflected differences in either properties of the ice nucleating agents associated with corn leaf surfaces or ineffective wetting of the leaf surface (Ashworth et al. 1985c).

Differences noted in the extent of supercooling between specimens having dry surfaces, which were apparently nucleated internally, and specimens with wetted surfaces which froze at significantly warmer temperatures, possibly arose from differences in the relative efficiencies of ice nucleating substances associated with the external and internal portions of the tissue. Presumably in the absence of surface moisture, the more efficient external ice nucleating components are isolated from significant amounts of water. This either prevents ice formation from being catalyzed, or prevents the subsequent spread of ice crystals into the internal tissues. Under these conditions, tissue water supercools until freezing is initiated by an internal ice nucleation event. An alternative possibility would be that ice nucleation is not catalyzed by a surface component of the leaf, but that atmospheric sources of ice nucleating agents or suspended ice crystals are responsible for initiating ice formation.

Neither of these alternatives has been firmly established. Marcellos and Single (1976) assayed the air space above a wheat canopy for stable ice nucleating agents active at temperatures above −5°C. Despite the fact that hoar frost would form in these fields at temperatures just below 0°C, atmospheric sources of ice nuclei were only detected on 6 of the 73 evenings assayed. Airborne ice crystals were also not detected above these fields, and it was concluded that atmospheric components were not responsible for catalyzing the freezing of dew on plant surfaces

(Marcellos and Single 1976; Single and Marcellos 1981). Whether freezing was initiated by components associated with the leaf surface could also not be established. Specimens submerged, or homogenized, in water supercooled to lower temperatures than comparable plants covered with a fine mist of water (Ashworth et al. 1985c). This observation was surprising since in all three instances, water should have been in contact with ice nucleation active agents associated with the leaf surfaces.

Despite the lack of a plausible explanation for the effect, the presence of moisture on the surface of plant tissues had a pronounced effect on the extent of supercooling. This effect must be considered in order to obtain meaningful results when attempting to simulate hoar frost conditions.

B. Ice Nucleating Agents Associated with Plant Tissues

Observations that ice generally forms in plants within a few degrees of the tissue melting point imply that freezing is initiated via heterogeneous ice nucleation, and that some component associated with the plant is responsible for catalyzing ice formation. The prime candidates for this role have been conveniently categorized as either epiphytic INA bacteria residing on the plant or as intrinsic components of the tissue. Each will be considered separately in this review. The reader is referred to several earlier reviews which examine INA bacteria in more detail (Lindow 1982a, 1983a, 1983b). A series of papers characterizing the ice nucleation active component of the bacteria have also appeared (Green and Warren 1985; Orser et al. 1985; Wolber et al. 1986; Govindarajan and Lindow 1988; Burke and Lindow 1990). This aspect will not be covered here.

1. Ice nucleation active bacteria. An examination of potential ice nucleating agents associated with precipitation led researchers to observe that decaying vegetation is a very effective ice nucleating agent (Schnell and Vali 1972, 1973, 1976). Subsequently, the active component of the leaf litter was isolated and identified as the bacterium *Pseudomonas syringae* van Hall (Maki et al. 1974). The relevance of this to the freezing of plant tissues was first reported by Arny et al. (1976). In a series of subsequent reports, Lindow and coworkers reported that strains of ice nucleation active (INA) bacteria were effective in catalyzing heterogeneous ice nucleation at temperatures associated with frost injury; that in the absence of INA bacteria, ice formation in frost sensitive plants would not be initiated until much lower temperatures ($-8°$ to $-10°C$); that the temperature at which frost injury was initiated was predictable based on the population and ice nucleation activity of the INA bacteria associated with the tissue; and that this was a general response common to a range of plant species (Lindow et al. 1978a, 1978b, 1982a, 1982b; Hirano et al. 1985). In addition to *P. syringae*, other species of bacteria exhibiting ice nucleation activity above −5°C have been identified. These include *Erwinia herbicola* (Löhnis) Dye (Lindow et al. 1978b); *Pseudomonas fluorescens* Migula (Maki and Willoughby 1978); and *Pseudomonas viridiflava* (Burkholder) Dowson (Paulin and Luisetti 1978). In each instance, not all strains of a species express ice nucleation activity. In *P. syringae* about half the strains appear active, with a lower proportion of INA strains noted in the other species (Lindow 1982a). INA bacteria appear to be widely distributed in nature. Lindow et al. (1978a) screened a range of plants collected at five U.S. locations for the presence of INA bacteria. INA strains of *P. syringae* and *E. herbicola* are widely distributed with respect to both host range and geographic distribution. Most plant species harbor detectable populations of INA bacteria, the exceptions being conifers and some crucifer species. Although strains of both these bacterial species are pathogenic to some hosts, both are primarily found as epiphytes. The ubiquitous nature of these species suggests that INA bacteria are a significant source of ice nucleation activity associated with plant tissues, and could contribute as a source of atmospheric ice nucleating agents as well (Lindow et al. 1978a; Lindow 1982a). Even though INA bacteria were reported to be associated with most plant tissues, the populations detected varied among individual leaves (Hirano et al. 1982), among species (Lindow et al. 1978a), among cultivars of the same species (Marshall 1988), and in response to environmental conditions (Lindow et al. 1978a).

Numerous reports have linked the presence of INA bacteria with frost injury to various crop plants (Arny et al. 1976; Lindow et al. 1978b, 1982b, 1985; Anderson et al. 1982; Lindow 1983a; Hirano et al. 1985; Marshall 1988). These reports demonstrated that when sufficient populations of INA bacteria were present on plant tissues, ice formation would be initiated at −3° to −5°C. In frost sensitive plants, injury was initiated at this temperature. In contrast, when population levels were low or undetectable, tissues would supercool to −8° to −10°C before freezing. In the absence of ice formation, frost sensitive plants would escape injury. These observations led researchers to consider INA bacteria as the causal agents of frost injury above −5°C (Lindow 1983a, 1983b) and propose that controlling INA bacteria populations would provide a means of frost control.

If INA bacteria are involved in frost injury to a particular crop, three separate conditions must be met. The first is that INA bacteria must catalyze ice nucleation over the same temperature range that frost injury occurs. Secondly, INA bacteria must be present in sufficient numbers to initiate freezing. Finally, in the absence of sufficient populations of INA bacteria, plant tissues would supercool to significantly lower temperatures before freezing was initiated. Failure to satisfy all of these criteria would indicate that INA bacteria do not play a role in frost injury to a particular species, and that frost control methods based on manipulating the activity and populations of these organisms would not be successful. It should also be remembered that the success or failure of a particular combination of host, bacterium, and environmental condition does not dictate the general success or failure of this approach. Rather, each situation should be evaluated independently.

The criteria that INA bacteria are effective at catalyzing ice formation at temperatures important to frost damage has been well documented. INA bacteria were reported to be among the most effective ice nucleating agents identified (Maki et al. 1974; Lindow et al. 1978). These organisms were active both in vitro and in vivo, and in the latter case, appear to be equally effective at initiating freezing in a range of plant tissues (Anderson et al. 1982; Lindow et al. 1978c, 1982a; Lindow 1982a).

Determining whether or not INA bacteria are present in sufficient populations on plants to catalyze ice formation has not been as easy as might be anticipated. Satisfying this criterion requires determining both the population of INA bacteria on tissues and the threshold population level needed to initiate ice formation. Populations of INA bacteria are not evenly distributed among leaves within a crop canopy. Hirano et al. (1982) noted that within a set of 24 individual maize leaves, INA bacterial populations varied 500-fold. Populations were not normally distributed, but instead were skewed and best represented by a lognormal distribution. Therefore, assaying populations from a combined sample of leaves will overestimate the median bacterial population found on individual leaves.

The second obstacle involves determining the threshold population of INA bacteria necessary to catalyze ice formation. Several reports noted that frost injury, a presumed reflection of the temperature of ice nucleation in frost sensitive plants, was directly proportional to the logarithm of the INA bacterial population (Lindow et al. 1978b, 1978c). Therefore, reductions in the population of INA bacteria should result in predictable reductions in both the temperature of ice nucleation and in frost injury. Unfortunately, the relationship between bacterial population and the temperature of ice nucleation appears to be more complex. Although the ice nucleation characteristic is a function of the bacteria genotype, not every cell within a population of INA bacteria will catalyze ice nucleation at a particular temperature and time (Maki et al. 1974). Therefore, the temperature at which ice formation is initiated is a function of both the INA bacterial population and the proportion of cells within that population which are active ice nucleating agents. This latter component, referring to the number of ice nuclei of bacterial origin expressed within a population of bacterial cells, has been called the nucleation frequency (Lindow 1982a, 1983b; Lindow et al. 1982a, 1982b). Reported values for bacterial nucleation frequencies have ranged from one cell in ten, to one in 10^8 cells being active catalysts for ice nucleation. Nucleation frequency appears to be both a product of the cell genotype and of the prevailing environment. Several reports have noted that bacterial ice nucleation frequencies varied among bacterial species and strains (Maki et al. 1974; Lindow et al. 1982a, 1982b; Hirano et al. 1985). When grown in culture such factors as the growth temperature, chemical composition, and the physical consistency of the media influences nucleation frequency (Maki et al. 1974; Lindow et al. 1982b; Rogers et al. 1987). Likewise, the nucleation frequency of bacteria colonizing a leaf is also influenced by the host and the prevailing environment (Hirano et al. 1985; O'Brien and Lindow 1988). The nucleation frequency of a particular bacterial strain in culture does not correlate with the nucleation frequency observed in plant tissues (O'Brien and Lindow 1988). This unfortunately complicates assessing the threshold population of INA bacteria necessary to incite frost injury.

A further complication involves the potential role of dead INA bacteria as active ice nucleating agents. Maki et al. (1974) reported that dead *Pseudomonas syringae* cells retained ice nucleation activity, and Lindow (1982a, 1983a, 1983b) has suggested that dead bacteria may serve as a significant source of ice nucleating activity on plant tissues. This situation causes two distinct problems. First of all, dead INA bacteria present investigators with the problem of being undetectable using standard bacteriological methods, yet potentially responsible for catalyzing ice formation. This makes it extremely difficult to establish that ice nucleation activity associated with plant tissues is not of bacterial origin. The second potential problem involves the use of bactericidal sprays as a potential frost control method. Killing INA bacteria just prior to an anticipated frost would not be effective if the dead organisms retained ice nucleation activity, and alternative strategies would be required. Experiments with INA strains of *P. viridiflava* noted that cells lost ice nucleation activity when killed by UV light exposure, desiccation, and streptomycin treatment (Anderson and Ashworth 1986). The reduction in activity observed varied with treatment. Cells treated with streptomycin and incubated at 21°C gradually lost ice nucleation activity over a 9-h period, whereas cells killed by UV radiation exhibited an immediate 3°C reduction in activity. Desiccated cells were less effective at catalyzing ice formation (−4° to −4.5°C compared to −2.5° to −3°C for controls) and maintained this difference for at least a 40-day period. It was also observed that bacterial cells killed by either streptomycin or UV radiation would lose further nucleation activity following exposure to warm (33°C) temperatures. Viable cells also lose activity when exposed to warm temperatures, but, unlike dead cells, would regain the capacity to catalyze ice formation at −3°C when transferred to inductive temperatures (4°C). Although it is not known whether fluctuating temperatures would inactivate dead INA bacterial cells in the field, it appears likely that nucleation activity would not persist with dead INA bacteria, and that this would not serve as a significant source of ice nucleation activity under field conditions (Anderson and Ashworth 1986).

The last criterion to establish if INA bacteria play a significant role in frost injury depends upon whether plant tissues in the absence of INA bacteria would supercool to significantly cooler temperatures before freezing. Levitt (1980) noted that, with the exception of water within xylem ray parenchyma cells and overwintering flower buds of some woody perennials, little supercooling occurs in nature. However, reports as to whether plants contain intrinsic ice nucleating substances are mixed. Many reports have indicated that a variety of plant species lack intrinsic ice nucleating agents active at temperatures warmer than $-5°$ to $-10°C$ (Arny et al. 1976; Marcellos and Single 1976, 1979; Lindow et al. 1982a; Rajashekar et al. 1983; Lindow and Connell 1984; Hirano et al. 1985). Instead, the primary ice nucleating agents associated with these tissues appear to be of bacterial origin. Tissues harboring significant epiphytic populations of INA bacteria froze at $-2°$ to $-4°C$, similar to the range over which frost injury occurred under field conditions (Arny et al. 1976; Lindow et al. 1978b, 1982a; Anderson et al. 1982; Hirano et al. 1984) In addition, both laboratory and field experiments employing treatments to either diminish the population of INA bacteria or inactivate the active catalytic site responsible for nucleation on frost sensitive plants have decreased the incidence and severity of frost damage (Lindow et al. 1978b, 1978c; Lindow 1982b, 1983a; Lindow and Connell 1984).

However, some independent studies failed to demonstrate a relationship between the population of INA bacteria and the temperature at which freezing was initiated in deciduous fruit trees tissues (Ashworth and Davis 1984; Gross et al. 1984; Ashworth et al. 1985; Andrews et al. 1986; Anderson et al. 1987; Cody et al. 1987; Proebsting and Gross 1988). Ice formed in the woody tissues of these plants at $-2°$ to $-3°C$ in both laboratory studies (Ashworth and Davis 1984; Ashworth et al. 1985a; Anderson and Ashworth 1987) and in the field (Ashworth et al. 1985b; Ashworth and Davis 1986). The temperature at which freezing was initiated was consistent throughout the year (Ashworth et al. 1985a) and among developmental stages (Ashworth and Davis 1984; Gross et al. 1984). In addition, the temperature of ice nucleation was similar regardless of whether or not detectable populations of the INA bacteria were present on the tissue (Gross et al. 1984; Ashworth et al. 1985a; Andrews et al. 1986; Proebsting and Gross 1988). The application of surface disinfestants, bactericides, chemicals inhibitory to bacterial ice nucleation, and antagonistic organisms which lower the population of INA bacteria on the tissue all failed to alter the temperature at which freezing was initiated or the sensitivity of the tissue to frost injury (Ashworth and Davis 1984; Cody et al. 1987; Proebsting and Gross 1988). The combined evidence suggests that the woody tissues of these species contain an intrinsic component active in catalyzing ice formation. This component limits the extent of supercooling, and once freezing is initiated within the stem tissues, ice could spread to the attached blossoms. This was demonstrated by Proebsting and coworkers who noted that excised blossoms and fruitlets would supercool more than specimens inoculated with INA bacteria or those left attached to a segment of the woody stem (Proebsting et al. 1982; Gross et al. 1984; Andrews et al. 1986). However, flowers and fruitlets left attached to the stem froze at the same temperature regardless of whether or not INA bacteria were detected. This led to conclusions that an intrinsic component of the woody stem tissue was active at catalyzing ice formation over a similar temperature range as INA bacteria, and that once freezing was initiated, ice crystals would spread into the blossoms and fruit via the vascular system.

2. Intrinsic ice nucleating agents. The inability to observe a relationship between INA bacterial populations and the temperature of ice nucleation has led some workers to speculate that an intrinsic component associated with *Prunus* stem tissues catalyzes ice nucleation (Ashworth et al. 1985a; Gross et al. 1985; Andrews et al. 1986; Anderson and Ashworth 1987; Gross et al. 1988; Proebsting and Gross 1988). This component has neither been identified nor well characterized. It appears to be a constitutive component of woody tissues active above $-3°C$. Since mature stem tissues supercool less than equivalent-sized younger stems (Ashworth et al. 1985; Gross et al. 1988), the component appears to develop during the maturation of the shoot. Once established, ice nucleation activity is stable through out the year (Ashworth and Davis 1984; Ashworth et al. 1985; Gross et al. 1988). Several characteristics of the intrinsic ice nucleating agent are distinct from those associated with INA bacteria. While the bacterial component active in ice nucleation has been identified as a protein (Lindow 1983; Wolber et al. 1986), the component associated with woody tissue does not appear to be a protein. Gross et al. (1988) observed that it was insensitive to pronase, p-hydroxymer-curibenzoate, and iodine, all three of which inactivated the bacterial ice nucleating protein. They also noted that the intrinsic component could be inactivated by soaking the tissue in water, a treatment which had no effect on the ice nucleation activity of bacterial cells. Other differences between the intrinsic ice nucleating component and that associated with INA bacteria include differences in heat stability and chemical stability when treated with bacterial ice nucleation inhibitors (Ashworth and Davis 1984; Gross et al. 1988). Further characterization of the intrinsic component will be complicated, since it appears to be present at low frequency and inactivated by homogenization (Ashworth and Davis 1984; Ashworth et al. 1985a).

While ice formation appears to be initiated first within stem tissues of

Prunus species, Kaku (1973) reported that leaf tissues of Veronica persica Poir., Pisum sativum L., Vicia faba L., Vicia sativa L. and Buxus microphylla Siebold & Zucc. were more effective than stem tissues in initiating ice formation. Of the five species examined, Veronica leaves contained the most effective ice nucleating agents, and freezing was initiated within the leaf tissues between −2° to −6°C. Experiments comparing the ice nucleation activity of the leaf blade, vein, and petiole tissue, and homogenates and supernates of leaf tissues indicate that the most effective ice nucleating agents are associated with the leaf blade. Homogenizing leaf tissue reduces the median nucleation temperature, and the supernatant from this treatment is much less effective than the crude extract. The results suggest that the most effective ice nucleating agents associated with Veronica leaves are located in or on the leaf blade, and that the agents are a structural rather than a soluble component (Kaku 1973). In contrast, experiments with Buxus published in the same study (Kaku 1973) noted that the most effective INA components were associated with the midrib, rather than the leaf blade or stem. Again, homogenization of the tissue lowered the temperature of ice nucleation, suggesting the importance of tissue structure. Alternatively, it is possible that lower ice nucleation temperatures observed for homogenized tissues may not have resulted from homogenization per se. In these experiments 15–20 µl subsamples of the tissue homogenates were assayed for ice nucleation, and the increased supercooling observed may be more a reflection of the reduced likelihood of ice nucleation in small samples compared to larger ones (Ashworth and Davis 1984; Ashworth et al. 1985a, 1985c).

Interestingly, Kaku (1971) observed that mature Buxus leaves nucleate at warmer temperatures than younger leaves. It is not clear whether this is due to quantitative or qualitative differences in ice nucleating agents. Since freezing appeared to be initiated in the midrib, it was speculated that the warmer ice nucleation in mature leaves could be related to maturation of the vascular system. This corroborated similar observations made on spruce needles which reported that ice formation was initiated within the steele (Salt and Kaku 1967).

The observed decrease in supercooling noted in both Buxus leaves (Kaku 1971) and Prunus stems (Ashworth et al. 1985; Gross et al. 1988) during tissue maturation suggest that intrinsic ice nucleating agents are synthesized during differentiation of the tissue. An alternative explanation would be the gradual colonization of tissue with INA bacteria over time. The studies by Kaku and coworkers (Salt and Kaku 1967; Kaku 1971, 1973) preceded recognition that INA bacteria may play a role in initiating ice formation in plant tissues (Arny et al. 1976). Therefore, since INA bacterial populations were not measured, interpretations concerning differences among species, among tissues (stems vs. leaves), and with maturation will be difficult to assess.

To establish whether plant tissues contain intrinsic ice nucleating agents which limit the extent of supercooling in the absence of INA bacteria will require their identification. Indirect evidence based on the inability to detect INA bacteria or on apparent differences in the sensitivity to inhibitors is suggestive but inconclusive. Presumably, plant tissues contain numerous substances which can act as heterogeneous ice nucleating agents. Collectively, these components would be effective over a range of temperatures. However, only those limiting supercooling in nature would be important, since unless sequestered within a tissue and isolated from the remainder of the plant, only those agents active at the warmest temperatures would be significant. Therefore, the question remains whether intrinsic components or INA bacteria initiate freezing within plant tissues. Intrinsic ice nucleating substances have been observed in insects (Zachariassen and Hammel 1976; Duman 1982; Zachariassen 1982); in the Afro-Alpine shrub, *Lobelia telekii* Schweinf. (Krog et al. 1979); and in lichens (Kieft 1988). Research to identify and characterize potential ice nucleating substances within crop plants is important. Resolving this issue will have a bearing on whether or not supercooling of crop plants can be effectively promoted by simply eliminating or inactivating INA bacteria.

It should also be noted that the inability to detect INA bacteria does not mean that the organisms are not present. All techniques have a lower limit of detection, and values below this level should not be automatically equated to zero. With ice nucleation frequency values of one cell in ten being active as an effective ice nucleating agent, it is certainly possible that populations below the detection level could still be responsible for initiating ice formation. However, if this is the case, then potential frost control methods based on reducing the populations of INA bacteria will need to be extremely effective. This may not be attainable.

INA bacteria undoubtedly comprise an important source of ice nucleation activity associated with plant tissues. However, on some species (Lindow et al. 1978a) and in some climates and production systems (Lindow 1978a; Gross et al. 1983; Ashworth et al. 1985a), INA bacterial populations are undetectable. Do these plants supercool to significantly lower temperatures before freezing is initiated in the field, or do other sources of ice nucleation activity initiate freezing? In addition, what is the relationship between the number of ice nucleating agents present and the freezing temperature? Since in most tissues water exists in a continuous network throughout the plant, a single ice nucleation event could initiate freezing throughout the plant. Presumably in many cases numerous ice nuclei form before ice crystals originating from the first nucleation event can spread throughout the entire tissue. But how many ice nuclei typically form on a leaf or flower blossom during a frost episode? Would blossoms containing 1, 100, or 10,000 ice nucleating agents active at $-3°C$ freeze at significantly different temperatures? Perhaps tissues containing high numbers of effective ice nuclei would be more likely to freeze first, but how much of a difference would be expected? It has been reported that INA bacteria account for 95% of the ice nucleating substances active above −5°C in some plant tissues (Lindow 1983b). While this is certainly significant, it does not follow that the other 5% were insignificant. Efforts to account for all the sources of ice nucleation activity on plant tissues are required, and it would be anticipated that species will vary in both their capacity to support epiphytic populations of INA bacteria and the relative contribution of intrinsic ice nucleating substances.

IV. ICE CRYSTALLIZATION

Freezing occurs in two steps involving both ice nucleation and crystallization. Although often considered together, each is a separate process having different temperature optima (Luyet 1968) As a result, conditions during freezing can influence the pattern of ice formation. For example, when solutions are cooled slowly to temperatures at which the ice nucleation rate is low, but the rate of crystal growth is high, a small number of large ice crystals are observed. In contrast, if solutions are rapidly cooled to temperatures where the rate of ice nucleation is high, and the rate of crystal growth is reduced, a large number of small ice crystals are noted (Luyet 1968). The former condition is likely to occur during a natural freeze episode when cooling rates are slow. The latter occurs in response to the rapid cooling rates often used in cryofixation techniques for microscopy or during cryopreservation protocols.

The rate of ice crystal growth in water and dilute aqueous solutions is primarily a function of the rate at which the latent heat of fusion is dissipated. Experiments by Pruppacher (1967a, 1967b) demonstrated that the growth rate of ice crystals increased exponentially with increased supercooling. Greater supercooling before ice nucleation increases the temperature gradient between the crystallizing solution and the surrounding environment, and since the dissipation of the heat of fusion is primarily a function of the temperature gradient, faster rates of ice crystal growth are noted. In more concentrated aqueous solutions, the rate of solute diffusion from the ice-water interface also influences the rate of crystal growth (Pruppacher 1967b). Since solutes are generally excluded from the crystal lattice, a concentration gradient of solutes could build up in front of the advancing ice front, leading to a local lowering of the solution melting point (Franks 1981). This condition would impede continued crystal growth until either solute diffusion dissipated the gradient, or the temperature was reduced below the melting point of the adjacent solution.

V. ICE CRYSTALLIZATION WITHIN PLANT TISSUES

In contrast to the freezing of a simple aqueous solution, ice crystals growing within plant tissues cannot develop freely in all directions. Tissue structure and compartmentalization impede the mixing of solutes, and both cell walls and membranes are obstacles to the spread of ice (Luyet 1968). Under conditions generally experienced in the field, ice formation will be initiated within either the extracellular space or the vascular tissues (Levitt 1980). Once initiated, ice crystals can rapidly propagate along continuous columns of water. Measurements of linear rates of ice growth in woody stem tissues of lemon (Lucas 1954), mulberry (Kitaura 1967a, 1967b), citrus (Yelenosky 1975), and peach (Anderson and Smith 1989) noted rates between 7 and 74 cm/minute. As expected, the rate of ice propagation increased as temperatures decreased. Kitaura (1967b) observed that ice propagated at 20 and 60 cm/minute in mulberry shoots supercooled to −1.5° and −3°C, respectively. In mulberry, ice apparently propagates through the vascular tissues. Studies on mulberry and peach show that ice spreads along the stem and into attached leaves and blossoms (Kitaura 1967; Anderson and Smith 1989). Peach blossoms froze within 30 seconds of the time that the ice front passed the point of blossom attachment. Results were similar at both bloom and post bloom, and among different peach cultivars. No apparent barriers to the linear propagation of ice in these tissues were apparent (Anderson and Smith 1989).

A. Factors Affecting the Spread of Ice Within Plants

As in the freezing of any aqueous solution, the rate at which a plant tissue freezes will be influenced by the speed at which the heat of fusion is dissipated. Heat dissipation will be influenced by the tissue water content, temperature gradients between the specimen and the surrounding environment, tissue surface:volume ratio, and the specimen surface characteristics. Tissue water content dictates the amount of heat which must be dissipated, while the other three factors influence the rate of heat transfer.

Although the initial rates of ice propagation through continuous columns of water are rapid and primarily a function of the dissipation of the latent heat of fusion, the extracellular water fraction includes only a small fraction of the available tissue moisture (Tyree and Jarvis 1982), and continued growth of ice crystals will depend upon the diffusion of intracellular water to the growing extracellular ice crystals. The rate of water loss from cells will depend upon the water potential gradient, cell wall and membrane permeability, and the cell surface area. In many tissues and experimental systems, cells will be surrounded by water and extracellular ice. Under these conditions, the movement of cellular water to extracellular ice crystals will be rapid and probably not rate limiting. Cells will maintain equilibrium with the extracellular ice, and the continued rate of extracellular ice crystal growth will be primarily a function of the cooling rate and the dissipation of the heat of fusion. However, if very rapid cooling rates (>3°C/minute) are experienced, cells will not remain in equilibrium with extracellular ice, and the intracellular solution will become supercooled (Steponkus 1984). Under these conditions, cells are susceptible to intracellular ice formation resulting from either intracellular ice nucleation (Levitt 1980) or seeding from extracellular ice (Morris and McGrath 1981; Dowgert and Steponkus 1983).

In many tissues, extracellular ice crystals are not in direct contact with the cell surface. Ice crystals are often unevenly distributed and in contact with only portions of a cell or further removed (Wiegand 1906; Quamme 1978; Sakai 1979, 1982; Ishikawa and Sakai 1982, 1985; Ashworth et al. 1988, 1989; Pearce 1988; Ashworth 1990). The distance between the intracellular solution and extracellular ice can be several cell diameters or more, and the diffusion of water could be through either the liquid or vapor phase. This aspect of plant cold hardiness research has received only limited attention, and the mode of water movement may vary among tissues, and species. In the absence of sufficient moisture within the intercellular spaces, water movement would occur predominantly through the vapor phase. Under these conditions, water movement would be significantly slower, and tissues may fail to remain in equilibrium with extracellular ice. Studies examining ice formation within apple bark noted that large ice crystals formed within the cortex, and the extent of cell collapse in response to freezing varied (Ashworth et al. 1988). Cells adjacent to the extracellular ice collapsed more than cells within the periderm, phloem, and cambium, which were more distant.

This situation was observed to an even greater extent in the freezing of overwintering buds. In these tissues, ice crystals are not uniformly distributed within the tissues but instead form in discrete areas within the bud scales and subtending tissues (Wiegand 1906; Dorsey 1934; Quamme 1978; Sakai 1979, 1982; Ishikawa and Sakai 1982, 1985; Ashworth et al. 1989; Ashworth 1990). Ishikawa and Sakai (1982) have described this condition as extraorgan freezing. In these buds, a water potential gradient is established between the ice crystals in the scales and subtending tissues, and the water within the bud primordium. To maintain equilibrium, water leaves the bud primordium or developing floral organ and diffuses to the adjacent ice crystals (Quamme 1978, 1983; Ishikawa and Sakai 1981, 1985; Andrews and Proebsting 1987). If either the cooling rate is sufficiently slow or the rate of water loss sufficiently rapid, bud tissues become dehydrated. In these instances, the intracellular solute content increases, and intracellular ice formation within the bud primordium or developing floral organ is avoided (Burke and Stushnoff 1979; Sakai 1979). However, if these conditions are not met, and the temperature continues to decline, water within the bud primordium is exposed to temperatures below its melting point and supercools. In instances where extensive supercooling occurs, intracellular ice nucleation and lethal freezing may occur. Sakai (1979) demonstrated that hardy conifer buds supercooled when exposed to rapid cooling rates. When cooled at rates more likely to be experienced in the field, the shoot primordia dehydrated and were able to survive temperatures well below the homogeneous ice nucleation temperature. Sakai (1979) observed that cold hardy conifer species generally had smaller shoot primordia than less hardy species, and speculated that smaller primordia size may facilitate dehydration of the tissue during extraorgan freezing. As a corollary, buds which exhibit the deep supercooling characteristic tend to be larger and more developed, and presumably the larger size impedes establishment of equilibrium even at slow cooling rates.

1. Tissue moisture content. Several studies have shown a relationship between tissue moisture content and the spread of ice. In conifers, ice propagates faster in hydrated needles than in needles allowed to dry (Salt and Kaku 1967; Kaku 1971). Both the linear and radial spread of ice are reduced. Kaku (1971) observed that ice formation was initiated within the steele of pine needles and subsequently spread outward. When ice formation is initiated in either needles having a high moisture content or in young developing needles, ice spreads rapidly. However, when the tissue moisture content has declined, or the needle has matured and the endoderms become suberized and lignified, ice propagates more slowly, and two distinct exotherms are observed in differential thermal analysis experiments.

A similar reduction in the rate of ice propagation with decreased moisture content is observed in bean leaves. Ice spreads rapidly in well-watered plants; but if leaves are wilted, ice spreads slowly, and small, isolated, frozen spots are visible on the leaves (Cary and Mayland 1970). Presumably the reduced rates of ice propagation result primarily from reduced moisture within the intercellular spaces.

2. Tissue structure. Studies examining the distribution of ice crystals within leaves (Nath and Fisher 1971; Pearce 1988), buds (Wiegand 1906; Dorsey 1934; Quamme 1978; Ishikawa and Sakai 1982, 1985; Ashworth et al. 1989; Ashworth 1990), and stem tissues (Wiegand 1906; Ashworth et al. 1988) note that ice crystals are not uniformly distributed among plant tissues. Instead, large ice crystals are segregated into specific locations within the various tissues. These observations have led to suggestions that tissue anatomy and morphology may influence the formation and spread of ice within plant tissues. However, despite the attractiveness of this concept, evidence linking particular tissue structural features with the pattern of ice propagation are limited.

Several studies have examined the distribution of ice crystals within overwintering buds using frozen tissue sections, isothermal freeze fixation and freeze-substitution techniques (Wiegand 1906; Dorsey 1934; Quamme 1978; Ishikawa and Sakai 1982, 1985; Ashworth et al. 1989; Ashworth 1990). Evidence of large ice crystals within the bud scale tissues and in the lower portions of the developing bud axis tissues suggests that once freezing is initiated within these tissues, water migrates from adjacent tissues to the growing ice crystals (Figure 6.5). In peach buds, the formation of these large extracellular ice crystals results in a mechanical disruption of the bud scale and axis tissues (Figures 6.6 and 6.7) (Ashworth et al. 1989). Upon thawing and refreezing, ice apparently forms in the same locations within the bud. An increase in mechanical injury has been noted as winter progresses and the buds are exposed to numerous freeze-thaw cycles.

In most studies of dormant flower buds, no evidence of ice formation within the developing floral organ was observed above the lethal temperature (Quamme 1978; Ishikawa and Sakai 1982, 1985; Ashworth et al. 1989). It has been suggested that the segregation of ice outside the developing floral tissue is critical for the survival of the delicate floral organ. These floral tissues do not appear to have any tolerance for the presence of ice crystals and extraorgan freezing and supercooling provide a mechanism of avoiding injury (Quamme 1978; Ashworth 1982; Ishikawa and Sakai 1982, 1985; Ashworth 1989). However, an examination of overwintering forsythia flower buds, which exhibit the deep supercooling characteristic, has noted that ice crystals form within the lower portions of the developing floral tissues (Figure 6.8). Evidence of large ice crystals has been noted just below the epidermal tissues of the hypanthium (Ashworth 1990). Ice crystals were not observed within the developing pistil, petals, or stamen tissues of dormant buds, and it was proposed that water in these tissues supercooled. Why ice did not continue to spread throughout forsythia bud tissues is not apparent.

Although it is clear that there is a segregation of ice when overwintering buds of woody species are frozen, it was not clear why ice forms preferentially in some portions of the tissue and not in others. Several explanations have been presented and these will be considered in turn. One suggestion is that since bud scales are present on the exterior of the bud, the scales cool faster than the internal tissues and are thereby more likely to freeze first (Ishikawa and Sakai 1985). Once freezing is initiated within the scales, water migrates from the internal tissues to the ice within the scales. This does not appear likely. Wiegand (1906) demonstrated that bud scales are poor insulators, and that temperature differences between the exterior and interior of the bud are minimal. In addition, this hypothesis did not explain why large ice crystals form in the subtending axis tissues (Sakai 1979, 1982; Ashworth et al. 1989). However, the suggestion that ice accumulates in external tissues due to

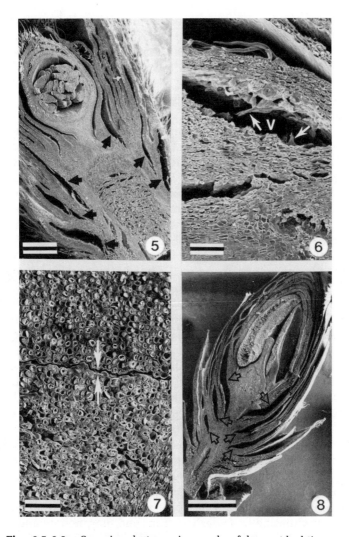

Figs. 6.5–6.8. Scanning electron micrographs of dormant bud tissues.

Figure 6.5. Peach flower bud freeze-fixed at −5°C. Note prominent voids within the bud scales (arrows) and smaller voids within the bud axis. No voids or tissue disruptions were observed within the floral organs. Scale = 500 ℞m (Ashworth et al. 1989).

Figure 6.6. Bud scale tissue from −5°C freeze-fixed peach flower bud. Evidence of disrupted cells (arrows) on periphery of internal void (V). Mechanical disruption of bud scale tissues by ice crystal growth apparent. Scale = 100 ℞m (Ashworth et al. 1989).

Figure 6.7. Axis tissue of conventionally fixed peach bud. Bud harvested from the field in late winter after numerous freeze-thaw cycles. Note long, thin split (arrows). This split was perpendicular to the axis of the bud and corresponded to the region where a large ice crystal formed during freezing. Scale = 100 ℞m (Ashworth et al. 1989).

Figure 6.8. Dormant forsythia flower bud freeze-fixed at −5°C. Note prominent voids within both the lower portion of the developing floral organ and the bud scales (arrows). Scale = 1mm (Ashworth 1990).

increased exposure should not be dismissed completely. Ice often accumulated just below the epidermal tissues in leaves (Nath and Fisher 1971), stem tissues (Wiegand 1906; Ashworth et al. 1988) and buds (Ashworth 1990). Since the dissipation of the latent heat of fusion often limits the rate of ice crystal growth, freezing should proceed faster in the external tissues than in adjacent interior tissues, and as a result may accumulate there.

An alternative proposal to explain the segregation of ice in buds suggests that the scale tissues contain intrinsic ice nucleating components (Ishikawa and Sakai 1981, 1982, 1985). Freezing within the bud would therefore be initiated within these tissues, and large crystals would grow as water was withdrawn from adjacent portions of the flower. Experiments with peach buds noted that the buds scales were more efficient at nucleating ice than the developing floral tissues (Quamme and Gusta 1987; Ashworth, unpublished results). However, neither the excised buds nor the separated component tissues contain effective nucleating agents active above −5°C. In addition, several studies have reported that woody plants begin to freeze near −2°C and that ice subsequently spreads from the woody tissues into the attached buds (Proebsting et al. 1982; Andrews et al. 1986). Therefore, it appears unlikely that ice nucleating agents within the bud play a role in the pattern of ice formation within these tissues.

It has also been suggested that differences in tissue water potentials lead to freezing in certain tissues, and a subsequent redistribution of moisture (Quamme 1978, 1983; Ishikawa and Sakai 1981, 1985; Quamme and Gusta 1987). While differences in tissue water potentials certainly drive the movement of water once freezing is initiated, it appears unlikely that these differences account for ice being preferentially initiated in one tissue region over another. Water initially supercools below the solution melting point of the individual tissue components until either seeded by external ice or heterogeneously nucleated.

An alternative suggestion has proposed that the segregation of ice within overwintering buds derives from bud morphology and vascular development (Ashworth 1989; Ashworth et al. 1989, 1990). In peach buds, xylem vessels form a continuous conduit between the woody tissues and the bud axis and scales. However, vascular tissues within the dormant flower are not fully differentiated, and xylem continuity between the flower bud and subtending stem tissues has not been established (Ashworth 1982, 1984). It has also been noted that ice formation is initiated in the woody tissues of peach between −2° and −3°C (Ashworth and Davis 1984; Ashworth et al. 1985a, 1985b), whereas freezing is not initiated in excised flower buds until lower temperatures are reached (Quamme and Gusta 1987). Therefore, as ice forms within woody stems, ice spreads via the vascular system and initiates freezing within the bud axis and scales, but not in the dormant floral organ. Once ice forms, a water potential gradient is established. This in turn causes water to move from the floral organ to the growing ice crystals within the bud axis and scales. The relationship between vascular differentiation and the distribution of ice crystals within overwintering flower buds has been observed not only in peach (Ashworth 1982; Ashworth et al. 1989) but in forsythia buds as well (Ashworth 1990; Ashworth et al. 1990). Further support for this hypothesis comes from observations that as both species deacclimate in the spring, vascular differentiation progresses and parallel changes in the distribution of ice crystals are noted (Ashworth 1982, 1990; Ashworth et al. 1989, 1990).

Direct observations of ice propagating through the xylem and into the bud tissues have not been made. However, experiments have demonstrated that ice initiated within stem tissues rapidly propagates into attached buds (Proebsting et al. 1982; Andrews et al. 1986; Anderson and Smith 1989; Ashworth et al. 1990). When a strip of bark was removed between a bud and the site of ice initiation on a woody stem, there was no delay in the spread of ice into the bud (Ashworth et al. 1990). Although this does not prove that ice propagates through the xylem vessels per se, it does demonstrate that ice could propagate through the wood and into attached buds. The most likely conduit for ice propagation through the stem would be the xylem vessels.

Although ice crystals form in the vicinity of vascular tissues within buds, the crystals do not surround the xylem elements as might be anticipated. Instead, ice often forms between the vascular tissues and the epidermis. If ice does enter buds through the xylem, it is not clear how ice propagates radially and what conditions influence its subsequent spread. If ice does not enter these tissues through the xylem, then alternative explanations must be developed. Hopefully further research will provide clarification.

3. Cell wall. The cell wall and membrane serve as barriers impeding the spread of extracellular ice into the intracellular solution (Levitt 1980). The proposed mechanisms of these barriers are that water within small pores transversing both the wall and membrane would have a significantly lower melting point than bulk water (Mazur 1969; George and Burke 1977; Levitt 1980; Ashworth and Abeles 1984). This would restrict the growth of ice through these structures. Instead, water would leave the cell, move through both the membrane and cell wall, and join the growing extracellular ice crystals.

Cell walls would not only be expected to affect the growth of ice across it, but also along its length. In many tissues, cells are densely packed and share common walls. In these instances, the intercellular spaces are essentially composed of cell wall materials, and the extracellular water would be associated with microcapillaries of the cell wall. These capillaries form a continuous network between the microfibrals and the molecules of the cell wall matrix (Preston 1974). The size and shape of these capillaries is probably irregular, and an estimated diameter of 4nm has been reported (Preston 1974). Water within such pores would have an estimated melting point between −15° to −25°C. The presence of water in the microcapillaries of the cell wall would not only depress the freezing point of the intercellular solution but would also impede the spread of ice throughout the tissue (Ashworth and Abeles 1984). When propagating through the wall, ice would spread first through the largest diameter capillaries, but its spread would be limited by the smallest diameter within a given capillary. Several papers have suggested that small diameter pores within cell walls may facilitate the supercooling of xylem ray parenchyma cells within many hardwood tree species (George and Burke 1977; George 1983; Ashworth and Abeles 1984). Studies by Wisniewski and coworkers (1987a, 1987b) have examined the porosity of woody plant cell walls in relation to freezing response. Using apoplastic tracers, they proposed that the porosity and permeability of the pit membranes rather than the entire cell wall would limit both the spread of ice and the movement of water in these tissues.

Although it appears likely that properties of the cell wall affect the spread of ice in plants, and that ice and water may move preferentially through specific regions of the cell wall, continued progress within this area will be difficult. These tissue are not amenable to cryomicroscopy, so direct observation of ice propagation is unlikely. Sectioning tissues to a manageable size for this purpose would eliminate much of the three dimensional tissue structure, and promote artifacts such as the spreading of ice along cut surfaces. The best approaches for future study may be further use of apoplastic tracers to determine potential pathways of ice and water movement, and low temperature scanning electron microscopy to determine the location of ice following freezing. In addition, modeling the physical features of the tissue along with the properties of water and ice may provide insights into the process and provide ideas for further testing. An innovative alternative approach has utilized cell wall digesting enzymes to alter wall structure and note the effects on tissue freezing (Wisniewski et al. 1991). Hopefully progress will continue in these areas.

VI. CONCLUDING REMARKS

A. The Influence of Ice Nucleation and Crystal Growth on Plant Survival

In the absence of ice formation, tissues are generally not injured by freezing temperatures. Therefore, strategies that promote supercooling often favor survival. This is particularly true with frost sensitive plants which have no tolerance to ice formation. Based on the premise that INA bacteria are the primary agents responsible for initiating ice formation in plants. Lindow and coworkers (Lindow 1982b, 1983; Lindow et al. 1982) have proposed that controlling the level and activity of these organisms would provide a measure of frost control. Methods involving the use of antibiotics and antagonistic bacteria to lower populations of INA bacteria, and bacterial ice nucleation inhibitors to reduce the ice nucleation activity of the bacteria, have been suggested (Lindow 1983a, 1983b; Lindow et al. 1983). Successful use of these techniques in laboratory and field tests have been reported (Lindow 1982, 1983; Lindow et al. 1983; Lindow and Connell 1984; Marshall 1988). As noted earlier, successful frost protection using these strategies requires that plants supercool significantly in the absence of INA bacteria. This does not always appear to be the case. Both laboratory and field trials with deciduous fruit trees failed to either establish a relationship between INA bacteria populations and the temperature at which tissues froze, or demonstrate frost protection (Ashworth and Davis 1984; Gross et al. 1984; Ashworth et al. 1985a; Andrews et al. 1986; Anderson et al. 1987; Cody et al. 1987; Proebsting and Gross 1988). If in these latter cases intrinsic ice nucleating components limited the extent of supercooling, then the question of whether the presence and activities of these components could be regulated becomes appropriate. In the woody species which have been investigated, the unidentified intrinsic ice nucleating agent(s) appeared to be a constitutive component of mature tissues (Ashworth et al. 1985; Gross et al. 1988). However, in some insect species, substances active in catalyzing ice formation were inducible (Zachariassen and Hammel 1976; Zachariassen 1982). If analogous situations exist within some plant tissues, then it may be possible to regulate levels of intrinsic ice nucleating substances and modify plant freezing temperatures.

The supercooling of water to extremely low temperature has been described as a survival mechanism for overwintering flower buds (George et al. 1974; Quamme 1974; Pierquet et al. 1977; Burke and Stushnoff 1979) and xylem ray parenchyma cells (Quamme et al. 1973; George and Burke 1977a; Pierquet et al. 1977; Burke and Stushnoff 1979; Ashworth et al. 1983) in many deciduous woody plants. In both types of tissues, death and ice nucleation occurred at the same temperature. In living xylem tissues that supercooled, it has been hypothesized that intracellular ice formation was responsible for cell death, and that each individual cell froze as a distinct unit (Hong and Sucoff 1980; Ashworth et al. 1983). It has not been established whether intracellular ice formation resulted from intracellular ice nucleation or seeding from external ice, although the former was believed to be the more likely (George and Burke 1977a).

In contrast, when ice formation occurred in supercooled overwintering buds, a single nucleation event led to the freezing of an entire flower, and each was killed as an individual unit (George et al. 1974;

Quamme 1974, 1978; Graham and Mullin 1976). Since freezing typically occurred above the homogeneous ice nucleation temperature, either heterogeneous ice nucleation or seeding from external ice was responsible for initiating ice formation. The former seems more likely since isolated primordia and floral organs supercooled to temperatures similar to that observed with the freezing of intact buds (George et al. 1974; Quamme 1978; Ashworth 1982). In both wood and bud tissues which exhibit the deep supercooling characteristic, low temperature survival is limited by the homogeneous ice nucleation temperature ($-40°C$). However, this point only provides a lower limit, and more often, particularly in overwintering flower buds or in xylem tissues which are not fully cold acclimated, ice nucleation and tissue death occurs above the homogeneous ice nucleation temperature. The factors limiting supercooling in these instances are unknown. Likewise, the features which determine the extent of supercooling and distinguish the hardiness of species and cultivars have not been identified.

Although increased supercooling is synonymous with increased survival in many instances, excessive supercooling is not always desirable. Several laboratory studies demonstrated that excessive supercooling caused increased freezing injury when ice formation was eventually initiated (Siminovitch and Scarth 1938; Gusta and Fowler 1977; Levitt 1980; Lindstrom and Carter 1983; Rajashekar et al. 1983). Presumably the increased damage resulted from rapid rates of ice crystal growth when supercooled water eventually froze. To avoid this problem, plant cold hardiness researchers often inoculate tissues with ice just below $0°C$. In instances where tissues are not seeded with ice, inappropriate results and misinterpretations can result. For example, differential thermal analysis and nuclear magnetic resonance experiments typically involve the freezing of small specimens within the instrument. In such studies, the likelihood of excessive supercooling is high. Experiments with peach flower buds demonstrated that specimens which supercooled excessively exhibited altered freezing behavior which sometimes led to misinterpretations regarding tissue cold hardiness (Ashworth and Davis 1987). Therefore, even when studying deep supercooling phenomena, researchers need to ensure that the temperature at which tissues initially freeze is representative of field conditions.

The importance of the ice nucleation temperature on tissue survival has led some researchers to propose that survival of frost tolerant plants might be enhanced by seeding with artificial ice nucleating agents (Rajashekar et al. 1983). In addition, Lindow (1982) speculated that the colonization of frost tolerant plant species with INA bacteria could be advantageous for both the survival of the host and the epiphytic bacteria. In both these instances, the promotion of ice nucleation would increase tissue freezing resistance and provide an adaptive advantage. Whether overwintering plants have evolved mechanisms to ensure nucleation either through an intrinsic ice nucleating agent or through a symbiotic relationship with INA bacteria needs to be determined.

The pattern of ice crystal growth is apparently linked to the survival of some plant tissues. This is particularly true in those species which exhibit extraorgan freezing (Quamme 1978; Sakai 1982; Ishikawa and Sakai 1982, 1985). Apparently some plants have evolved mechanisms to influence the spread of ice. Interestingly, the patterns of ice formation change with deacclimation (Ashworth et al. 1989; Ashworth 1990). It is not clear whether this change is a cause or an effect of the reduced freezing resistance which occurs during deacclimation. Factors which control the spread of ice in organized tissues need to be elucidated.

B. Summation

The nucleation and subsequent spread of ice are the initial processes involved in a freezing stress. These processes are sometimes overlooked or considered inconsequential in studies of plant cold hardiness, and this can be a mistake. Although apparently simple and straightforward, these processes are complex, and numerous aspects are not well understood. Both ice nucleation and crystallization have an important impact on plant injury and survival at cold temperatures and therefore warrant further research.

LITERATURE CITED

Anderson, J. A., and E. N. Ashworth. 1985. Ice nucleation in tomato plants. *J. Am. Soc. Hort. Sci.* 110:291–296.

_____. 1986. The effects of streptomycin, desiccation, and UV radiation on ice nucleation by *Pseudomonas viridiflava*. *Plant Physiol.* 80:956–960.

Anderson, J. A., E. N. Ashworth, and G. A. Davis. 1987. Nonbacterial ice nucleation in peach shoots. *J. Am. Soc. Hort. Sci.* 112:215–218.

Anderson, J. A., D. W. Buchanan, R. E. Stall, and C. B. Hall. 1982. Frost injury of tender plants increased by *Pseudomonas syringae* van Hall. *J. Am. Soc. Hort. Sci.* 107:123–125.

Anderson, J. A., and M. W. Smith. 1989. Ice propagation in peach shoots and flowers. *HortScience* 24:480–482.

Andrews, P. K., E. L. Proebsting Jr., and D. C. Gross. 1983. Differential thermal analysis and freezing injury of deacclimating peach and sweet cherry reproductive organs. *J. Am. Soc. Hort. Sci.* 108:755–759.

_____. 1986. Ice nucleation and supercooling in freeze-sensitive peach and sweet cherry tissues. *J. Am Soc. Hort. Sci.* 111:232–236.

Arny. D. C., S. E. Lindow, and C. D. Upper. 1976. Frost sensitivity of *Zea mays* increased by application of *Pseudomonas syringae*. *Nature* 262:282–284.

Ashworth, E. N. 1982. Properties of peach flower buds which facilitate supercooling. *Plant Physiol.* 70:1475–1479.

_____. 1984. Xylem development in *Prunus* flower buds and the relationship to deep supercooling. *Plant Physiol.* 74:862–865.

_____. 1989. Properties of peach flower buds which facilitate supercooling. p. 153–157. In: P. H. Li (ed.), *Low Temperature Stress Physiology in Crops*. CRC Press, Boca Raton, FL.

_____ . 1990. The formation and distribution of ice within forsythia flower buds. *Plant Physiol.* 92:718–725.

Ashworth, E. N., and F. B. Abeles. 1984. Freezing behavior of water in small pores and the possible role in the freezing of plant tissues. *Plant Physiol.* 76:201–204.

Ashworth, E. N., J. A. Anderson, and G. A. Davis. 1985a. Properties of ice nuclei associated with peach trees. *J. Am. Soc. Hort. Sci.* 110:287–291.

Ashworth, E. N., J. A. Anderson, G. A. Davis, and G. W. Lightner. 1985b. Ice formation in *Prunus persica* under field conditions. *J. Am. Soc. Hort. Sci.* 110:322–324.

Ashworth, E. N., and G. A. Davis. 1984. Ice nucleation within peach trees. *J. Am. Soc. Hort. Sci.* 109:198–201.

_____ . 1986. Ice formation in woody plants under field conditions. *HortScience* 21:1233–1234.

_____ . 1987. Influence of ice nucleation temperature on the freezing of peach flower buds. *HortScience* 22:923–925.

Ashworth, E. N., G. A. Davis, and J. A. Anderson. 1985c. Factors affecting ice nucleation in plant tissues. *Plant Physiol.* 79:1033–1037.

Ashworth, E.N, G. A. Davis, and M. E. Wisniewski. 1989. The formation and distribution of ice within dormant and deacclimated peach flower buds. *Plant Cell & Environ.* 12:521–528.

Ashworth, E. N., P. Echlin, R. S. Pearce, and T. L. Hayes. 1988. Ice formation and tissue response in apple twigs. *Plant Cell & Environ.* 11:703–710.

Ashworth, E. N., and D. J. Rowse. 1982. Vascular development in dormant *Prunus* flower buds and its relationship to supercooling. *HortScience* 17:790–791.

Ashworth, E. N., D. J. Rowse, and L. A. Billmyer. 1983. The freezing of water in woody tissues of apricot and peach and the relationship to freezing injury. *J. Am. Soc. Hort. Sci.* 108:299–303.

Ashworth, E. N., T. J. Willard, and S. R. Malone. 1990. The relationship between vascular development and the spread of ice within forsythia flower buds. *Plant Physiol.* (suppl.) 93:86.

Bigg, E. K. 1953. The supercooling of water. *Proc. Phys. Soc.* B66:688–694.

Burke, M. J., L. V. Gusta, H. A. Quamme, C. J. Weiser, and P. H. Li. 1976. Freezing and injury in plants. *Annu. Rev. Plant Physiol.* 27:507–528.

Burke. M. J., and S. E. Lindow. 1990. Surface properties and size of the ice nucleation site in ice nucleation active bacteria: Theoretical considerations. *Cryobiology* 27:80–84.

Burke, M. J., and C. Stushnoff. 1979. Frost hardiness: A discussion of possible molecular causes of injury with particular reference to deep supercooling of water. p. 199–225. In: H. Mussell and R. C. Staples (eds.), *Stress Physiology in Crop Plants.* Wiley, New York.

Cary, J. W., and H. F. Mayland. 1970. Factors influencing freezing of supercooled water in tender plants. *Agron. J.* 62:715–719.

Chandler, W. H. 1954. Cold resistance in horticultural plants: A review. *Proc. Am. Soc. Hort. Sci.* 64:552–572.

Cody, Y. S., D. C. Gross, E. L. Proebsting, Jr., and R. A. Spotts. 1987. Suppression of ice nucleation-active *Pseudomonas syringae* by antagonistic bacteria in fruit tree orchards and evaluations of frost control. *Phytopathology* 77:1036–1044.

Dorsey. M. J. 1934. Ice formation in the fruit bud of the peach. *Proc. Am. Soc. Hort. Sci.* 31:22–27.

Dowgert. M. F., and P. L. Steponkus. 1983. Effect of cold acclimation on intracellular ice formation in isolated protoplasts. *Plant Physiol.* 72:978–988.

Duman, J. G. 1982. Insect antifreezes and ice-nucleating agents. *Cryobiology* 19:613–627.

Fletcher, N. H. 1970. *The Chemical Physics of Ice.* Cambridge University Press.

Franks, F. 1981. Biophysics and biochemistry of low temperatures and freezing. p. 3–19. In: G. J. Morris and A. Clark (eds.), *Effects of Low Temperatures on Biological Membranes.* Academic Press, London.

_____ . 1982. The properties of aqueous solutions at subzero temperatures. p. 215–338. In:

F. Franks (ed.), *Water, a Comprehensive Treatise*. Vol. 7. Plenum Press, New York.

_____ . 1987. Nucleation: A maligned and misunderstood concept. *Cryo-Letters* 8:53–55.

George, M. F. 1983. Freezing avoidance by deep supercooling in woody plant xylem: Preliminary data on the importance of cell wall porosity. p. 84–95. In: D. D. Randall, D. G. Blevins, R. L. Larson, and B. J. Rapp. (eds.), *Current Topics in Plant Biochemistry and Physiology*. University of Missouri Press, Columbia.

George, M. F., M. J. Burke, and C. J. Weiser. 1974. Supercooling in overwintering azalea flower buds. *Plant Physiol.* 54:29–35.

George. M. F., and M. J. Burke. 1977a. Cold hardiness and deep supercooling in xylem of shagbark hickory. *Plant Physiol.* 59:319–325.

_____ . 1977b. Supercooling in overwintering azalea flower buds. *Plant Physiol.* 59:326–328.

Graham. P. R., and R. Mullin. 1976. The determination of lethal freezing temperatures in buds and stems of deciduous azalea by a freezing curve method. *J. Am. Soc. Hort. Sci.* 101:3–7.

Gokhale, N. 1965. Dependence of freezing temperature of supercooled water drops on rate of cooling. *J. Atmos. Sci.* 22:212–216.

Govindarajan, A. G., and S. E. Lindow. 1988. Size of bacterial ice-nucleation sites measured in situ by radiation inactivation analysis. *Proc. Natl. Acad. Sci.* (USA) 85:1334–1338.

Green, R. L., and G. J. Warren. 1985. Physical and functional repetition in a bacterial ice nucleation gene. *Nature* 317:645–648.

Gross, D. C., Y. S. Cody, E. L. Proebsting Jr., G. K. Radamaker, and R. A. Spotts. 1983. Distribution, population dynamics, and characteristics of ice nucleation active bacteria in deciduous fruit tree orchards. *Appl. Env. Microbiol.* 46:1370–1379.

Gross. D. C., E. L. Proebsting Jr., and P. K. Andrews. 1984. The effects of ice nucleation active bacteria on temperatures of ice nucleation and freeze injury of *Prunus* flower buds at various stages of development. *J. Am. Soc. Hort. Sci.* 109:375–380.

Gross, D. C., E. L. Proebsting Jr., and H. MacCrindle-Zimmerman. 1988. Development, distribution, and characteristics of intrinsic, nonbacterial ice nuclei in *Prunus* wood. *Plant Physiol.* 88:915–922.

Gusta. L. V., and D. B. Fowler. 1977. Factors affecting the cold survival of winter cereals. *Can. J. Plant Sci.* 57:213–219.

Guy, C. L. 1990. Cold acclimation and freezing stress tolerance: Role of protein metabolism. *Annu. Rev. Plant Physiol. Plant Mol. Biol.* 41:187–223.

Hallett, J. 1968. Nucleation and growth of ice crystals in water and biological systems. p. 23–52. In: J. Hawthorn and E. J. Rolfe (eds.), *Low Temperature Biology of Foodstuffs*. Pergamon Press, New York.

Hirano, S. S., L. S. Baker, and C. D. Upper. 1985. Ice nucleation temperature of individual leaves in relation to population sizes of ice nucleation active bacteria and frost injury. *Plant Physiol.* 77:259–265.

Hirano. S. S., E. V. Nordheim, D. C. Arny, and C. D. Upper. 1982. Lognormal distribution of epiphytic bacterial populations on leaf surfaces. *Appl. Environ. Microbiol.* 44:695–700.

Hong, S., and E. Sucoff. 1980. Units of freezing of deep supercooled water in woody xylem. *Plant Physiol.* 66:40–45.

Ishikawa, M., and A. Sakai. 1981. Freezing avoidance mechanisms by supercooling in some *Rhododendron* flower buds with reference to water relations. *Plant & Cell Physiol.* 22:953–967.

_____ . 1982. Characteristics of freezing avoidance in comparison with freezing tolerance: A demonstration of extraorgan freezing. p. 325–340. In: P. H. Li and A. Sakai (eds.), *Plant Cold Hardiness and Freezing Stress*. Vol. 2. Academic Press, New York.

_____ . 1985. Extraorgan freezing in wintering flower buds of *Cornus officinalis* Sieb et Zucc. *Plant Cell & Environ.* 8:333–338.

Kaku, S. 1971. Changes in supercooling and freezing processes accompanying leaf maturation in *Buxus*. *Plant & Cell Physiol.* 12:147–155.

———. 1973. High ice nucleating ability in plant leaves. *Plant & Cell Physiol.* 14:1035–
1038.

Kaku, S., and R. W. Salt. 1968. Relationship between freezing temperature and length of
conifer needles. *Can. J. Bot.* 46:1211–1213.

Kieft, T. L. 1988. Ice nucleation activity in lichens. *App. Environ. Microbiol.* 54:1678–1681.

Kitaura, K. 1967a. Freezing and injury of mulberry trees by late spring frost. *Bull. Sericult.
Expt. Sta.* 22:202–323.

———. 1967b. Supercooling and ice formation in mulberry trees. p. 143–156. In: E.
Asahina (ed.), *Cellular Injury and Resistance in Freezing Organisms.* Inst. Low Temp.
Sci., Hokkaido Univ., Sapporo, Japan.

Krog, J. O., K. E. Zachariassen, B. Larsen. and O. Smidsrod. 1979. Thermal buffering in
afro-alpine plants due to nucleating agent-induced water freezing. *Nature* 282:300–30l.

Levine, J. 1950. Statistical explanation of spontaneous freezing of water droplets. National
Advisory Committee for Aeronautics Tech. Note 2234.

Levitt, J. 1980. *Responses of Plants to Environmental Stresses.* Vol. 1 (2nd ed.). Academic
Press, New York.

Li, P. H. 1984. Subzero temperature stress physiology of herbaceous plants *Hort. Rev.*
6:373–416.

———. 1987. *Plant Cold Hardiness.* Alan R. Liss, New York.

———. 1989. *Low Temperature Stress Physiology in Crops.* CRC Press, Boca Raton, FL.

Li, P. H., and A. Sakai. 1978. *Plant Cold Hardiness and Freezing Stress.* Vol. 1. Academic
Press, New York.

———. 1982. *Plant Cold Hardiness and Freezing Stress.* Vol. 2. Academic Press, New York.

Lindow, S. E. 1982a. Epiphytic ice nucleation-active bacteria. p. 335–362. In: M.S. Mount
and G. H. Lacy (eds.), *Phytopathogenic Prokaryotes.* Vol. 1. Academic Press, New York.

———. 1982b. Population dynamics of epiphytic ice nucleation active bacteria on frost
sensitive plants and frost control by means of antagonistic bacteria. p. 395–416. In: P. H.
Li and A Sakai (eds.), *Plant Cold Hardiness and Freezing Stress.* Vol. 2. Academic Press,
New York.

———. 1983a. Methods of preventing frost injury caused by epiphytic ice-nucleation-
active bacteria. *Plant Dis.* 67:327–333.

———. 1983b. The role of bacterial ice nucleation in frost injury to plants. *Annu. Rev.
Phytopath.* 21:363–384.

Lindow, S. E., D. C. Arny, and C. D. Upper. 1978a. Distribution of ice nucleation-active
bacteria on plants in nature. *Applied Env. Microbiol.* 36:831–838.

———. 1978b. *Erwinia herbicola*: A bacterial ice nucleus active in increasing frost injury to
corn. *Phytopathology* 68:523–527.

———. 1982a. Bacterial ice nucleation: A factor in frost injury to plants. *Plant Physiol.*
70:1084–1089.

———. 1983. Biological control of frost injury: An isolate of *Erwinia herbicola*
antagonistic to ice nucleation active bacteria. *Phytopathology* 73:1097–1106.

Lindow, S. E., D. C. Arny, C. D. Upper, and W. R. Barchet. 1978c. The role of bacterial ice
nuclei in frost injury to sensitive plants. p. 249–263. In: P. H. Li and A. Sakai (eds.), *Plant
Cold Hardiness and Freezing Stress.* Vol. 1. Academic Press, New York.

Lindow, S. E., and J. H. Connell. 1984. Reduction of frost injury to almond by control of ice
nucleation active bacteria. *J. Am. Soc. Hort. Sci.* 109:48–53.

Lindow, S. E., S. S. Hirano, W. R. Barchet, D. C. Arny, and C. D. Upper. 1982b. Relationship between ice nucleation frequency of bacteria and frost injury. *Plant Physiol.*
70:1090–1093.

Lindstrom, O. M., and J. V. Carter. 1983. Assessment of freezing injury of cold-hardened
undercooled leaves of *Solanum commersonii*. *Cryo-Letters* 4:361–370.

Lucas, J. W. 1954. Subcooling and ice nucleation in lemons. *Plant Physiol* 29:245–251.

Luyet, B. J. 1968. The formation of ice and the physical behavior of the ice phase in aqueous solutions and in biological systems. p. 53–77. In: J. Hawthorne and E. J. Rolfe (eds.), *Low Temperature Biology of Foodstuffs.* Pergamon Press, New York.

Maki. L. R., E. L. Galyan, M. Chang-Chien, and D. R. Caldwell. 1974. Ice nucleation induced by *Pseudomonas syringae. Applied Microbiol.* 28:456–459.

Maki, L. R., and K. J. Willoughby. 1978. Bacteria as biogenic sources of freezing nuclei. *J. Appl. Meteor.* 17:1049–1053.

Marcellos, H., and W. V. Single. 1976. Ice nucleation on wheat. *Agr. Met.* 16:125–129.

_____ . 1979. Supercooling and heterogeneous nucleation of freezing in tissues of tender plants. *Cryobiology* 16:74–77.

Marshall, D. 1988. A relationship between ice-nucleation-active bacteria, freeze damage, and genotype in oats. *Phytopathology* 78:952–957.

Mathias, S. F., F. Franks, and K. Trafford. 1984. Nucleation and growth of ice in deeply undercooled erythrocytes. *Cryobiology* 21:123–132.

Mazur, P. 1969. Freezing injury in plants. *Annu. Rev. Plant Physiol.* 20:419–448.

Michelmore, R. W., and F. Franks. 1982. Nucleation rates of ice in undercooled water and aqueous solutions of polyethylene glycol. *Cryobiology* 19:163–171.

Modlibowska, I. 1962. Some factors affecting supercooling of fruit blossoms. *J. Hort. Sci.* 37:249–261.

Morris, G. J., and J. J. McGrath. 1981. Intracellular ice nucleation and gas bubble formation in *Spirogyra. Cryo-Letters* 2:341–352.

Nath, J., and T. C. Fisher. 1971. Anatomical study of freezing injury in hardy and nonhardy alfalfa varieties treated with cytosine and guanine. *Cryobiology* 8:420–430.

Nemethy, G. 1968. The structure of water and of aqueous solutions. p. 1–21. In: J. Hawthorne and E. J. Rolfe (eds.), *Low Temperature Biology of Foodstuffs.* Pergamon Press, New York.

O'Brien, R. D., and S. E. Lindow. 1988. Effect of plant species and environmental conditions on ice nucleation activity of *Pseudomonas syringae* on leaves. *Appl. and Environ. Microbiol.* 54:2281–2286.

Orser, C., B. J. Staskawicz, N.J. Panopoulos, D. Dahlbeck, and S. E. Lindow. 1985. Cloning and expression of bacterial ice nucleation genes in *Escherichia coli. J. Bacteriol.* 164:359–366.

Paulin, J. P., and J. Luisetti. 1978. Ice nucleation activity among phytopathogenic bacteria. p. 725–731. In: *Station de Pathologie Végétale et Phytobacteriologie,* Proc. 4th Intl. Conf. on Plant Pathogenic Bacteria. Vol. II. Institute National Récherche Agronomique. Beaucouze, France.

Pearce, R. S. 1988. Extracellular ice and cell shape in frost-stressed cereal leaves: A low-temperature scanning-electron-microscopy study. *Planta* 175:313–324.

Pellett, H. M., and J. V. Carter. 1981. Effect of nutritional factors on cold hardiness of plants. *Hort. Rev.* 3:144–171.

Pierquet, P., C. Stushnoff, and M. J. Burke. 1977. Low temperature exotherms in stem and bud tissues of *Vitis riparia* Michx. *J. Am. Soc. Hort. Sci.* 102:54–55.

Preston, R. D. 1974. *The Physical Biology of Plant Cell Walls.* Chapman and Hall, London.

Proebsting, E. L. Jr., P. K. Andrews, and D. Gross. 1982. Supercooling young developing fruit and floral buds in deciduous orchards. *HortScience* 17:67–68.

Proebsting, E. L. Jr., and D. C. Gross. 1988. Field evaluations of frost injury to deciduous fruit trees as influenced by ice nucleation-active *Pseudomonas syringae. J. Am. Soc. Hort. Sci.* 103:57–61.

Pruppacher, H. R. 1967a. Interpretation of experimentally determined growth rates of ice crystals in supercooled water. *J. Chem. Physics* 47:1807–1813.

_____ . 1967b. Some relations between the structure of the ice-solution interface and the free growth rate of ice crystals in supercooled aqueous solutions. *J. Colloid Interface Sci.* 25:285–294

Quamme, H. A. 1974. An exothermic process involved in the freezing injury to flower buds of several *Prunus* species. *J. Am. Soc. Hort. Sci.* 99:315–318.

———. 1978. Mechanism of supercooling in overwintering peach flower buds. *J. Am. Soc. Hort. Sci.* 103:57–61.

Quamme, H. A., and L. V. Gusta. 1987. Relationship of ice nucleation and water status to freezing patterns in dormant peach flower buds. *HortScience* 22:465–467.

Quamme, H. A., C. J. Weiser, and C. Stushnoff. 1973. The mechanism of freezing injury in xylem of winter apple twigs. *Plant Physiol.* 51:273–277.

Rajashekar, C. B., P. H. Li, and J. V. Carter. 1983. Frost injury and heterogeneous ice nucleation in leaves of tuber-bearing *Solanum* species. *Plant Physiol.* 71:749–755.

Rasmussen, D. H., and A. P. MacKenzie. 1972. Effect of solute on ice-solution interfacial free energy: Calculation from measured homogeneous nucleation temperatures. p. 126–145. In: H. Jellinck (ed.), *Water Structure at the Polymer Interface*. Plenum Press, New York.

Reid, D. S., A. T. Foin, and C. A. Lem. 1985. The effect of solutes on the temperature of heterogeneous nucleation of ice from aqueous solutions. *Cryo-Letters* 6:189–198.

Rieger. M. 1989. Freeze protection for horticultural crops. *Hort. Rev.* 11:45–109.

Sakai, A. 1982. Extraorgan freezing of primordial shoots of winter buds of conifer. p. 199–209. In: P. H. Li and A. Sakai (eds.), *Plant Cold Hardiness and Freezing Stress*. Vol 2. Academic Press, New York.

———. 1984. Cryopreservation of apical meristems. *Hort. Rev.* 6:357–372.

Sakai, A., and W. Larcher. 1987. *Frost Survival of Plants*. Springer-Verlag, Berlin.

Salt, R. W. 1961. Principles of insect cold-hardiness. *Annu. Rev. Ent.* 6:55–74.

Salt, R. W., and S. Kaku. 1967. Ice nucleation and propagation in spruce needles. *Can. J. Bot.* 45:1335–1346.

Siminovitch, D., and G. W. Scarth. 1938. A study of the mechanism of frost injury to plants. *Can. J. Res. C* 16:467–481.

Single, W. V., and H. Marcellos. 1981. Ice formation and freezing injury in actively growing cereals. p. 17–33. In: C. R. Olein and M. N. Smith (eds.), *Analysis and Improvement of Plant Cold Hardiness*. CRC Press, Boca Raton, FL.

Steffens, K. L., R. Arora, and J. P. Palta. 1989. Relative sensitivity of photosynthesis and respiration to freeze-thaw stress in herbaceous species. *Plant Physiol.* 89:1372–1379.

Steponkus, P. L. 1984. Role of the plasma membrane in freezing injury and cold acclimation. *Annu. Rev. Plant Physiol.* 35:543–584.

Thomas, D. A., and H. N. Barber. 1974. Studies on leaf characteristics of a cline of *Eucalyptus urnigera* from Mount Wellington, Tasmania. I. Water repellency and the freezing of leaves. *Austral. J. Bot.* 22:501–512.

Tyree, M. T., and P. G. Jarvis. 1982. Water in tissues and cells. p. 35–77. In: A. Pirson and M. H. Zimmerman (eds.), *Encyclopedia of Plant Physiology*. Vol. 12B, *Physiological Plant Ecology II*. Springer-Verlag, Berlin.

Vali, G. 1969. The characteristics of freezing nuclei. *Proc. 7th Intern. Conf. Condensation and Ice Nuclei.* Prague and Vienna, p. 387–393.

———. 1971. Quantitative evaluation of experimental results on the heterogeneous freezing nucleation of supercooled liquids. *J. Atmos Sci.* 28:402–409.

Vali. G., and E. J. Stansbury. 1966. Time-dependent characteristics of the heterogeneous nucleation of ice. *Can. J. Physics* 44:477–502.

Vonnegut, B. 1948. Variation with temperature of the nucleation rate of supercooled liquid tin and water drops. *J. Colloid Sci.* 3:563–569.

White, W. C., and C. J. Weiser. 1964. The relation of tissue desiccation, extreme cold, and rapid temperature fluctuations to winter injury of American arborvitae. *J. Am. Soc. Hort. Sci.* 85:554–563.

Wiegand, K. M. 1906. Some studies regarding the biology of buds and twigs in winter. *Bot. Gaz.* 41:373–424.

Wisniewski, M., E. Ashworth, and K. Schaffer. 1987a. The use of lanthanum to characterize cell wall permeability in relation to deep supercooling and extracellular freezing in woody plants. Intergeneric comparisons between *Prunus, Cornus* and *Salix. Protoplasma* 139:105–116.

_____. 1987b. The use of lanthanum to characterize cell wall permeability in relation to deep supercooling and extracellular freezing in woody plants. II. Intrageneric comparisons between *Betula lenta* and *Betula papyrifera. Protoplasma* 141:160–168.

Wisniewski, M., G. Davis, and K. Schaffer. 1991. Mediation of deep supercooling of peach and dogwood by enzymatic modifications in cell wall structure. Planta (in press).

Wolber, P. K., C. A. Deininger, M. W. Southworth, J. Vanderkerckhove, M. van Montagu, and G. J. Warren. 1986. Identification and purification of a bacterial ice-nucleation protein. *Proc. Natl. Acad. Sci.* (USA) 83:7256–7260.

Yelonosky, G. 1975. Cold hardening in citrus stems. *Plant Physiol.* 56:540–543.

_____. 1983. Ice nucleation active (INA) agents in freezing of young citrus trees. *J. Am. Soc. Hort. Sci.* 108:1030–1034.

_____. 1985. Cold hardiness of citrus. *Hort. Rev.* 7:201–238.

Zachariassen, K. E. 1982. Nucleating agents in cold-hardy insects. *Comp. Biochem. Physiol.* 73A:557–562.

Zachariassen, K. E., and H. T. Hammel. 1976. Nucleating agents in heamolymph of insects tolerant to freezing. *Nature* 262:285–287.

_____. 1988. The effect of ice-nucleating agents on ice-nucleating activity. *Cryobiology* 25:143–147.

7

Responses of Fruit Crops to Flooding*

Bruce Schaffer
University of Florida, IFAS,
Tropical Research and Education Center
18905 S.W. 280 Street, Homestead, FL 33031

P. C. Andersen
University of Florida, IFAS,
Agricultural Research and Education Center
P.O. Route 4, P.O. Box 63, Monticello, FL 32344

R. C. Ploetz
University of Florida, IFAS,
Tropical Research and Education Center
18905 S.W. 280 Street, Homestead, FL 33031

I.   Introduction
II. Plant Tolerance
III. Chemical and Physical Changes in Soil
IV. Plant Responses
    A. Plant Metabolism
    B. Growth, Anatomical, and Morphological Responses
    C. Water Relations and Photosynthesis
    D. Plant Nutrition
V.   Effects of Flooding on Soilborne Diseases
    A. Effects on the Pathogen
    B. Host Resistance
    C. Disease-Flood Interactions
    D. Host Responses
VI. Conclusions

I. INTRODUCTION

Plants vary considerably in their ability to tolerate waterlogged soil (Kramer 1833; Kozlowski 1984b). Poor soil aeration associated with flooding results in changes in soils that are usually deleterious to growth

*The authors wish to thank F. S. Davies, J. H. Crane, D. J. Mitchell, and T. L. Davenport for suggestions and critical review of the manuscript. Portions of work reported by B. Schaffer and R. C. Ploetz were supported by grants from the Florida Avocado Administrative Committee and the Cooperative State Research Service, USDA, Agreement nos. 88-34135-3567 and 89-34135-4573. Florida Agricultural Experiment Station Journal Series No. R-00613.

and development of many plant species (Kozlowski 1984a). An understanding of the effect of flooding on the physiology and growth of fruit crops may aid in the selection of cultivars adapted to flood-prone or poorly drained sites, or suggest cultural methods to increase flood tolerance.

In a previous review, Rowe and Beardsell (1973) examined the effects of waterlogged soil on fruit trees. Tremendous diversity in flood tolerance exists, and knowledge of the flood tolerance of a particular species or cultivar is of primary importance in flood-prone or poorly drained sites. This review provides an updated account of the relative tolerances of various fruit crops to flooding.

Flood responses among fruit crops are directly related to the dramatic changes in O_2 availability and the chemical and physical states of soil that occur after flooding. These changes can have profound effects on plant growth and development.

Plants respond to flooded soil in many different ways. Plant metabolic processes and functions are affected by flooding and an outline of these changes is presented. Investigations on responses of fruit crops to flooding have focused primarily on growth, water relations, leaf gas exchange, and plant nutrition. Thus, a substantial portion of this review is devoted to detailing this diverse and large body of reasearch.

The last section of this review describes the influence of flooding on soilborne diseases of fruit crops. Studies of the responses of these crops to flooding have generally neglected the possible confounding effects of root disease. Flooding may provide a favorable environment for pathogen activities, reduce host resistance, and cause more severe disease to develop than would occur under nonflooded conditions. An awareness of disease-flood interactions should assist attempts to combat flood-associated production problems.

In this review, primary emphasis has been given to research on woody fruit crops conducted since 1973. However, work on herbaceous and forest tree species that has relevance to an understanding of the responses of fruit crops to flooding is also discussed.

II. PLANT TOLERANCE

Plant tolerance to waterlogged soil conditions may be influenced by a number of factors: (1) soil type, porosity, and chemistry; (2) degree and duration of anaerobiosis; (3) soil microbe and pathogen status; (4) vapor pressure deficits and root zone and air temperatures; (5) plant age, stage of development, or season of the year; and (6) plant preconditioning responses as a result of prior climatic and edaphic conditions. The importance of each of these variables may differ with plant species. It is particularly difficult to assess intraspecific and interspecific tolerance from experiments performed under different climatic and edaphic conditions.

The tolerance of fruit trees to waterlogged conditions is mainly determined by the rootstock and not the scion (Ford 1964; Rowe and Catlin 1971; Rowe and Beardsell 1973; Andersen et al. 1984a, 1984b), although specific foliar symptoms of flooding injury may vary with the scion (Andersen et al. 1984b). Studies which have encompassed many fruit crop species and cultivars have provided useful information concerning the relative waterlogging sensitivity of fruit crops under a given set of conditions. "Relative tolerance" to waterlogged conditions should be viewed with some caution due to the many factors cited above. Nevertheless, a sufficient variation in interspecific as well as intraspecific flood tolerance occurs in fruit crops to justify flood tolerance as a selection criterion (Rowe and Beardsell 1973).

Rowe and Beardsell (1973) compiled a ranking of waterlogging tolerance of fruit trees as follows: extremely tolerant, quince (*Cydonia oblonga* Mill.); very tolerant, pear (*Pyrus* spp.); moderately tolerant, apple (*Malus* × *domestica* L.), *Citrus* spp., and plums (*Prunus domestica* L. and *Prunus cerasifera* L.); sensitive, plum (*Prunus salicina* Lindl.); very sensitive, cherry (*Prunus avium* L. and *Prunus mahaleb* L.), apricot (*Prunus armeniaca* L.), and peach [*Prunus persica* (L.) Batsch and *Prunus davidana* L]. The present review will further delineate differential sensitivity to waterlogging.

The waterlogging tolerance of six pear species, quince, willow (*Salix discolor* Muhl.), apple, and peach were compared after continuous flooding for up to two years (Andersen et al. 1984b). Willow was differentiated from other species in that it thrived under flooded conditions, whereas pear species and quince merely survived in a state characterized by some leaf chlorosis and reduced shoot and leaf growth. The relative tolerance of pear rootstocks based on survival, growth, and stomatal conductance is indicated in Table 7.1. All species survived one month of continuous flooding. *Pyrus betulaefoli* Bunge survived more than 20 consecutive months of soil flooding and only a few seedlings of *Pyrus calleryana* Decne, quince, and *Pyrus communis* L. cv. Barlett died. *Pyrus dimorphophylla* (Mak.), *Pyrus pyrifolia* (Burm.) Nak., *Pyrus ussuriensis* (Max.), *Pyrus pashia* (D. Don), *Pyrus communis* 'Old Home' × 'Farmingdale 97', *Pyrus amydaliformes* (Vill.), and apple (Table 7.2) were notably less tolerant (Andersen 1983; Andersen et al. 1984b).

The high degree of tolerance exhibited by *P. betulaefolia* and *P. calleryana* has been supported by field observations (West and Lombard 1966). Quince, another common rootstock for pear, has been reported to be both tolerant (Anonymous 1961; Bini 1963) and sensitive (Westwood and Lombard 1966) to flooded soil conditions. Although many pear rootstocks are considered very tolerant to soil waterlogging, one month of spring soil flooding reduced growth of all species except *P. betulaefolia*

Table 7.1.   Relative tolerance of species of *Pyrus* (pear) and *Cydonia* (quince) to soil waterlogging.

Species	Tolerance Level	Reference
P. betulaefolia	Extremely tolerant	Andersen 1983; Andersen et al. 1984b
P. calleryana	Very tolerant	Andersen 1983; Andersen et al. 1984b
C. oblonga 'Emla C'	*Very tolerant*	*Andersen 1983; Andersen et al. 1984b*
C. oblonga 'Provence BA29'	*Very tolerant*	*Andersen 1983; Andersen et al. 1984b*
P. faurei	Very tolerant	Andersen 1983; Andersen et al. 1984b
P. communis (Bartlett sdlg)	Very tolerant	Andersen 1983; Andersen et al. 1984b
P. dimorphophylla	Moderately tolerant	Andersen 1983; Andersen et al. 1984b
P. pyrifolia	Moderately tolerant	Andersen 1983; Andersen et al. 1984b
P. ussuriensis	Moderately tolerant	Andersen 1983; Andersen et al. 1984b
P. pashia	Moderately tolerant	Andersen 1983; Andersen et al. 1984b
P. communis 'OH × F 97'	Moderately tolerant	Andersen 1983; Andersen et al. 1984b
P. amydaliformes	Moderately sensitive	Andersen 1983; Andersen et al. 1984b

Table 7.2.   Relative tolerance of apple rootstocks to flooded soil conditions.

Seedling or Clone	Tolerance Level	Reference
M.1, M.3	Moderately tolerant	Remy and Bidabe 1962; Saunier 1962
M.7	Moderately tolerant	Remy and Bidabe 1962; Rom and Brown 1979
M.6, M.13, M.14, M.15, M.16, Crab C sldg, Jonathan sdlg.	Moderately tolerant	Remy and Bidabe 1962
M.2	Moderately tolerant	Saunier 1966
M.26	Moderately tolerant	Rom and Brown 1979
M.26	Moderately sensitive	Salesses et al. 1970
MM.106, M.111	Moderately sensitive	Andersen et al. 1984a
M.9	Moderately sensitive	Cummins and Aldwinkle 1982; Remy and Badabe 1962
M.4	Moderately sensitive	Remy and Bidabe 1962
M.2, MM.104, MM.109	Sensitive	Remy and Bidabe 1962
MM.106	Sensitive	Cummins and Aldwinkle 1982; Rom and Brown 1979
M.111, Delicious sdlg.	Very Sensitive	Rom and Brown 1979
M.26	Very Sensitive	Cummins and Aldwinkle 1982
M.779, M.789, M.793, Northern Spy sdlg.	Very sensitive	Remy and Bidabe 1962

and *P. calleryana*; one month of flooding in early fall reduced growth of all species the following spring (Andersen 1983; Andersen et al. 1984b).

Apple rootstocks may survive from one month up to one year of soil flooding depending on tree size, season in which trees are flooded, and soil pathogen status (Rowe and Beardsell 1973; Rom and Brown 1979; Cummins and Aldwickle 1983; Olien (1987). Less genetic variation exists among apple rootstock (all *Malus domestica*) than pear rootstocks, which include several genera and many species. Hence, less variation in waterlogging tolerance may be expected for apple than pear. A fairly extensive collection of apple germplasm has been tested in reference to waterlogging sensitivity (Table 7.2); however, conflicting tolerance assessments are found in the literature. For example, 'Delicious' apple on MM.111 rootstock was reported to be more sensitive to waterlogging than on MM.106 and M.27 rootstocks; M.26 was the most tolerant rootstock (Rom and Brown 1979). In another study, MM.106 and MM.111 were equally tolerant of flooding (Andersen 1983). M.26 rootstocks also have been rated moderately sensitive (Salesses et al. 1990) and very sensitive (Cummins and Aldwinkle 1982) to flooded soil conditions. Similarly, M.2 has been reported to be both tolerant (Saunier 1966) and susceptible (Remy and Bidabe 1962) to soil flooding.

Very few studies have been conducted on the waterlogging tolerance of bearing fruit trees. Soil waterlogging for six weeks in the spring, summer, or fall reduced vegetative growth and yield of 'Macspur' apple scions on M.26 rootstock, but the greatest yield decline resulted from spring flooding (Olien 1987). In contrast to a previous study for 'Stayman Winesap' apples (Childers and White 1950), fruit quality of 'Macspur' on M.26 was not affected appreciably by flooding. Olien (1987) pointed out that a decline in apple tree productivity rather than tree death is likely to be the consequence of soil flooding.

A substantial amount of information exists concerning differential waterlogging sensitivity of *Prunus* because there is a wealth of genetic diversity within this genus, and many different species are graft-compatible (Table 7.3). Plum and prune rootstocks (*Prunus cerasifera* J. F. Ehrh., *Prunus domestica* L., and *Prunus japonica* Thunb.) are the most tolerant species and may survive up to several months of soil waterlogging (Salesses et al. 1970). Japanese plum (*Prunus salicina* Lindl.) is considered to be less tolerant than the other plums. Differences in flood tolerance may exist among clones of a single species of plum rootstocks (Table 7.3). Waterlogging tolerance of cherry rootstocks from most to least has been reported as follows: *Prunus cerasus* L. 'Stockton Morello' > *Prunus avium* 'Mazzard' > *Prunus mahaleb* 'Mahaleb' (Norton et al. 1963b). Common peach rootstocks *Prunus persica* (L.) Batsch 'Halford', 'Lovell', and 'Nemaguard', or *Prunus tomentosa* Thunb. and other related species [*Prunus davidana* (Carriere) Franch., *Prunus mume* Siebold and Zucc., and *Prunus subhirtella* L.] are all intolerant of flooded

Table 7.3. Relative tolerance of Prunus species to flooded soil conditions.

Species	Cultivar	Tolerance Level	Reference
Prunus japonica		Moderately tolerant	Mizutani et al. 1979
P. cerasifera (Marianna)	S2544-2	Moderately tolerant	Salesses et al. 1970
P. cerasifera	GF 8-1	Moderately tolerant	Salesses et al. 1970
P. domestica	Damas GF 1869, Damas de Toulouse	Moderately tolerant	Bernhard 1970
P. domestica	Cirule 43, Brompton, St. Julian A, St. Julian GF 355-2	Moderately sensitive	Bernhard 1970
P. cerasifera	P34, P855, P936, P938	Moderately sensitive	Salesses et al. 1970
P. cerasifera × P. salicina	GF 31	Moderately sensitive	Salesses et al. 1970
P. salicina	S37, S300, S2540, S2541	Sensitive	Salesses et al. 1970
P. cerasus	Stockton Morello	Sensitive	Norton et al. 1963
P. salicina	S573, 2508, 2514, 2538	Very sensitive	Salesses et al. 1970
P. avium	Mazzard	Very sensitive	Norton et al. 1963
P. davidana	GF308	Extremely sensitive	Salesses et al. 1970
P. mahaleb	Mahaleb, St. Lucie 39	Extremely sensitive	Saunier 1966; Norton et al. 1963b
P. persica[z]	sdlgs of Lovell, Halford, Nemaguard, Siberian C, New, Rutgers Red Leaf, Harrow Blood	Extremely sensitive	Mizutani et al. 1979
P. mume		Extremely sensitive	Mizutani et al. 1979
P. tomentosa		Extremely sensitive	Mizutani et al. 1979
P. subhirtella		Extremely sensitive	Mizutani et al. 1979
P. pauciflora		Extremely sensitive	Mizutani et al. 1979
P. dulcis		Most sensitive	Norton et al. 1963a
P. armeniaca		Most sensitive	Mizutani et al. 1979

[z]Reported differential tolerance of Prunus persica: 'Lovell' and 'Halford', both very sensitive (Andersen 1983; Andersen et al. 1984b); 'Lovell' and 'Halford' more sensitive than 'Nemaguard' (Rom and Brown 1979); sensitivity of 'Lovell'> 'Harrow Blood'> 'Siberian C'> 'New'> 'Rutgers Red Leaf' (Chaplin et al. 1974).

soil conditions (Mizutani et al. 1979). Peach rootstocks may be killed after only 2–5 days of flooding if soil temperatures are high (Rowe and Catlin 1971; Andersen 1983). Chaplin et al. (1974) reported that survival of peach seedlings under flooded soil conditions from high to low was as follows: 'Rutgers Red Leaf', 'New', 'Siberian C', 'Harrow Blood', 'Lovell'. In another study 'Halford' and 'Lovell' were more sensitive than 'Nemaguard' (Rom and Brown 1979). Almond (Prunus dulcis Mill.) (Norton et al. 1963a) and apricot are among the most sensitive of all deciduous tree species to soil waterlogging, and should be planted only in well-drained, sandy soils.

Little information is available concerning differential tolerance of pecan [Carya illinoensis (Wagenh.) C. Koch.] to soil flooding since these trees are grown on various seedling rootstocks obtained from different cultivars or an unknown seed source. Pecan trees, which are native to the Mississippi flood-plains, are often exposed to flooded soil conditions in both their native and cultivated range. Pecan seedlings flooded during the winter, spring, and summer all survived four weeks of flooding. Spring flooding reduced growth and altered concentrations of mineral elements in the leaf more than summer flooding; dormant flooding had little or no effect (Smith and Bourne 1989). Loustalot (1945) reported that 35 days of flooding resulted in extensive root damage of pecan seedlings. Mature 'Stuart' pecan trees have displayed leaf chlorosis, necrosis, and defoliation when flooded in the field (Alben 1958).

Waterlogging tolerance of walnut seedling rootstocks has been characterized. Plant survival after four weeks of soil flooding was greater for pecan than for black walnut (Juglans nigra L.) (Louks and Keen 1973). Chinese wingnut (Pterocarya stenoptera D. C.) was more tolerant of soil flooding than Juglans rootstocks (Catlin et al. 1977). Northern California black walnut (Juglans hindsii Deps.), English walnut (Juglans regia L.), and black walnut had a 50% survival rate after 7–14 days of flooding at 23°C (Catlin et al. 1977; Catlin and Olson 1986). Paradox hybrids (J. hindsii × J. regia) were more flood tolerant than either parent (Catlin et al. 1977).

Blueberry plants (Vaccinium spp.) appear to be moderately tolerant of waterlogged soil conditions (Crane and Davies 1989). Rabbiteye blueberry (Vaccinium ashei Reade) has been reported to be more tolerant than highbush blueberry (Vaccinium corymbosum L.) (Kender and Brightwell 1966). However, Crane and Davies (1989) observed little difference in flood tolerance among Vaccinium species when flooded for up to 55 days. Rabbiteye blueberries may survive 25–35 days of summer flooding and more than 117 days of spring flooding (Crane and Davies 1988a, 1988b). Abbott and Gough (1985) found that 'Bluecrop' highbush blueberry plants survived up to 30 months of flooding. Similarly, cranberry (Vaccinium macropcarpan Ait.) may survive extended flooding during the dormant season, but succumb to flooding quickly during the growing season (Bergman 1943). Crane and Davies (1989) suggested that the differential tolerance often reported for Vaccinium species may be a result of damage caused by Phytophthora cinnamomi Rands, a fungal root pathogen, rather than physiological factors.

Grape (Vitis spp.) and fig (Ficus carica L.) are considered moderately tolerant and intolerant of soil waterlogging, respectively (Rowe and Beardsell 1973; West and Taylor 1984). A severe leaf disorder of kiwi [Actinidi deliciosa (A. Chev.) C. F. Liang et A. R. Ferguson var. deliciosa]

Table 7.4. Tolerance of citrus rootstocks to flooded soil conditions.

Species	Common Name	Tolerance Level	Reference
Poncirus trifoliata	trifoliate orange	Moderately tolerant	Ford 1964
Citrus jambhiri	rough lemon	Moderately tolerant	Ford 1964; Syvertsen et al. 1983
C. sinensis × P. trifolia	carrizo citrange	Moderately sensitive	Ford 1964
C. reticulata	cleopatra and others	Moderately sensitive	Ford 1964
C. sinensis	sweet orange	Moderately sensitive	Ford 1964
C. auratium	sour orange	Sensitive	Ford 1964, 1969; Syvertsen et al. 1983
C. aurantiifolia	Rangpur lime	Sensitive	Hagin et al. 1985

has been attributed to low oxygen concentrations in nonflooded soils (Smith et al. 1990). Save and Seranno (1986) also reported that kiwi is sensitive to soil waterlogging.

Citrus rootstocks vary in their ability to tolerate soil flooding. Waterlogging tolerances of citrus rootstocks, which are not infected by *Phytophthora cinnamomi*, are listed in Table 7.4. Trifoliate orange [*Poncirus trifoliata* (L.) Raf.] and rough lemon (*Citrus jambhiri* Lush) are considered to be the most tolerant; sweet orange [*Citrus sinensis* (L.) Osbeck] and mandarin (*Citrus reticulata* Blanco) intermediate; and sour orange (Citrus aurantium L.) and sweet lime [*Citrus aurantiifolia* (Christin.) Swingle] least tolerant (Gardner 1961a, 1961b; Ford 1964, 1969; Syvertsen et al. 1983). Ford (1964) reported that 14 days of soil flooding damaged roots of citrus rootstocks.

Relatively little information is available concerning waterlogging sensitivity of tropical fruit crops. Mango (*Mangifera indica* L.) (Larson et al. 1991), carambola (*Averrhoa carambola* L.) (Joyner and Schaffer 1989), and guava (*Pisidium guajava* L.) (Jawana 1961) are moderately flood-tolerant, whereas avocado (*Persea americana* Mill.) (Ploetz and Schaffer 1987, 1988, 1989; Schaffer and Ploetz 1989), loquat [*Eriobotrya japonica* (Thung.) Lindl.], and papaya (*Carica papaya* L.) (Jawanda 1961) are reported to be sensitive to waterlogging.

III. CHEMICAL AND PHYSICAL CHANGES IN SOIL

Upon flooding, oxygen in soil pores is displaced by water; aerobic organisms in the soil consume available O_2 and become quiescent or die. Facultative and obligate anaerobes, which ultimately predominate in flooded soil (Takai et al. 1956; Takeda and Furusaka 1970), use NO_3^-, Mn^{+4}, SO_4^{-2}, CO_2, N_2, H^+, and various organic molecules as electron acceptors in the processes of oxidative phosphorylation and electron transport when O_2 is limiting. Therefore, N_2, Mn^{+2}, Fe^{+2}, H_2S, CH_4, NH_4^+ and H_2 often predominate, respectively (Ponnamperuma 1972, 1984). These chemical changes and various excretory by-products of bacteria serve to alter the oxidation reduction status of the soil (Ponnamperuma 1972, 1984).

Redox potential (Eh) is a quantitative measurement of the electrochemical status of the soil redox system (Ponnamperuma 1972, 1984). Soil redox potentials of aerobic soils typically range from 300 to 800 mV, and anaerobic soils from −450 to 200 mV, although the Eh of flooded and nonflooded soils may overlap (Nichols and Turner 1957; Cho and Ponnamperuma 1971; Ponnamperuma 1972; Larson et al. 1991). The rate and degree of soil reduction is dependent on soil pH, temperature, organic matter content, and the identity of the electron acceptors. The chemical reduction of constituents in a flooded soil proceeds in a sequence described by the thermodynamics of the redox system (i.e., $O_2 >$ $NO_3^- > Mn^{+4} > Fe^{+3} > SO_4^{-2} > CO_2$) (Takai and Kamura 1966; Cho and Ponnamperuma 1971; Ponnamperuma 1972, 1984). Oxygen is the first element in the soil to become chemically reduced in a flooded soil. Next, nitrate reduction proceeds when O_2 falls to a low level. The presence of NO^-_3 retards the reduction of elements such as Mn^{+4}, Fe^{+3}, SO_4^{-2} that have a less positive Eh. Unfortunately, Eh measurements are seldom taken adjacent to plant roots. In nonflooded soils Eh (and pH) are typically reduced in the rhizosphere compared to the remainder of the soil (Fischer et al. 1989). Conversely, in flooded soils, the Eh may actually increase in the rhizosphere if the plant is capable of internal oxygen transport (Armstrong 1968, Laan et al. 1989).

Soil redox potential may be a useful indicator of plant survival in a given soil type (Pezeshki et al. 1989). For example, Eh of soil planted with rabbiteye blueberry declined to −200 mV within 24 h of flooding (Crane and Davies 1988a, 1988b). However, in another study with containerized highbush blueberry, Eh of the potting media declined to only 75 mV after 30 months of continuous flooding (Abbott and Gough 1987b).

Oxygen diffusion rates (ODRs) in soils have often been quantified by the platinum microelectrode technique (Lemon and Erickson 1952, 1955; Stolzy 1964; McIntyre 1970; Mann and Stolzy 1972; Sojka and Stolzy 1980). Both the theory of measurement (Stolzy 1964; McIntyre 1970) and plant response to soil ODR (Sojka and Stolzy 1980) have been reviewed. In brief, O_2 diffuses from the soil to the platinum surface of the electrode and is electrochemically reduced, producing an electric current. The technique is not adequate for measuring ODR in nonsaturated soils with a moisture level much less than field capacity since hydraulic continuity must exist among the platinum surface, soil particles, and the calomel electrodes. Measurements of soil ODR, like measurements of soil Eh, suffer from a significant pH effect and possible buildup of contaminants on the electrode surface. Devitt et al. (1989) recommended that elec-

trodes be kept in place less than 2 months to minimize the buildup of contaminants. Furthermore, the chemical reaction(s) which occur at the
platinum surface are not known with certainty (McIntyre 1970). Nevertheless, measurements of soil ODR have provided quantitative flux rates
of soil O_2 that have been correlated with various plant responses (Sojka
and Stolzy 1980). An ODR of 20×10^{-8} g cm^{-2} min^{-1} inhibited root growth
of numerous herbaceous species (Lemon and Erickson 1952, 1955; Stolzy
and Letey 1964) and was associated with reduced stomatal conductance
of several woody plant species (Sojka and Stolzy 1980; Andersen et al.
1984b). Succession of species in plant communities growing on wet sites
were associated with ODRs between 5 and 25×10^{-8} g cm^{-2} min^{-1} (Poel
1960).

Flooding increases the pH of acidic soils and depresses the pH of basic
soils (Ponnamperuma 1972). Upon soil flooding, soil pH first declines for
several days (due to CO_2 production by microorganisms), then increases
to a stable value of 6.5–7.2 after several weeks in most soils, regardless of
the initial pH (Ponnamperuma 1972; Larson et al. 1991). The subsequent
increase in pH is believed to be affected by the concentrations of Fe^{+3} and
organic matter (Ponnamperuma 1972).

Nutrient availability increases shortly after soil flooding, but eventually declines (Cho and Ponnamperuma 1971; Ponnamperuma 1972).
Increased nutrient availability is a function of increased solubility of
reduced forms of iron (Fe^{+2}) and maganese (e.g., Mn^{+3}, $MnHCO^{+3}$), and
displacement of K^+, Ca^{+3}, and Mg^{+2} from soil colloids by Fe^{+2} and Mn^{+2}
(Ponnamperuma 1984). Nitrate in flooded soil is quickly assimilated and
incorporated into cellular material or utilized as an alternative to O_2 as an
electron acceptor. Soil flooding inhibits the nitrification process, and
much of the NO_3^- is lost by volatilization as N_2 gas (Ponnamperuma
1972).

The solubility of Fe and Mn increases significantly in flooded soil, and
Fe usually stabilizes the decline of Eh and pH (Bloomfield 1951; Drew and
Sisworo 1979; Larson et al. 1991). Iron toxicity has been implicated as a
possible cause of flooding injury in *Erica* species (Jones 1972).

In anaerobic soils, SO_4^{-2} (and the sulfur containing amino acids) are
degraded to H_2S, thiols, NH_3, and fatty acids at an Eh of −0.150 mV
(Ponnamperuma 1972). Hydrogen sulfide, which is produced by soil bacteria of the genus *Desulfovibria*, inhibits root growth, root respiration,
nutrient uptake, and leaf photosynthesis (Ponnamperuma 1972; Rowe
and Beardsell 1973; Koch and Mendelssohn 1989). Culbert and Ford
(1972) reported that rapid flooding injury of citrus was directly related to
H_2S formation in a sandy soil. If the soil contains sufficient amounts of
Fe^{-2}, injurious levels of H_2S can be avoided by precipitation of FeS
(Ponnamperuma 1972).

The net result of soil flooding on other soil nutrients is to increase the
availability of PO_4^{-2}, Co, Cu, and Mn. Availability of Zn usually is not

changed in flooded soil (Ng and Bloomfield 1962; Ponnamperuma 1972).

Organic matter decomposition proceeds slowly in flooded soils. Many substances produced by the anaerobic metabolism of bacteria may accumulate to toxic levels; these include sulfur and hydrocarbon gases such as hydrogen sulfide, mercaptans, methane, ethane, and butane; alcohols such as methanol and ethanol; organic acids such as formic, acetic, and propionic acids which eventually are converted to methane; and various aldehydes, fatty acids, and phenolic componds (Ponnamperuma 1972, 1984; Drew and Lynch 1980).

The deleterious effects of excess CO_2 and insufficient O_2 in flooded soil warrant particular attention. Carbon dioxide is highly soluble in water. High CO_2 concentrations occur in flooded soil mainly as a result of respiration by soil organisms. For example, flooding a sandy loam for two weeks changed O_2 concentration from 20% to 1% and CO_2 concentration from 0.34% to 3.4% (v/v) (Drew and Sisworo 1979). Several reviews have concluded that low O_2 and not excess CO_2 is likely the major source of damage associated with soil flooding (Grable 1966; Rowe 1966); Rowe and Beardsell 1973).

IV. PLANT RESPONSES

A. Plant Metabolism

Research involving effects of flooding on metabolism of fruit crops has been limited. Therefore, information from work with herbaceous and forest tree species was referenced. Knowledge at the biochemical level derived from other plant species may be applicable to fruit species; however, research on metabolic responses of fruit crops to flooding is required for confirmation.

Metabolic characteristics associated with flood tolerance of plants have been proposed, including: (1) control over the Pasteur effect (i.e., an increase in the rate of carbohydrate catabolism under low oxygen due to a shift from aerobic to anaerobic glycolysis) and the control of alcohol dehydrogenase activity; (2) diversification of the end products of glycolysis and transfer of the O_2 debt from the root to the shoot; (3) an affinity of terminal cytochrome oxidases for O_2; (4) a favorable energy balance [e.g., adequate levels of adenosine triphosphate (ATP)] and an adequate supply and transport of carbohydrates; (5) maintenance of proper membrane function; (6) minimization or prevention of cytoplasm acidosis; and (7) regulation of phytohormones.

1. Regulation of the rate of glycolysis. In flood-tolerant plants, flooding has been hypothesized to accelerate glycolysis (the Pasteur effect) and to increase alcohol dehydrogenase (ADH) activity (Crawford 1966, 1967,

1969, 1972, 1975, 1976; Crawford and McManmon 1968; McManmon and Crawford 1971; Francis et al. 1974; Crawford and Baines 1977). It was proposed that differential tolerance to soil flooding is due to differing inductive properties of glycolytic enzymes, particularly ADH. The net result of increased glycolytic and ADH activities is an increase in the ethanol concentration in flood-intolerant plants. Many investigators have studied the relationship of ADH activity and flood tolerance. For example, *Senecio* species that are native to dry but not wet habitats, had increased rates of ethanol production (ADH activity) with soil waterlogging (Crawford 1966). Moreover, numerous, often unrelated, plant species judged to be tolerant to soil flooding exhibited low ADH activity and a low rate of ethanol production. Plants judged to be flood-intolerant had a high ADH activity and increased ethanol production under flooded conditions (Crawford 1967,1972; McManmon and Crawford 1971; Chirkova 1975; Crawford and Baines 1977). Electrophoresis analysis of ADH isozymes from flood-tolerant strains of corn (*Zea mays* L.) (Marshall et al. 1973), *Trifolium subterraneum* L. (Francis et al. 1974), and *Lupinis augustifolius* L. (Marshall et al. 1974) displayed different banding patterns than flood-intolerant strains.

During the 1970s, evidence accumulated disputing the theory that low or no induction of ADH activity is a characteristic of flood-tolerant plant species. Many flood-tolerant species have been shown to exhibit increased glycolytic rates and ADH activity when exposed to root hypoxia (Grineva 1963; John and Greenway 1976; Smith and Ap Rees 1979a, 1979b; Rumpho and Kennedy 1981; Jackson et al. 1982; Tripepi and Mitchell 1984). For example, flooding resulted in increased ADH activity for flood-tolerant cultivars of barley (*Hordeum vulgare* L.), yet root and shoot growth declined more for sensitive species (Wignarajah et al. 1976). Similarly, intermediate concentrations of O_2 (3–13%) rather than low O_2 (<3%) produced the greatest stimulation of ADH activity in roots of barley (Wignarajah et al. 1976) and corn (Wignarajah and Greenway 1976). Mendelssohn et al. (1981) determined that ADH activity is greatest at intermediate levels of O_2 for *Spartina* marshgrass.

The hypothesis that flood-intolerant plants are damaged as a result of uncontrolled glycolysis when exposed to anaerobiosis assumes that ethanol accumulates to toxic levels (Crawford 1966; McManmon and Crawford 1971). Studies involving several plant species have shown that ethanol concentration increased in the roots of flooded plants (Jackson et al. 1982; Drew 1983). Mizutani et al. (1982) concluded that differential sensitivity of *Prunus* species to soil flooding was correlated with ethanol accumulation. Similarly, apple rootstocks exhibited increased ethanol concentrations with high root temperature induced anaerobiosis (Gur et al. 1972). Rowe (1966) observed increased rates of ethanol production in peach, plum, and pear with the onset of anaerobiosis, although rates fell drastically after 4 h. Johnson et al. (1989) reported that pre-exposure to hypoxic conditions resulted in increased ADH activity, adenylate charge ratio and survival of corn roots.

Ethanol may not necessarily be deleterious to plants and may not be a primary determinant of flooding injury. Many plants have the ability to metabolize ethanol (Keneflick 1962; Cossins and Beevers 1963; Rowe 1966; Crawford and Baines 1977;) and ethanol can be detected in trunks and roots of both nonflooded and flooded trees (Crawford and Baines 1977). Additionally, exogenous ethanol has not been satisfactorily demonstrated to be toxic to plants, even when the concentrations applied are above endogenous levels (Beletskya 1977; Drew and Lynch 1980; Jackson et al. 1982; Jackson and Drew 1984; Barta 1988). Toxic levels of ethanol may be dissipated by diffusion into the rooting medium (Chirkova and Gutman 1972; McKee and Mendelssohn 1987) or by entry into xylem fluid (Fulton and Erickson 1964; Barta 1984). Beletskya (1977) demonstrated that acetylaldehyde, which also may accumulate in flooded plants, is much more toxic than ethanol on a equivalent molar basis.

2. Regulation of glycolytic end products. The requirement for proton disposal and regeneration of NAD^+ for continued glycolysis may be accomplished by production of molecules other than ethanol such as lactate, malate, alanine, aspartate, glutamate, glycerol, and shikimate (Crawford 1976). Of these, malate is viewed as a nontoxic substitute for ethanol; concentration of this compound tends to increase under hypoxic conditions. Crawford and colleagues proposed that a malic enzyme, which catalyzes the conversion of malate to pyruvate, is present in flood-intolerant plants and is absent from flood-tolerant plants (Crawford 1966, 1967, 1969, 1972, 1975, 1976; Crawford and McManmon 1968; McManmon and Crawford 1971; Francis et al. 1974; Crawford and Baines 1977). Malate is formed from phosphoenol pyruvate via dark fixation of CO_2 with the participation of phospheonol pyruvate carboxylase and malate dehydrogenase (McManmon and Crawford 1971). McManmon and Crawford (1971) suggested that flood-intolerant species can convert phosphoenol pyruvate to pyruvate directly or via a pathway involving oxaloacetate and malate as intermediates. However, in flood-tolerant plants, the direct pathway does not occur, and malate accumulates. Crawford (1972) observed a greater malate concentration in the xylem fluid of birch (*Betula pubescens* Ehrh.) trees grown on wet sites as opposed to trees grown on drier sites. He argued that starch utilization with the production of malate transferred the O_2 debt from the root to the shoot. Davies et al. (1974b) challenged the "malic enzyme theory" by documenting the presence of malic enzyme in 27 of 28 plant species [including those previously classified as lacking the enzyme by McManmon and Crawford (1971)]. Davies et al. (1974b) reported that malic enzyme exhibited allosteric properties in all species except those in the Graminaceae. Moreover, Keely (1978) reported that malate is an unlikely alternative end product to ethanol since malate concentrations and ethanol were not inversely correlated in flooded plants. The full significance of the increased malate concentration found in roots under flooded conditions remains to be resolved (Crawford and Tyler 1969; Davies et al. 1974a; Chirkova 1978; McKee and Mendelssohn 1987).

Another method of NAD+ regeneration may be accomplished by lactate production. The increase in lactate dehydrogenase activity under flooded conditions often is one to three orders of magnitude less than the increase in ADH (Smith and Ap Rees 1979a, 1979b; Rumpho and Kennedy 1981; McKee and Mendelssohn 1987). Hence, lactic fermentation is viewed as a minor pathway of carbohydrate metabolism under anoxic and hypoxic conditions. (The significance of lactic acid production is discussed in the section dealing with regulation of cytoplasm pH.)

Plants may utilize various transaminase enzymes to alter the concentration of pyruvate or other compounds of the glycolytic pathway. Amino acid metabolism is altered greatly under conditions of low O_2. The concentration of free amino acids and the rate of interconversion increase under hypoxic conditions, with alanine increasing to the greatest concentrations (Dubinina 1961; Streeter and Thompson 1971; Smith and Ap Rees 1979; Bertani and Brambilla 1982; Reggiani et al. 1985,1988; McKee and Mendelssohn, 1987). The change in free amino acid concentration in rice was due to protein degradation, and the change in amino acid profile was due to the influence of altered organic acid levels (Regiani et al. 1988). Other amino acids reported to increase with hypoxia include gamma-aminobutyric acid (Dubiniha 1961; Streeter and Thompson 1971; Reggiani et al. 1988), proline (Dubinina 1961; Wample and Bewley 1975), glycine, and serine (Guinn and Brinkerhoff 1970). The concentrations of the amides (glutamine and asparagine) and their respective acids decreased under hypoxic conditions (Streeter and Thompson 1971; Reggiani et al. 1988). Alanine accumulation may be due to the availability of acidic amino acids for transamination (via glutamate-pyruvate transaminase) and to the increased pool of pyruvic acid (Reggiani et al. 1988). The increased concentrations of alanine, gamma-aminobutyric acid, and proline may be adaptive responses to flooding since these amino acids reduce cytoplasm pH less than lactate, glutamate, and aspartate.

3. Affinity of cytochrome oxidases for O_2.   Root respiration characteristics of flood-tolerant and flood-sensitive species may differ with respect to: (1) the ability to maintain capacity for aerobic respiration after prolonged periods of hypoxia, and (2) the affinity of terminal oxidases for O_2 and differing proportions of cyanide-sensitive and an alternative, cyanide-resistant respiratory pathway.

Inherently high or low respiration rates of roots in air were not correlated with differential tolerance of pear rootstocks, although a very flood-tolerant species, Pyrus betulaefolia, maintained the capacity for aerobic respiration after 26 days of flooding (Andersen et al. 1985). Similarly, flood-tolerant red maple (Acer rubrum L.) and bald cypress [Taxodium distichum (L.) Rich] maintained the capacity for aerobic respiration longer than sugar maple (Acer saccharum Marsh).

The cyanide-resistant respiratory pathway is located in mitochondria (Siedow and Berthold 1986). The pathway conserves only up to one third the amount of ATP as the cyanide-sensitive pathway in the oxidation of pyruvate, and the remainder of the energy is given off as heat (Moore and Rich 1985). The hypothesis that under hypoxic conditions the ratio of cyanide-sensitive to cyanide-insensitive respiration may correlate with waterlogging sensitivity has not been supported by results from several studies (Lambers 1976,1980; Carpenter and Mitchell 1980a, 1980b; Lambers and Smakman 1980). The affinity of cytochrome oxidase for O_2 (Km $<0.1\mu$M) is actually greater than that of the alternative oxidase (Km = 1 to μ2M) (Siedow and Berthold 1986). Thus, the importance of the cyanide-insensitive pathway probably is reduced under conditions of hypoxia.

4. Energy balance and supply of carbohydrates. A reduced level of ATP, ADP, or adenine energy-charge ratio (ATP + 1/2 ADP; ATP + ADP + AMP) has been reported in hypoxic roots of several plant species (Saglio et al. 1980; Tripepi and Mitchell 1984; McKee and Mendelssohn 1987;). The production of ATP during anoxia is dependent upon glycolysis and alcohol fermentation (Saglio et al. 1980; Mendelssohn et al. 1981; Tripepi and Mitchell 1984; McKee and Mendelssohn 1987).

The maintenance of the normal structure of mitochondria is dependent on a sufficiently high adenine energy charge ratio derived from oxidative phosphorylation (Luzikov et al. 1973). Mitochondria of plant roots subjected to flooded conditions were reduced in size and number and developed tubular inclusions (Oliveira 1977). Alterations in mitochondrial ultrastructure and function are reversible if the period or degree of hypoxia is not excessive.

The inherent inefficiency of anaerobic metabolism has prompted work concerning the adequacy and utilization of energy reserves (carbohydrates) in the root and the transport of carbohydrates under flooded conditions. Root viability and root adenine energy-charge ratios have been extended under hypoxic conditions with exogenous carbohydrate applications (Vartapetian et al. 1977; Saglio et al. 1980). Several investigators have found that soluble carbohydrates actually increase in hypoxic roots, which indicates that the lack of available substrates was not limiting metabolism (Benjamin and Greenway 1979; Spek 1981; Papenhuijzen 1983; Barta 1988a).

5. Maintenance of membrane integrity. Membrane integrity and proper membrane function are contingent upon an adequate supply of available energy. Anoxia induces the efflux of K^+, Cl^-, as well as amino and organic acids, from the roots of barley (Marschberm and Mengel 1966; Hiatt and Lowe 1967) and electrolytes from roots of rabbiteye blueberry (Crane and Davies 1987). At first glance, solute efflux under conditions of low O_2 may be ascribed to an increase in membrane permeability. Alternatively, it is known that hypoxia may induce depolarization of membrane potential (Pitman 1969; Buwalda et al. 1988). Buwalda et al. (1988) demonstrated that a depolarzation of root cell membranes (in reference to sorbitol and K^+) of wheat (*Triticum aestivum* L.) occurred after only three minutes of hypoxia, and an increase in root membrane permeability did not occur even after 10 days. They suggested that a subsequent increase in root permeability may be either a cause or an effect of cell death.

Crane and Davies (1987) noted that electrolyte leakage from roots of rabbiteye blueberry occurred 6 days after flooding, later than decreases in stomatal and hydraulic conductances in the sequence of flooding symptomatology, and well after observed reductions in stomatal conductance and root hydraulic conductance. Similarly, the release and hydrolysis of cyanogenic glycosides in roots of flooded *Prunus* species (Rowe 1966; Rowe and Catlin 1971), and the release of phenolic compounds in three species of walnut rootstocks exposed to flooding (Catlin et al. 1977), were not viewed as the primary events of flooding injury.

6. Regulation of cytoplasm pH. Perhaps the most significant advance in our understanding of biochemical adaptations associated with hypoxia has been in relation to the regulation of cytoplasmic acidosis (Roberts et al. 1984, 1985). Davies et al. (1974a) suggested that cytoplasm pH determines the identity of the glycolytic end product via a high pH optimum of lactate decarboxylase. This leads to lactic acid production and a low pH optimum of pyruvate decarboxylase, and subsequent ethanol production. Lactic acid production predominates in alkaline cytoplasm with the onset of hypoxia. The resultant acidification activates pyruvate decarboxylase, and ethanol ultimately is formed (Davies et al. 1974). Roberts et al. (1984) determined that cytoplasm acidification in maize root tips exposed to hypoxia was due to lactic acid production. Supporting evidence for this included the finding that a maize mutant lacking a locus for ADH activity could not synthesize ethanol; pH was not stabilized and irreversible cell injury occurred.

The change in amino acid profile in roots under hypoxic conditions reported earlier in this review may be considered an adaptive response to flooding since concentrations of amino acids such as alanine, gamma-aminobutyric acid, and proline, which normally increase due to flooding, tend to decrease pH less than the amino acids such as glutamate/glutamine or asparatate/asparagine, which predominate in a nonflooded state.

7. Regulation of phytohormones. Although significant advances have been made in studying the effects of flooding on the biochemistry of ethylene production, fewer data are available concerning other plant hormones. Moreover, most of the published reports deal with herbaceous crops. Thus, we present a brief overview of effects of soil flooding on phytohormones; a more comprehensive review can be found in Reid and Bradford (1984).

It is well known that ethylene may induce leaf epinasty, adventitious rooting, stem hypertrophy, and aerenchyma production in certain species of flooded plants (Bradford and Yang 1981; Kawase 1981; Reid and Bradford 1984). Increased ethylene concentrations in waterlogged plants are due to accelerated production in the plant, decreased diffusion out of the plant, and absorption of microbially produced ethylene. In the late 1970s, evidence accumulated for a factor which emanated from the root, was transported in the xylem fluid, and promoted shoot ethylene production in flooded plants (Jackson and Campbell 1976; Jackson et al. 1978). Adams and Yang (1977, 1979) elucidated the pathway of ethylene biosynthesis and identified s-adenosylmethionine (SAM) and l-aminocyclopropane-l-carboxylic acid (ACC) as intermediates between methionine and ethylene. Under anaerobic conditions, the biosynthesis of ethylene proceeds only to ACC since the final conversion of ethylene requires O_2 (Yang et al. 1980). Although ethylene is known to be an important factor in flooding symptomology of some herbaceous and woody species, the significance in the symptomology of fruit trees remains to be resolved. Andersen (1983) determined the ACC concentration in xylem fluid and ethylene evolution from excised roots of flooded and nonflooded tomato, willow, peach, quince, and four pear species after 20 and 40 days of flooding. ACC was detected in xylem fluid of flooded tomato and willow, but not in pear, peach, or quince. Ethylene was found to emanate from flooded and nonflooded roots of most species and was attributed to wound ethylene. Ethylene evolution was not detected from roots of flooded rabbiteye blueberry plants (Crane 1987).

Waterlogging has been reported to enhance abscisic acid (ABA) levels in leaves of several plant species (Hiron and Wright 1973; El-Beltagy and Hall 1974; Hall et al. 1977; Sivakumaran and Hall 1978; Jackson and Hall 1987), including *Juglans* species (Shaybany and Martin 1977). There has been considerable interest in the role of ABA in stomatal closure and root hydraulic conductance of flooded plants (see water relations and photosynthesis section). Increased ABA concentrations have been detected in roots of pea (*Pisum sativum* L.) 22 h after the onset of soil flooding; and 14 h later ABA was detected in the leaf (Zhang and Davies 1987). Zhang and Davies (1987) provided evidence that elevated leaf ABA content was a consequence of increased transport out of the root, while Jackson and Hall (1987) maintained that it was a result of decreased transport of ABA out of the leaf. Flore et al. (1989) found that ABA concentration in leaves of normal ('Alisa Craig') and an ABA-deficient ('Flacca') tomato manifested increased ABA concentrations after 24–48 h. Increased leaf ABA was attributed to decreased translocation out of the leaf under flooded conditions; ABA was undetected in xylem fluid. The physiological importance of ABA in plants with flooded root zones requires further research.

There is even less information concerning the effects of flooding on growth-promoting phytohormones such as auxins, cytokinins, and gibberellins. The limited available data generally are from work conducted before 1980. Soil flooding increases auxin concentration in the shoot (Phillips 1964a, 1964b; Hall et al. 1977; Wample and Reid 1979). Reduced auxin transport to the roots of flooded plants may explain elevated shoot auxin concentrations. Wample and Reid (1979) found that flooding inhibited transport into the root and metabolic breakdown of ^{14}C-labelled IAA. They suggested that increased ethylene concentration may stimulate auxin production, which in turn promotes adventitious root formation (see section on growth, anatomical, and morphological responses).

Root meristems are a significant source of cytokinins and gibberellins (Moore 1979). Cytokinins have been reported in xylem fluid of flooded tomato (Reid and Crozier 1971) and sunflower (Burrows and Carr 1969). Jackson and Campbell (1979) found that an application of benzyladenine plus gibberellic acid increased the transpiration of waterlogged tomato and often induced wilting. In contrast, Flore et al. (1989) found that flooding did not influence cytokinin concentrations in the xylem fluid of normal and ABA-deficient ('Flacca') tomato plants. Very little is known about the biosynthesis, metabolism, and transport of these hormones under hypoxic conditions.

B. Growth, Anatomical, and Morphological Responses

Reductions in both shoot and root growth are common responses to waterlogging (DeWit 1978; Kozlowski 1982). For flood-sensitive plants, flooding often results in damage and death of root tips. Many plants, including fruit crops, develop anatomical or morphological structures in response to flooding such as hypertrophied lenticels and adventitious roots. In this section, specific examples of growth reductions and anatomical and morphological changes of fruit crops in response to flooding are discussed.

Several researchers have quantified the effects of flooding or reduced soil O_2 on fruit crop growth. Heinicke (1932) observed that apple trees flooded for one week during the spring had restricted root growth and developed small leaves which often desiccated on hot days. Childs (1941) noted that shoot growth of apple abruptly declined when the O_2 in the soil decreased below 12%, and then gradually declined as O_2 concentrations were further decreased to 1.5–2%, where another abrupt growth decline occurred. In addition, roots grown in soil with 1.5–2% O_2 were more brittle, longer, less lignified, and had greater mortality than roots grown at greater O_2 concentrations (Childs 1941). Boynton (1940) reported that when soil O_2 concentration dropped below 10% due to flooding, so few rootlets were formed that shoot growth of apple trees was reduced. Olien (1987) observed that current season's shoot and trunk growth of apple trees were reduced following 6-week flooding periods. This decrease was greater for trees flooded in the spring and summer than for trees flooded in the fall. Summer waterlogging of apple also decreased leaf and root dry weights (Olien 1987). Andersen et al. (1984b) observed that, in addition to reducing shoot growth of apple, flooding reduced shoot growth of several peach, quince, and pear species. Some pear species appeared to be more flood-tolerant than others with respect to shoot growth. Working with rough lemon (*Citrus jambhiri*) and sour orange (*Citrus aurantium*), Syvertsen et al. (1983) observed that three weeks of continuous flooding resulted in a cessation of shoot growth and a sloughing off of roots. In contrast, Larson et al. (1989, 1991) reported that root growth of mango trees ceased after two weeks of flooding, whereas shoot growth of flooded plants continued to increase during the flooding period. Thus, flooding of lemon, sour orange, and mango resulted in increased shoot:root ratios. Joyner and Schaffer (1989) observed that leaf, stem, and root dry weights of carambola decreased with increased flood duration; flooding reduced shoot and root growth in roughly the same proportion. After several weeks of flooding, removal of mango (Larson et al. 1989) and carambola (Joyner et al. 1988; Joyner and Schaffer 1989) from flooding resulted in resumption of normal shoot and root growth.

Flooding has been shown to reduce shoot and root growth of highbush blueberry (Herath and Eaton 1968; Abbott and Gough 1987b; Crane and Davies 1989) and shoot growth of rabbiteye blueberries (Davies and Wilcox 1983; Crane and Davies 1988b, 1989), cranberry (Bergman 1943), and grape (West and Taylor 1984).

Another common response to flooding is an inhibition of leaf expansion resulting in smaller leaves, reduced leaf number, and increased leaf abscission (Kozlowski 1984b). These responses have been reported for several fruit crops including apple (Boynton 1940), mango (Larson et al. 1989), avocado (Ploetz and Shaffer 1989), carambola (Joyner and Schaffer 1989), cranberry (Bergman 1943), highbush blueberry (Herath and Eaton 1967;Abbott and Gough 1987b; Crane and Davies 1989), rabbiteye blueberry (Crane and Davies 1988a, 1988b, 1989), peach, apple, quince, pear (Andersen 1984b), and pecan (Wazir et al. 1988). Abbott and Gough (1987b) observed that leaves of flooded highbush blueberries had a thinner epidermal layer, a greater percentage of intercellular spaces in the spongy mesophyll, and a disrupted palisade layer compared to leaves of nonflooded plants.

Flooding has been shown to affect flowering, fruit set, yield, and fruit quality of several fruit crops. Both summer and winter flooding reduced fruit set, yield, and fruit size of cranberry (Bergman 1943). Percent fruit set and yield of rabbiteye blueberries increased after 25 days of flooding, whereas after 5 days of flooding the number of flower buds was less for flooded than for nonflooded plants (Crane and Davies 1985). Abbott and Gough (1987a, 1987b) noticed that flooded highbush blueberry plants had smaller flower buds and fruits than nonflooded plants, and that flooding delayed flower bud development by about one week. Flooding of highbush blueberries resulted in a reduced soluble solid content of the fruit (Abbott and Gough 1987c). Waterlogging of apple trees has been associated with poor fruit set (MacDaniels and Heinicke 1929) and yield (Olien 1987). Olien (1987) found that decreased apple yield due to flooding was followed by an increased return bloom. In contrast, Joyner and Schaffer (1989) observed that carambola trees flooded for 6–18 weeks tended to have greater flowering and fruit set than trees flooded for shorter time periods or nonflooded trees. In addition, carambola trees intermittently flooded for 2, 3, or 6 weeks (trees were removed from flooding for 3 weeks between flooding periods) had greater fruit set than nonflooded trees (Joyner and Schaffer 1989).

Several morphological and anatomical adaptations of plants to flooding have been reviewed by Kawase (1981). Two important adaptations are the development of adventitious roots and the formation of hypertrophied lenticels. Many flood-tolerant woody plant species respond to waterlogging by developing adventitious roots, which facilitate increased O_2 absorption (Sena Gomes and Kozlowski 1980a, 1980b; Coutts 1982; Kozlowski 1982). Andersen et al. (1984b) observed the formation of adventitious roots by waterlogged quince and apple; flooded pear species did not form these structures. Adventitious roots not only increase water-absorbing efficiency, but have been correlated with near normal stomatal conductance under flooded conditions (Sena Gomes and Kozlowski 1980a). The ability to form adventitious roots often is considered to be a major factor conferring flood tolerance to woody plants (Hook and Brown 1973).

Alternatively, adventitious rooting may be a stress symptom rather than an adaptation to flooding (Hall and Smith 1955), since plants with a wide range of flood tolerance develop these structures (Pereira and Kozlowski 1977; Kawase 1981). Interspecific comparisons of flood tolerance based upon the development of adventitious rooting are tenuous. For example, moderately flood-tolerant apple and quince formed adventitious roots with flooding; however, no pear species formed adventitious roots whether they were moderately or extremely flood tolerant (Andersen 1984b).

A common anatomical response of flood-tolerant woody plants to flooding is the formation of hypertrophied (swollen) lenticels on the stem below and above the water line, and increased aerenchyma formation in the cortex of stems and shoots (Hook et al. 1970; Coutts and Armstrong 1976; Hook and Scholtens 1978; Coutts 1982; Abbott and Gough 1987a; Justin and Armstrong 1987; Sena Gomes and Kozlowski 1988; Larson et al. 1990). Flooding resulted in lenticel hypertrophy of peach, apple, quince, pear (Andersen et al. 1984b), mango (Larson et al. 1990, 1991), carambola (Joyner and Schaffer, unpublished), and highbush blueberry (Abbott and Gough 1987a). Although formation of hypertrophied lenticels and increased aerenchyma were observed on flooded highbush blueberry plants (Abbott and Gough 1987a), development of these structures was not observed on rabbiteye blueberry plants, even after 100 days of flooding (Crane and Davies 1988b).

A typical cross-section of a hypertrophied lenticel of a flooded mango tree, 60 days after initiating flooding, and a normal lenticel from the stem of a nonflooded mango tree are shown in Figure 7.1. Hypertrophied lenticels are characterized by increased intercellular air space, due either to cell separation (schizogeny) or cell lysis (lysigeny); cell enlargement; a more spherical cell structure; and the disappearance of the lenticel-closing layer (Esau 1965; Coutts and Armstrong 1976; Hook and Scholtens 1978). Hypertrophies may extend into the subtending cortex and phloem tissues (Hook et al. 1970).

The increased intercellular space associated with hypertrophied lenticels indicate that they may function as organs for increased O_2 absorption and transport to the roots (Hook et al. 1970; Kawase 1981). In addition, hypertrophied lenticels may serve as excretory sites for toxic metabolities such as ethanol and acetylaldehyde formed as a result of anaerobic respiration in the roots (Chirkova and Gutman 1972; Chirkova 1978). Larson et al. (1989) observed that when hypertrophied lenticels of

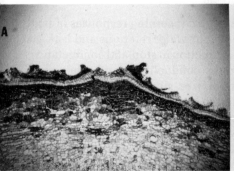

Figure 7.1. Cross-sections of A) normal lenticel from the stem of a nonflooded mango, and B) a hypertrophied lenticel on the stem of a flooded mango tree (100×) (from Larson et al., 1991).

flooded mango trees were sealed, leaves became chlorotic from the veins outward, suggesting the possibility that toxic substances were transported to the foliage when excretion of these metabolites through lenticels was blocked.

Kawase (1981) attributed hypertrophied lenticel formation and lysigenous aerenchyma development in flooded plants to lack of O_2 rather than hydration, since bubbling O_2 through the water in which plants were submerged prevented the development of such structures. However, recent studies with mango (Larson et al., unpublished) indicate that stems must be hydrated for lenticels to hypertrophy. Formation of hypertrophied lenticels and lysigenous aerenchyma was attributed to increased cellulase activity resulting from increased ethylene concentration in the stems of waterlogged plants (Kawase 1981). A further indication of the role of ethylene in lenticel hypertrophy is that exposing apple scions to ethylene resulted in the formation of hypertrophied lenticels (Janick 1975). For hydroponically grown mango trees, ethylene evolution at the air-water interface was greater from stems of trees grown in a non-oxygenated solution compared with trees grown in an oxygenated solution. Trees grown in the non-oxygenated solution formed hypertrophied lenticels earlier than trees in the oxygenated solution (Larson and Schaffer, unpublished). Thus, evidence suggests that lenticel hypertrophy is dependent on several factors, including stem hydration, lack of O_2 in the root zone, and ethylene.

In general, growth responses to flooding such as damage and death of root tips, reduced root and shoot growth, leaf abscission, a reduction in flower initiation and fruit production, and eventual shoot and root mortality are similar for all flood-sensitive fruit species. Many flood-tolerant fruit crops exhibit anatomical or morphological features which appear to be related to their ability to tolerate prolonged flooding.

C. Water Relations and Photosynthesis

During the past 20 years, much research concerning responses of fruit crops to waterlogging has focused on plant-water relations and leaf gas exchange. These responses, such as a reduction in stomatal conductance, photosynthesis, root and stem hydraulic conductivity, and often eventual reductions in leaf water potential, are similar for most species studied. The specific time course required to elicit these responses, however, varies among species. This presumably is due to inherent differences among species, differences in methodologies used to measure these responses, or the unique biotic and environmental conditions of each study, such as plant age and condition, air temperature, and relative humidity. Despite the large body of work conducted on water relations and photosynthesis of flooded fruit crops, cause-and-effect relationships leading to the sequence of events that take place under hypoxic conditions have not been elucidated. This section documents the effects of soil waterlogging on plant-water relations and leaf gas exchange of fruit crops; possible mechanisms for such responses based on recent investigations are discussed.

1. Stomatal conductance, transpiration, and hydraulic conductivity. One of the earliest physiological responses to flooding is a reduction of stomatal conductance (Kozlowski 1982; Kozlowski and Pallardy 1984); this has been observed for many herbaceous species (Moldau 1973; Wenkert et al. 1981; Bradford and Hsiao 1982; Jackson and Hall 1987), forest tree species (Regehr et al. 1973; Pereira and Kozlowski 1977; Kozlowski and Pallardy 1979; Tang and Kozlowski 1982; Andersen et al. 1984a), and fruit crops (Andersen et al. 1984a, 1984b; Davies and Wilcox 1984; Davies and Flore 1985, 1986a, 1986b, 1986c; Beckman et al. 1987; Crane and Davies 1987, 1988, 1989; Ploetz and Shaffer 1987, 1988, 1989; Schaffer and Ploetz 1987, 1989; Joyner and Schaffer 1988, 1990;Smith and Agar 1988; Wazir et al. 1988; Larson et al. 1989; Ploetz et al. 1989; Vu and Yelenoski 1991). A reduction of stomatal conductance, however, is not a consistent response to flooding. For example, Sena-Gomes and Kozlowski (1988) observed that stomatal conductance of rubber tree seedlings (Hevea brasieliensis Muell. Arg.) was not significantly affected by flooding.

The duration of flooding required to decrease stomatal conductance varies widely among fruit crop species. Syvertsen et al. (1983) observed a decrease of stomatal conductance of sour orange seedlings four days after flooding, whereas stomatal conductance of rough lemon seedlings did not decrease until eight days after root submergence. Effects of flooding on stomatal conductance may also vary with season of flooding. Olien (1989) reported that leaf conductance of apple trees in an orchard decreased during spring and summer waterlogging but not after fall waterlogging. Andersen (1984b) reported that leaf conductance of spring-flooded pear trees declined more rapidly than that of trees flooded in the fall.

Reductions of stomatal conductance due to waterlogging appears to be a reversible process. Stomatal conductance of highbush and rabbiteye blueberries declined after 1–7 days of flooding (Davies and Wilcox 1984; Davies and Flore 1985, 1986a, 1986b, 1986c; Crane and Davies 1987, 1988b). However, 18 days after plants were removed from flooding, stomatal conductance of rabbiteye blueberries recovered to that of the nonflooded controls, whereas stomatal conductance of highbush blueberries remained lower than that of the controls (Davies and Flore 1986b). Larson et al. (1989) observed that removing mango trees from flooding resulted in stomatal conductance rates returning to those of the controls. Similarly, Moon and Schaffer (unpublished) observed that stomatal conductance of Tahiti lime (Citrus × 'Tahiti'), which declined within three days of flooding, returned to preflood rates three days after removal from flooding. Smith and Agar (1988) observed that stomatal conductance of pecan trees decreased after one day of flooding. Eight days after plants were removed from flooded conditions, stomatal conductance rates returned to those of the nonflooded controls. However, during the course of their experiment, stomatal conductance of pecan trees flooded for 15 days did not return to that of the controls when plants were removed from flooding. Joyner and Schaffer (1989) observed that stomatal conductance rates of carambola trees nearly returned to that of the nonflooded controls three weeks after removal from flooding. However, after repeated flooding cycles, stomatal conductance recovered to a lesser degree than it did after the previous flooding cycles. Stomatal conductance of rabbiteye blueberries recovered to that of non-flooded controls after two flooding periods of 2–7 days (Crane and Davies 1988).

Decreased stomatal conductance under flooded conditions reduces transpiration of herbaceous crops (Kramer 1951; Grineva 1961; Bradford and Yang 1981; Wenkert et al. 1981; Bradford and Hsiao 1982; Jackson and Hall 1987) and forest species (Regehr et al. 1973; Coutts 1981; Kozlowski and Pallardy 1984; Sena Gomes and Kozlowski 1988). Among fruit crops, flood-induced reductions of transpiration have been reported for apple (Childers and White 1942; Olien 1989), pecan (Loustalot 1945; Smith and Ager 1988), citrus (Phung and Knipling 1979), avocado (Ploetz and Schaffer 1987), mango (Larson et al. 1989), carambola (Joyner et al. 1988; Joyner and Schaffer 1989), lime (Schaffer and Moon 1990) and blueberry (Davies and Flore, 1986b; Crane and Davies 1989).

The factors responsible for stomatal closure and hence reduced stomatal conductance under flooded conditions are currently a matter of speculation (Sojka and Stolzy 1980; Bradford and Yang 1981; Bradford and Hsiao 1982; Andersen et al. 1984a). It is unlikely that reduced stomatal conductance of flooded fruit trees is due to leaf dehydration since leaf water potential for several plant species (Pereira and Kozlowski 1977), including pecan (Smith and Agar 1988), blueberies (Davies and Wilcox 1983; Davies and Flore 1986a; Crane and Davies 1989), and pear (Andersen et al. 1984a), usually does not decline during early stages of flooding.

Kozlowski and Pallardy (1984) postulated that reduced conductance under flooded conditions may be due to an efflux of K^+ ions from the guard cells associated with stomatal closure. They based their hypothesis on the fact that low soil O_2 tension often results in reduced K^+ content in leaves (Hammond et al. 1955; Harris and van Bavel 1957). In addition, Harris and van Bavel (1957) observed that low O_2 concentrations in soil resulted in low leaf potassium concentrations and stomatal closure of tobacco (Nicotiana tabacum L.). It is possible that such a mechanism exists for fruit crop species. For example, microscopic observation of highbush blueberry leaves showed that flooded plants had fewer open stomata than nonflooded plants (Abbott and Gough 1987c).

Reductions in stomatal conductance under flooded conditions may result from a chemical signal from the roots to the shoot (Reid et al. 1969; Bradford and Yang 1981). Although no specific substance has been identified in fruit crops, Davies and Flore (1985) reported that a root exudate from flooded highbush blueberry plants caused a reduction in photosynthesis and stomatal conductance when applied to nonflooded plants. They suggested that a substance translocated from the roots under flooded conditions could inhibit net CO_2 assimilation which might affect stomatal conductance by increasing the substomatal CO_2 concentration. Schaffer and Ploetz (unpublished) approach-grafted pairs of avocado plants and measured stomatal conductance of leaves corresponding to the flooded and nonflooded side of the graft union. They observed a significant decrease in stomatal conductance on both flooded and nonflooded sides of they graft union relative to nonflooded pairs, suggesting the transmission of a graft-transmissible substance from the flooded to the nonflooded portion of the pair that induced stomatal closure.

Phytohormones, particularly ABA and cytokinins, have been implicated in eliciting stomatal closure under waterlogged conditions (Davies and Kozlowski 1975a, 1975b; Bradford and Yang 1981; Bradford and Hsiao 1982; Kozlowski and Pallardy 1984; Reid and Bradford 1984). This is based on the fact that flooding results in a buildup of ABA in leaves (Hiron and Wright 1973; Shaybany and Martin 1977; Jackson and Hall 1987; Zhang and Davies 1987), and ABA buildup has been correlated with stomatal closure (Cummins et al. 1971; Mansfield and Davies 1981; Zhang and Davies 1987; Zhang et al. 1987).

There is evidence that reduced cytokinin levels under flooded conditions also may play a role in stomatal closure (Reid and Bradford 1984). Cytokinins, which are produced in roots, have been implicated in controlling stomatal opening (Livne and Vaadia 1965; Meidner 1967). Burrows and Carr (1969) noted decreased cytokinin concentrations in the xylem fluid of flooded sunflower (Helianthus annus L.). Therefore, the possibility exists that reduced stomatal conductance of flooded fruit trees may result from altered cytokinin or ABA concentrations or shifted ABA:cytokinin ratios. Further investigations are needed to elucidate the involvement of these hormones in stomatal closure of flooded plants.

Ethylene buildup in flooded plants also has been implicated in stomatal closure of some herbaceous plant species (Bradford and Yang 1981). However, there is no experimental evidence for this theory in waterlogged fruit trees (see section on regulation of phytohormones).

It has been suggested that decreased stomatal conductance of fruit crops due to waterlogging may be due to a decrease in root hydraulic conductivity (Syvertsen et al. 1983; Andersen et al. 1984a; Davies and Flore 1986a). Davies and Flore (1986a) reported a decrease in root hydraulic conductivity of highbush blueberry plants within 1–2 days of flooding. Since stomatal conductance usually decreased within 4–7 days of flooding, they postulated that reduced root hydraulic conductivity may be responsible for subsequent stomatal closure. Crane and Davies (1987) observed a concomitant decrease in root hydraulic conductivity and stomatal conductance of rabbiteye blueberries 4–6 days after flooding. Stem hydraulic conductivity of blueberry decreased following a decrease in root hydraulic conductivity. Root hydraulic conductivity of flooded blueberries decreased prior to an increase in electrolyte leakage from roots (Crane and Davies 1987). Thus, decreased root hydraulic conductivity does not appear to be associated with damage to root cells during the early stages of flooding. Syvertsen et al. (1983) reported a 49% decrease in root hydraulic conductivity of sour orange associated with an approximately 47% decrease in stomatal conductance. According to Andersen et al. (1984a), the simultaneous reductions in root hydraulic conductivity and stomatal conductance preceded reductions in leaf water potential or root osmotic potential of flooded pear species. In addition, several studies with highbush and rabbiteye blueberries indicated that water potential was similar between flooded and nonflooded plants during the first 3–5 weeks of flooding (Davies and Wilcox 1983; Davies and Flore 1986b; Crane and Davies 1989). Therefore, stomatal closure under flooded conditions has not been demonstrated to be induced by water stress resulting from reduced root hydraulic conductivity (Crane and Davies 1989).

2. Photosynthesis. An early response of fruit crops to waterlogging is a reduction of net CO_2 assimilation. This has been reported for apple (Childers and White 1942), pecan (Loustalot 1945; Smith and Agar 1988), highbush blueberry (Davies and Flore 1985; Davies and Flore 1986a, 1986b), rabbiteye blueberry (Davies and Flore 1986b, 1986c), grape (Striegler et al. 1987), rough lemon, orange (Phung and Knipling 1976; Vu and Yelenoski 1991), 'Tahiti' lime (Schaffer and Moon 1990), tart cherry (Beckman et al. 1987), avocado (Ploetz and Schaffer 1987, 1988, 1989; Schaffer and Ploetz 1987, 1988; Ploetz et al. 1989), mango (Larson et al. 1989), and carambola (Joyner et al. 1988; Joyner and Schaffer 1989). Several stomatal and nonstomatal factors limit photosynthesis under flooded conditions.

The decline in net CO_2 assimilation by fruit crops associated with flooding is affected by flood duration as well as environmental factors and disease. For example, the optimum temperature for net CO_2 assimilation of rabbiteye blueberries is from 20° to 25°C (Davies and Flore 1986b). However, net CO_2 assimilation of flooded rabbiteye blueberries compared to nonflooded plants was not lower at temperatures below 20°C, and decreased at 20–25°C (Davies and Flore 1986b; Crane and Davies 1989). Davies and Flore (1986b) also noted that increasing the vapor-pressure deficit (VPD) from 1.0 to 3.0 kPa decreased net CO_2 assimilation along with stomatal and residual conductance of nonflooded highbush blueberries plants, but had less of an effect on flooded plants (as indicated by a reduced slope of the regression line) (Fig. 7.2). The response of blueberry gas exchange to increasing VPD was attributed to stomatal factors since stomatal conductance and net CO_2 assimilation responded similarly to changes in VPD (Fig. 7.2) (Davies and Flore 1986b). Increasing external CO_2 concentration in the vicinity of the leaves within a 100 to 400 μmol mol^{-1} range increased net CO_2 assimilation for rabbiteye and highbush blueberries. However, increases in net CO_2 assimilation generally were 1.4 to 2.4 times greater for nonflooded than for flooded plants (Davies and Flore 1986b). Although light-response curves for net CO_2 assimilation of rabbiteye blueberries were similar for flooded and nonflooded plants, net CO_2 assimilation was lower from 2 to 5 days after

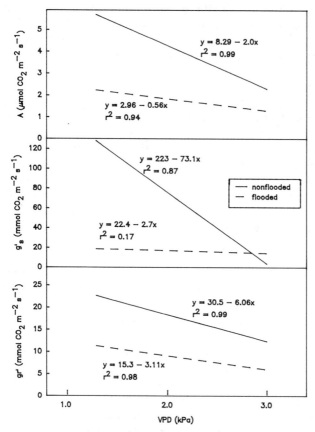

Figure 7.2. Effect of vapor-pressure deficit (VPD) on net CO_2 assimilation (A), stomatal conductance (g'_s) and residual conductance (g'_r) of nonflooded and flooded highbush blueberry plants (from Davies and Flore 1986b).

flooding for flooded than for nonflooded plants at each photosynthetic photon flux level (Davies and Flore 1986c).

Several theories have been put forth to explain the reduction in net CO_2 assimilation resulting from flooding. The most obvious explanation is the concomitant reduction in stomatal conductance that usually occurs when net CO_2 assimilation is reduced (Davies and Flore 1985, 1986a, 1986b, 1986c; Smith and Agar 1988; Joyner and Schaffer 1989; Larson et al. 1989; Ploetz and Schaffer 1989; Schaffer and Ploetz 1989). However, this theory assumes that leaf conductance decreases prior to a decrease in net CO_2 assimilation, a sequence that lacks experimental verification. Larson et al. (1991) measured net CO_2 assimilation, stomatal conductance, and internal partial pressure of CO_2 at hourly intervals for flooded mango trees. Both net CO_2 assimilation and stomatal conductance decreased simultaneously after three days. In addition, internal partial pressure of CO_2 for mango trees increased as net CO_2 assimilation and stomatal conductance decreased (Fig. 7.3). If reductions in net CO_2 assimilation were caused by reductions of stomatal conductance, then internal partial pressure of CO_2 would be expected to decrease as net CO_2 assimilation and stomatal conductance decreased, indicating a stomatal limitation to sufficient CO_2 entering the leaf for maximum assimilation (Farquhar and Sharkey 1982). A concurrent increase in internal partial pressure of CO_2 with decreasing net CO_2 assimilation and stomatal conductance also has been noted for flooded avocado (Schaffer and Ploetz, 1989). In contrast to these observations, Davies and Flore (1986a, 1986c) reported simultaneous reductions in net CO_2 assimilation, stomatal conductance, and internal partial pressure of CO_2 24 h after blueberry plants were flooded, thereby supporting their theory that stomatal conductance limits net CO_2 assimilation. However, they observed that a subsequent decrease in residual conductance increased internal partial pressure of CO_2 a few days after flooding was imposed.

Reductions in net CO_2 assimilation under flooded conditions also have been attributed to nonstomatal factors. Childers and White (1942) could not correlate the observed decrease in net CO_2 assimilation of flooded apple trees with any change in stomatal activity. Beckman et al. (1987) extracted a translocatable photosynthetic inhibitor from the xylem sap of flooded tart cherry (Prunus cerasus L. 'Montmorency' grafted onto 'Mahaleb' rootstock). Nonflooded plants given this exudate exhibited a significant decrease in net CO_2 assimilation within two hours, but stomatal conductance did not decrease, suggesting a nonstomatal limitation to net CO_2 assimilation under flooded conditions. Vu and Yelenoski (1991) observed a flood-induced decrease in chlorophll concentration and activity of rebulose bisphosphate carboxylase-oxygenase (Rubisco) of sweet orange grafted onto rough lemon or sour orange rootstocks. The decrease in Rubisco activity was concomitant with a decrease in net CO_2

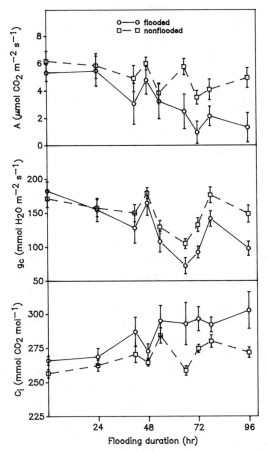

Figure 7.3. Net CO_2 assimilation (A), stomatal conductance of CO_2 (g_c), and internal partial pressure of CO_2 (Ci) for flooded and nonflooded mango trees (from Larson et al. 1991).

assimilation. Long-term flooding decreased carboxylation efficiency of highbush blueberries (Davies and Flore 1986a) and quantum-use efficiency of rabiteye blueberry plants (Davies and Flore 1986c), indicating an eventual direct response of the photosynthetic apparatus to flooding.

D. Plant Nutrition

Flooding of fruit crops generally results in reduced uptake of most mineral elements. The effects of flooding on nutrient absorption differ between flood-intolerant and flood-tolerant species. Flooding decreases the absorption of mineral nutrients in flood-intolerant species, whereas nutrient absorption in flood-tolerant plants actually may increase under flooded conditions (Kozlowski and Pallardy 1984). Increased nutrient uptake for waterlogged, flood-tolerant plants has been attributed to morphological adaptations such as the development of adventitious roots that proliferate in the upper, well-aerated portion of the soil (Hook et al. 1971; Sena Gomes and Kozlowski 1980a, 1988), and to the formation of aerenchymatous tissue in the roots, which increases O_2 movement from the shoot to roots and increases root metabolism (Armstrong 1968; Coutts and Armstrong 1976).

Macronutrient concentrations in several fruit crops decrease as a result of reduced O_2 to the roots. Reduced soil O_2 decreased total N, P, and K concentrations in citrus (Labanauskas et al. 1972), apple (Heinike et al. 1939; Olien 1989), avocado (Labanauskas et al. 1968; Slowick et al. 1979), peach (Kepka and Morris 1967), and highbush blueberry (Herath and Eaton 1967). Labanauskas et al. (1965) observed that whole-plant concentrations of N, P, K, Ca, and Mg in 'Bessie' sweet orange seedlings decreased when O_2 supply to the roots was decreased to less than 2%. Labanauskas et al. (1970) decreased O_2 supply to the root zone of potted sweet orange by displacing O_2 with nitrogen. They observed that whole-plant concentrations of N, P, K, Ca, and Mg decreased. Stolzy et al. (1975) observed that displacing O_2 from the root zone of containerized lemon [Citrus limon (L.) Burm. f.] and navel orange [Citrus sinensis (L.) Osbeck 'Atwood' navel], both grafted on sweet orange rootstock, resulted in reduced N, P, and K concentrations in the leaves and reduced K and Mg in the roots. Total plant concentrations of N, P, K, Ca, and Mg were lower for avocado plants grown with 2% soil O_2 than for plants grown with 21% O_2; however, leaf contents of N, P, Ca, and Mg were greater for plants grown at low O_2 (Slowick et al. 1979). Wazir et al. (1988) reported that 31 days of flooding reduced K, Ca, and Mg concentrations in the leaves, trunks, and roots of 'Dodd' pecan seedlings.

Micronutrient concentrations also are affected by flooding or reduced O_2 supply to the roots. Iron, Mn, and B concentrations in highbush blueberry (Herath and Eaton 1967), sweet orange (Labanauskas et al. 1970), and avocado (Labanauskas et al. 1968; Slowick et al. 1979) decreased when soil O_2 concentration decreased. Decreasing the soil O_2 supply to less than 2% decreased Zn, Mn, B, and Fe concentrations in 'Bessie' sweet orange seedlings (Labanauskas et al. 1966). Low soil O_2 (2%) decreased Zn, Mn, and Cu concentrations in leaves of avocado trees, whereas leaf Fe concentration increased. However, for avocado, Fe concentration (on a whole-plant basis) decreased when soil O_2 decreased (Slowick et al. 1979). Concentrations of Zn, Fe, and Mn in leaves, stems, and roots of pecan seedlings decreased as a result of soil waterlogging (Wazir et al. 1988).

Sodium content often increases due to flooding or low soil O_2 (Hopkins et al. 1949; Labanauskas et al. 1970). West and Taylor (1984) found that Na concentration in shoots of grapes (Vitis vinefera L.) increased seven days after waterlogging commenced. Chloride concentrations also often increase in fruit crops under low soil O_2 conditions. Low soil O_2 elevated Cl content of 'Bessie' sweet orange seedlings (Labanauskas et al. 1970), apple (West 1978), and grapes (West and Taylor 1984). Reduced soil O_2 concentration also increased concentrations of Na and Cl in the roots and stems of avocado, whereas Na and Cl concentrations in leaves were not affected (Slowick et al. 1979).

In addition to decreasing whole-plant macro- and micronutrient concentrations, reduced soil O_2 may alter the partitioning of nutrients within plant tissues. For example, Labanauskas et al. (1966) found that reduced soil O_2 resulted in decreased concentrations of K, Ca, Cl, and Cu in the roots and N, P, K, Cu, and B in shoots of 'Osbeck' citrus seelings. However, N, Na, Zn, Mn, B, and Fe in the roots and Mg, Na, Mn, and Fe in the shoots increased as result of low soil O_2. Wazir et al. (1988) found that flooding reduced N concentration in leaves and trunks of 'Dodd' pecan seedlings, whereas N concentration in the roots increased. Flooding also increased the P concentration in leaves of these plants but did not affect P concentration in roots.

Effects of flooding or low soil O_2 on plant nutrient uptake may fluctuate seasonally. Herath and Eaton (1968) observed that N, P, K, Ca, Mg, and Fe concentrations were lower in leaves of flooded than in leaves of nonflooded 'Bluecrop' blueberry plants during September and October. However, no differences were observed in leaf Ca, Mg, and Fe contents between treatments during August. In field studies with apple trees, Olien (1989) observed that N, P, K, and Cu contents in leaves were reduced by summer waterlogging. However, spring or fall waterlogging did not result in decreases of these elements. Olien (1989) also observed that spring and summer waterlogging of apple elevated leaf concentrations of Fe, whereas spring waterlogging increased Mn in the leaves.

Changes in plant nutrient concentrations under flooded or low soil oxygen conditions cannot arbitrarily be attributed to a dilution factor, because the dry weight of plants grown under these conditions usually is decreased (Labanauskas et al. 1970). Reduced nutrient uptake under hypoxic conditions could be attributed to several factors including root mortality and reductions in root respiration, water uptake, and hydraulic conductivity. In addition, changing nutrient availability in the soil under flooded conditions (see section on chemical and physical changes in soil) is involved with reduced mineral uptake under hypoxic conditions. Labanauskas et al. (1970) observed that hypoxia increased soil concentrations of N, P, Ca, and Mg. However, under low soil O_2 conditions, concentrations of these elements decreased within the plant. Therefore, decreased absorption of these elements under reduced soil O_2 was presumed to be due to decreased root metabolism and not decreased availability in the soil. The concomitant increase in mineral concentrations in roots with decreasing shoot mineral concentrations under flooded conditions has been attributed to the fact that nutrient translocation under flooded conditions was inhibited more than mineral absorption under low soil O_2 conditions (Shapiro et al. 1956; Kozlowski and Pallardy 1984).

Most research on flooding and mineral nutrition conducted to date indicates that flooding of fruit crops generally results in decreased plant concentrations of all macro- and microelements with the exception of Na and Cl which are often found in greater concentrations in flooded than in nonflooded plants. However, Larson et al. (1991) observed that flooding increased solubility of Fe and Mn in oolitic limestone soils. Mango trees growing in these soils under flooded conditions had greater leaf Fe and Mn concentrations than nonflooded trees (Larson et al., unpublished).

V. EFFECTS OF FLOODING ON SOILBORNE DISEASES

Numerous reviews over the last 20 years have been wholly or partially devoted to describing the influence of flooded soil on the development of plant disease (Cook and Papendick 1972; Schoenweiss 1975; Duniway 1979,1983; Drew and Lynch 1980; Stolzy and Sojka 1984; Waller 1986; Bruehl 1987). The scope of this section is somewhat narrower than that found in some of the reviews listed above, since it is limited to the effects of flooding on soilborne diseases.

Drew and Lynch (1980) indicated that infection and the development of plant disease may increase or decrease in waterlogged soils depending on: (1) the growth and activities of the pathogen under such conditions; (2) changes in host resistance to disease; and (3) activities of microorganisms antagonistic to the pathogen.

Flooding increases the severity of diseases caused by a diverse array of pathogens, but those caused by Oomycetes are probably the most frequently affected (Duniway 1979). The dependence of these fungi on high soil moisture to complete many stages of their life cycle and their need for free water to release zoospores and ensure the mobility of these propagules are prominent reasons for these associations.

Alva et al. (1985) indicated that it was important, in studies of disease-flood interactions, to distinguish the separate effects of each factor on the host. Their work demonstrated that flooding and phytophthora root rot of alfalfa, caused by *Phytophthora megasperma* Drechs. f. sp. *medicaginis* Kuan & Erwin, each independently influenced host productivity. Work on other phytophthora root rots has demonstrated the need to define conditions under which these pathogens cause disease (Sidebottom and Shew 1985; Wilcox and Mircetich 1985b; Bowers 1989). For example, *P. cryptogea* Pethb. & Laff. and *P. megasperma* each caused severe root rot (81–99% of the root system rotted) of 'Mahaleb' and 'Mazzard' cherry seedlings if plants were flooded for 48 h after inoculation (Wilcox and

Mircetich 1985b). However, disease severity was low (2–7%) when plants were not flooded, and contrary to host predisposition noted in other pathosystems (see disease-flood interaction section), flooding for up to 44 h prior to inoculation had no significant effect on disease severity.

Recent work has also demonstrated the importance of specific species and isolates of *Phytophthora* when considering the etiology of these diseases under flood conditions (Wilcox and Mircetich 1985a; Bielenin and Jones 1988; Brown and Mircetich 1988; Duncan and Kennedy 1989; Luz 1989; Wilcox 1989). Bielenin and Jones (1988) recovered *P. megasperma*, *P. cactorum* (Leb. & Cohn) Schröt., *P. cryptogea*, *P. syringae* (Kleb.) Kleb., and *P. cambivora* (Petri) Buism. from dying tart cherry trees in heavy, poorly drained soils in Michigan, but only isolates of *P. cryptoges* and *P. cambivora* were consistently virulent on seedlings of 'Mahaleb' cherry under flood conditions. Isolates of *P. megasperma* varied widely in their virulence to root systems of 'Mahaleb' cherry and isolates of *P. cactorum* and *P. syringae* did not cause significant root rot, but did cause stem cankers under flood conditions. Duncan and Kennedy (1989) reported that *P. megasperma* var. *megasperma*, *P. cactorum*, *P. syringae*, *P. cambivora*, and *P. erythroseptica* Pethb. caused significant root necrosis on the susceptible red rasberry 'Glen Moy' under flood and nonflood conditions, and that flooding increased disease severity. However, *P. megasperma* killed plants, whether or not they were flooded, and *P. drechsleri* Tucker caused significant disease only if plants were flooded for four days.

A. Effects on the Pathogen

A beneficial attribute of flooding is its ability to control certain soilborne pathogens (Strandberg 1987). Palti (1981) listed diseases caused by nematodes, parasitic plants, and a variety of fungi that have been controlled by flooding fallow fields. He indicated that reduced survival of pathogens under these situations was caused by O_2 deprivation. Anaerobic conditions in flooded soils also result in the generation of H_2S and possibly other compounds that reduce the survival of important nematode pests (Good 1987). Although the use of flooding is constrained by the availability of adequate water and the requirement for relatively flat land, it is a commonly used and effective method for controlling many diseases.

Detrimental effects of flooding on disease development are more frequently recognized. Flooding can directly influence the occurrence of disease through the movement of inoculum. Inoculum of *Fusarium oxysporum* Schlecht. f. sp. *cubense* (E. F. Smith) Snyd. & Hans., incitant of fusarial wilt of banana, is spread by flood water (Stover 1962). In addition, plant pathogens are often present in surface waters, such as streams and lakes, used for irrigation purposes. Shokes and McCarter (1979) recovered species of Rhizoctonia, Fusarium, Pythium, and Phytophthora from surface water in Georgia. The use of infested irrigation water can contribute to the spread of disease caused by pathogens capable of surviving these conditions (Strandberg 1987).

Most plant pathogens grow optimally under high moisture conditions, but few tolerate prolonged periods of flooding (Strandberg 1987). Those pathogens which benefit from short-term flooding may require water-saturated conditions for the production of infective propagules or for the optimal germination and infection by these propagules. They also may have a relatively greater tolerance of hypoxic or anoxic conditions than their hosts, or they may have a greater tolerance of these conditions than their microbial cohorts in soil. Water in flooded or very moist soil plays an extremely important role in the activities of soilborne pathogens, and a discussion of these relationships is relevant to the effects of flooded soil on soilborne disease.

1. Soil water potential. When considering the influence of soil water potential on plant pathogens, matric potential is the most important component (Cook and Papendick 1972). Therefore, the effect of soil water on activities of pathogens has often been assessed by determining the precise responses of pathogens to soil matric potentials (Duniway 1979).

a. *Oomycetes.* Saturated soils and soils near field capacity have profound influences on the formation, survival, and germination of different propagules of oomyceteous plant pathogens. Oomycetes form several different propagules in soil and and host tissue. Thick-walled, resting structures, chlamydospores (asexual) and oospores (sexual), play important roles in the survival of these fungi. Depending on the species involved, these propagules also can serve as primary inoculum sources for many of the diseases they cause (Duniway 1983). Sporangia are asexual structures that generally are more ephemeral than either chlamydospores or oospores, but which are important in the epidemiology of diseases incited by these fungi. Sporangia may germinate directly by producing germ tubes or indirectly by producing motile zoospores.

Oospores and chlamydospores. Flooding or high soil moisture generally increases the germination of oospores and chlamydospores of Oomycetes. Stanghellini and Burr (1973) reported that more oospores of Pythium aphanidermatum (Edson) Fitzp. germinated in moist soil (matric potential of −1 kPa) than in drier soils (−10, −100, and −150 kPa). Based on results in nutrient-amended and non-amended soil, they concluded that enhanced germination in moist soil resulted from an increase in available nutrients. Sterne et al. (1977) reported that chlamydospore germination and growth of *Phytophthora cinnamomi* was uniformly high at matric potentials of 0, 5, and −10 kPa, but declined at −25 kPa. They also indicated that increased availability of nutrients in moist soil played a key role in the germination of chlamydospores of this pathogen. Thus, increased availability of nutrients rather than soil water per se may be a reason that these pathogens often cause more disease in moist soil.

Most diseases caused by Oomycetes require multiple infections, which usually are incited by zoospores, for significant disease to develop (MacKenzie et al. 1983). Since oospores and chlamydospores actually may play minor roles in the epidemiology of these diseases after the host has been infected, much work on the effects of soil water potential on these pathogens has keyed on sporangia and zoospores.

Sporangia. The formation of sporangia by species of Phytophthora occurs over a wide range of matric potentials. Duniway (1983) suggested that many species of Phytophthora are capable of forming sporangia at matric potentials that encompass those found in agricultural soils throughout the growth cycle of a given crop.

Depending on the species involved and the substrate on which these pathogens are tested, flooding or saturation of soil (water potentials more negative than 0 or 0 Kpa, respectively) may reduce or increase the number of sporangia that are produced (Pfender et al. 1977; Bernhardt and Grogan 1982; Kuan and Erwin 1982). The optimal formation of sporangia for most species of Phytophthora, however, occurs between matric potentials of 0 to −25 kPa; these responses probably reflect the requirements of these fungi for high moisture conditions and adequate aeration in soil (Duniway 1975a, 1975b; Gisi et al. 1980).

Duniway (1979) concluded that sporangia of Phytophthora species shift from indirect to direct germination as water potential decreases. Although the precise water potentials at which this change takes place depend on the species involved and the relative contribution of the matric and osmotic components of the water potential, most species will germinate directly only in soils drier than about −10 kPa. Zoospore release is probably dependent on sporangium contact with liquid water (Duniway 1979, 1983). Sporangia of Phytophthora cryptogea released zoospores only at soil matric potentials greater than −5 kPa; release was maximal at 0 kPa (Duniway 1976). Wilcox and Mircetich (1985a) also indicated that once sporangia were flooded, zoospore release was a fairly rapid process. Sporangia of Phytophthora cryptogea began releasing zoospores within an hour of inundation, and those of P. cambivora and P. drechsleri were released within 3–6 h. Duniway (1983) observed that the high water potentials at which Oomycetes release zoospores coincide with the water potentials at which diseases caused by these fungi are most severe. This relationship, at least partially, accounts for the prevalence of diseases caused by these fungi in saturated or flooded soils.

Under saturated soil conditions, the indirect germination of sporangia of species of Phytophthora is influenced by temperature and usually is favored over direct germination within a wide range of temperatures.

Duniway (1983) summarized reports on temperatures at which indirect germination of sporangia of the following species predominated over direct germination under flooded conditions: *P. cactorum*, 15°–27°C; *P. capsici* Leonian, 10°–24°C; *P. cambivora*, 9°–27°C; *P. drechsleri*, 9°–30°C; *P. cryptogea*, 8°–30°C; and *P. megasperma*, 12°–28°C or 8°–21°C, depending on the isolate.

Once zoospores are released, their active disperal depends on free water and is influenced by soil texture. Duniway (1976) demonstrated that zoospores of *Phytophthora cryptogea* could swim horizontally 25–35 mm in a coarsely textured soil at matric potentials more negative than −0.1 kPa, but that their active movement was limited to about 5 mm in a loam soil of finer texture held at the same matric potentials. He suggested that soil pore size had a direct impact on zoospore movement, and that larger, less tortuous pores would better accommodate the active movement of these propagules. It was also determined that water-filled channels greater than 60 μm in diameter are required for prolonged zoospore movement (Duniway 1976, 1979).

Zoospores, either motile or nonmotile (encysted), also are capable of passive movement in water. Depending on the distance the water containing the zoospores moves, it is probable that zoospores could be moved substantially further via passive means than would occur through active means. The passive movement of zoospores (and other propagules) in flood water probably has more of an effect on the long-distance spread of these pathogens, whereas their short-distance movement to infection sites on the host probably are more frequently accomplished via motile zoospores.

2. Pathogens other than Oomycetes. Cook and Papendick (1972) reported that the prevalence of most root diseases in moist soil is due, in part, to the detrimental effect of drier soil on pathogen growth. If O_2 is not limiting, growth of most fungal and bacterial plant pathogens usually is optimal near matric potentials of 0 kPa (Cook and Baker 1983). Although the influence of soil saturation or flooding on Oomycetes is well known, little work has been conducted with other pathogens under these conditions.

Flooding influences the interaction of plant pathogens with antagonistic microbes (Drew and Lynch 1980). After 5 days incubation, mortality of *Criconemella xenoplax* (Raski) Luc & Raski, a nematode pest of peach, increased from 0 to 2, 10, and 37% at soil osmotic potentials of 0, −600, −1200, and −1800 kPa, respectively (Jaffee and Zehr 1983). In the presence of the fungal nematode parasite *Hirsutella rhossilensis* Minter & Brady, nematode mortality of 100% was observed at all but the 0 kPa treatment, but no mortality was detected at 0 kPa. The results indicate that saturated soil may interact adversely with peach by promoting the survival of an important peach pest.

3. Soil aeration. Most pathogens that cause greater disease in flooded than in nonflooded soil are less sensitive to low O_2 than their hosts. For example, Covey (1970) determined that mycelial growth of Phytophthora cactorum, an apple pathogen that causes its greatest damage in wet soil, was still about 72 and 37% of the maximum rate of O_2 concentrations of 5 and 0.2%, respectively. In contrast, Boynton (1940) reported that apple roots died at O_2 concentrations less than 3%, that growth of existing roots required O_2 levels of at least 5–10%, and that new roots were initiated only at O_2 concentrations greater than or equal to 12%.

Most plant pathogens can be adversely affected by reduced levels of O_2 and elevated CO_2 which develop in flooded soil. Mitchell and Zentmyer (1971) investigated effects of various concentrations of O_2 and CO_2 on the formation of oospores and sporangia of six different species of Phytophthora. The effects were complex, but oospore production generally was greater at low O_2 concentrations (1 or 5%) versus ambient (21%) in the absence of CO_2. However, oospore production generally decreased as CO_2 concentrations increased to 30%, regardless of O_2 concentration. Production of sporangia decreased either when O_2 was reduced or CO_2 increased. Klotz et al. (1962, 1963) studied the effect of O_2 concentration on germination of zoospores of Phytophthora parasitica Dastur and P. citrophthora (R. E. S. & E. H. Sm.) Leonian and endoconidia of Thielaviopsis basicola (Berk. & Broome) Ferraris, each which cause diseases that are favored by wet soil. Their work demonstrated that zoospores germinated sooner (0.5 vs. 4 h) and at much lower O_2 concentrations (0.1 vs. 6%) than endoconidia. Davison and Tay (1986) indicated that low soil O_2 did not affect the germination percentage of Phytophthora cinnamomi zoospores, but their subsequent growth was adversely affected under such conditions.

In general, plant parasitic nematodes appear to be very sensitive to O_2 levels in soil (See Stolzy and Sojka 1984 for a summary of work on these pathogens). For example, reproductive rates of Hemicycliophora arenaria Raski, the sheath nematode, increased lineary when O_2 diffusion rates increased from 10 to 40 $\mu g\ cm^{-2}\ min^{-1}$, and rates of less than 20 $\mu g\ cm^{-2}\ min^{-1}$ resulted in a decline in nematode numbers after one month (Van Gundy and Stolzy 1963). Under continuous exposure to 5% O_2, reproduction of this nematode was reduced; reproduction was completely inhibited at 4% O_2 (Cooper et al. 1970).

Although most phytopathogenic organisms have an obligate requirement for O_2, some are facultative anaerobes (e.g., those in the genus Erwinia), whereas others are anaerobes (e.g., those in the genus Clostridium). Bruehl (1987) indicated that the failure of early workers to use anaerobic conditions when studying Clostridium species led them to overlook the importance of these bacteria as plant pathogens.

B. Host Resistance

Host predisposition, or the tendency of nongenetic factors to reduce host resistance prior to infection, has been recognized for more than 100 years as an important factor in the development of certain plant diseases (Schoeneweiss 1975). Although flooding is assumed to predispose hosts to many different diseases, experimental evidence in this regard is sparse. Hosts have been inoculated during or after flooding in most studies of flood-disease interactions. Thus, it is difficult to distinguish effect(s) of host predisposition by flooding from the enhanced activities of pathogens that may occur in flooded soil or from other chemical or biological factors in soil that are altered by flooding.

Hosts are not always predisposed to greater levels of disease by flooding. Wilcox and Mircetich (1985b) reported that extreme root rot of 'Mahaleb' cherry was caused by *Phytophthora cryptogea* if plants were flooded for 48 h after inoculation, but not if they were flooded for about the same length of time before inoculation. Davison and Tay (1987) reported a significant increase in the severity of root rot of jarrah (*Eucalyptus mariginata* Sm.), caused by *Phytophthora cinnamomi*, when plants were flooded for four days during or after inoculation, but not when plants were flooded before inoculation.

C. Disease-Flood Interactions

1. Zoospore attraction and infection of the host. Zoospores of plant pathogenic species of *Pythium* and *Phytophthora* are attracted to host root exudates (Zentmyer 1980; Stolzy and Sojka 1984). The ability of these pathogens to follow gradients of exudates originating at infection sites (e.g., root wounds or the zone of elongation) allows them to find and infect host roots in soil. The establishment of gradients of exudates in saturated soil and the complete dependence of zoospore motility on free water each can influence the development of diseases caused by pythiaceous fungi in flooded soil.

Few studies have quantified the influence of flooding on host infection by pathogens. Results from studies on diseases caused by species of *Phytophthora* may be difficult to interpret if, as often is the case, inoculation is achieved with motile zoospores added to hosts in standing water. In these situations, the host is flooded in both the "nonflooded" and "flooded" treatments. In studies using various inoculum densities of sorghum seed colonized by *Phytophthora cinnamomi* to infest soil, infection of avocado roots was greatest at each inoculum density tested when soil was flooded (Ploetz and Schaffer, unpublished).

2. Disease development.

a. Host resistance. Host resistance to disease may or may not be overcome in flooded soil, and effects of flooding on host resistance varies depending on host species or cultivar as well as the pathogen involved. Duncan and Kennedy (1989) reported that 'Glen Moy', a susceptible cultivar of red rasberry, was severely affected by Phytophthora megasperma var. megasperma, P. cactorum, P. syringae, P. cambivora, P. drechsleri, P. megasperma, and P. erythroseptica in flooded soil, but that the resistant 'Tayberry', was unaffected by P. cambivora, P. drechsleri, and P. megasperma under these conditions.

b. Edaphic factors. Biesbrock and Hendrix (1970) suggested that increased root necrosis of peach in saturated soil, caused by Pythium vexans de Bary, was due to enhanced production of zoospores by the pathogen under such conditions. In contrast, root necrosis caused by Pythium irregulare Buisman was not increased under saturated conditions because sporangia of this species do not form zoospores and only germinate directly, a process that is not as sensitive to soil moisture as the formation and release of zoospores.

Sterne et al. (1977b) reported nine-fold increases in root rot severity of Persea indica L., caused by Phytophthora cinnamomi, when soil matric potentials in a sandy loam soil were increased from −25 kPa to 0 kPa. However, less than two-fold increases in disease severity were observed when the same matric potentials were tested in a clay soil. Since disease severity could be increased in the sandy loam soil at −25 kPa by amending the soil with glucose and asparagine, these workers concluded that nutrient availability, rather than matric potential, influenced disease severity in the drier soil. In a related study they determined that soil matric potential, rather than osmotic potential, had a greater influence on disease severity in this pathosystem. In addition, it was concluded that matric potentials that favored disease probably reflected soil water conditions favorable for germination of chlamydospores and infection by zoospores (Sterne et al. 1977a).

The duration, timing, and frequency of flooding episodes can affect the development of diseases that are influenced by flooding. Stolzy et al. (1965) indicated that the length of time that the soil was saturated was more important than the frequency of saturation in determining the severity of root-rot of sweet orange caused by Phytophthora citrophthora. Little or no root decay was caused by this pathogen unless soil was flooded.

Matheron and Mircetich (1985) reported that flooding duration influenced root-rot severity in the walnut species Juglans hindsii and the hybrid J. hindsii × regia when disease was caused by Phytophthora cryptogea and P. citrophthora in the former and P. citricola Sawada in the latter host. Mean root necrosis on these walnut rootstocks increased from 13% to 43% under nonflooded conditions to as high as 100% with biweekly flooding events of 48 h. Disease severity was high (82–100%) under both nonflooded and flooded conditions when disease was caused by either P. citricola on J. hindsi or P. cinnamomi on both rootstocks. They concluded that careful irrigation practices could minimize disease in walnut orchards infested with either *P. cryptogea* or *P. citrophthora*, but that water management alone would not help control root rot caused by *P. cinnamomi*.

The O_2 status of soil not only influences the host and pathogen, but also affects disease development. Stolzy et al. (1965) indicated that a low supply of O_2 prevented growth and regeneration of sweet orange roots. They suggested that it was unlikely that this host could recover from infection by *Phytophthora citrophthora* under flooded or saturated soil conditions. However, reduced O_2 can result in root damage, even in the absence of disease. For example, when O_2 diffusion rates were either high or low (about 33 and 10 μg cm^{-2} min^{-1}, respectively) root necrosis of avocado was affected little by inoculation with *Phytophthora cinnamomi*, even though reduced diffusion rates increased the severity of root necrosis (Stolzy et al. 1967).

Low O_2 in soil also may affect the development of diseases caused by nonpythiaceous pathogens. Miller and Burke (1975) reported that the severity of bean root rot, caused by *Fusarium solani* (Mart.) Sacc. F. sp. *phaseoli* (Burkholder) Snyd. & Hans., increased as the length of time roots were deprived of O_2 increased. Periods of anoxia as short as one day increased disease severity, but in the absence of the pathogen host roots were not injured by hypoxic conditions.

D. Host Responses

1. Water stress, stomatal responses, and photosyntheses. Ayres (1981) and Hall (1986) recently reviewed the effects of root-infecting pathogens on host water status, and their chapters should be consulted for an in depth treatment of the subject. Water stress induced by root-infecting pathogens may or may not be conspicuous. For example, diseases caused by species of *Verticillium* and *Fusarium oxysporum* may cause hosts to wilt rapidly, whereas woody plants infected by *Phytophthora cinnamomi* may develop water stress slowly and without comspicuous wilt symptoms (Hall 1986).

Results from work on pathogen-induced water stress indicate that the type and sequence of host responses to such stress are similar among diverse pathosystems (Hall 1986). Unfortunately, these studies usually have been conducted on diseased plants in the absence of flooding; very few studies have monitored host responses to disease under flooded conditions (Davidson and Tay 1987; Ploetz and Schaffer 1987, 1989; Schaffer and Ploetz 1989).

Hall (1986) composed the following sequence of events for soilborne diseases that affect host water status in the absence of flooding: (1) infection and subsequent death of root tissue; (2) increased resistance to water flow, leading to reductions in xylem water potential and relative water content of the plant, then to reduced stomatal conductance, stomatal aperture and transpiration; and (3) reduced growth and productivity of the host if these changes persist.

a. *Host water status*. In Hall's (1986) scenario, in the absence of flooding xylem water potential is reduced (becomes more negative) before stomatal conductance and transpiration are reduced. In contrast, under flooded conditions in the absence of disease, xylem water potential usually is reduced after conductance and transpiration have declined. Thus, the sequence by which these parameters decline apparently differs in diseased and flooded plants.

Reduced water content in hosts with root disease, noted either as lower xylem water potential or relative water content, may result from reduced capacity of the host to absorb (Sterne et al. 1977a, 1977b) or transport (Duniway 1975b) water. The former effect may simply be the result of reduced root surface in diseased plants. Although water transport deficiencies may result from vascular obstruction (Hall 1986), this may not always be the case (Davison and Tay 1987).

To our knowledge, no published accounts exist that describe the influence of flooding on xylem water potential of diseased plants. In one unpublished experiment, the leaf water potential of avocado plants with phytophthora root rot was significantly reduced four days after flooding compared to nonflooded healthy or nonflooded root-rotted plants. The water potential of nonflooded plants with root rot or healthy nonflooded and flooded plants did not differ (Fig. 7.4; Ploetz and Schaffer, unpublished).

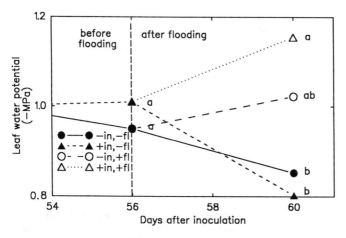

Figure 7.4.   Leaf water potential of avocado ('Simmonds' scions on 'Waldin' rootstocks) 56 days after planting in soil infested (+in) or noninfested (−in) with *Phytophthora cinnamomi* and either flooded (+fl) or nonflooded (−fl). Water potentials were determined immediately prior to flooding and four days after flooding (from Ploetz and Schaffer, unpublished). Different letters indicate significant difference (P ≤ 0.05).

b. *Stomatal and photosynthetic responses.* Host transpiration usually is reduced by diseases that affect host water status. Sterne et al. (1977a, 1977b) reported that root disease of the avocado cultivar Bacon, caused by *Phytophthora cinnamomi*, resulted in a six-fold reduction of transpiration in the field. In greenhouse studies, Ploetz and Schaffer (1987) investigated the effects of this disease, flooding, and the disease-flood interaction on transpiration and stomatal conductance of 'Simmonds' avocado scions grafted on rootstock of either 'Lula' or 'Waldin'. Although flooding and disease each reduced these responses by about 50%, transpiration and stomatal conductance rates were near zero in diseased plants that were flooded, indicating an interaction between flooding and disease. Davison and Tay (1987) indicated that leaf conductance was affected only when jarrah seedlings infected by *Phytophthora cinnamomi* were flooded; conductance was not affected when infected plants were not flooded. Thus, limited research indicates that disease and flooding severely impact stomatal conductance and transpiration.

Net CO_2 assimilation of avocado with phytophthora root rot was dramatically reduced under flood conditions (Ploetz and Schaffer 1987, 1989; Schaffer and Ploetz 1989). Photosynthesis of noninfected avocado plants was not reduced after seven days of flooding. Under nonflood conditions, root-rotted plants with high levels of root necrosis (greater than 50%) exhibited a 65% decrease in net CO_2 assimilation. Under flood conditions, however, as little as 20% root necrosis resulted in almost complete inhibition of net CO_2 assimilation (Fig. 7.5). Stomatal conductance and net CO_2 assimilation of avocado were positively correlated under

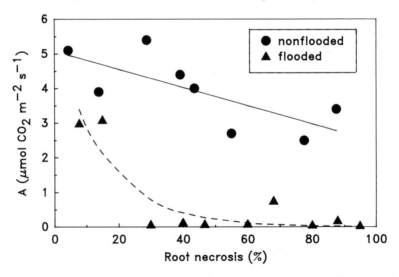

Figure 7.5. Net CO_2 assimilation (A) and percent root necrosis for flooded and nonflooded avocado plants. The regression line for flooded plants is $y = 5.6e^{-65x}$, $r^2 = 0.83$ and for nonflooded plants is $y = 5.1 - 0.26x$, $r^2 = 0.53$ (from Schaffer and Ploetz 1989).

both flood and nonflood conditions (Schaffer and Ploetz 1989). Since intercellular concentrations of CO_2 increased under those conditions, it was hypothesized that reduced photosynthetic capacity in these plants, rather than reduced stomatal conductance, may have caused a reduction in net CO_2 assimilation. Regardless of which parameter is reduced first, it is clear that reduced photosynthesis and stomatal conductance are two of the earliest indicators of host stress. Under these conditions, each variable was reduced about three days before the appearance of more conspicuous symptoms, such as wilting and defoliation.

2. Host growth and biomass accumulation. Hall (1986) suggested that host growth and biomass accumulation are among the last of host attributes reduced by disease. When diseased avocado plants are flooded, reduced growth remains one of the last symptoms of host stress to become apparent (Ploetz and Schaffer 1989). Root and shoot biomass of diseased plants may be reduced further by flooding (Alva et al. 1985; Bielenin and Jones 1988; Duncan and Kennedy 1989; Ploetz and Schaffer 1989).

Flooding exacerbated reductions in root and shoot biomass of red raspberry infected with *Phytophthora drechsleri* and *P. cambivora*; reductions caused by *P. megasperma*, which was highly virulent in flooded and nonflooded soil, were not increased by flooding. When seedlings of 'Mahaleb' cherry, artifically inoculated with isolates of *P. megasperma*, *P. cryptogea*, and *P. syringae*, were flooded at 2-week intervals, root dry weights were usually reduced relative to the noninoculated control (Bielenin and Jones 1988).

VI. CONCLUSIONS

The majority of work over the past 20 years on the responses of fruit crops to flooding has focused on plant susceptibility, growth, water relations, and photosynthesis. Although research has been conducted on biochemical and metabolic responses of herbaceous plants to flooding, this area has been virtually overlooked for fruit crops.

Although it is difficult to give a precise time course that applies to all fruit crops for the series of events which occur during flooding, responses of these crops can be loosely divided into early, mid, and late responses. Crane and Davies (1989) defined these three periods for rabbiteye blueberry as Phases 1, 2, and 3 (Table 7.5). The exact time required for each phase varies among species and cultivars and depends on environmental factors such as air and soil temperature and relative humidity. However, the sequence of events resulting from waterlogging generally is similar among fruit crops. The earliest responses (Phase 1) are presumably biochemical. In addition, other early responses include reductions in net CO_2 assimilation, stomatal conductance, transpiration,

Table 7.5. Time course of physiological responses of rabbiteye blueberry plants to flood duration.[z]

Time (days after onset of flooding)	Plant response[y,x]
	Phase 1
2	g_s ($\downarrow$); g_s' ($\downarrow$); A ($\downarrow$); Φ ($\downarrow$)
3	RLp ($\downarrow$); ψ_w (no change)
5	Flower bud number ($\downarrow$) after summer flooding
	Phase 2
6	Leaf expansion and shoot elongation cease
8	RLp ($\downarrow\downarrow$)
9	SLp ($\downarrow$); EL ($\uparrow$); g_r' ($\downarrow$); Φ ($\downarrow$)
10	A ($\downarrow$) and may become zero or negative; g_s less responsive to environment
11	New shoots and/or leaves may wilt
13	Leaves senesce and abscise
14	g_s ($\downarrow\downarrow$); A ($\downarrow\downarrow$); ψ_w (no change)
	Phase 3
16	RLp ($\downarrow\downarrow$); SLp ($\downarrow\downarrow$); EL ($\uparrow\uparrow$)
17	Fruit shriveling may begin
20–45	Continued decline in plant vigor and a reduction in yields
46–120	ψ_w ($\downarrow\downarrow$); eventual plant death

[z]Adapted from Crane and Davies (1989)

[y]Phytophthora root rot, caused by *Phytophthora cinnamomi*, may rapidly accelerate symptoms of flood stress.

[x]Symbols used to describe plant responses of flooding are: g_s = stomatal conductance to water; g_s' = stomatal conductance to carbon dioxide; A = carbon assimilation; Φ = quantum yield; RLp = root hydraulic conductivity; ψ_w = leaf xylem water potential; SLp = stem hydraulic conductivity; EL = electrolyte leakage; g_r' = residual conductance to carbon dioxide; $\downarrow$, decrease; $\downarrow\downarrow$, greater decrease; $\uparrow$, increase; and $\uparrow\uparrow$, greater increase.

root hydraulic conductivity, and shoot and root growth.

During Phase 2 or the mid stage, these variables continue to decline and net CO_2 assimilation and stomatal conductance may shut down completely. Additionally, root electrolyte leakage continues to increase, and visible leaf chlorosis, epinasty, senescense, and abscission occur. At this point, leaves may wilt.

During Phase 3 or the late stage, root and stem hydraulic conductivities continue to decrease, plant water potential decreases (becomes more negative), shoot and root growth decline, and fruit shriveling and a reduction in yield may occur; ultimately, mortality occurs.

Although the sequence of flood responses is similar for all fruit crops, the specific time required to elicit each response varies for each species based on inherent flood tolerance. For example, avocado, a flood-sensitive species, exhibited a similar sequence of responses to flooding as rabbiteye blueberry, a relatively flood-tolerant species. However, the time required to elicit these responses was accelerated for avocado (even in the absence of root rot) compared to rabbiteye blueberry (Ploetz and Schaffer 1987, 1988, 1989; Schaffer and Ploetz 1987, 1989).

It may be difficult to visually distinguish a plant's response to flooding from its response to soilborne disease, especially in field situations. Some researchers have indicated that flood-associated declines previously attributed to flooding may have, in fact, been due to root disease exacerbated by flooding (Crane and Davies 1989). Although the importance of these associations is not always understood or appreciated, such relationships may be fairly common, and in those situations that have been investigated, they often are very destructive. The continued study of flood-disease interactions is clearly warranted.

LITERATURE CITED

Abbott, J. D., and R. E. Gough. 1985. Flooding highbush blueberry plants. HortScience 20:88–89 (Abstr.).

_____. 1987a. Prolonged flooding effects on anatomy of highbush blueberry. HortScience 22:622–625.

_____. 1987b. Growth and survival of the highbush blueberry in response to root zone flooding. J. Am. Soc. Hort. Sci. 112:603–608.

_____. 1987c. Reproductive response of the highbush blueberry to root-zone flooding. HortScience 22:40–42.

Adams, D. O., and S. F. Yang. 1977. Methionine metabolism in apple tissue. Implication of S-adenosylmethionine as an intermediate in the conversion of methionine to ethylene. Plant Physiol. 60:892–896.

_____. 1979. Ethylene biosynthesis: Identification of l-aminocyclopropane-l-carboxylic acid as an intermediate in the conversion of methionine to ethylene. Proc. Nat. Acad. Sci. (USA) 76:170–174.

Alben, A. O. 1958. Waterlogging of subsoil associated with scorching and defoliation of Stuart pecan trees. Proc. Am. Soc. Hort. Soc. 72:219–223.

Alva, A. K., L. E. Lanyon, and K. T Leath. 1985. Influence of fungal-soil water interactions of phytophthora root rot of alfalfa. Biol. Fert. Soils 1:91–96.

Andersen, P. .C. 1983. Effects of flooding on peach, apple, quince and several pear species. Ph.D. Thesis, Oregon State University, Corvallis, OR.

Andersen, P. C., P. B. Lombard, and M. N. Westwood. 1984a. Effect of root anaerobiosis on the water relations of several Pyrus species. Physiol. Plant. 62:245–252.

_____. 1984b. Leaf conductance, growth, and survival of willow and deciduous fruit tree species under flooded soil conditions. J. Am. Soc. Hort. Sci. 109:132–138.

Andersen, P. C., J. M. Montano, and P. B. Lombard. 1985. Root anaerobiosis, root respiration, and leaf conductance of peach, willow, quince, and several pear species. HortScience 20:248–250.

Anonymous. 1961. Pear rootstocks in California. Cal. Agr. Expt. Sta. Ext. Ser. Leafl. 136.

Armstrong, W. 1968. Oxygen diffusion from the roots of woody species. Physiol. Plant. 21:579–543.

Ayres, P. G. 1981. Effect of disease on plant water relations. p. 131–148. In: Ayres, P. G. (ed.), Effects of Disease on the Physiology of the Growing Plant. S. E. B. Seminar Series Vol. 11. Cambridge Univ. Press.

Barta, A. L. 1984. Ethanol synthesis and loss from flooded roots of Medicago sativa L. and Lotus corniculatus L. Plant Cell. Environ. 7:187–191.

_____. 1988. Response of field grown alfalfa to root waterlogging and shoot removal. I. Plant injury and root content of carbohydrates and minerals. Agron. J. 80:889–892.

Beckman, T. G., J. A. Flore, and R. L. Perry. 1987. Sensitivity of various growth indices and the production of a translocatable photosynthesis inhibitor by one year-old cherry trees during flooding. HortScience 22:1141 (Abstr.).

Beletskya, E. K. 1977. Changes in metabolism of winter crops during their adaptation to flooding. Soviet Plant Physiol. 23:750–756.

Benjamin, L. R., and H. Greenway. 1979. Effects of a range of O_2 concentrations on porosity of barley roots and on their sugar and protein concentrations. Ann. Bot. 43:383–391.

Bergman, H. F. 1943. The relationship of ice and snow cover on winter-flooded cranberry bogs to vine injury from oxygen deficiency. p. 1–24. In: H. J. Franklin, H. F. Bergman, and N. E. Stevens (eds.), Weather in cranberry culture. Mass. Agr. Expt. Sta. Bul.

Bernhard, R. 1970. Patrones de melocotonero. Inf. Tech. Econ. Agr. 1:89–104

Bernhardt, E. A., and R. G. Grogan. 1982. Effect of soil matric potential on the formation and indirect germination of sporangia of Phytophthora parasitica, P. capsici, and P. cryptogea. Phytopathology 72:507–511.

Bertani, A., and I. Brambilla. 1982. Effect of decreasing oxygen concentration on some aspects of protein and amino-acid metabolism in rice roots. Z. Pflanzenphysiol. 107:193–200.

Bielenin, A., and A. L. Jones. 1988. Prevalence and pathogenicity of Phytophthora spp. from sour cherry trees in Michigan. Plant Dis. 72:473–476.

Biesbrock, J. A., and F. F. Hendricks. 1970. Influence of soil water and temperature on root necrosis of peach caused by Pythium spp. Phytopathology 60:880–882.

Bini, G. 1963. Studies on the resistance of peach and pear trees to pear asphyxia. Riv. Ortoflorofruttic. Ita. 47:22–36.

Bloomfield, C. 1951. Experiments on the mechanism of gley formation. J. Soil. Sci. 2:196–211.

Bowers, J. H. 1989. The effects of inoculum density and soil-water matric potential on infection of pepper by Phytophthora capsici and the epidemiology of Phytophthora blight in the field. Ph.D. Thesis. Univ. Florida, Gainesville, FL.

Boynton, D. 1940. Soil atmosphere and the production of new rootlets by apple tree root systems. Proc. Am. Soc. Hort. Sci. 37:19–26.

Bradford, K. J., and T. C. Hsiao. 1982. Stomatal behavior and water relations of waterlogged tomato plants. Plant Physiol. 70:1508–1513.

Bradford, K. J., and S. F. Yang. 1981. Physiological responses of plants to waterlogging. HortScience 16:25–30.

Brown, G. T., and S. M. Mircetich. 1988. Effects of flood duration on the development of Phytophthora root and crown rots of apple. Phytopathology 78:846–851.

Bruehl, G. W. 1987. Soilborne Plant Pathogens. MacMillan, New York.

Burrows, W. J., and D. J. Carr. 1969. The effects of flooding the root system of sunflower plants on the cytokinin content in the xylem sap. Physiol. Plant. 22:1105–1112.

Buwalda, F., C. J. Thompson, W. Steigener, E. G. Barret-Lennard, J. Gibbs, and H. Greenway. 1988. Hypoxia induces membrane depolarization and potassium loss from wheat roots but does not increase their permeability to sorbitol. J. Expt. Bot. 39:1169–1183.

Carpenter, J. R., and C. A. Mitchell. 1980a. Root respiration characteristics of flood-tolerant and intolerant tree species. J. Am. Soc. Hort. Sci. 105:684–687.

_____. 1980b. Flood-induced shift of electron flow between cyanide-sensitive and alternative respiratory pathways in roots of tolerant and intolerant tree species. J. Am. Soc. Hort. Sci. 105:688–690.

Catlin, P. B., G. C. Martin, and E. A. Olsson. 1977. Differential sensitivity of Juglans hindsii, J. regia, Paradox hybrid, and Pterocarya stenoptera to waterlogging. J. Am. Soc.

Hort. Sci. 102:101–104.

Catlin, P. B., and E. A. Olsson. 1986. Response of eastern black walnut and northern California black walnut seedlings to waterlogging. HortScience 21:1379–1380.

Chaplin, E. E., G. W. Schneider, and D. C. Martin. 1974. Rootstock effect on peach tree survival on a poorly drained soil. HortScience 9:28–29.

Childers, N. F., and D. G. White. 1942. Influence of submersion of the roots on transpiration, apparent photosynthesis, and respiration of young apple trees. Plant Physiol. 17:603–618.

_____. 1950. Some physiological effects of excess soil moisture on Stayman Winesap apple trees. Ohio Agr. Expt. Sta. Res. Bul. 694.

Childs, W. H. 1941. Photosynthesis, transpiration, and growth of apple trees as influenced by various concentrations of oxygen and carbon dioxide in the soil atmosphere. Proc. Am. Soc. Hort. Sci. 38:179–180.

Chirkova, T. V. 1975. Metobolism of ethanol and lactate in tissues of woody plants differing with respect to their resistance to oxygen deficiency. Soviet Plant Physiol. 22:834–838.

_____. 1978. Some regulatory mechanisms of plant adaptation to temporal anaerobiosis. p. 137–154. In: D. D. Hook and R. M. M. Crawford (eds.), Plant Life in Anaerobic Environments. Ann Arbor Sci. Press, Ann Arbor, MI.

Chirkova, T. V., and T. S. Gutman. 1972. Physiological role of branch lenticels in willow and poplar under conditions of root anaerobiosis. Soviet Plant Physiol. 19:289–295.

Cho, D. Y., and F. N. Ponnamperuma. 1971. Influence of soil temperature on the chemical kinetics of flooded soils and the growth of rice. Soil. Sci. 112:184–194.

Cook, R. J., and K. F. Baker. 1983. The Nature and Practice of Biological Control of Plant Pathogens. Am. Phytopathol. Soc., St. Paul, MN.

Cook, R. J. and R. I. Papendick. 1972. Influence of water potential of soils and plants on roots disease. Annu. Rev. Phytopathol. 10:349–374.

Cooper, A. F., S. D. Van Gundy, and L. H. Stolzy. 1970. Nematode reproduction in environments in fluctuating aeration. J. Nematol. 2:305–315.

Correll, J. C., and R. W. Schneider. 1982. The effect of frequency of irrigation on severity of Fusarium yellows of celery. Phytopathology 72:948 (Abstr.).

Cossins, E. A., and H. Beevers. 1963. Ethanol metabolism in plant tissues. Plant Physiol. 38:375–380.

Coutts, M. P. 1982. The tolerance of tree roots to waterlogging. V. Growth of woody roots of sitka spruce and lodgepole pine in waterlogged soil. New Phytol. 90:467–476.

Coutts, M. P. and W. Armstrong. 1976. Role of oxygen transport in the tolerance of trees to waterlogging. p. 361–385. In: M. G. R. Cannel and F. T. Last (eds.), Tree Physiology and Yield Improvement, Academic Press, New York.

Covey, R. P., Jr. 1970. Effect of oxygen tension on the growth of Phytophthora cactorum. Phytopathology 60:359–359.

Crane, J. H. 1987. Soil temperature and flooding effects on young rabbiteye blueberry survival, growth, and ethylene levels. Proc. Fla. State Hort. Soc. 100:301–305.

Crane, J. H., and F. S. Davies. 1985. Responses of rabbiteye blueberries to flooding. Proc. Fla. State Hort. Soc. 98:153–155.

_____. 1987. Hydraulic conductivity, root electrolyte leakage, and stomatal conductance of flooded and unflooded rabbiteye blueberry plants. HortScience 22:1249–1252.

_____. 1988a. Flooding duration and seasonal effects on growth and development of young rabbiteye blueberry plants. J. Am. Soc. Hort. Sci. 113:180–184.

_____. 1988b. Periodic and seasonal flooding effects on survival, growth, and stomatal conductance of young rabbiteye blueberry plants. J. Am. Soc. Hort. Sci. 113:488–493.

_____. 1989. Flooding responses of Vaccinium species. HortScience 24:203–210.

Cräwford, R. M. M. 1966. The control of anaerobic respiration as a determining factor in the distribution of the genus Senecio. J. Ecol. 54:403–413.

_____. 1967. Alcohol dehydrogenase activity in relation to flood tolerance in roots. J. Expt.

Bot. 18:458–464.

⸻ . 1969. The physiological basis of flooding tolerance. Ber. Deutsch Bot. Ges. 82:111–114.

⸻ . 1972. Physiologische okologie: Ein vergleich der anpassung von pflanzen und tieren an sauerstoffarme umgebung. Flora 161:209–223.

⸻ . 1975. Metabolic adaptations to anoxia in plants and animals. Proc. 12th Internatl. Bot. Congress, Leningrad. [cited in: p. 387–440, M. G. R. Cannel and F. T. Last (eds.), Tree Physiology and Yield Improvement. Academic Press, New York].

⸻ . 1976. Tolerance of anoxia and the regulation of glycolysis in tree roots. p. 387–401. In: M .G .R. Cannell and F. T. Last (eds.), Tree Physiology and Yield Improvement. Academic Press, New York.

Crawford, R. M. M., and M. A. Baines. 1977. Tolerance of anoxia and the metabolism of ethanol in tree roots. New Phytol. 79:519–526.

Crawford, R. M. M., and M. McManmon. 1968. Inductive responses of alcohol and malic dehydrogenase in relation to flooding tolerance in roots. J. Expt. Bot. 19:435–441.

Crawford, R. M. M., and P. D. Tyler. 1969. Organic acid metabolism in relation to flooding tolerance in roots. J. Ecol. 57:235–244.

Culbert, D. L., and H. W. Ford. 1972. The use of a multi-celled apparatus for anaerobic studies of flooded root systems. HortScience 7:29–31.

Cummins, J. N., and H. S. Aldwickle. 1983. Breeding apple rootstocks. Plant Breed. Rev. 1:294–376.

Cummins, W. R., H. Kende, and K. Raschke. 1971. Specificity and reversibility of the rapid stomatal response to abscisic acid. Planta 99:347–351.

Davies, D. D., S. Grego, and P. Kenworthy. 1974a. The control of the production of lactate and ethanol by higher plants. Planta. 118:297–310.

Davies, D. D., K. H. Nascimento, and K. K. Patil. 1974b. The distribution and properties of NADP enzyme in flowering plants. Phytochemistry 13:2417–2425.

Davies, F. S., and J. A. Flore. 1985. Root conductivity, CO_2 response and quantum yield of flooded and unflooded highbush blueberry plants. HortScience 20:89 (Abstr.).

⸻ . 1986a. Flooding, gas exchange and hydraulic conductivity of highbush blueberry. Physiol. Plant. 67:545–551.

⸻ . 1986b. Gas Exchange and flooding stress of highbush and rabbiteye blueberries. J. Am. Soc. Hort. Sci. 111:565–571.

⸻ . 1986c. Short-term flooding effects on gas exchange and quantum yield of rabbiteye blueberry (Vaccinium ashei Reade). Plant Physiol. 81:289–292.

Davies, F. S., and D. Wilcox. 1983. Waterlogging of containerized rabbiteye blueberry in Florida. J. Am. Soc. Hort. Sci. 109:520–524.

Davies, W. J., and T. T. Kozlowski. 1975a. Effects of applied abscisic acid and plant water stress on transpiration of woody angiosperms. For. Sci. 22:191–195.

⸻ . 1975b. Effect of applied abscisic acid and silicone on water relations and photosynthesis of woody plants. Can. J. For. Res. 5:90–96.

Davison, E. M., and F. C. S. Tay. 1987. The effect of waterlogging on infection of Eucalyptus marginata seedlings by Phytophthora cinnanmomi. New Phytol. 105:585–594.

Devitt, D. A., L. H. Stolzy, W. W. Miller, J. E. Campana, and P. Sternberg. 1989. Influence of salinity, leaching fraction, and soil type on oxygen diffusion rate measurements and electrode poisoning. Soil Sci. 148:327–335.

De Witt, M. C. J. 1978. Morphology and function of roots and shoot growth of crop plants under oxygen deficiency. p. 333–350. In: D. D. Hook and R. M. M. Crawford (eds.), Plant Life in Anaerobic Environments. Ann Arbor Sci. Press, Ann Arbor, MI.

Drew, M. C. 1983. Plant injury and adaptation to oxygen deficiency in the root environment: A review. Plant and Soil 75:179–199.

Drew, M. C., and J. M. Lynch. 1980. Soil anaerobiosis, microorganisms, and root function. Annu. Rev. Phytopathol. 18:301–314.

Drew, M. C., and E. J. Sisworo. 1979. The development of waterlogging damage in young barley plants in relation to plant nutrient status and changes in soil properties. *New Phytol.* 82:301–314.

Dubinina, I. M. 1961. Metabolism of roots under various levels of aeration. *Soviet Plant Physiol.* 8:305–312.

Duncan, J. M., and D. M. Kennedy. 1989. The effect of waterlogging on *Phytophthora* root root of red rasberry. *Plant Pathol.* 38:161–168.

Duniway, J. M. 1975a. Formation of sporangia by *Phytophthora drechsleri* in soil at high matric potentials. *Can. J. Bot.* 53:1270–1275.

_____ . 1975b. Limiting influence of water potential on the formation of sporangia by *Phytophthora drechsleri* in soil. *Phytopathology* 65:886–891.

_____ . 1976. Movement of zoospores of *Phytophthora cryptogea* in soils of various textures and matric potentials. *Phytopathology* 66:877–882.

_____ . 1979. Water relations of water molds. *Annu. Rev. Phytopathol.* 17:431–460.

_____ . 1983. Role of physical factors in the development of *Phytophthora* diseases. p. 175–187 In: D. C. Erwin, S. Bartnicki-Garcia, and P. H. Tsao (eds.), Phytophthora: *Its Biology, Taxonomy, Ecology, and Pathology*. Amer. Phytopathol. Soc. St. Paul, MN.

El-Beltagy, A. S., and M. A. Hall. 1974. Effect of water stress upon endogenous ethylene levels in *Vicia faba*. *New Phytol.* 73:47–60.

Esau, K. 1965. *Plant Anatomy*. 2nd ed. Wiley, New York.

Farquhar, G. D., and T. D. Sharkey. 1982. Stomatal conductance and photosynthesis. *Annu. Rev. Plant Physiol.* 33:317–345.

Fischer, W. R., H. Flessa, and G. Schaller. 1989. pH values and redox potentials in microsites of the rhizosphere. *Z. Pflanzen. Bodenk.* 152:191–195.

Flore, J. A., J. R. Lenton, and M. B. Jackson. 1989. Gas exchange and hormone content of normal ('Alisa Craig') and ABA deficient ('Flacca') tomato under flooding conditions. ASHS 1989 Annu. Mtg., Tulsa, Okla., Prog. and Abstracts, p. 126, Abst. 506.

Ford, H. W. 1964. The effect of rootstock, soil type, and soil pH on citrus root growth in soils subject to flooding. *Proc. Fla. State Hort. Soc.* 77:41–45.

_____ . 1969. Water management of wetland Citrus in Florida. *Proc. 1st Intl. Citrus Symp.* 3:1759–1770.

Francis, C. M., and A. C. Devitt, and P. Steele. 1974. Influence of flooding on the alcohol dehydrogenase activity of roots of *Trifolium subterraneum* L. *Aust. J. Plant Physiol.* 1:9–13.

Fulton, J. M., and A. E. Erickson. 1964. Relation between soil aeration and ethyl alcohol accumulation in xylem exudate of tomatoes. *Proc. Amer. Soc. Soil. Sci.* 28:610–614.

Gardner, F. E. 1961a. Evaluation of citrus rootstocks for Florida citrus. *Citrus and Veg. Mag.* 24:12.

_____ . 1961b. Evaluation of citrus rootstocks for Florida citrus, II. *Citrus and Veg. Mag.* 24:16.

Gisi, U., G. A. Zentmyer, and L. J. Klure. 1980. Production of sporangia by *Phytophthora cinnamomi* and *P. palmivora* in soils at different matric potentials. *Phytopathology* 70:301–306.

Good, J. M. 1987. The effect of flooding on nematode populations. p. 35–39. In: Agricultural flooding of organic soils. Univ. Florida, IFAS Tech. Bul. 870.

Grable, A. R. 1966. Soil aeration and plant growth. *Adv. Agron.* 18:57–106.

Grineva, G. M. 1963. Alcohol formation and excretion by plant roots under anaerobic conditions. *Soviet Plant Physiol.* 10:360–376.

Guinn, G., and L. A. Brinkerhoff. 1970. Effect of root aeration on amino acid levels in cotton plants. *Crop. Sci.* 10:175–178.

Gur, A., B. Bravdo, and Y. Mizrahi. 1972. Physiological responses of apple trees to supraoptimal root temperature. *Physiol. Plant.* 27:130–138.

Hagin, J., A. Lifshitz, and S. P. Monselise. 1965. The influence of soil aeration on the growth of citrus. *Isreal J. Agric. Res.* 15:59–64.

Hall, M. A., J. A. Kapuya, S. Sivakumaran, and A. John. 1977. The role of ethylene in the response of plants to stress. *Pestic. Sci.* 8:217–223.

Hall, R. 1986. Effects of root pathogens on plant water relations. p. 241–265. In: P. G. Ayres and L. Boddy (eds.), *Water, Fungi, and Plants*. Brit. Mycol. Soc. Symp. no. 11. Cambridge Univ. Press.

Hall, T. F., and G. E. Smith. 1955. Effects of flooding on woody plants. West Sandy dewatering project, Kentucky Reservoir. *J. For.* 53:383–385.

Hammond, L. C., W. H. Allaway, and W. E. Loomis. 1955. Effects of oxygen and carbon dioxide levels upon absorption of potassium by plants. *Plant Physiol.* 30:155–161.

Harris, D. G., and C. H. M. Van Bavel. 1957. Nutrient uptake and chemical composition of tobacco plants as affected by the composition of the root atmosphere. *Agron. J.* 49:176–181.

Heinicke, A. J. 1932. The effect of submerging the roots of apple trees at different times of the year. *Proc. Am. Soc. Hort. Sci.* 29:205–207.

Heinicke, A. J., D. Boynton, and W. Reuther. 1939. Cork experimentally produced in Northern Spy apples. *Proc. Am. Soc. Hort. Sci.* 37:47–52.

Herath, H. M. E., and G. W. Eaton. 1968. Some effects of water table, pH, and nitrogen fertilization upon growth and nutrient-element content of highbush blueberry plants. *Proc. Am. Soc. Hort. Sci.* 92:274–283.

Hiatt, A. J., and R. H. Lowe. 1967. Loss of organic acids, amino acids, potassium, and chlorine from barley roots treated anaerobically and with metabolic inhibitors. *Plant Physiol.* 42:1731–1736.

Hiron, W. P., and S. T. C. Wright. 1973. The role of endogenous abscissic acid in the response of plants to stress. *J. Expt. Bot.* 24:769–781.

Hook, D. D., and C. L. Brown, 1973. Root adaptations and relative flood tolerance of five hardwood species. *For. Sci.* 19:22–9.

Hook, D. D., C. L. Brown, and P. P. Kormanik. 1970. Lenticel and water root development of swamp tupelo under various flooding conditions. *Bot. Gaz.* 131:217–224.

――――. 1971. Inductive flood tolerance in swamp tupelo (*Nyssa sylvatica* var. *biflora* (Walt.) Sarg.). *J. Expt. Bot.* 22:78–79.

Hook, D. D., and J. R. Scholtens. 1978. Adaptations and flood tolerance of tree species. p. 299–331. In: D. D. Hook and R. M. M. Crawford (eds.), *Plant Life in Anaerobic Environments*. Ann Arbor Sci. Press, Ann Arbor, MI.

Jackson, M. B., and D. J. Campbell. 1976. Waterlogging and petiole epinasty in tomato: the role of ethylene and low oxygen. *New Phytol.* 76:21–29.

――――. 1979. Effects of benzyladenine and gibberellic acid on the responses of tomato plants to anaerobic root environments and to ethylene. *New Phytol.* 82:331–340.

Jackson, M. B., and M. C. Drew. 1984. Effects of flooding on growth and metabolism of herbaceous plants. p. 47–128. In: T. T. Kozlowski (ed.), *Flooding and Plant Growth*. Academic Press, Inc., New York.

Jackson, M. B., K. Gales, and D. J. Campbell. 1978. Effect of waterlogged soil conditions on the production of ethylene and on water relationships in tomato plants. *J. Expt. Bot.* 29:183–193.

Jackson, M. B., and K. C. Hall. 1987. Early stomatal closure in waterlogged pea plants is mediated by abscisic acid in the absence of foliar water deficits. *Plants, Cell and Environ.* 10:121–130.

Jackson, M. B., B. Herman, and A. Goodenough. 1982. An examination of the importance of ethanol in causing injury to flooded plants. *Plant, Cell and Environ.* 5:163–172.

Jaffee, B. A., and E. I. Zehr. 1983. Effects of certain solutes, osmotic potential, and soil solutions on parasitism of *Criconemella xenoplax* by *Hirsutella rhossiliensis*. *Phytopathology* 73:544–546.

Janick, J. 1975. Ethylene effects on apple scions. *HortScience* 10:70–72.

Jawanda, J. S. 1961. The effect of waterlogging on fruit trees. *Punjab. Hort. J.* 1:150–152.

John, C. D., and H. Greenway. 1976. Alcoholic fermentation and activity of some enzymes in rice roots under anaerobiosis. *Austral. J. Plant Physiol.* 3:325–336.

Johnson, J., B. G. Cobb, and M. C. Drew, 1989. Hypoxic induction of anoxia tolerance in root tips of *Zea mays* . Plant Physiol. 91:837–841.

Jones, R. 1972. Comparative studies of plant growth and distribution in relation to waterlogging. V. The uptake of iron and manganese by dune slack plants. *J. Ecol.* 60:131–140.

Joyner, M. E. B., and B. Schaffer. 1989. Flooding tolerance of 'Golden Star' carambola trees. *Proc. Fla. State Hort. Soc.* 102:236–239.

Joyner, M. E. B., B. Schaffer, and P. G. Webb. 1988. Flooding tolerance and recovery of carambola trees. *HortScience* 23:818 (Abstr.).

Justin, S. H. F. W., and W. Armstrong. 1987. The anatomical characteristics of roots and plant response to soil flooding. *New. Phytol.* 106:465–495.

Kawase, M. 1981. Anatomical and morphological adaptation of plants to waterlogging. HortScience 16:8-12.

Keeley, J. E. 1978. Malic acid accumulation in roots in response to flooding: Evidence contrary to its role as an alternative to ethanol. *J. Expt. Bot.* 29:1345–1349.

Kender, W. J., and W. T. Brightwell. 1966. Environmental relationships. p. 75–93. In: P. Eck and N. F. Childers (eds.), *Blueberry Culture*. Rutgers Univ. Press, New Brunswick, NJ.

Keneflick, D. G. 1962. Formation and elimination of ethanol in sugar beet discs. *Plant Physiol.* 37:434–439.

Kepka, M., and J. R. Morris. 1967. The effect of nitrogen and aeration on nutrient uptake and growth of peach seedlings in solution culture. *Roczn. Nauk. Roln. Sev. A.* 92:653–664.

Klotz, L. J., L. H. Stolzy, and T. A. DeWolfe. 1962. A method for determining the oxygen requirement of fungi in liquid media. *Plant Dis. Reptr.* 46:606–608.

_____ . 1963. Oxygen requirement of three root-rotting fungi in a liquid medium. *Phytopathology* 53:302–305.

Koch, M. S., and I. A. Mendelssohn. 1989. Sulphide as a soil phytotoxin: Differential responses in two marsh species. *J. Ecol.* 77:565–587.

Kozlowski, T. T. 1982. Water supply and tree growth. Part II. Flooding. *For. Abstr.* 43:145–161.

_____ . 1984a. Extents, causes, and impacts of flooding. p. 1–7. In: T. T. Kozlowski (ed.), *Flooding and Plant Growth*. Academic Press, Inc., New York.

_____ . 1984b. Responses of woody plants to flooding. p. 129–163. In: T. T. Kozlowski (ed.), *Flooding and Plant Growth*. Academic Press, Inc., New York.

Kozlowski, T. T., and S. G. Pallardy. 1984. Effect of flooding on water, carbohydrate, and mineral relations. p. 165–193. In: T. T. Kozlowski (ed.), *Flooding and Plant Growth*. Academic Press, Inc., New York.

Kramer, P. J. 1952. Causes of injury to plants resulting from flooding of the soil. *Plant Physiol.* 26:722–736.

_____ . 1983. Development of root systems. p. 170–178. In: P. J. Kramer (ed.), *Water Relations of Plants*. Academic Press, Inc., New York.

Kuan, T. L., and D. C. Erwin. 1982. Effect of soil matric potential of *Phytophthora* root rot of alfalfa. *Phytopathology* 72:543–548.

Laan, P., A. Smolder, C. W. P. M. Blom, and W. Armstrong. 1989. The relative roles of internal aeration, radial oxygen losses, iron exclusion and nutrient balances in flood tolerance of *Rumex* species. Acta. Bot. Neerl. 38:131–145.

Labanauskas, C. K., J. Letey, L. H. Stolzy, and N. Valoras. 1966. Effects of soil-oxygen and irrigation on the accumulation of macro and micro-nutrients in citrus seedlings (*Citrus sinensis* var. 'Osbeck'). *Soil Sci.* 101:378–384.

Labanauskas, C. K., L. H. Stolzy, and M. F. Handy. 1972. Concentrations and total amount of nutrients in citrus seedlings (*Citrus sinensis* 'Osbeck') and in soil as influenced by differential soil oxygen treatments. *Proc. Soil Sci. Soc. Am.* 36:454–457.

Labanauskas, C. K., L. H. Stolzy, L. J Klotz, and T. A. DeWolfe. 1965. Effects of soil temperature and oxygen on the amounts of macronutrients and micronutrients in citrus seedlings (Ctirus sinensis var. 'Bessie'). Soil Proc. Sci. Soc. Am. 29:60–64.

———. 1970. Soil oxygen diffusion rates and mineral accumulations in citrus seedlings (Citrus sinensis var. 'Bessie'). Soil Sci. 111:386–392.

Labanauskas, C. K., L. H. Stolzy, G. A. Zentmyer, and T. E Szusziwicz. 1968. Influence of soil oxygen and soil water on the accumulation of nutrients in avocado seedlings (Persea americana Mill.). Plant and Soil 29:391–406.

Lambers, H. 1976. Respiration and NADH-oxidation of the roots of flood-tolerant and flood-intolerant Senecio species as affected by anaerobiosis. Physiol. Plant. 37:117–122.

———. 1980. The physiological significance of cyanide-resistant respiration in higher plants. Plant, Cell and Environ. 3:293–302.

Lambers, H., and G. Smakman. 1978. Respiration of the roots of flood-tolerant and flood-intolerant Senecio species: Affinity for oxygen and resistance to cyanide. Physiol. Plant. 42:163–166.

Larson, K. D., D. A. Graetz, and B. Schaffer. 1991. Flood-induced chemical transformations in calcareous agricultural soils of South Florida. Soil Sci. (in press).

Larson, K. D., B. Schaffer, and F. S. Davies. 1989. Flooding, carbon assimilation and growth of mango trees. ASHS 1989 Annu. Mtg., Tulsa Okla, Prog. and Abstr. p. 126.

———. 1990. Lenticel hypertrophy of flooded mango trees. HortScience (Abstr.) 25:1001.

———. 1991. Flooding, leaf gas exchange and growth of mango in containers. J. Am. Soc. Hort. Sci. 116:156–160.

Larson, K. D., F. S. Davies, and B. Schaffer. 1991. Floodwater temperature and stem lenticel hypertrophy in Mangifera indica L. Amer. J. Bot. (in press).

Lemon, E. R., and A. E. Erickson. 1952. The measurement of oxygen diffusion in the soil with a platinum microelectrode. Proc. Soil Sci. Soc. Am. 16:160–163.

———. 1955. Principle of the platinum microelectrode as a method of characterizing soil aeration. Soil Sci. 79:383–392.

Livine, A., and Y. Vaadia. 1965. Simulation of transpiration rate in barely leaves of kinetin and gibberellic acid. Physiol. Plant. 18:658–664.

Loucks, W. L., and R. A. Keen. 1973. Submersion tolerance of selected seedling trees. J. For. 71:496–497.

Loustalot, A. J. 1945. Influence of soil moisture conditions on apparent photosynthesis and transpiration of pecan leaves. J. Agr. Res. 71:519–533.

Luz, E. D. M. N. 1989. The roles of five species of Phytophthora in infection and disease of roots, stems, and pods of Theobroma cacao L. Ph.D. Thesis Univ. Florida, Gainesville, FL.

Luzikov, N. V., A. S. Zubatov, and E. I. Rainina. 1973. Formation and degradation of mitochondria in the cell. I. Increasing stability of mitochondria during aerobic growth of Saccharomyces cerevisiae. Bioenergetics 5:129–149.

MacDaniels, L. H., and A. J. Heinicke. 1929. Pollination and other factors affecting the set of fruit, with special reference to the apple. Cornell Agr. Expt. Sta. Bul. 497.

MacKenzie, D. R., V. J. Elliot, B. A. Kidney, E. D. King, M. H. Royer, and R. L. Theberge. 1983. Application of modern approaches to the study of the epidemiology of diseases caused by Phytophthora. p. 303–313. In: D. C. Erwin, S. Bartnicki-Garcia, and P. H. Tsao (eds.), Phytophthorr:Its Biology, Taxonomy, Ecology, and Pathology. Amer. Phytopathology Soc. St. Paul, MN.

Mann, L. D., and L. H. Stolzy. 1972. An improved construction method for platinum electrodes. Proc. Soil Sci. Soc. Amer. 36:358–384.

Mansfield, T. A. and W. J. Davies. 1981. Stomata and stomatal mechanisms. p. 315–346. In: L. G. Paleg and D. Aspinall (eds.), The Physiology and Biochemistry of Drought Resistance in Plants. Academic Press, New York.

Marschberm, H., and K. Mengel. 1966. Der Einfluss von Ca-und-H-ioen bei unterschiedlichen Stoffwechselbedingungen auf die Membranpermeabilitat junger Gerstenwurzeln. Z. Pflazen. Bodenk. 112:39–49.

Marshall, D. R., P. Broue, and R. N. Oram. 1974. Genetic control of alcohol dehydrogenase in narrow leaved lupins. *J. Hered.* 65:198–207.

Marshall, D. R., P. Broue, and A. J. Pryor. 1973. Adaptive significance of alcohol dehydragenase isoenzymes in maize. *Nature New Biol.* 244:16–24.

Matheron, M. E., and S. M. Mircetich. 1985. Influence of flooding duration on development of *Phytophthora* root and crown rot of *Juglans hindsii* and Paradox walnut rootstocks. *Phytopathology* 75:973–976.

McIntyre, D. S. 1970. The platinum microelectrode method for soil aeration measurement. *Adv. Agron.* 22:235–283.

McKee, K. L., and I. A Mendelssohn. 1987. Root metabolism in the black mangrove (*Avicennia germinans* L.): Response to hypoxia. *Environ. Expt. Bot.* 27:147–156.

McManmon, M., and R. M. M. Crawford. 1971. A metabolic theory of flooding tolerances: The significance of enzyme distribution and behavior. *New. Phytol.* 70:299–306.

Meidner, H. 1967. The effect of kinetin on stomatal opening and rate of intake of carbon in mature primary leaves of barley. *J. Expt. Bot.* 18:556–561.

Mendelssohn, I. A., K. L. McKee, and W. H. Patrick, Jr. 1981. Oxygen deficiency in *Spartina alterniflora* roots: Metabolic adaption to anoxia. *Science* 214:439–441.

Miller, D. E., and D. W. Burke. 1975. Effect of soil aeration on *Fusarium* root rot of beans. *Phytopathology* 65:519–523.

Mitchell, D. H. and G. A. Zentmyer. 1971. Effects of oxygen and carbon dioxide tensions on sporangium and oospore formation by *Phytophthora* spp. *Phytopathology* 61:807–812.

Mizutani, F., M. Yamada, A. Sugiura, and T. Tomana. 1979. Differential water tolerance among *Prunus* species and the effect of waterlogging on the growth of peach scions on various rootstocks. *Engeigaku Kenkyu Shuroku* (Stud. Inst. Hortic., Kyoto Univ.) 9:28–35.

Mizutani, F., M. Yamada, and T. Tomana. 1982. Differential water tolerance and ethanol accumulation in *Prunus* species under flooded conditions. *J. Japan. Soc. Hort. Sci.* 51:29–34.

Moore, A. L., and P. R. Rich. 1985. Organization of the respiratory chain and oxidative phosphorylation. p. 134–172. In: R. Douce and D. A. Day (eds.), *Higher Plant Cell Respiration*. Springer-Verlag Press, Berlin.

Moore T. C. 1979. *Biochemistry and Physiology of Plant Hormones*. Springer-Verlag Press, New York.

Mouldau, H. 1973. Effects of various water regimes on stomatal and mesophyll conductances of bean leaves. *Photosynthetica* 7:1–7.

Ng, S. K., and C. Bloomfield. 1962. The effect of flooding and aeration on the mobility of certain trace elements in soils. *Plant and Soil* 16:108–135.

Nichols, W. E., and R. C. Turner. 1957. The pH of noncalcareous near neutral soils. *Can. J. Soil. Sci.* 37:96–101.

Norton, R. A., C. J. Hansen, H. J. O'Reilley, and W. H. Hart. 1963a. Rootstocks for peaches and nectarines in California. Calif. Agr. Exp. Sta. Ext. Leaf. 157.

———. 1963b. Rootstocks for plums and prunes in California. Calif. Agr. Exp. Sta. Ext. Leav. 159.

Olien, W. C. 1987. Effect of seasonal soil waterlogging on vegetative growth and fruiting of apple trees. *J. Am. Soc. Hort. Sci.* 112:209–214.

———. 1989. Seasonal soil waterlogging influences water relations and leaf nutrient content of bearing apple treas. *J. Am. Soc. Hort. Sci.* 114:537–542.

Oliveira, L. 1977. Changes in the ultrastructure of mitochondria of roots of *Triticale* subjected to anaerobiosis. *Protoplasma* 91:267–280.

Palti, J. 1981. *Cultural Practices and Infectious Crop Diseases*. Springer-Verlag Press, Berlin.

Papenhuijzen, C. 1983. Effect of interruption of aeration of the root medium on distribution of dry matter, sugar, and starch in young plants of *Phaseolus vulgaris*. *Acta. Bot. Neerl.* 32:63–67.

Pereira, J. S., and T. T. Kozlowski. 1977. Variation among woody angiosperms in response to flooding. *Physiol. Plant.* 41:184–192.

Pezeshki, S. R., R. D. Delaune, and W. H. Patrick, Jr. 1989. Effect of fluctuating rhizosphere redox potential on carbon assimilation of *Spartina alterniflora*. *Oecologia* 80:132–135.

Pfender, W. F., R. B. Hine, and M. E. Stanghellini. 1977. Production of sporangia and release of zoospores by *Phytophthora megasperma* in soil. *Phytopathology* 67:657–663.

Phillips, I. D. J. 1964a. Root-shoot hormone relations. I. The importance of aerated root system in the regulation of growth hormone levels in the shoot of *Helianthus annuus*. *Ann. Bot. N.S.* 28:17–45.

_____. 1964b. Root-shoot hormone relations. II. Changes in endogneous auxin concentration produced by flooding of the root system in *Helianthus annuus*. *Ann. Bot. N.S.* 28:37–45.

Phung, H. T., and E. B. Knipling. 1976. Photosynthesis and transpiration of citrus seedlings under flooded conditions. *HortScience* 11:131–113.

Pitman, M. G. 1969. Adaptation of barley roots to low oxygen supply and its relation to potassium and sodium uptake. *Plant Physiol.* 44:1233–1240.

Ploetz, R. C., J. L. Ramos, J. L. Parrado, B. Schaffer, and S. P. Lara. 1989. Performance of clonal avocado rootstocks in Dade County, Florida. *Proc. Fla. State Hort. Soc.* 102:234–236.

Ploetz, R. C., and B. Schaffer. 1987. Effects of flooding and Phytophthora root rot on photosynthetic characteristics of avocado. *Proc. Fla. State Hort. Soc.* 100:290–294.

_____. 1988. Damage thresholds for flooded and nonflooded avocado with *Phytophthora* root rot. *Phytopathology* 78:1544 (Abstract).

_____. 1989. Effect of flooding and Phytophthora root rot on net gas exchange and growth of avocado. *Phytopathology* 79:204–208.

Poel, L. W. 1960. A preliminary survey of soil aeration conditions in a Scottish hill grazing. *J. Ecol* 48:733–736.

Ponnamperuma, F. N. 1972. The chemistry of submerged soils. *Adv. Agron.* 24:29–96.

_____. 1984. Effect of flooding on soils. p. 9–45. In: T. T. Kozlowski (ed.), *Flooding and Plant Growth*. Academic Press, London.

Regehr, D. L., F. A. Bazzaz, and W. R. Boggess. 1975. Photosynthesis, transpiration and leaf conductance of *Populus deltoides* in relation to flooding and drought. *Photosynthetica* 9:52–61.

Reggiani, R., I. Brambilla, and A. Bertani. 1985a. Effect of exogenous nitrate on anaerobic metabolism in excised rice roots. I. Nitrate reduction and pyridine nucleotide pools. *J. Expt. Bot.* 36:1193–1199.

_____. 1985b. Effect of exogenous nitrate on anaerobic metabolism in excised rice roots. II. Fermentative activity and adenylic energy charge. *J. Expt. Bot.* 36:1698–1704.

Reggiani, R., C. A. Cantu, I. Brambilla, and A. Bertani. 1988. Accumulation and interconversion of amino acids in rice roots under anoxia. *Plant Cell Physiol.* 29:981–987.

Reid, D. M., and K. J. Bradford. 1984. Effects of flooding on hormone relations. p. 195–219. IN: T. T. Kozlowski (ed.), *Flooding and Plant Growth*. Academic Press, N.Y.

Reid, D. M., and A. Crozier. 1971. Effect of waterlogging on the gibberillin content and growth of tomato plants. *J. Expt. Bot.* 22:39–48.

Remy, P., and B. Bidabe. 1962. Root asphyxia and collar rot in pome fruit trees. The influence of the rootstock. *Congr. Pomol.*, 92nd Session p. 17–28.

Roberts, J. K. M., F. H. Andrade, and I. C. Anderson. 1985. Further evidence that cytoplasmic acidosis is a determinant of flooding intolerance in plants. *Plant Physiol.* 77:492–494.

Roberts, J. K. M., J. Callis, O. Jardetzky, V. Walbot, and M. Freeling. 1984. Cytoplasmic acidosis as a determinant of flooding intolerance in plants. *Proc. Nat. Acad. Sci.* (USA) 81:6029–6033.

Roberts, J. K. M., J. Callis, D. Wemmer, V. Walbot, and O. Jardetzky. 1984. Mechanism of cytoplasmic pH regulation in hypoxic maize root tips and its role in survival under hypoxia. *Proc. Nat. Acad. Sci. (USA)* 81:3379–3383.

Rom, R. C., and S. A. Brown. 1979. Water tolerance of apples on clonal rootstocks and peaches on seedling rootstocks. *Compact Fruit Tree* 12:30–33.

Rowe, R. N. 1966. Anaerobic metabolism and cyanogenic glycoside hydrolysis in differential sensitivity of peach, plum and pear roots under water-saturated condition. Ph.D. Thesis. Univ. Calif., Davis.

Rowe, R. N., and D. V. Beardsell. 1973. Waterlogging of fruit trees. *Hort. Abstr.* 43:534–544.

Rowe, R. N., and P. B. Catlin. 1971. Differential sensitivity to waterlogging and cyangenesis by peach, apricot, and plum roots. *J. Am. Soc. Hort. Sci.* 96:305–308.

Rumpho, M. E., and R. A. Kennedy. 1981. Anaerobic metabolism in germinating seeds of *Echinochloa crus-galli* (Barnyard Grass) (metabolic and enzyme studies). *Plant Physiol.* 68:165–168.

Saglio, P. H., P. Raymond, and A. Pradet. 1980. Metabolic activity and energy charge of excised maize root tips under anoxia. *Plant Physiol.* 66:1053–1057.

Salesses, G., R. Saunier, and A. Bonnet. 1970. L'asphyxie radiculaire chez les arbres fruitiers. *Bull. Tech. Inf.* 251:1–12.

Saunier, R. 1966. Method de determination de la resistance a l'asphyzie radiculaire de certains porte greffes d'argres fruiters. *Ann. Amel. Plantes* 16:367–384.

Save, R., and L. Serrano. 1986. Some physiological growth responses of kiwi fruit (*Actinida chinensis*) to flooding. *Physiol. Plant.* 66:75–78.

Schaffer, B., and P. Moon. 1990. Influence of rootstock on flood tolerance of Tahiti lime. *Proc. Fla. State Hort. Soc.* (in press).

Shaffer, B., and R. C. Ploetz. 1987. Effects of phytophthora root rot and flooding on net gas exchange of potted avocado seedlings. *HortScience* 22:1141 (Abstr.).

_____ . 1989. Gas exchange characteristics as indicators of damage thresholds for phytophthora root rot of flooded and nonflooded avocado trees. *HortScience* 14:653–655.

Schoeneweiss, D. F. 1975. Predisposition, stress, and plant dieseace. *Annu. Rev. Phytopathol.* 13:193–211.

Sena Gomes, A. R. and T. T. Kozlowski. 1980a. Growth responses and adaptations of *Fraxinus pennsylvanica* seedlings to flooding. *Plant Physiol.* 66:267–271.

_____ . 1980b. Responses of *Melaleuca quinquenervia* seedlings to flooding. *Physiol. Plant.* 49:373–377.

_____ . 1988. Physiological and growth responses to flooding of seedlings of *Hevea brasiliensis*. *Biotropica* 20:286–293.

Shapiro, R. E., G. S. Taylor, and G. W. Volk. 1956. Soil oxygen contents and ion uptake by corn. *Proc. Soil. Sci. Soc. Am.* 20:193–197.

Shaybany, B., and G. C. Martin. 1977. Abscisic acid identification and its quantitation in leaves of *Juglans* seedlings during waterlogging. *J. Am. Soc. Hort. Sci.* 102:300–302.

Shokes, F. M., and S. M. McCarter. 1979. Occurrence, dessimination, and survival of plant pathogens in surface irrigation ponds in southern Georgia. *Phytopathology* 69:510–516.

Sidebottom, J. R., and H. D. Shew. 1985. Effect of soil type and soil matric potential on infection of tobacco by *Phytophthora parasitica* var. *nicotianae*. *Phytopathology* 75:1439–1443.

Siedow, J. N., and D. A. Berthold. 1986. The alternative oxidase: A cyanide-resistant respiratory pathway in higher plants. *Physiol. Plant.* 66:569–573.

Sivakumaran, S., and M. A. Hall. 1978. Effects of age and water stress on endogenous levels of plant growth regulators in *Euphorbia lathyrus* L. *J. Expt. Bot.* 29:195–205.

Slowick, K., C. K. Labanauskas, L. H. Stolzy, and G. A. Zentmyer. 1979. Influences of rootstocks, soil oxygen, and soil moisture on the uptake and translocation in young avocado plants. *J. Am. Soc. Hort. Sci.* 104:172–175.

Smith, A. M., and T. Ap Rees. 1979a. Pathways of anaerobiosis on carbohydrate oxidation by roots of *Pisum sativum* Phytochemistry. 18:1453–1458.

———. 1979b. Pathways of carbohydrate fermentation in the roots of marsh plants. *Planta* 146:327–334.

Smith, G. S., J. G. Buwalda, T. G. A. Green, and C. J. Clark. 1989. Effect of oxygen supply and temperature at the root on the physiology of kiwifruit vines. *New Phytol.* 113:431–437.

Smith, M. W., and P. L. Ager. 1988. Effects of soil flooding on leaf gas exchange of seedling pecan trees. *HortScience* 23:370–372.

Smith, M. W., and R. D. Bourne. 1989. Seasonal effects of flooding on greenhouse-grown seedling pecan trees.*HortScience* 24:81–83.

Sojka, R. E., and L. H. Stolzy. 1980. Soil-oxygen effects of stomatal response. *Soil Sci.* 130:350–358.

Spek, L. Y. 1981. Influence of nitrate and aeration on growth and chemical composition of *Zea mays* L. *Plant and Soil* 63:115–118.

Strangellini, M. E., and T. J. Burr. 1973. Effects of soil water potential on disease incidence and oospore germination of *Pyhtium aphidermatum*. *Phytopathology* 63:1496–1498.

Sterne, R. E., G. A. Zentmyer, and M. R. Kaufman. 1977a. The effect of matric and osmotic potential of soil on Phytophthora root disease of *Persea indica*. *Phytopathology* 67:1491–1494.

———. 1977b. The influence of matric potential, soil texture, and soil amendment on root disease caused by *Phytophthora cinnamomi*. *Phytopathology* 67:1495–1500.

Stolzy, L. H. 1964. Measurement of oxygen diffusion rates with the platinum microelectrode. I. Theory and equipment. *Hilgardia* 35:545–554.

Stolzy, L. H., C. K. Labanauskas, L. J. Klotz, and T. A. DeWolfe. 1975. Nutritional responses of root rot of *Citrus limon* and *Citrus sinensis* under high and low soil oxygen supplies in the presence and absence of *Phytophthora* spp. *Soil Sci.* 119:136–142.

Stolzy, L. H., and J. Letey. 1964. Measurement of oxygen rates with the platinum microelectrode. III. Correlation of plant responses to soil oxygen diffusion rates. *Hilgardia* 35:567–576.

Stolzy, L. H., J Letey, L. J. Klotz, and C. K. Labanauskas. 1965. Water and aeration as factors in root decay of *Citrus sinensis*. *Phytopathology* 55:270–275.

Stolzy, L. H., and R. E. Sojka. 1984. Effects of flooding on plant diseases. p. 221–264. In: T. T. Kozlowski (ed.), *Flooding and Plant Growth*. Academic Press, New York.

Stolzy, L. H., G. A. Zentmyer, L. J. Klotz, and C. K. Labanauskas. 1967. Oxygen diffusion, water, and *Phytophthora cinnamomi* in root decay and nutrition of avocados. *Proc. Am. Soc. Hort. Sci.* 90:67–76.

Stover, R. H. 1962. Fusarial wilt (Panama disease) of banana and other *Musa* species. Kew, Commonw. Mycol. Inst. Phytopathol. Papers no. 4.

Strandberg, J. O. 1987. The effect of flooding on plant pathogen populations. p. 41–56. In: Agricultural Flooding of organic soils. Univ. Florida, IFAS Tech. Bul. 870. 63 p.

Streeter, J. G., and J. F. Thompson. 1971. Anaerobic accumulation of γ-aminobutyric acid and alanine in radish leaves (*Raphanus sativus* L.). *Plant Physiol.* 49:572–578.

Striegler, R. K., G. S. Howell, and J. A. Flore. 1987. Effect of flooding on the physiology and growth of selected grapevine cultivars. *HortScience* 22:1141 (Abstr.).

Syvertsen, J. P., R. M. Zablotowicz, and M. L. Smith, Jr. 1983. Soil temperature and flooding effects on two species of citrus. I. Plant growth and hydraulic conductivity. *Plant and Soil* 72:3–12.

Takai, Y., and T. Kamura. 1966. The mechanism of reduction in waterlogged paddy soil. *Folia Microbiol.* 11:304–313 (cited in *Adv. Agron.* 24:29–95).

Takai, Y., T. Koyoma, and T. Kamura. 1956. Microbial metabolism in the reduction process of paddy soils (part 1). *Soil Plant Food* (Tokyo) 2:63–66.

Takeda, K., and C. Furusaka. 1970. On bacteria isolated anaerobically from paddy field soil

I. Succession of facultative anaerobes and strict anaerobes. *J. Agr. Chem. Soc. Jap.* 44:343–348.

Tang, Z. C., and T. T Kozlowski. 1982. Physiological, morphological, and growth responses of *Platanus occidentalis* seedlings to flooding. *Plant and Soil* 66:243–255.

Tripepi, R. R., and C. A. Mitchell. 1984. Metabolic response of river birch and European birch roots to hypoxia. *Plant Physiol.* 76:31–35.

Van Gundy, S. D., and L. H. Stolzy. 1963. The relationship of oxygen diffusion rates to the survival, movement and reproduction of *Hemicycliophora arenaria. Nematologia* 9:605–612.

Vartapetian, B. B., I. N. Andreeva, G. I. Kozlova, and L. P. Agapova. 1977. Mitchondrial ultrastructure in roots of mesophyte and hydrophyte at anoxia and after glucose feeding. *Protoplasma* 91:243–256.

Vu, J. C. V. and G. Yelenosky. 1991. Photosynthetic responses of citrus trees to flooding. *Physiol. Plant.* 81:7–14.

Waller, J. M. 1986. Drought, irrigation, and fungal diseases of tropical crops. p. 175–187. In: *Water, Fungi, and Plants.* Brit. Mycol. Soc. Symp. no. 11. Cambridge Univ. Press.

Wample, R. L., and J. D. Bewley. 1975. Proline accumulation in flooded and wilted sunflower and the effects of benzyladenine and abscisic acid. *Can. J. Bot.* 53:2893–2896.

Wample, R. L., and D. M. Reid. 1979. The role of endogenous auxins and ethylene in the formation adventitious roots and hypocotyl hypertrophy in flooded sunflower plants (*Helianthus annuus*). *Physiol. Plant.* 45:219–226.

Wazir, F. K., M. W. Smith, and S. W. Akers. 1988. Effects of flooding on phosophorous levels in pecan seedlings. *HortScience* 23:595–597.

Wenkert, W., N. R. Fausey, and H. D. Watters. 1981. Flooding responses in *Zea mays* L. *Plant and Soil* 62:351–366.

West, D. W. 1978. Water use and sodium chloride uptake by apple trees. II. The responses to soil oxygen deficiency. *Plant and Soil* 50:51–65.

West, D. W., and J. A. Taylor. 1984. Responses of six grape cultivars to the combined effects of high salinity and rootzone waterlogging. *J. Am. Soc. Hort. Sci.* 109:844–851.

Westwood, M. N. 1978. *Temperate Zone Pomology.* W. H. Freeman and Co., San Francisco, CA.

Westwood, M. N., and P. B. Lombard. 1966. Pear rootstocks. *Ann. Rep. Oregon State Hort. Soc.* 58:61–68.

Wignarajah, K., and H. Greenway. 1976. Effect of anaerobiosis on activities of alcohol dehydrogenase and pyruvate decarboxylase in roots of *Zea Mays. New Phytol.* 17:575–584.

Wignarajah, K., H. Greenway, and C. D. John. 1976. Effects of waterlogging on growth and activity of alcohol dehydrogenase in barley and rice. *New Phytol.* 77:585–594.

Wilcox, W. F. 1989. Identity, virulence, and isolation frequency of seven *Phytophthora* spp. causing root rot of raspberry in New York. *Phytopathology* 79:93–101.

Wilcox, W. F., and S. M. Mircetich. 1985a. Influence of soil water matric potential on the development of *Phytophthora* root and crown rots of Mahaleb cherry. *Phytopathology* 75:648–653.

———. 1985b. Effects of flooding duration on the development of *Phytophthora* root and crown rots of cherry. *Phytopathology* 75:1451–1455.

Yang, S. F., D. O. Adams, C. Lizada, Y. Yu, K. J. Bradford, and A. C. Cameron. 1980. Mechanism and regulation of ethylene biosynthesis. p. 239–248. In: F. Skoog (ed.), *Plant Growth Substances 1979.* Academic Press, New York.

Zhang, J., and W. J. Davies. 1987. ABA in roots and leaves of flooded pea plants. *J. Expt. Bot.* 38:649–659.

Zhang, J., U. Schurr, and W. J. Davies. 1987. Control of stomatal behaviour by abscisic acid which apparently originates in the roots. *J. Expt. Bot.* 38:1174–1181.

Zentmyer, G. A. 1980. *Phytophthora cinnamomi and the Diseases It Causes.* Am. Phytopathol. Soc., St. Paul, MN.

8

Developmental Morphology and Anatomy of Grape Flowers*

Jean M. Gerrath
Department of Horticultural Science, University of Guelph
Guelph, Ontario, N1G2W1 Canada

I. INTRODUCTION

Reproductive development is central to studies of grape (*Vitis* L.) production. One aspect which has not been emphasized until recently has been flower development. This area has been neglected by both horticulturists and botanists. Most production studies do not require a detailed understanding of the early stages of floral development, thus reducing the likelihood that horticulturists will undertake such studies. For many other crop species, horticulturists could use botanical works to provide the basis for floral anatomical and developmental studies to supplement their own interests. However, flowers of Vitaceae, the grape family, are not particularly variable in morphology at maturity, so are not useful in systematic studies (Lavie 1970; Jackes 1984). As a result, less is known about anatomical and morphological floral development in grapes than might be expected. This review concentrates on anatomical developmental studies that have been carried out since the very thorough review of reproductive anatomy in grapes by Pratt in 1971. This paper uses her

*The author gratefully acknowledges the help and encouragement of her former supervisor, Dr. U. Posluszny. Financial support was provided by a Natural Sciences and Engineering Research Council of Canada Postdoctoral Fellowship. Facilities were generously provided by the Horticulture Department, University of Guelph.

work, as well as her excellent review of vegetative anatomy (Pratt 1974), as the base from which to begin, and acts as a supplement to it.

A number of general works have discussed floral structure in the last 20 years. Rives (1972) gives a broad review of grapevine floral initiation and development. The ampelography by Galet (1970) and Morton's translation of it into English (Galet 1979) provide some information on mature flower structure, as does the more recent book by Huglin (1986). The review by Lavee (1985) provides a more physiological approach.

The only review of grapevine flower development per se is by Srinivasan and Mullins (1981a), and although it contains a great deal of anatomical information, it also takes a physiological approach. The recent review by Gerrath and Posluszny (1988c) places the floral development of both wild and cultivated grapes in context with what is known about floral development in the rest of the Vitaceae. The implementation within the last 20 years of imaging tools such as scanning electron microscopy and epi-illumination light microscopy (Polsuszny et al. 1980) has enabled three-dimensional observations of developing flowers. These improvements, coupled with a renewed appreciation of the importance of development as one of the key aids to understanding mature structure, have resulted in a new look at grape floral development.

Two major topics are discussed in this review. First, inflorescence initiation and development and its relationship to tendril initiation is discussed. Second, floral development, including initiation and development of the floral structures as well as sex determination and pollination, is covered.

II. FLORAL DEVELOPMENT

A. Inflorescence Initiation and Early Development

Shoots of the grape family are unique because the shoot apical meristem produces two types of primordia, the leaf primordium (L) and the uncommitted primordium (UP), as illustrated in Figure 8.1. The fact that the uncommitted primordium is leaf-opposed at maturity has resulted in considerable interest in how to interpret it morphologically. Normally, plants possessing leaf-opposed structures grow sympodially. That is, a terminal determinate structure, often an inflorescence, becomes laterally displaced as the axillary bud of the subtending leaf takes over the function of shoot elongation (Bugnon 1953; Hallé et al. 1978). A number of ontogenetic studies (Moens 1956; Millington 1966; Tucker and Hoefert 1968; Shah and Dave 1970; Gerrath 1988; Gerrath and Posluszny 1988a, 1989a, 1989c) however, have found that the vitaceous shoot is monopodial, in the sense that the uncommitted primordium is a lateral appendage from its inception. Thus reviewers such as Pratt (1974)

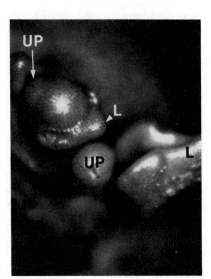

Figure 8.1.   Oblique view of shoot apex of V. *vinifera* 'Riesling', illustrating the fact tnat the apex produces both leaf (L) and uncommitted (UP) primordia. Epi-illumination light microscopy. × 99.

and Srinivasan and Mullins (1981a) have concluded that the vitaceous shoot can be viewed as a monopodium.

However, because the uncommitted primordium (a) does not consistently bear the same close spatial relationship to its subjacent leaf that an axillary bud does, (b) is spatially separate from the meristematic tissue that will form axillary buds, unlike the situation found in *Ampelopsis brevipedunculata,* a species with serial accessory buds (Gerrath 1989c), and (c) consistently possesses an abaxial first bract instead of the expected adaxial bract (see Bugnon 1953 for an excellent discussion of this point), it has remained difficult to imagine how it could be derived historically from a monopodial shoot in which the uncommitted primordium corresponds to an axillary bud. For these reasons, Gerrath and Posluszny (1989c) concluded that although the vitaceous shoot is developmentally monopodial, it is likely that the ancestor of the Vitaceae grew sympodially. This complementary view (Rutishauser and Sattler 1985), in which two apparently conflicting theories are held simultaneously depending on the aspect one is emphasizing, confirms Goebel's (1905) statement that the grape shoot should be considered sympodial from a comparative morphological point of view, but monopodial from a developmental point of view. Since the real ancestor can only be discovered through fossil evidence it is unlikely that this question will ever by satisfactorily resolved.

Among present-day members of the family, it is known that at maturity, species of *Cayratia* have axillary or pseudoaxillary inflorescences as well as leaf-opposed tendrils (Jackes 1987), whereas some species of *Cyphostemma* possess apparently terminal inflorescences and lack leaf-opposed appendages (Descoings 1960). However, the ontogeny of these genera is unknown. The shoot of *Leea guineensis*, a member of the Leeaceae, the family most closely related to the Vitaceae, has terminal inflorescences and is therefore sympodial (Lacroix and Gerrath 1990). Recent experimental work by Fournioux and Bessis (1990) has shown that in vitro shoots of cultivar 'Pinot noir' undergo "sympodization," apparently a reflection of a smaller apical meristem. They present an exhaustive listing of authors who consider the vitaceous shoot to be sympodial but do not speculate about its historical origin.

Srinivasan and Mullins (1980c) carried out a histochemical study on the shoot apex, using mitotic indices as well as acid phosphatase and peroxidase studies to determine changes in the apex as primordia are formed. They found a high number of nuclei at the 4C DNA level before uncommitted primordia formation, many 2C nuclei in the developing inflorescence, and a rise in the number of 4C nuclei once the tendril or inflorescence primordium is fully formed. They found, not unexpectedly, that the cells of the peripheral zone, which are the ones involved in the initiation of leaf and uncommitted primordia, were mitotically and enzymatically the most active.

It has long been known that the structure which forms a tendril can also develop into an inflorescence. Barnard (1932) referred to the primordium as an "anlagen," and that term has often been used since (Pratt 1971; Srinivasan and Mullins 1976, 1981a; Lavee 1985; Swanepoel and Archer 1988). However, because "anlage" is simply the German word for "primordium" and therefore does not distinguish between the two types of primordia produced at the apex, Posluszny and Gerrath (1986) used "uncommitted primordium" as an English term to replace the less precise "anlage." This is also more in keeping with the findings by Srinivasan and Mullins (1978, 1979, 1980a, 1980b) that show that the primordium is initially uncommitted, and can be induced to follow a particular developmental pathway as the result of the application of exogenous hormones. Indeed, on occasion they use both terms (1980c), as do Palma and Jackson (1989). Alleweldt (1966) did not use the term "anlage," but referred to it as the "undifferenziertes primordium."

Uncommitted primordia may occur opposite every leaf in cultivars such as 'Concord' with *V. labrusca* parentage. In this instance the shoots are said to have a continuous tendril pattern. However, in most cultivars of *Vitis* the uncommitted primordium is found opposite two of every three leaves (an interrupted pattern). Figure 8.2 illustrates a typical interrupted shoot plan in longitudinal section, in which there is no uncommitted primordium (UP) opposite leaf 4 (L_4). Figure 8.3 shows the basic

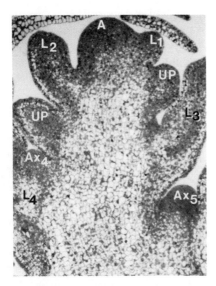

Figure 8.2.   Near median longitudinal section of a shoot apex (A) of *V. riparia,* showing the interrupted shoot pattern. There is no uncommitted primordium opposite every third leaf (eg., L_4). Axillary buds (Ax) form at every node. Toluidine blue O stain. × 99.

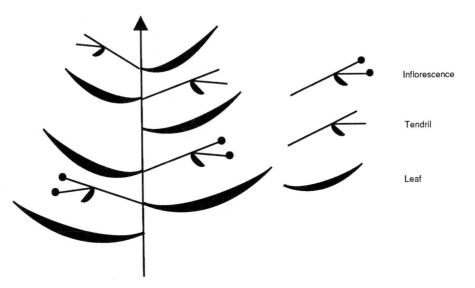

Figure 8.3.   Scheme of the interrupted shoot plan of a typical grape (*Vitis*) shoot in which tendrils of inflorescences arise opposite two of every three nodes. This shoot illustrates two cycles of nodes (six leaves); the lower showing inflorescences, the upper, tendrils.

interrupted shoot plan typical of shoots of most grape cultivars. Because every third node lacks a tendril, grape shoots are modular, in that they only repeat themselves every three nodes (Fournioux and Bessis 1973). Millington (1966) designated the nodes as lower tendril nodes, upper tendril nodes, or tendril-less nodes, depending on their position within the three-node cycle.

Gerrath and Posluszny (1988a) classified the shoot apex of *V.riparia* (a species with an interrupted pattern) into four categories. Type 1 is an apex that is forming a leaf at a lower tendril node. Type 2 is an apex initiating a leaf at either an upper tendril node or a tendril-less node. Type 3 is an apex giving rise to an uncommitted primordium at a lower tendril node, and Type 4 is an apex at the stage at which it forms an uncommitted primordium at an upper tendril node. In shoots with a continuous pattern the apex is, by definition, Type 2 when it gives rise to a leaf and Type 4 when it initiates an uncommitted primordium (Gerrath and Posluszny 1989c). They found that in *V.riparia* an uncommitted primordium of Type 3 is initiated high on the apical flank, approaching an apical bifurcation, whereas a Type 4 uncommitted primordium arises relatively lower on the flank. Thus, Gerrath (1988) concluded that it will be important to specify exactly what shoot apex type is being examined in future histological studies of grape shoot apices.

Most recent work on inflorescence initiation and development has focussed on what chemically determines the fate of the uncommitted primordium and at what stage the primordium's fate is determined. In a series of important papers in which they describe work on in vitro material, Srinivasan and Mullins (1978, 1979, 1980a, 1980b) and Mullins (1980) found that gibberellic acid stimulates initiation of the uncommitted primordium and its development into a tendril. They also found that cytokinins stimulate branching of the uncommitted primordium, which leads to inflorescence development. They concluded that the tendril could, therefore, be viewed as a weakly differentiated inflorescence. Thus they were able to confirm the statement made by Albertus Magnus over 700 years ago that ". . . the tendril is to be interpreted as a bunch of grapes incompletely developed" (see Arber 1938, p. 5). The close relationship of the tendril and the inflorescence has also been strengthened by the observations by Latiff (1981) that in many other members of the grape family it is normal to have structures that are part inflorescence, part tendril. Palma and Jackson (1989) also confirmed the close relationship between tendril and inflorescence, and found that in vivo exogenous applications of auxins enhanced inflorescence and flower production, whereas applications of gibberellic acid had the opposite effect.

The uncommitted primordium arises as a knob-like structure (Fig. 8.4, UP). Subsequently an abaxial bract forms on the primordium, and the primordium itself bifurcates to form an inner (main) and an outer (lateral)

Figure 8.4. Scanning electron micrograph of an overwintering bud of V. *riparia*, showing the shoot apex (A), the leaves (L), the stipules (St), a tendril (T), and two inflorescences (I). The inflorescences have formed branches (I_b), but there is no evidence of any floral primordia. The center "inflorescence" is an arm of the right inflorescence. × 116.

arm. Olivain and Bessis (1987) found that the first morphological sign that the uncommitted primordium will develop into an inflorescence is an enlargement of the inner arm, so as to appear as a dome. Gerrath and Posluszny (1988a, 1988b, 1989a, 1989b, 1989c) also found that in Vitis *riparia, Parthenocissus inserta* and *Ampelopsis brevipedunculata* the proportion of inner arm to outer arm is greater in an inflorescence primordium when compared to a tendril primordium. Posluszny and Gerrath (1986) found that the inflorescence primordium can be distinguished from the tendril primordium in Vitis cultivar 'Ventura' by the club-shaped appearance of the primordium even before the subtending bract is formed.

Pratt (1971) noted that most inflorescence development takes place in the inner arm in V. *vinifera*, an observation confirmed by Srinivasan and Mullins (1976, 1981a). This is consistent with the observation by Gerrath and Posluszny (1988b) that inflorescence branches form first on the inner arm in V. *riparia*.

Although the mature inflorescence pattern has been well described by Snyder (1933b), the actual pattern of initiation was first described by Gerrath and Posluszny (1988b). They found that in V. *riparia* the inflorescence branch primordia arise in a 2/5 spiral phyllotaxy, and although others have illustrated this same apparent spiral pattern (Scholefield and Ward 1975; Srinivasan and Mullins 1976, 1981a; Posluszny and Gerrath

1986) for cultivars of *Vitis*, they did not actually mention the pattern in their text.

For most cultivars of *V. vinifera* and *V. labrusca*, and in *V. riparia*, subsequent orders of branch primordia form as pairs at right angles to the previous pair, with each new primordium being subtended by a bract (Payer 1857; Snyder 1933b; Pratt 1971; Posluszny and Gerrath 1986). Ultimately the highest orders of primordia develop directly into floral primordia. Thus, the basic module on which the inflorescence is based is usually that of a three-flowered cyme or dichasium, in which the central flower is not subtended by its own bract (Pratt 1971). Srinivasan and Mullins (1981a) report that in 'Shiraz' the basic floral unit is five. Scholefield and Ward (1975) illustrate a different pattern for 'Sultana'. They found that the highest order branch primordium apex apparently subdivides into three floral primordia separated by approximately 120° with each flower subtended by its own bract.

The lateral branches of *Vitis* are complex in two ways. First, normally they elongate from an overwintering bud that was formed during the previous growing season. However, each branch is not entirely preformed, and the leaves of its distal portion grow out during the season of their formation. Second, a bud complex forms in the axil of every leaf (Fig. 8.2), i.e., higher orders of buds form in the axils of the prophylls at the base of the first order, or true axillary bud (Pratt 1971, 1979; Gerrath and Posluszny 1988a). Putting these two facts together, first order buds of the distal portion of the shoot normally grow the year they are initiated (the "prompt buds" or "summer laterals") and may or may not possess inflorescences. Second and higher orders of buds normally remain dormant and become overwintering buds (Fig. 8.4) on which the bulk of the next year's fruit crop forms. Thus first order buds in the basal region of the branch do not grow out during the year they were formed, since they were part of an overwintering bud. Figure 8.4 illustrates a dormant winter bud, showing the apex (A), several leaf primordia (L), and their associated stipules (St). It also shows a tendril (T) primordium and two partly formed inflorescence (I) primordia, which have given rise to inflorescence branches (I_b).

Because grapes undergo a period of dormancy after inflorescence initiation, there has been considerable work done to determine whether or not floral primordia have formed prior to dormancy in overwintering buds. The results of these studies are variable and are summarized in Table 8.1. Aglaoglu (1971) and Alleweldt and Balkema (1965) observed that floral primordia do form in the fall, before dormancy, and Gerrath and Posluszny (1988b) reported that calyces were discerned in early November, in some functionally male plants of *V. riparia*. All other workers, however, reported floral initiation to take place post-dormancy, in the spring of the year following inflorescence initiation (Barnard and Thomas 1933; Snyder 1933a, 1933b; Winkler and Shemsettin 1937;

Table 8.1.   Time of initiation of floral primordia.

Author	Cultivar or species	Location	Time of Floral Primordium Formation
Dormant Buds			
Aglaoglu (1971)	*V. vinifera*	Ankara, Turkey	August, prior to dormancy
Allweldt and Balkema (1965)	*Vitis* spp.	Various	Prior to dormancy
Barnard and Thomas (1933)	'Sultana'	Merbein, Australia	Spring
Carolus (1971)	'Merlot'	Bordeaux, France	No flower primordia prior to dormancy
Considine and Knox (1979b)	'Gordo'	Mildura, Australia	At bud burst in spring
Gerrath and Posluszny (1988b)	*V. riparia*	Ontario, Canada	Variable; usually in spring; some form on male plants prior to dormancy
Snyder (1933a, 1933b)	'Concord'	Iowa, U.S.A.	Bud swell in spring
Srinivasan and Mullins (1981a)	'Shiraz'	Sydney, Australia	Spring
Winkler and Shemsettin (1937)	'Sultanina'	California, U.S.A.	Just before and for a short time after bud burst in spring
Prompt Buds			
Madhava Rao and Mukherjee (1970)	'Pusa Seedless'	Delhi, India	45–53 days after bud break
Olivain and Bessis (1987)	'Pinot'	Dijon, France	Early to mid-June, post bud break

Carolus 1971; Considine and Knox 1979b; Srinivasan and Mullins 1981a). Madhava Rao and Mukherjee (1970) and Olivain and Bessis (1987) assessed development in summer laterals, and therefore their observations were on inflorescences that initiated and developed during one growing season.

B. Initiation of Floral Organs

Flowers of grape are fairly constant in their structure. Their general floral plan is illustrated in Figure 8.5. Typically the flower of *Vitis* consists of five sepals which are joined at the base, five petals joined at their tips (connivent) which alternate with the sepals, five petal-opposed stamens which arise in common with the petals, and a two-loculed superior ovary with four anatropous bitegmic ovules. In addition, there is a floral disc at the base of the ovary.

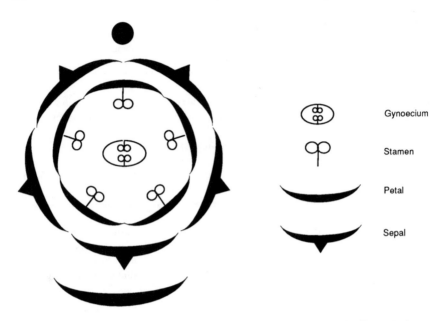

Figure 8.5. Floral diagram of a typical hermaphrodite grape flower. The flower is shown in relation to its subtending bract (arc below), and the shoot axis (circle above).

Because there are so many orders of inflorescence branch formation, and a sepal primordium ordinarily arises in the same position as the bract of an inflorescence branch, it is not easy to distinguish between an inflorescence branch primordium and a floral primordium. Gerrath and Posluszny (1988b) found that in V. riparia the stage at which one can recognize a flower primordium is when at least three sepal primordia have formed. Although there are five sepals at initiation, their individuality can be masked by the development of a basal ring primordium. Thus one might well speak of a calyx ring (Fig. 8.6, K) as Srinivasan and Mullins (1976) found in some cultivars, or five separate sepals in others (Scholefield and Ward 1975). Pratt (1971) indicates that there is considerable dispute over whether the calyx has five papillae, or is a ring. Most likely these differences are the result of differences in timing of development of the ring meristem that determines how discrete the sepals remain (Gerrath 1988).

The petals and stamens are opposite one another, an unusual condition in flowers, but a diagnostic feature in the Rhamnales, to which the Vitaceae belong (Cronquist 1981). Beille (1901) noted that the petals and stamens arise as five common petal-stamen primordia (Fig. 8.6, CA) and that subsequently each subdivided into separate petal (Fig. 8.6, C) and stamen (Fig. 8.6, A) primordia. Subsequent workers have not particularly remarked on this although Considine and Knox (1979b), using sectioned material state that petals are formed before stamens. However, detailed work by Gerrath and Posluszny (1988b) confirmed Beille's observation, and extended it to other members of the Vitaceae (Gerrath and Posluszny 1988c).

Further stages of floral development are obscured by the growth of the individual petals over the floral apex. In fact, the petals of Vitis push through the calyx, thus eventually performing a protective function for the reproductive floral organs (Fig. 8.7). As stamen development proceeds the floral apex expands sufficiently to leave a flat portion in the centre on which the gynoecium will arise (Fig. 8.8). There are few studies of the initiation and early development of the gynoecium, and these have not yielded consistent results. Gerrath and Posluszny (1988b) found that the gynoecium arose as a ring meristem in V. riparia (Fig. 8.9, G). However, they also found that the gynoecium of the hybrid cultivar 'Ventura' apparently was initiated as five separate primordia which subsequently formed a ring (Posluszny and Gerrath 1986). Considine and Knox (1979b) described the gynoecium of V. vinifera as arising from two growth centers on the central floral disc. Aglaoglu (1971), working with sections, stated that the gynoecium may be two- or three-carpellate, and Dorsey (1912) found the carpel number to range from two to four. By the stage that growth of a portion of the inner gynoecial wall is redirected to form the septa (Fig. 8.10, Se) the gynoecium appears to form two units. The septa may touch (Fig. 8.11), but workers did not report that they fuse (Aglaoglu 1971; Gerrath and Posluszny 1988b).

Ovule initiation and development appear to be consistent in the genus, with two anatropous bitegmic ovules per locule (Fig. 8.11, O) formed at the base of the septa (Gerrath and Posluszny 1988b) or at the base of the pericarp (Considine and Knox 1979b).

The pistil develops as the result of apical growth of the gynoecial ring meristem (Considine and Knox 1979b), and eventually a pistil with a short style and a discoid stigma forms (Fig. 8.12). As the flower approaches anthesis, there is heavy vascularization of the ovary wall surrounding the ovules (Fig. 8.13). Considine et al. (1986) reported that most of this vascular tissue consists of phloem, and that there is only a limited amount of primary xylem development. The stigma is of the wet type (Considine and Knox 1979b).

The petals of Vitis are unusual, in that they become laterally joined at maturity and dehisce basally as a unit (the calyptra, Figs. 8.14, 8.15). Snyder (1933a) reported that the interdigitating cells which join the individual petals together to form the cap in 'Concord' were the product of extensive cell divisions. However, Srinivasan and Mullins (1976, 1981a) and Gerrath and Posluszny (1988b) observed that the cells are interlocking but not fused, and did not report any callus-like formation or extensive cell division in the region, thus confirming the results of Winkler and Shemsettin (1937). At anthesis the calyptra dehisces basally

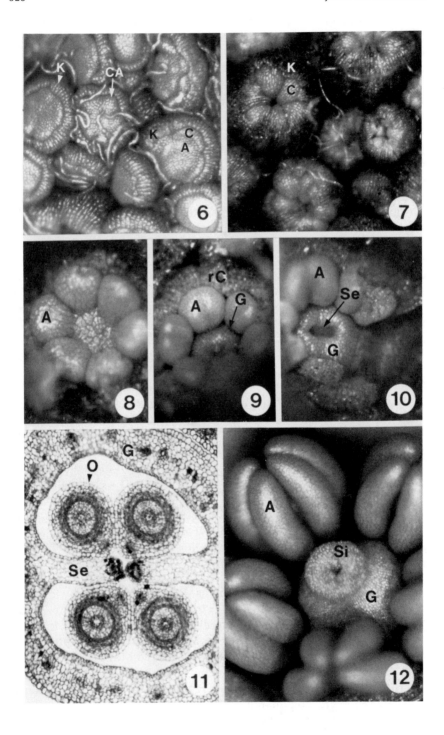

and is pushed off as the filaments (A) straighten (Fig. 8.15). Dorsey (1912) attributed the shedding of the calyptra to a "drying-up" process, not to a straightening of the filament. There does not seem to be any experimental evidence regarding the mechanism of calyptra dehiscence. The floral disc (D), sometimes referred to as a nectary (Pratt 1971; Lavee 1985), develops from the base of the ovary (Gerrath and Posluszny 1988b) and may be variously lobed or may appear as a unit, depending on the cultivar.

Shortly after anthesis the ovaries begin to expand as fruit formation begins (Fig. 8.16). Considine and Knox (1979a, 1979b) carried out a detailed study of the development and histochemistry of the pistil and the fruit wall in *V. vinifera* 'Gordo'. They delineated three phases of pericarp differentiation: ring meristem formation just after pistil initiation; cell proliferation by anticlinal divisions; and a maturation phase characterized by periclinal cell division and differentiation. Jona and Botta (1988) reported that the peak of mitosis occurred after fertilization, with the nucellar tissues being the most active. Growth of the berry itself was mainly the result of mesocarp cell enlargement. The main components of the cell wall were pectins, which appeared to increase during the mitotic phase of tissue growth.

Figures 8.6–8.12.   Stages of grape (*Vitis riparia*) floral development. **Figures 8.6–8.10.** Epi-illumination light microscopy. **Figure 8.6.** Inflorescence showing the initiation of sepals (K), followed by the formation of a common petal-stamen primordium (CA), and the subsequent subdivision of this primordium into petals (C) and stamens (A). × 99. **Figure 8.7.** Inflorescence illustrating that the petals push through the sepals, leaving the sepals to form a collar at the base of the flower. × 64. **Figure 8.8.** Top view of a floral apex illustrating the maximum stage of development of the stamens (A) before gynoecium initiation. × 168. **Figure 8.9.** Oblique view of a flower illustrating the initiation of the gynoecium (G) as a ring primordium. The petals have been removed (rC). × 99. **Figure 8.10.** Face view of a flower in which the septa (Se) are beginning to form on the inner gynoecial wall (G). × 99. **Figure 8.11.** Cross section through an ovary, showing the gynoecial wall (G), the septa (Se), and the presence of two ovules (O) per locule in the gynoecium. × 64. Safranin-alcian green stain. **Figure 8.12.** Oblique view of the stamens (A) and gynoecium (G) of a pistillate flower 10 days before anthesis. The discoid papillate stigma (Si) has formed. Externally the anthers are indistinguishable from those of a male flower of this species. Epi-illumination. × 99.

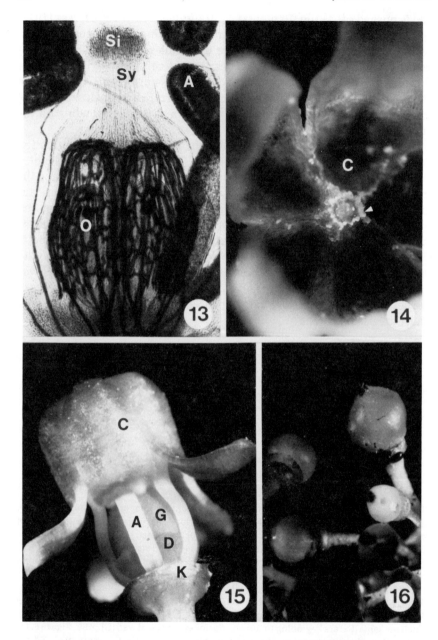

Figures 8.13–8.16. Later stages of grape floral development. **Figure 8.13.** Longitudinal view of a cleared pistil of *V. vinifera* 'Riesling', showing the heavy vascularization of the gynoecial wall, surrounding the ovules (O). The short style (Sy) and the papillate stigma (Si) are also visible. The anthers (A) may also be seen. Safranin stain. × 40. **Figures 8.14–8.16.** Living material. **Figure 8.14.** View of the inner surface of the calyptra (fused petals, C) of *V. labrusca* 'Concord', showing the interlocking cells and the papillate cells (arrowhead)

C. Sex Determination

One of the interesting and potentially important facts about the genus *Vitis* is that, with the important exception of *V vinifera*, few species have hermaphrodite flowers. Most species appear to be functionally dioecious based on floral structure, although this has rarely been confirmed experimentally (Kevan et al. 1985, 1988). There is considerable variation in terminology used for the two basic floral types. Thus, flowers that have undeveloped stigmas and styles and erect filaments may be called "male," "female sterile," "functionally staminate," or "staminate," whereas those with well-developed stigmas and styles and recurved filaments may be referred to as "functionally pistillate," "imperfect hermaphrodite," or "fruit producing morph" (Stout 1921; Oberle 1938; Kevan et al. 1985). For the purposes of this paper, the terms "pistillate" and "staminate" will be used when referring to the flowers and the plants will be designated as "male" and "female." The word "functionally" should be inserted as understood before both the floral and plant terms.

Oberle (1938) states that *V. vinifera* may have staminate, pistillate, or hermaphrodite flowers, although Baranov (1927) reports that there are no true staminate flowers of *V. vinifera* in the wild. As a result, it is not surprising that there are variations in sexual expression in flowers of the many cultivars that have been produced. Interest in sex determination has centred on three main questions: first, the description of floral development in functionally dioecious species; second, the genetic determination of sex; and third, the environmental and hormonal control of sex expression.

It is generally recognized that unisexual flowers of *Vitis* possess rudimentary organs of the opposite sex (Stout 1921; Lavee 1985). Thus, although they appear bisexual, they are functionally unisexual. Differences between the sexes are ones of quantitative development, not qualitative differences (Gerrath and Posluszny 1988b). Watanabe et al. (1983) and Watanabe (1984) have described floral development in the dioecious species *V. coignetiae*. Gerrath and Posluszny (1988b), following floral development in *V. riparia*, confirmed the observations by Dorsey (1912) for this species (as *V. vulpina*) that the staminate flowers have upright stamens, the anthers of which produce an abundance of viable pollen grains, but the pistil is aborted or rudimentary. In contrast, the pistillate flower has a normal flask-shaped pistil (containing ovules), a papillate stigma, but reflexed filaments and non functional pollen.

ringing the point at which all the petals join. × 40. **Figure 8.15.** Flower of 'Concord', at anthesis, showing the calyptra (C) being pushed off as the filaments (A) elongate. The basal floral disc (D) and the ovary (G) can be seen between the filaments. The sepals (K) remain as a small collar around the base of the flower. × 20. **Figure 8.16.** Young berries of *V. riparia*, less than a week after anthesis, showing the swelling pericarp. × 8.

Dorsey (1914) carried out the most careful study of the degeneration of pollen in female plants of 'Brighton' and in V. riparia, and found a variety of stages at which development failed to continue. In general, pollen sterility seems due to disintegration subsequent to mitosis in the microspore nucleus. There is little information about degeneration of the pistil in staminate flowers. However, Watanabe et al. (1983) and Watanabe (1984) illustrate the pistil of staminate flowers as flattened (the style sank and atrophied, and the ovaries shrank and became hidden by the basal floral disc). Gerrath and Posluszny (1988b) found that the gynoecium of staminate flowers of V. riparia initiates ovules that develop at least one integument. However, they did not observe the formation of a megaspore mother cell, meiosis, or any signs of ovule development past this stage. Thus, by anthesis, the central portion of the ovule disintegrates, leaving a mass of dead tissue. The style and stigma also fail to develop in staminate flowers of V. riparia.

The genetic determination of sex has recently been studied by Carbonneau (1983a, 1938b), who has suggested that sex phenotype is the result of epistatic relationships between alleles of two linked polymorphic loci in a dominance hierarchy. In the first gene, the staminate phenotype is dominant over the hermaphrodite phenotype, which dominates over the pistillate phenotype. The second gene is associated with modifications of expression of the first gene. Negi and Olmo (1971a) suggest that sex determination is heterogametic, with the male dominant. Antcliff (1980) proposes a model for the inheritance of sex in Euvitis, in which there are two alleles. One is a staminate allele, the other a pistillate allele. In this scheme staminate flowers are homozygous for the staminate allele, pistillate flowers are homozygous for pistillate allele, and hermaphrodites are heterozygous. The phenomenon of temporal sex change, in which a plant changes its sex over time, is apparently exceedingly rare in grape, as there are no published reports since Barrett (1966) reported that a normally functionally male plant of V. riparia produced hermaphrodite flowers.

The greatest interest has been in the manipulation of sexual expression by exogenous application of growth regulators. Thus, Negi and Olmo (1966) reported that cytokinins stimulated the production of hermaphrodite flowers in male plants of V. vinifera, and Iizuki and Hashizu (1968) reported that cytokinins stimulated production of pistillate flowers on plants that normally produce staminate flowers. Negi and Olmo (1971b) also found that cytokinin application is best made at the time of megaspore mother cell formation in both V. vinifera and V. riparia. In 1972, they concluded that in V. vinifera subsp. sylvestris there is a dominant staminate gene that inhibits cytokinin synthesis at a certain stage of pistil development, leading to maleness. Moore (1970), however, who has also made exogenous applications of cytokinins, concluded that floral sex in Vitis is controlled by the relative levels of two inhibitors, and speculated that cytokinins may act either by suppressing female inhibitor production, by altering the level of the critical threshold, or that cytokinin is the inhibitor itself. Milyaeva et al. (1984) reported that inflorescences of 'Kohmish Black Seedless' could be induced to set seed when dipped in cytokinins, or a 0.01% concentration of (2-chloroethyl) phosphoric acid (ethephon) at the end of flowering. Srinivasan and Mullins (1918b) reported that cytokinin can induce flowering in four-week grape seedlings. However, in this instance, all the flowers were pistillate, thus masking the genetic sex of the seedlings.

How closely the embryological development of seedless grapes resembles that of male plants is a question that has not been particularly explored. The recent thorough review of seedlessness in grapes by Ledbetter and Ramming (1989) precludes the necessity to cover this aspect of grape floral development in this paper.

D. Pollen Morphology and Pollination

The structure of pollen has been examined from a number of points of view, but most of them focus on identification. The study by Reille (1967) is the only one in which an attempt has been made to relate pollen morphology to the systematics of the family. He concluded that the pollen of *Vitis* is so homogeneous that the two sections of *Vitis*, Muscadinia and Euvitis, should not be separated on the basis of pollen characters. Several studies, using scanning electron microscopy, have found that differences in exine structure were sufficient to use pollen morphology to distinguish cultivars (Tompa-Kasirskaya and Kozma 1978; Ahmedullah 1983; Tompa-Kasirskaya 1984, 1985; Castelli et al. 1986a). Tompa-Kasirskaya (1985) noted that some of the cultivars of *V. riparia* had inaperturate pollen and were thus functionally female. This is similar to the reports by Kevan et al. (1985) and Gerrath and Posluszny (1988b) that in naturally occurring populations of *V. riparia* the pollen of the staminate flowers is tricolporate, whereas the pollen of the pistillate flowers is inaperturate. Kevan et al. (1988) have also found similar results in *V. aestivalis*, but in this instance pollen of the pistillate flowers is significantly smaller than that of the staminate. They also suggest that sterile pollen cultivars of *V. vinifera* may be a consequence of dioecious ancestors in the species. Watanabe et al. (1983) report that the pollen from pistillate flowers of *V. coignetiae* was shrivelled or collapsed (Watanabe 1987) and failed to germinate.

There has been little work on pollen cytology in grapes since the work of Dorsey (1914). Goldy (1985) reports that meiosis takes place in the spring when the developing buds are 15 cm long, that meiosis begins at noon, and that although meiosis is not synchronous within an inflorescence, it is so within a flower. Corriveau and Coleman (1988) showed by DAPI screening that the pollen of grape is binucleate when shed and that plastid DNA is maternally inherited.

The actual pollination of grapes has not been much studied in cultivation. Free (1970) states that pollination in grapes is accomplished partly by wind, partly by insects. Branties (1978) found that in wild plants of V. vinifera subsp. silvestris wind is probably not an important vector because the pollen is large enough that it sinks rapidly and the plants are not close to each other. He reported that the insect visitors were beetles, Halyctids, honeybees, and Syrphids. He found that the visitors were attracted to and consumed the pollen, and although the floral disc might act as an odor attractor, it apparently produced no nectar, and insect visitors were not seen to lick it. Crane and Walker (1984) also considered that insect pollination may be more important than previously thought. However, Kevan et al. (1985) concluded that V. riparia is essentially wind pollinated, although supplemented by insect vectors, which feed on the pollen of both sexes.

Booth (1902) reported that the anthers burst before the pistil is receptive, but that the pollen may remain viable for up to three weeks. Castelli et al. (1986b) report that there is evidence that pollen grains of some cultivars germinate before anthesis, thus allowing for the possibility of cleistogamy; stigma receptivity was not discussed. Thus, it would appear that dichogamy (separation of sexes in timing of maturity) does not readily occur in grape. Randhawa and Negi (1965) also found that pollen dehiscence can take place before anthesis: the four cultivars studied ('Alamwick', 'Bharat Early', 'Black Muscat', and 'Pusa Seedless') were self-pollinating, self-fertile, with some cleistogamy. The question of self-fertility still seems to be controversial (Free 1970; Crane and Walker 1984).

Kassemeyer and Staudt (1983) found that fertilization takes place within 24 h of pollination in cultivars of V. vinifera ('Weisser Bergunder' and 'Gewerztraminer'). They also reported that the whole mitotic cycle of the zygote takes 20 days; the endosperm becomes cellular toward the end of this period.

III. SUMMARY AND CONCLUSIONS

Work on floral development in grape over the past 20 years has filled many gaps in our knowledge of this area. It has contributed a clearer idea of the details of inflorescence and flower ontogeny as well as pollen morphology and pollination. However, there are still a number of questions which have not been adequately addressed. Although the developmental morphology of the inflorescence is better understood, there is little detailed information about the anatomy, especially the vascular relationships within the inflorescence. Nor is there much information about the vascular relationship between the inflorescence and the shoot itself, especially as the inflorescence develops. There is little information about floral vasculature, especially the pistil and the developing fruit. Embryological information is still not particularly complete, despite the interest in development of seedless grape cultivars. Pollen structure is more thoroughly understood than before, but there are many aspects of the pollination biology of grapes which have not been fully explored. These include questions about the degree of cleistogamy present in cultivars, whether or not flowers are self-fertile, the relative importance of biotic and abiotic pollination vectors, and the variability in floral disc function between cultivars.

This review deals exclusively with work done on the section Euvitis. This is because there is still no work done on the floral anatomy and development in *V. rotundifolia* Section Muscadinia. This seems to be a major oversight, both from a botanical and horticultural point of view. In the same vein, it is also important, especially if hybrids continue to be produced, that we know more about the floral development of other species of *Vitis*. Thus, although certain aspects of grape floral development are better understood than they were 20 years ago, much work remains to be done.

LITERATURE CITED

Aglaoglu, Y. S. 1971. A study of the differentiation and the development of floral parts in grapes (*V. vinifera* L.). *Vitis* 10:20–26.

Ahmedullah, M. 1983. Pollen morphology of selected *Vitis* cultivars. *J. Am. Soc. Hort. Sci.* 108:155–160.

Alleweldt, G. 1966. Die differenzierung der Blütenorgane der Rebe. *Wein-Wissenschaft* 21:393–402.

Alleweldt, G., and G. H. Balkema. 1965. Über die anlage von infloreszenz-und Blüten-primordien in den Winterknospen der Rebe. *Z. Acker-Pflanzenbau* 123:59–74.

Antcliff, A. J. 1980. Inheritance of sex in *Vitis*. *Ann. Amélior. Plantes.* 30:113–122.

Arber, A. 1938. *Herbals. Their Origin and Evolution.* 3rd Ed. Reprinted 1986, Introduction and annotation by W. T. Stearn. Cambridge Univ. Press, New York.

Baranov, P. 1927. The true female flower of the grape. *Trans. Irrig. Expt. Sta. Ak-Kavak.* 4:1–78, 119–137.

Barnard, C. 1932. Fruit bud studies. I. The Sultana. An analysis of the distribution and behaviour of the buds of the Sultana vine, together with an account of the differentiation and development of the fruit buds. *J. Counc. Sci. Ind. Res. Austral.* 5:47–52.

Barnard, C., and J. E. Thomas. 1933. Fruit bud studies. II. The Sultana: Differentiation and development of the fruit buds. *J. Counc. Sci. Ind. Res. Austral.* 6:285–294.

Barrett, H. C. 1966. Sex determination in a progeny of a self-pollinated staminate clone of *Vitis. Proc. Am. Soc. Hort. Sci.* 88:338–340.

Beille, L. 1901. Recherches sur le developpement floral des Disciflores. *Actes Soc. Linneenne Bordeaux* 56:231–407.

Booth, N. O. 1902. A study of grape pollen. New York Agr. Expt. Sta. Bull. 224.

Branties, N. B. M. 1978. Pollinator attraction of *Vitis vinifera* subsp. *sylvestris*. *Vitis* 17:229–233.

Bugnon, F. 1953. Recherches sur la ramification des Ampélidacées. Publications de l'Université de Dijon 11. Presses Universitaires de France, Paris.

Carbonneau, A. 1983a. Male and female sterility in the genus Vitis I. Modelling of their inheritance. Agronomie, Paris 3:635–644.

———. 1983b. Male and female sterility in the genus Vitis. II. Consequences for genetics and breeding. Agronomie, Paris 3:645–649.

Carolus, M. 1971. Description des stades du développment des primordia inflorescentiels durant l'organogenese des bourgeons latents de la Vigne (V. vinifera L.) Conn. Vigne Vin. 2:163–173.

Castelli, S., P. L. Pisani, and E. Rinaldelli. 1986. Recerche sulle caratteristiche del polline di cloni di alcuni vitigna aduva da vino. Rev. Ortofloro frutticoltura Ital., Firenze 69:287–297.

Castelli, S., A. Gemmiti, P. L. Pisani, and F. Barnato. 1986. Some observations on cleistogamy in the grapevine (Vitis vinifera L.) (in Italian). Vignevini, Bologna 13:39–41.

Considine, J. A., J. Farrant, A. Lang, and B. G. Coombe. 1986. Vascular anatomy, development, and function in Vitis spp. HortScience 21:165.

Considine, J. A., and R. B. Knox. 1979a. Development and histochemistry of the cells, cell walls, and cuticle of the dermal system of fruit of the grape, Vitis vinifera L. Protoplasma 99:347-365.

———. 1979b. Development and histochemistry of the pistil of the grape, Vitis vinifera. Ann. Bot. 43:11–22.

Corriveau, J. L., and A. W. Coleman. 1988. Rapid screening method to detect potential biparental inheritance of plastid DNA and results for over 200 angiosperm species. Am. J. Bot. 75:1443–1458.

Crane, E., and P. Walker. 1984. Pollination Directory for World Crops. International Bee Research Assn. London.

Cronquist, A. 1981. An Integrated System of Classification of Flowering Plants. Columbia Univ. Press, New York.

Descoings, B. 1960. Un genre méconnu de Vitacées: Comprehension et distinction des genres Cissus L. et Cyphostemma (Planch.) Alston. Not. System. (Paris) 16:113–125.

Dorsey, M. J. 1912. Variations in the floral structures of Vitis. Bull. Torrey Bot. Club 39:37–52.

———. 1914. Pollen development in the grape with special reference to sterility. Univ. Minn. Agr. Exp. Bull. 144:1–60.

Fournioux, J.-C., and R. Bessis. 1973. Étude du parcours caulinaire des faisceaux conducteurs foliares permettant la mise en évidence d'une rythmicité anatomique chez le vigne (V. vinifera L.) Rev. Gen. Bot. 80:177–185.

———. 1990. Élements du contrôle de la morphogenèse de la vigne in vitro: sympodisation. Can. J. Bot. 68:841–851.

Free, J. B. 1970. Insect Pollination of Crops. Chapt. 13, Vitaceae. Academic Press, New York.

Galet, P. 1970. Précis de Viticulture. Paul Déhan, Montpellier.

———. 1979. A practical ampelography. Grapevine Identification. Transl. Lucie T. Morton. Comstock Publ. Assoc. Cornell U. Press. Ithaca, N.Y.

Gerrath, J. M. 1988. Morphological and anatomical development in the Vitaceae. Ph. D. Thesis, University of Guelph, Guelph.

Gerrath, J. M., and U. Posluszny. 1988a. Morphological and anatomical development in the Vitaceae. I. Vegetative development in Vitis riparia. Can. J. Bot. 66:209–224.

———. 1988b. Morphological and anatomical development in the Vitaceae. II. Floral development in Vitis riparia. Can. J. Bot. 66:1334–1351.

———. 1988c. Comparative floral development in some members of the Vitaceae. p. 121–131. In: P. Leins, S. C. Tucker, and P. K. Endress (eds). Aspects of Floral Development. J. Cramer, Berlin.

———. 1989a. Morphological and anatomical development in the Vitaceae. III. Vegetative development in Parthenocissus inserta. Can. J. Bot. 67:803–816.

_____. 1989b. Morphological and anatomical development in the Vitaceae. IV. Floral development in *Parthenocissus inserta*. *Can. J. Bot.* 67:1356–1365.

_____. 1989c. Morphological and anatomical development in the Vitaceae. V. Vegetative and floral development in *Ampelopsis brevipedunculata*. *Can. J. Bot.* 67:2371–2386.

Goebel, K. 1905. *Organography of Plants*. Part II. Special organography. Reprinted 1969. Hafner Publ. Co., New York.

Goldy, R. G. 1985. Relationship of morphological development to meiotic occurrence in various small fruits. *HortScience* 20:255–256.

Hallé, F., R. A. A. Oldeman, and P. B. Tomlinson. 1978. *Tropical Trees and Forests—An Architectural Analysis*. Springer Verlag. Berlin.

Huglin, P. 1986. *Biologie et écologie de la vigne*. Editions Payot Lausanne Techniques et Documentation, Paris.

Iizuka, M., and T. Hashizuma. 1968. Induction of female organs in staminate grape by 6-substituted adenine derivatives. *Japan. J. Genetics* 43:393.

Jackes, B. R. 1984. Revision of the Australian Vitaceae. 1. *Ampelocissus* Planchon. *Austrobaileya* 2:81–86.

_____. 1987. Revision of the Australian Vitaceae. 2. *Cayratia* Juss. *Austrobaileya* 2:365–379.

Jona, R., and R. Botta. 1988. Fruit set and early berry development in two grapevine cultivars. *Israel J. Bot.* 37:203–216.

Kassemeyer, H-H., and G. Staudt. 1983. The mitotic cycle of the zygote nuclei in *Vitis vinifera*. *Ber. Dtsch. Bot Ges.* 95:449–456.

Kevan, P. G., D. C. A. Blades, U. Posluszny, and J. D. Ambrose. 1988. Pollen dimorphism and dioecy in *Vitis aestivalis*. *Vitis* 27:143–146.

Kevan P. G., R. W. Longair and R. M. Gadawski. 1985. Dioecy and pollen dimorphism in *Vitis riparia* (Vitaceae), *Can. J. Bot.* 63:2263–2267.

Lacroix, C. R., and J. M Gerrath. 1990. Developmental morphology of *Leea guineensis*. 1. Vegetative development. *Bot. Gaz.* 151:204–209.

Latiff, A. 1981. Inflorescence structure and evolution in the Vitaceae. Abstract 08-24-06. XIII International Botanical Congress, Sydney.

Lavee, S. 1985. *Vitis vinifera*. In: A. H. Halevy (ed.), *C. R. C. Handbook of Flowering*. Vol. IV. C. R. C. Press Inc., Boca Raton, FL.

Lavie, P. 1970. Contribution à l'étude caryosystematique des Vitacées. Thèse, Univ. de Montpellier, Montpellier.

Ledbetter, C. A., and D. W. Ramming. 1989. Seedlessness in grapes. *Hort. Rev.* 11:159–184.

Madhava Rao, V. N., and S. K. Mukherjee. 1970. Studies on pruning of grape III. Fruit bud formation in Pusa Seedless grapes (*Vitis vinifera* L.) under Delhi conditions. *Vitis* 9:52–59.

Millington, W. F. 1966. The tendril of *Parthenocissus inserta*: Determination and development. *Am. J. Bot.* 53:74–81.

Milyaeva, E. L., N. K. Smirnova, and M. Kh. Chailakhyan. 1984. Hormonal regulation of development of the female reproductive sphere in seedless grapes. *Dokl. Akademii Nauk SSSR.* 267:252–256.

Moens, P. 1956. Ontogenèse des vrilles et différentiation des ampoules adhésives chez quelques végétaux (*Ampelopsis, Bignonia, Glaziovia*). *La Cellule* 57:369–401.

Moore, J. N. 1970. Cytokinin-induced sex conversion in male clones of *Vitis* species. *J. Am. Soc. Hort. Sci.* 95:387–393.

Mullins, M. G. 1980. Regulation of flowering in the grapevine (*Vitis vinifera* L.). pp. 323–330. In: F. Skoog (ed.). *Plant Growth Substances*. Springer Verlag, Berlin.

Negi, S. S., and H. P. Olmo. 1966. Sex-conversion in a male *Vitis vinifera* L. by a kinin. *Science* 152:1624–1625.

_____. 1971a. Conversion and determination of sex in *Vitis vinifera* L. (*sylvestris*). *Vitis* 9:265–279.

_____. 1971b. Induction of sex conversion in male *Vitis*. *Vitis*. 10:1–19.

_____ . 1972. Certain embryological and biochemical aspects of cytokinin S. D. 8339 in converting sex of a male *Vitis vinifera sylvestris*. *Am. J. Bot.* 59:851–857.

Oberle, G. D. 1938. A genetic study of variations in floral morphology and function in cultivated forms if *Vitis.*. New York State Agr. Expt. Sta. Tech. Bull. 250.

Olivain, C., and R. Bessis. 1987. L'organogenèse inflorescentielle dans les bourgeons anticipés de vigne (*Vitis vinifera* L. cépage Pinot). *Vitis* 26:98–106.

Palma, B. A., and D. I. Jackson. 1989. Inflorescence initiation in grapes—Response to plant growth regulators. *Vitis.* 28:1–12.

Payer, J. B. 1857. Traité d'organogénie comparée de la fleur: orders des Ampélidées. Librairie de Victor Masson, Paris. 157–160.

Posluszny, U., and J. M. Gerrath. 1986. The vegetative and floral development of the hybrid grape cultivar 'Ventura'. *Can. J. Bot.* 64:1620–1631.

Posluszny, U., M. G. Scott, and R. Sattler. 1980. Revisions in the technique of epi-illumination light microscopy for the study of floral and vegetative apices. *Can. J. Bot.* 58:2491–2495.

Pratt, C. 1971. Reproductive anatomy of cultivated grapes—A review. *Am. J. Enol and Vitic.* 22:92–109.

_____ . 1974. Vegetative anatomy of cultivated grapes—A review. *Am. J. Enol. and Vitic.* 25:131–150.

_____ . 1979. Shoot and bud development during the prebloom period of *Vitis*. *Vitis* 18:1–5.

Randhawa, G. S., and S. S. Negi. 1965. Further studies on flowering and pollination in grapes. *Indian J. Hort.* 22:287–308.

Reille, M. 1967. Contribution à l'étude palynologique de la famille des Vitacées. *Pollen et Spores* 9:279–303.

Rives, M. 1972. L'initiation florale chez la vigne. *Conn. Vigne Vin.* 2:127–146.

Rutishauser, R., and R. Sattler. 1985. Complementarity and heuristic value of contrasting models in structural botany. I. General considerations. *Bot. Jahrb. Syst.* 107:415–455.

Scholefield, P. B., and R. C. Ward. 1975. Scanning electron microscopy of the developmental stages of the Sultana inflorescence. *Vitis* 14:14–19.

Shah, J. J., and Y. S. Dave. 1970. Morphohistogenic studies on tendrils of Vitaceae. *Amer. J. Bot.* 57:363–373.

Snyder, J. D. 1933a. Flower bud development in the Concord grape. *Bot. Gaz.* 94:771–779.

_____ . 1933b. Primordial development of the inflorescence of the Concord grapes. *Proc. Am. Soc. Hort. Sci.* 30:247–252.

Srinivasan, C., and M. G. Mullins. 1976. Reproductive anatomy of the grapevine (*Vitis vinifera* L.). Origin and development of the anlage and its derivatives. *Ann. Bot.* 40:1079–1084.

_____ . 1978. Control of flowering in the grapevine (*Vitis vinifera* L.): Formation of inflorescences in vitro by isolated tendrils. *Plant Physiol.* 61:127–130.

_____ . 1979. Flowering in *Vitis*: Conversion of tendrils into inflorescences and bunches of grapes. *Planta* 145:187–192.

_____ . 1980a. Effects of temperature and growth regulators on formation of anlagen, tendrils and inflorescences in *Vitis vinifera* L. *Ann. Bot.* 45:439–446.

_____ . 1980b. Flowering in *Vitis*: Effects of genotype on cytokinin-induced conversion of tendrils into inflorescences. *Vitis* 19:293–300.

_____ . 1980c. Flowering in the grapevine (*V. vinifera*): Histological changes in apices during the formation of the anlage and its derivatives. *Z. Pflanzenphysiol.* 97:299–308.

_____ . 1981a. Physiology of flowering in the grapevine—A review. *Am. J. Enol. Vitic.* 32:47–63.

_____ . 1981b. Induction of precocious flowering in grapevine seedlings by growth regulators. *Agronomie* 1:1–5.

Stout, A. B. 1921. Types of flowers and intersexes in grapes with reference to fruit development. New York Agr. Expt. Sta. Geneva, Bull. 82.

Swanepoel, J. J., and E. Archer. 1988. The ontogeny and development of Vitis vinifera L. 'Chenin blanc' inflorescence in relation to phenological stages. Vitis 27:133–141.

Tompa-Kasirskaya, A. 1984. Scanning electronmicroscopic characteristics of pollen taken from vine infra- and interspecific hybrids. (In Hungarian, English summary). Különlenyomat a Kertészeti Egyetem 48:127–135.

_____. 1985. Sculpture and structure of the pollen sporodermis with different grape-flower types. (In Hungarian, English summary). Különlenyomat a Kertészeti Egyetem 49:69–80.

Tompa-Kasirskaya, A., and P. Kosma. 1978. Study of vine pollen with scanning electron microscope. (In Hungarian, English summary). Különlenyomat a Kertészeti Egyetem 42:25–39.

Tucker, S. C., and L. L. Hoefert. 1966. Ontogeny of the tendril in Vitis vinifera. Am. J. Bot. 55:1110–1119.

Watanabe, S. 1984. Micro structures of pistils in apples and other deciduous fruit trees. Bull. Yamagata Univ. (Agr. Sci.) 9:327–349.

_____. 1987. Pollen-pistil interaction in deciduous fruit trees: Scanning electron microscopic observations. J. Yamagata Agr. For. Soc. 44:41–49.

Watanabe, S., B. Nanba, and Y. Orihara. 1983. A morphological study of floral organs in Vitis coignetiae. J. Yamagata Agr. For. Soc. 43:13–30.

Winkler, A. J., and E. M. Shemsettin. 1937. Fruit bud and flower formation in the Sultanina grape. Hilgardia 10:589–611.

Developmental Physiology of Rabbiteye Blueberry*

R. L. Darnell
Fruit Crops Department, University of Florida
Gainesville, FL 32611

G. W. Stutte
Department of Horticulture, University of Maryland
College Park, MD 20742

G. C. Martin
Pomology Department, University of California
Davis, CA 95616

G. A. Lang
Department of Horticulture, Louisiana Agricultural Experiment Station,
Louisiana State University
Baton Rouge, LA 70803

J. D. Early
Pomology Department, University of California
Davis, CA 95616

*University of Florida Journal Series no. R-00865. LAES Journal Series no. 90-28-4458.

I. INTRODUCTION

A. Approach to Review and Model Development

This review examines the influence of biotic and abiotic factors on the development of rabbiteye blueberry (*Vaccinium ashei* Reade), emphasizing reproduction. The annual cycle of plant development is divided into the following stages (Fig. 9.1):

1. *Vegetative development following harvest:* growth of vegetative apices and accumulation of carbon and nutrient reserves.

2. *Flower bud initiation:* floral induction and the transition of apices from vegetative to reproductive.

3. *Dormancy:* non-growth of the vegetative meristems and the differentiation of reproductive structures.

4. *Anthesis:* pollination and fertilization.

5. *Fruit development:* renewed growth of the vegetative structures, concomitant with growth and maturation of reproductive structures.

After a morphological description of each stage, the literature on blueberry developmental physiology is reviewed. Following this, we

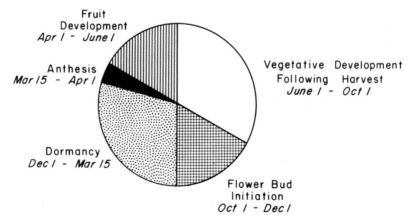

Figure 9.1.   Annual developmental cycle of rabbiteye blueberry in north-central Florida.

integrate the morphological, physiological, and climatological interactions that affect reproduction within each stage of the annual cycle. This integration focuses on reproductive development, vegetative growth, carbon exchange and partitioning, and nitrogen partitioning of mature (ca. six-year-old) rabbiteye blueberry plants. Since our knowledge of specific physiological processes in rabbiteye blueberry is incomplete, we have integrated data from other fruit crops to provide the basis for development of a preliminary model that will facilitate identification and design of future research.

The climatic conditions we consider in this preliminary model are those of north-central Florida, since the rabbiteye blueberry industry is expanding rapidly in this and other states along the Gulf of Mexico. The inherent climatological variation in the Gulf region and the significant variation in fruit production that occurs in this region offer the opportunity to explore yield limitations due to interactions among morphology, physiology, and climatology. Climatology is characterized at Gainesville, Florida (latitude 29°N), where chilling accumulation averages 300–400h below 7°C. Limitations to plant growth and development due to poor soil types, poor nutrient status, or insect and disease pressures will not be considered.

B. Taxonomy

Blueberries are native to North America, belonging to the section *Cyanococcus* of the genus *Vaccinium* in Ericaceae, the heath family (Camp 1945). Cultivated blueberries belong to four species, *V. corymbosum* L. (highbush), *V. angustifolium* Ait. and *V. myrtilloides* Michx. (both lowbush), and *V. ashei* Reade (rabbiteye). Although the rabbiteye blueberry has the longest history of cultivation, with documentation of transplanted native seedlings near Crestview, Florida, in the 1890s (Mowry and Camp 1928), this species' prominence in commercial blueberry production has occurred only in the past 30 years.

C. Geographic Range

The rabbiteye blueberry grows in the southeastern United States, principally southern Georgia, northern Florida, North and South Carolina, southern Mississippi and Louisiana, and southeast Alabama (Eck and Childers 1966; Krewer et al. 1986). It has various habitats and is commonly found along stream banks and lakeshores, in open woods, or in abandoned agricultural areas (Camp 1945). It is cultivated primarily in the piedmont and coastal plain regions of eastern Texas through the Gulf states, and along the Atlantic Coast from north-central Florida to southeastern Virginia (Fig. 9.2). Commercial production in the Northern Hemisphere occurs primarily between latitudes 30°N and 34°N. Climatic

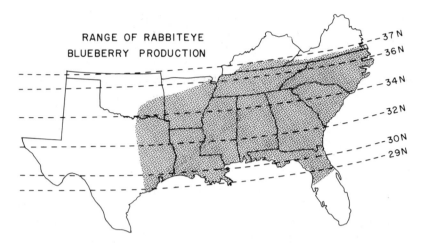

Figure 9.2.   Geographic boundaries of North American rabbiteye blueberry production.

factors present the major limitation to the geographic range of rabbiteye production; however, edaphic factors such as alkaline soils and irrigation water sources high in soluble salts are often the most limiting factors in western regions. In some areas where both highbush and rabbiteye can be grown, the latter is preferred due to the greater tolerance of rabbiteyes to heat, drought, and upland mineral soils (Eck and Childers 1966).

D. Commercial Production

Rabbiteye blueberries are produced on a commercial scale in 10 southeastern states (Table 9.1), constituting the primary blueberry species cultivated in Florida, Georgia, Alabama, Mississippi, Louisiana, and Texas. Each of these states also has limited planting of low chill southern highbush blueberries (tetraploid interspecific *V. corymbosum* hybrids). Both rabbiteye and northern highbush blueberries are grown in Arkansas, North Carolina, and Oklahoma, where rabbiteye plantings are limited primarily to small commercial and "pick-your-own" operations. Arkansas rabbiteyes are grown generally south of Little Rock (Schaller 1983), and North Carolina rabbiteyes are grown primarily in the piedmont and coastal plain regions (Mainland 1986, 1987), extending north into the coastal plain soils of Virginia (Borchers 1979). Rabbiteye blueberries comprise approximately two-thirds of the blueberries cultivated in Tennessee (D. E. Deyton, personal communication) and have been planted in limited experimental trials in Maryland (R. F. Korcak, personal communication), southern Kentucky (G. R. Brown, personal communication), southern Illinois (J. W. Courter, personal communica-

Table 9.1. Estimated 1986 area of rabbiteye blueberry cultivation and 1990 production (Nelson and Holbein 1986).

State	1986 Area (ha)	1990 Production (metric tons)
Georgia	1017	2722
Florida	358	794
Mississippi	253	567
Texas	200	454
Alabama	103	227
South Carolina	72	159
North Carolina	63	136
Louisana	51	113
Tennessee	32	68
Arkansas	10	—
Total	2159	5262

tion), as well as in California and other western states (D. Finch, personal communication).

The earliest-ripening rabbiteye fruit production occurs in Florida, where cultivars such as 'Beckyblue', 'Climax', and 'Bonita' ripen from May 20 through June 20 (Lyrene 1989b). The time of blueberry harvest occurs later in more northern latitudes, with the latest commercial harvests occurring from mid-June through mid-August in North Carolina (Mainland 1987).

E. Climatic Considerations

Rabbiteye blueberries are grown primarily where the mean annual temperature is between 15° and 21°C. Cultivation is generally limited to areas where mean winter minimum temperatures are greater than −9° to −12° (Fig. 9.3A). Buds of rabbiteye blueberry are more susceptible to low temperature damage than those of highbush blueberry (Patten et al. 1989), and extreme low temperature limits production in northern (Borchers 1979) and mountainous (Schaller 1983; Mainland 1986, 1987) regions.

While low winter temperatures limit rabbiteye blueberry production in nothern locations, lack of sufficient chilling temperatures can limit production in southern locations. Rabbiteyes are not recommended for culture in Florida south Ocala (Lyrene 1989b) or in south Texas, due to the potential for insufficient winter chilling. Mean January temperatures greater than 13°C (Fig. 9.3B) indicate the geographic regions that have a high probability for insufficient chilling. In addition, Lyrene (1989b) noted that production of late-ripening rabbiteye cultivars in Florida may be hindered by the increased potential for heavy rains after June 15, which can cause problems with fruit cracking and mechanical harvest.

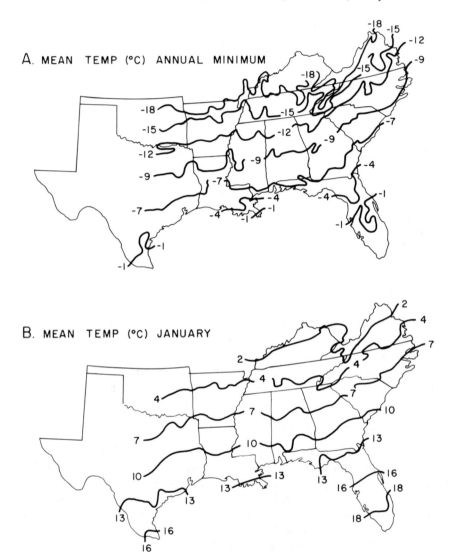

Figure 9.3.   Climatic norms (1898–1940) for the region of rabbiteye blueberry production. A. Mean annual minimum temperatures. B. Mean annual January temperatures. (Source: Climate and Man, Yearbook of Agriculture. 1941. USDA Printing Office, Washington, D.C.)

II. VEGETATIVE GROWTH FOLLOWING FRUIT HARVEST

A. General Literature Review

1. Vegetative morphology. The commercial blueberry is a deciduous perennial plant with simple leaves arranged alternately on the stem (Eck 1966). Leaf morphology varies among species and may be used as an aid to identification. Leaf shape ranges from elliptic in lowbush to ovate in highbush to spatulate or oblanceolate in rabbiteye blueberry. The abaxial leaf surface of highbush blueberry is commonly covered with fine hairs. Leaves of rabbiteye blueberry have minutely stalked glands on the abaxial surfaces; leaves of both lowbush and highbush blueberry are nonglandular.

Blueberry roots are very fine, fibrous, and lack root hair. There is typically no taproot (Eck and Childers 1966). The root distribution in rabbiteye blueberry depends on soil type. Austin (1982) found that rabbiteye blueberry root systems penetrated to 100 cm in a well-drained sandy loam.

Root and shoot growth of blueberries occurs in flushes throughout the growing season. Abbott and Gough (1987a) reported that highbush blueberries in Rhode Island undergo a major flush of shoot growth following fruit harvest. A flush of root growth begins during mid-August and continues through September, during the period of flower bud initiation, but before the onset of dormancy. There appears to be a cycling pattern between root and shoot growth during late summer, but the cycles are not absolute. The August/September flush of root growth may be due to a renewed availability of carbon for support of root growth in the absence of a fruit sink, although no supporting data were presented. There are no data on growth flushes of rabbiteye blueberry; however, observation indicates that several flushes of shoot growth are common.

2. Factors influencing growth.
  a. Carbon exchange and partitioning. No specific data describing carbon exchange and partitioning in blueberry following fruit harvest exist. In general, highbush blueberries exhibit leaf photosynthetic rates of 11–12 μmol CO_2 m^{-2} s^{-1} (Moon et al. 1987), higher than rates in rabbiteye blueberries, which average 5–7 μmol CO_2 m^{-2} s^{-1} (Teramura et al. 1979; Davies and Flore 1986a, 1986b). This difference is due to decreased stomatal conductance in rabbiteye blueberries (Davies and Flore 1986a, 1986b), which may partially account for their greater tolerance to high temperatures and drought conditions. However, optimum temperatures for photosynthesis were similar in both species, ranging from 20–30°C (Davies and Flore 1986a, 1986b; Moon et al. 1986). Photosynthetic rates

for both species increased with increasing photosynthetic photon flux (PPF) up to about 700–800 μmol m^{-2}s^{-1}. Dark respiration in highbush blueberry averaged 1.0–1.6 μmol CO_2 m^{-2} s^{-1}, about 10–15% of CO_2 assimilation (Moon et al. 1987). Respiration rates for rabbiteye blueberry ranged from 0.8 to 1.1 μmol CO_2 m^{-2} s^{-1} (Teramura et al. 1979).

There are no descriptive reports of carbon allocation between roots and shoots of rabbiteye blueberry, nor are there long-term studies of carbon partitioning in this species. Pritts and Hancock (1985) examined the biomass partitioning in wild populations of V. corymbosum and found that roots and shoots were the strongest vegetative carbon sinks until blueberry plants were about 20 years old, after which carbon allocation to leaves exceeded that to shoots and roots.

b. Nutrient partitioning. Blueberries are calcifuges and grow best on acid soils. A wide range of optimal soil pH has been reported. Korcak (1989) reported pH optima ranging from 3.9 to 6.1 for rabbiteye and 4.3 to 6.7 for highbush. Recommended pH is generally between 4.5 and 5.5. Cain and Eck (1966) and Korcak (1989) have summarized the numerous reports on leaf nutrients, sampling procedures, and deficiency symptoms in blueberry. Relative to other fruit crops, blueberries require low concentrations of all essential elements for adequate growth. The requirement for leaf Ca is especially striking, 0.3% for blueberry vs. 1.7% for apple (Korcak 1989).

The effects of ammonium vs. nitrate nitrogen on blueberry growth has recently been reviewed (Korcak 1989), and will be considered only briefly. Ammoniacal nitrogen is considered to be the most suitable nitrogen source for blueberries since it is more available in acid soils than the nitrate form (Townsend 1967, 1969; Spiers 1978a, 1979). Townsend (1967) found that fertilization with ammonium nitrogen produced more vegetative growth and a more extensive root system than nitrate nitrogen under field conditions. However, there is not universal agreement that blueberry growth is best with ammonium compared to nitrate nitrogen. Nitrogen source did not affect yield or leaf dry weight in a two-year solution culture study on highbush blueberry growth (Hammett and Ballinger 1972). Leaf nitrogen was slightly higher with ammonium nitrogen, which may be explained by the lower pH of ammonium solutions compared to nitrate solutions. Nitrogen source (ammonium vs. nitrate) did not affect shoot growth or final dry weight in half-high blueberries (V. corymbosum L. × V. angustifolium) grown hydroponically for seven months (Rosen et al. 1989). Under field conditions, Austin and Bondari (1986) and Korcak (1986) found little or no effect of nitrogen source on growth or yield of blueberry. These results indicate that blueberry can use both forms of nitrogen effectively.

Nitrogen source influences the uptake of other elements (Cain 1952; Townsend 1967). Increased nitrate rates increased root calcium, phosphorus, and manganese concentrations (Peterson et al. 1988).

Increased ammonium rates increased plant phosphorus and sulfur and shoot iron concentrations, and decreased shoot calcium concentrations and root manganese concentrations (Peterson et al. 1988). Cain (1952) suggested that ammonium lowers plant pH, making iron more soluble and thus more available to the plant. Townsend (1967) reported that pH of soil effluents from nitrate fertilized plants was above pH 6.0. Precipitation of phosphorus and iron out of soil solution at the higher pH may reduce potential uptake and result in lower iron and phosphorus leaf concentrations with nitrate fertilization (Townsend 1967).

The critical leaf nitrogen level for development of deficiency symptoms in rabbiteye blueberry leaves is about 1.5% (Spiers 1983). High levels of nitrogen fertilization may not dramatically increase levels of leaf nitrogen (Eck 1977), but may result in excessive vegetative growth that reduces final yield. There has been no research to determine the regulatory mechanisms of nitrogen transport and metabolism in blueberry.

Leaf nutrient concentrations vary with sampling position on the shoot. Trevett et al. (1968) found that the concentrations of N, P, K, Mg, Fe, Zn, Cu, Mo, and B were generally higher in tip leaves than in basal leaves, while concentrations of Mn and Ca were lower. Concentrations of N, P, K, Ca, and Mg were generally higher in leaves from one-year wood than from two-year wood.

No comprehensive studies of seasonal nutrient partitioning in blueberry have been conducted, although several studies of the seasonal changes in elemental content of lowbush, highbush, and rabbiteye blueberry have been performed to optimize leaf sampling for diagnostic purposes (Bailey et al. 1962; Ballinger 1966; Townsend and Hall 1970; Eaton and Meehan 1971; Daughty et al. 1984; Clark et al. 1989). Although a general picture of seasonal changes in elemental content of mature, mid-shoot leaves on fruiting shoots can be described, differences in cultivar, soil nutrient content, irrigation conditions, plant age, and sampling procedures make definitive statements difficult. Furthermore, studies that have been based on calendar dates are not readily correlated to developmental events.

In general, the nitrogen and phosphorous content of blueberry leaves declines throughout fruit development, then stabilizes following harvest (Ballinger 1966; Townsend and Hall 1970; Clark et al. 1989; Retamales and Hanson 1990). There is a trend towards declining potassium levels in leaves as fruit mature, but the concentrations are highly variable in all studies, making generalizations difficult. Calcium leaf content generally increases during the growing season, with the greatest increase following harvest (Ballinger 1966; Townsend and Hall 1970; Eaton and Meehan 1971; Bailey et al. 1982). Clark et al. (1989) reported only minor changes in leaf calcium content during the growing season under Arkansas conditions. There is little variation in leaf magnesium content during the growing season (Bailey et al. 1962; Ballinger 1966; Clark et al. 1989). Few generalizations can be made about seasonal changes in microelements due to the limited number of studies and the variability in concentrations found among those studies.

c. *Water relations.* In rabbiteye blueberry, root dry weight production was enhanced by irrigation, incorporation of peat into the soil, and mulching (Spiers 1986). Mulching resulted in the greatest increase in root dry weight but the shallowest root system compared to the other two treatments. Spiers explained these results in terms of improved water relations, but no data on soil moisture, water use, or temperature were presented. Spiers' conclusions are similar to those of Gough (1980), who reported that root growth in highbush blueberry under sawdust mulch occurred primarily in the decomposed portion of mulch at the soil-mulch interface. Creech (1988) found that rabbiteye blueberries grew as well under polypropylene fabric mulch as under pine bark mulch. This was attributed to enhanced lateral movement of drip irrigation water into the root zone.

Davies and Johnson (1982) found that moderate water stress (i.e., stomatal conductance 80% of well-watered plants) and severe water stress (i.e., stomatal conductance 50% of well-watered plants) decreased leaf area and stem and leaf dry weight in one-year-old rabbiteye blueberry. Root dry weight was unaffected by moderate water stress, decreasing significantly only under severe water stress. Korcak (1986) did an extensive study of the adaptability of blueberry species to various soil types. He grew non-fruiting blueberry plants in five different soil types with irrigation and various soil amendments. Incorporation of peat into the soil increased overall canopy growth and root dry weight of all species. This effect on growth appears to have been independent of water availability (continuous drip irrigation was applied) or temperature effect of a mulch (the peat was incorporated). However, there were no direct measurements of soil moisture or temperature. Korcak (1986) suggested that peat affects the cation exchange capacity within the soil, improving nutrient uptake.

B. Model Development: June 1 to October 1

1. Climatology. During the period of vegetative development following harvest in Gainesville, Florida, the average maximum and minimum air temperatures range from 32.7° and 20.6°C, respectively, in mid-June, increasing slightly in July and August to 33.1° and 21.8° before decreasing to 31.7° and 20.8° by mid-September (Fig. 9.4). Average soil temperatures (at 10 cm) range from 30.6° to 27.2° in mid-June decreasing to 29.2° to 26.8° by mid-September, and the daylength decreases from 14:00 to 11:50 h.

VEGETATIVE DEVELOPMENT FOLLOWING HARVEST

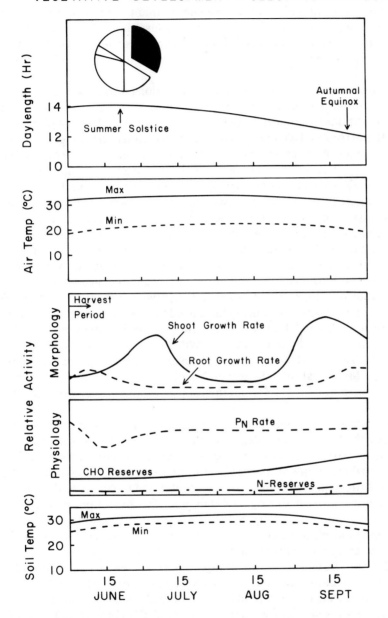

Figure 9.4. Environmental conditions (daylength, air and soil temperatures) and presumed physiological and morphological events during vegetative development following harvest of rabbiteye blueberry in north-central Florida. Climatological data based on 10-year average (1979–1989). P_N rate = whole plant net photosynthetic rate; CHO reserves, N-reserves = whole plant carbohydrate and nitrogen reserves, respectively.

2. Reproductive development. Although flower bud initiation may be beginning in some cultivars by the end of this period, the majority of initiation occurs when the photoperiod is less than 12 h. Flower bud formation and subsequent yield, however, are dependent on the development of the vegetative canopy during the period following fruit harvest.

3. Vegetative growth. Environmental conditions during the long north-central Florida summer, when supplemented with irrigation, promote ideal conditions to support meristem growth after fruit harvest (Fig. 9.4). Abbott and Gough (1987a) found that highbush blueberry underwent two major flushes of shoot and root growth under Rhode Island conditions. The long growing season in Florida results in at least three major flushes of growth. The first cycle takes place prior to harvest (see section VI and Fig. 9.12), with the remaining two cycles after harvest (Fig. 9.4). The second cycle of shoot growth occurs just after harvest and may be initiated by the decreased reproductive demand for carbon and nutrients following harvest. A limited period of root growth may also occur soon after fruit harvest; however, high soil temperatures during summer in north-central Florida and competition with vegetative growth may limit feeder root development. Average soil temperatures (at 10 cm) in the north-central Florida area during the period following harvest range from 26.8° to 30.6°C, much higher than the optimum soil temperatures of 14°–18°C reported for highbush root growth in Rhode Island (Abbott and Gough 1987a). The final and largest flush of root growth probably occurs after the final flush of late-season shoot growth (Fig. 9.4), when soil temperatures are favorable, competition is minimal, and resources are available.

Although environmental conditions seem to favor continuous vegetative growth after harvest, a depression in shoot growth rate occurs during the hottest part of the summer (Fig. 9.4). Shoot growth resumes in late summer when temperature, daylength, and light intensity are less. As there is little expected competition for water and nutrients with roots or reproductive growth during mid-summer, the reduction in potential canopy growth is probably a result of adverse, possibly interactive environmental effects such as evapotranspiration demands, high light intensity, and high temperatures on physiological processes.

In Florida, Davies et al. (1979) reported xylem water potential (ψ) for rabbiteye blueberry ranged from −0.2 MPa predawn to −1.04 MPa in the afternoon in well-watered fields. In rabbiteye blueberries, stomatal closure occurred when leaf ψ approached −2.2MPa (Davies and Johnson 1982), which is similar to values reported for apple and peach (Landsberg and Jones 1981). Stomatal conductance decreased rapidly as leaf ψ decreased (Davies and Johnson 1982), which would lead to a decrease in net photosynthesis. Other environmental factors also may affect stomatal conductance. In apricot, stomatal conductance was dependent more on relative humidity and leaf temperature (Schulze et al. 1974) than on water potential or internal CO_2 concentration (Schulze et al. 1975). The potential effect of high relative humidity during summers in north-central Florida on rabbiteye stomatal conductance has yet to be examined.

The lack of root hairs in blueberries limits their potential for water absorption under field conditions (Coville 1910). This feature, in concert with the rabbiteye blueberry's stomatal sensitivity to leaf ψ, may result in the development of a smaller canopy under moderate water stress (Davies and Johnson 1982). This effect on vegetative growth would be consistent with that reported in apple (Landsberg and Jones 1981), where cycles of low water potential reduced shoot extension, but not shoot number. Consequently, it is apparent that those blueberry cultivars with a smaller ratio (large above-ground canopy relative to root system) are likely to experience detrimental water balance conditions even under well-watered conditions. Rabbiteye cultivars with large canopy development and limited feeder root density are especially susceptible to water stress (Davies et al. 1979). Thus, management of canopy size may be important to reproductive development, reserve accumulation, and the ability to support a large crop the next season. The root:shoot ratios of most cultivars of rabbiteye blueberry have not been determined, but characterization may be necessary before developing cultural practices to optimize the bearing surface of the shoot system.

Strategies to develop root system enlargement prior to episodes of shoot elongation may lead to better development of future reproductive apices. It is likely that reduction of soil temperatures may increase root growth. Although mulching and drip irrigation encourage confinement of what is typically a shallow rooted plant (Gough 1980), these practices may also serve to decrease root temperatures and therefore increase root growth (Patten et al. 1988). Furthermore, a deep, well-drained soil that is well irrigated is critical for maximum root growth (Doehlert 1937; Patten et al. 1988).

4. Carbon exchange and partitioning. The high summer light intensity in north-central Florida should not limit meristem growth since photosynthesis in rabbiteye blueberry is saturated at 700–800 μmol m^{-2} s^{-1} PPF (Teramura et al. 1979). Even in the shaded portion of the canopy these authors reported PPF of 250–300 μmol m^{-2} s^{-1}, which is sufficient for 50–60% of maximum photosynthesis. There are no data on high light intensity-induced photoinhibition (Bjorkman 1981) in rabbiteye blueberry.

Optimum leaf temperatures for photosynthesis vary during the growing season for many crops. Lange et al. (1974) found that optimal temperatures for photosynthesis in apricot ranged from 24°C in late March to 36° in mid-August. Sour cherry exhibited a maximum photosynthetic temperature optimum at 25°C, with the variations in optima dependent on light intensity and vapor pressure deficit (Sams and Flore 1982). The photosynthetic temperature optima for the subarctic *Vaccinium* species, *V. ulisinosum* L. and *V. vitis-idaea* L., varied by 5° to 10°C during the growing season (Karlsson 1985b). There is no information available on variation in photosynthetic rate of rabbiteye blueberry during the growing season. The possibility exists that the mid-season depression in rabbiteye blueberry shoot growth rate (Fig. 9.4) may be, at least partially, due to reduced net photosynthesis resulting from evapotranspiration-induced low ψ, high light intensity-induced photoinhibition, and/or supraoptimal temperatures.

At fruit harvest, the plant adjusts to removal of the major sink. The amount of adjustment depends on the fruit growth rate at the time of harvest, the overall fruit load, and the proportion of fruit removed. Logic suggests that removal of sinks, such as fruit, would result in a decrease in net photosynthesis. This seems to be the case for many annuals, e.g., eggplant (Lenz 1979), pepper (Hall and Brady 1977), and soybean (Mondal et al. 1978). The results are variable for perennial plants. There is no information on photosynthetic adjustment following fruit harvest in blueberry. Chalmers et al. (1975) reported that total daily photosynthesis of peach trees increased during stage III of fruit development, followed by a significant decline after fruit harvest. Total daily photosynthesis was higher in fruiting vs. non-fruiting citrus (Lenz 1979) and grape (Downton et al. 1987). No consistent fruit effect on photosynthesis has been reported for sour cherry (Sams and Flore 1983) or sweet cherry (Roper et al. 1988). These contrasting results may be due to any of several aspects of experimental protocol: plant species, stage of plant development, environmental conditions, and method of gas exchange measurement. Gucci (1988) suggested that perennial fruit trees, in contrast to annuals, may utilize their expansive vegetative surface as a well-buffered reservoir for adjusting carbon and water relations following fruit removal.

However, there is general agreement among investigators that carbon assimilation rates are highest during the most rapid fruit growth phase (Chalmers et al. 1975; Fujii and Kennedy 1985; Roper et al. 1988). In blueberry, the most rapid growth phase occurs during stage III. Spiers (1981) reported that rabbiteye blueberries increased in volume at 29 cc per day during the last 20 days of growth, which was three times greater than the volume gain at stages I and II. There is a similar three-fold increase in dry weight gain during stage III (G. A. Lang, unpublished data). Based on this, it is likely that the photosynthetic rate in rabbiteye blueberry decreases, at least temporarily, following harvest (Fig. 9.4). Gucci (1988) reported that defruited plum trees exhibited a temporary (ca. six days) decrease in net photosynthesis; however, as vegetative growth increased in response to defruiting, net photosynthesis increased six-fold compared to fruiting trees.

The time required for photosynthetic recovery by rabbiteye blueberry following fruit harvest is unknown; however, the long growing season between harvest and terminal bud set in the north-central Florida area, coupled with favorable temperatures and high light intensity, promote sufficient carbon assimilation to support new root growth and canopy development for flower bud initiation, as well as support carbohydrate reserve build-up. Carbohydrate reserves probably increase gradually during the four months between harvest and leaf senescence (Fig. 9.4), as has been reported for a number of fruit crops (Ryugo 1988).

5. Nitrogen partitioning. Nitrogen is probably the element that has the greatest influence on reproductive development in blueberry. High nitrogen levels in the soil can result in excessive vegetative growth and a prolonged growing season. This increased demand for nutrients by non-reproductive structures may limit reproductive development the following season. In several fruit crops, nitrogen applications can be timed to optimize reproductive growth. Summer applications of nitrogen increased both the number and density of pear flower buds (Taylor et al. 1975). Nitrogen taken up by almond trees following harvest was partitioned primarily into subsequently formed flower buds (Weinbaum et al. 1984). Fall applications of nitrogen had little effect on vegetative growth of apple, but improved flower quality by extending stigma receptivity and ovule longevity (Williams 1965). Thus, nitrogen accumulation during the period following harvest influences not only the bearing surface and flower density, but may affect the duration of the effective pollination period the following spring. Consequently, nitrogen application to rabbiteye blueberries following harvest would be expected to increase vegetative growth, thus increasing the bearing surface and possibly flower bud density and floret quality.

Little experimental data exist to establish the timing of accumulation and utilization of nitrogen reserves in blueberry during the vegetative growth period following harvest. At the end of the growing season, nitrogen concentrations in rabbiteye blueberry are probably highest in leaf tissue and lowest in stem tissue, as reported for highbush blueberry (Retamales and Hanson 1989). Total nitrogen levels in rabbiteye blueberry probably remain relatively low and constant during the early part of the vegetative growth period following harvest. Towards the end of the period, the levels of total nitrogen in crown and root tissue probably begin to increase slowly (Fig. 9.4). This would be analagous to highbush blueberry, where crown and root tissues combined accounted for more than half of the total plant nitrogen at the end of the season (Retamales and Hanson 1989).

III. FLOWER BUD INITIATION

A. General Literature Review

1. Reproductive apex morphology. Blueberry flower buds are simple, containing a racemic inflorescence of 1–16 flowers (Eck 1988), depending on the species and cultivar. Flower buds are initiated in late summer and/or early fall (Aalders and Hall 1964; Gough et al. 1978), usually on the upper part of current season's growth (Gough et al. 1976; Eck 1988). Initiation is basipetal (Lyrene 1984), and within the cluster, florets are initiated and differentiated acropetally (Aalders and Hall 1964; Gough et al. 1978).

Flower development has been described in lowbush and highbush, but not rabbiteye, blueberry. Bell and Burchill (1955) reported that floret primordia appeared in June in lowbush blueberries under Nova Scotia conditions. Soon thereafter, flower parts began differentiating, starting with scales on the outside, followed by sepals, petals, stamens, and finally carpels. The actual reproductive cells differentiated by early August. By the end of August, microspore and megaspore mother cell formation began. Development continued during autumn, with the embryo sac undergoing cell division. A similar progression of anatomical events, but with slightly different timing, occurs in highbush blueberry (Gough et al. 1978). Meristem flattening and appearance of sepal primordia began in late July to late August under Rhode Island conditions. By October, all flower parts were visible, and the ovaries were highly meristematic. By the first of November, ovule primordia were apparent, microspore mother cell formation began and some megaspore mother cell activity was evident. Cell division ceased by early December, and did not begin again until spring.

2. Factors influencing flower bud initiation. Several cultural and environmental factors influence flower bud initiation in blueberries; however, as in any flowering species, the physiological and biochemical mechanisms of flower bud initiation remains unknown.

a. *Cultivar.* The number of flower buds initiated per shoot is dependent upon cultivar (Eck 1988). Lyrene (1984) found that the average number of flower buds on the distal 10 cm of shoots ranged from 2.2 to 6.4 for 10 rabbiteye cultivars. In a later study, Lyrene (1989a) reported an average range of 1.8 to 8.5 flower buds per shoot on 24 cultivated-type *V. ashei* clones and 1.9 to 9.0 flower buds per shoot on 22 native *V. ashei*, indicating that the clonal variation in flower bud number is inherent in the population and not a result of artificial selection pressure.

b. *Apical dominance.* Clonal variation in flower bud number may be related to the "strength" of apical dominance. The conversion of vegetative to floral buds in blueberry usually occurs at the shoot apex and proceeds basipetally for a certain number of nodes. Blueberries have the ability to compensate for flower bud removal by initiating new flower buds. Lyrene (1984) found that if rabbiteye blueberry shoots were pruned in late October to remove already formed flower buds, new flower buds were initiated in leaf axils below the cut. Clones that initiated an average of seven flower buds, which were then removed, reinitiated an average of six more flower buds farther down the shoot. Clones that initiated an average of two flower buds reinitiated an average of one flower bud upon removal of the first flower buds. These results indicate that flower buds formed distally on the shoot may inhibit the conversion of buds farther down on the same shoot, suggestive of apical dominance.

c. *Photoperiod.* A substantial amount of data indicates that flower bud initiation in blueberry is promoted under short days. Aalders and Hall (1964) and Hall and Ludwig (1961) reported that, in general, initiation in lowbush blueberries required photoperiods less than 12 h. There was clonal variation, however, with some clones able to produce flower buds at 14 and 16 h, although the number of flower buds produced was usually not as extensive (Hall and Ludwig 1961). At least six weeks of the required photoperiod was needed for normal flower bud formation. Concomitant with an increase in flower bud production was a decrease in vegetative growth (measured as total new shoot growth). The photoperiod response of highbush blueberries is similar to lowbush. Eight weeks of 8-, 10-, or 12-h photoperiods resulted in flower bud initiation in highbush cultivars, whereas fewer flower buds were formed under 14- and 16-h photoperiods (Hall et al. 1963). Vegetative growth was greatest under 16-h photoperiods and least under the 8-h photoperiod. Surprisingly, more flower buds were initiated under 10-h photoperiods than under 8-h. This was attributed to the greater amount of vegetative growth produced under 10-h photoperiods, which increased shoot growth relative to the 8-h treatment, providing more shoots on which to produce flower buds.

Photoperiod effects in rabbiteye blueberries are less well documented. Defoliating shoots of field-grown 'Aliceblue' plants in late August or mid-September markedly decreased flower bud initiation on those shoots (P.M. Lyrene, unpublished data). Partial defoliation, where every other leaf is removed, also decreased initiation, although not as dramatically. Interestingly, alternate leaf removal resulted in flower bud formation only at those nodes which still retained leaves; those nodes at which leaves had been removed generally formed only vegetative buds. This suggests that the effect of leaf removal is a localized one, and may argue against carbohydrate supply as a major factor in flower bud initiation. However, the possibility of localized carbohydrate effects cannot be ruled out.

Further studies suggest that flower bud initiation in some rabbiteye cultivars may be promoted directly by short daylengths. 'Beckyblue' plants, exposed to gradually decreasing daylengths from 10 to 8 h over a five-week period starting October 1, initiated twice as many flower buds as those plants exposed to natural daylengths (i.e., gradually decreasing daylengths from 12 to 11 h) during that same time period (R.L. Darnell, unpublished data). After equal amounts of artificial chilling, the two groups of plants were placed in the greenhouse to observe subsequent growth. Plants exposed to shortened photoperiods the previous fall had a greater percentage of floral budbreak (based on the number of flower buds formed within each treatment) and a shorter, more concentrated bloom period than did the control plants. Fruit set percentage decreased in the short photoperiod treatment (probably due to the three-fold increase in the number of pollinated flowers); however, the number of harvested fruit was double that of the control. Furthermore, fresh weight per berry increased under the short fall photoperiod treatment, despite the fact that fruit number was higher. This suggests that short photoperiods increase not only the number of flower buds initiated, but also flower bud development. Interestingly, 'Climax' rabbiteye blueberry did not respond to the photoperiod treatments in any way, suggesting that the critical daylength (or nightlength) for flower bud initiation in rabbiteye blueberry is cultivar dependent.

The interaction of photoperiod with temperature on flower bud initiation in rabbiteye blueberries has not been examined. In lowbush blueberry, temperature did not appear to alter the photoperiodic requirement for initiation (Hall and Ludwig 1961; Aalders and Hall 1964).

d. *Light intensity and carbohydrate supply.* Shading reduces flower bud initiation, probably by limiting carbohydrate production. Hall (1958) reported that the total number of flower buds in lowbush blueberry decreased with decreasing light intensities. In a later study, Hall and Ludwig (1961) found that lowbush blueberries grown under 60% full sun from the end of April through October initiated as many flower buds as plants grown in full sun. However, plants grown under 50% full sun produced significantly fewer flower buds. A critical light intensity for flower bud initiation in rabbiteye blueberries has not been determined.

There is indirect evidence that flower bud initiation in rabbiteye blueberries is affected by carbohydrate supply. Lyrene (1989a) reported a significant positive correlation between "healthy" leaves and high initiation in both cultivated and native populations. Leaf health was rated subjectively, based on the extent of leaf diseases present in mid-December (presumably after the photoperiod requirement for initiation was met for most cultivars). Furthermore, Crane and Davies (1988a) surmised that the reduction in initiation in rabbiteye blueberries flooded for five days resulted from a decrease in leaf area and/or a reduction in carbon assimilation observed.

e. *Mineral nutrition.* Application of nitrogen to rabbiteye blueberries late in the fall increased vegetative shoot growth as well as flower bud initiation per unit shoot length (R.K. Reeder and R.L. Darnell, unpublished data). Two-year-old potted 'Aliceblue' plants fertilized three times in October with 9 g 9N-9P-17K soluble fertilizer initiated 0.62 flower buds/cm new shoot length compared to 0.41 flower buds/cm new shoot length in plants which did not receive October fertilization. Flower bud initiation on the old growth (i.e., current season's growth occurring prior to the October fertilizer treatments) was 10-fold higher for the fertilized plants than for the non-fertilized (0.25 vs. 0.02 flower buds/cm shoot length).

B. Model Development: October 1 to December 1

1. Climatology. During the flower bud initiation period in Gainesville, Florida, the average maximum and minimum air temperatures are 28.3° and 15.9°C, respectively, in mid-October, decreasing to 24.4° and 11.1°, respectively, by mid-November (Fig. 9.5). Average soil temperatures (at 10 cm) range from 25.3° to 23.1°C in mid-October, decreasing to 20.8° to 18.7° by mid-November, and the daylength decreases from 11:50 to 10:25 h.

2. Reproductive development. There is no information on the anatomy and morphology of flower bud initiation in rabbiteye blueberries. However, the sequence of events is probably similar to that described previously for lowbush and highbush blueberries. Initial differentiation may begin later under central Florida conditions, due to the longer photoperiods in late summer and fall compared to more northern areas. Later differentiation, coupled with earlier anthesis for rabbiteye blueberries in this area, suggests that flower bud differentiation in central Florida may continue unabated throughout the winter, as long as winter temperatures remain conducive for growth.

Genetic, environmental, and cultural factors which influence flower bud initiation in rabbiteye blueberry—such as cultivar, apical dominance, carbohydrate supply, and mineral nutrition—are discussed in the general literature review. In the rabbiteye model, photoperiod and possibly photoperiod × temperature interactions might be major factors influencing initiation. Preliminary results of photoperiod effects on initiation in 'Beckyblue' and 'Climax' have been discussed above. However, there is no information on photoperiod × temperature interactions. Although the literature on lowbush suggests this interaction is negligible, literature on other plants indicates significant interaction between photoperiod and temperature. Furthermore, the variations in timing of blueberry floral anatomical development observed at different latitudes (Bell and Burchill 1955; Gough et al. 1978) might be explained by photoperiod/temperature interactions. Pettersen (1972) reported that

FLOWER BUD INITIATION

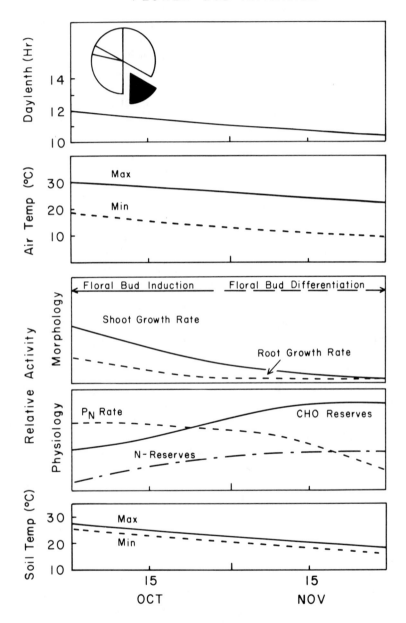

Figure 9.5. Environmental conditions (daylength, air and soil temperatures) and presumed physiological and morphological events during flower bud initiation of rabbiteye blueberry in north-central Florida. Climatological data based on 10-year average (1979–1989). P_N rate = whole plant net photosynthetic rate; CHO reserves, N-reserves = whole plant carbohydrate and nitrogen reserves, respectively.

flower bud initiation in several rhododendron cultivars was enhanced under short day (10 h) conditions when temperatures were above 20°C. However, at 15° to 16°C, cultivars were day-neutral with respect to initiation. More recently, Boyle (1990) found that short day enhancement of flower bud initiation in dwarf Easter cactus occurred when plants were grown at 10°C; however, initiation was day-neutral in plants grown at 18°. Durner et al. (1984) reported that initiation in June-bearing strawberry cultivars was inhibited under long days (16 h), or night interruption (9-h light period with a 3-h interruption in the middle of the night), compared to short days (8 h) at a constant 21°C. However, under a diurnal temperature regime of 18°C day/14° night, initiation was day-neutral.

In field situations, the average late summer/early fall temperatures in the central Florida area probably do not fluctuate enough from year to year to cause significant annual variation in the photoperiodic/temperature influences on flower bud initiation in rabbiteye blueberries. It is apparent, however, that cultivars vary in the magnitude of their initiation response to photoperiod. This suggests that careful attention must be given to cultivar selection to ensure that daylength does not limit initiation in areas of relatively low latitude, such as north-central Florida.

3. Vegetative growth. As daylength decreases below 12 h and flower bud initiation increases, there is a concomitant decrease in shoot growth rate (Fig. 9.5). Hall and Ludwig (1961) reported that shoot growth in highbush blueberry decreased as daylength decreased from 16 to 8 h. This response was particularly pronounced as temperatures decreased from 21° to 10°C. Shoot growth of rabbiteye blueberry in the north-central Florida area also decreased during flower bud initiation in response to decreasing photoperiods (R.L. Darnell, unpublished data). However, the relatively high air temperatures during this period (Fig. 9.5) suggest that the decrease in shoot growth rate may not be as rapid as in northern production areas.

Root growth rate of rabbiteye blueberries also probably decreases during the flower bud initiation period, following the increase presumed to occur at the end of the vegetative development period following harvest (Fig. 9.5). Periodicity in root growth has been reported for highbush blueberries (Abbott and Gough 1987a) as well as other fruit crops (Atkinson 1980). Although optimum root growth of highbush blueberry occurred at soil temperatures between 14° and 18°C (Abbott and Gough 1987a), total shoot growth of highbush blueberry increased as soil temperature increased up to 30° (Bailey and Jones 1938). Average soil temperatures in the north-central Florida area exceed 18°C during the flower bud initiation period (Fig. 9.5). The effects of soil temperatures on rabbiteye blueberry root and shoot growth have not been determined; however, decreased soil temperatures through the use of surface mulch (Patten et al. 1988) might be expected to enhance root growth.

4. Carbon exchange and partitioning. Little information is available on carbon exchange or partitioning during flower bud initiation in blueberry. Canopy photosynthetic rate probably begins decreasing as shoot growth rate decreases and leaf age increases (Fig. 9.5). This general effect of leaf age on photosynthetic rate has been reported for apples and cherries (Barden 1978; Sams and Flore 1982). Although photosynthetic rate is assumed to decrease, carbon reserves probably continue to accumulate during the initiation period. Worley (1979) reported that total carbohydrates in pecan trees increased during fall, coincident with the induction/initiation of pistillate flower buds. Early defoliation reduced flower formation in pecan (Worley 1979), as it did in blueberry (R.L. Darnell, unpublished data). Environmental conditions in the north-central Florida area (especially daylength and temperature) enable sufficient carbon assimilation to support both initiation and reserve accumulation as long as leaves remain healthy. Carbohydrate reserves in rabbiteye blueberry probably increase throughout initiation, reaching a maximum at the end of this period (Fig. 9.5).

As in other fruit crops (Hansen and Grauslund 1973; Tromp 1983), carbohydrate storage in rabbiteye blueberry probably occurs in roots and wood of shoots, primarily as starch, although soluble carbohydrates have also been identified as important storage compounds in several tree fruit species (Loescher et al. 1990). Starch concentration is usually higher in the roots than in the above-ground parts of woody plants, and thus roots are considered the major site of carbohydrate storage (Tromp 1983; Loescher et al. 1990). The preferential accumulation of starch reserves in roots, rather than shoots, of many fruit trees may reflect the maintenance of higher metabolic rates during winter due to higher soil, as opposed to air, temperatures (Tromp 1983). Partitioning of carbohydrate reserves in blueberry has not been examined. However, the absence of freezing air temperatures in the north-central Florida area during the flower bud initiation period suggests that starch synthesis and storage may occur to a similar extend in both roots and shoots. Furthermore, the fibrous, shallow root system of the blueberry plant probably does not have the capacity to store reserves to the same extent as plants with more extensive root systems.

5. Nitrogen partitioning. Although carbohydrates comprise the majority of reserve materials in fruit crops, nitrogen reserves are also important. Taylor and van den Ende (1969) found a positive correlation between spring growth of mature peach trees and the level of storage nitrogen present prior to growth resumption.

Nitrogen in leaf tissues is remobilized to other plant parts prior to leaf senescence (Millard and Thomson 1989). Remobilization of leaf nitrogen into shoot tissue of peach and apple began after harvest and continued until leaf abscission under Pacific Northwest conditions (Rogers et al.

1953; Batjer and Westwood 1958; Titus and Kang 1982). However, leaf nitrogen was not remobilized until just prior to leaf senescence in peaches in the mid-Atlantic area (G.W. Stutte, unpublished data). The majority of nitrogen remobilization from apple leaves occurred three to four weeks before leaf senescence (Oland 1963; Tromp 1970). The extent of remobilization appears to vary with species and environment, ranging from 23% to 50% in senescing apple leaves (Titus and Kang 1982) and 45% to 56% in senescing peach leaves (Taylor and van den Ende 1969; Castagnoli et al. 1990). If leaf senescence in the fall was delayed and/or if plants were defoliated suddenly (e.g., in response to a freeze or disease infestation), nitrogen remobilization from leaves was substantially reduced (Tromp 1970; Castagnoli et al. 1990). This decreases the nitrogen storage pool, and may result in reductions in subsequent vegetative development (Millard and Thomson 1989).

Total nitrogen reserves in rabbiteye blueberry probably continue to increase during flower bud initiation, as shoot growth rate decreases and the requirement for immediate utilization of nitrogen decreases (Fig. 9.5). This has been observed in apple (Tromp 1970) and peach (Taylor and van den Ende 1969). The timing and extent of nitrogen remobilization from leaves of rabbiteye blueberry are unknown; however, remobilization probably begins near the end of the flower bud initiation period under north-central Florida conditions, as leaf senescence begins. However, total leaf abscission in this area rarely occurs except in response to a severe freeze, which seldom occurs before mid-December. Thus, nitrogen remobilization from non-senescing leaves may continue throughout initiation and the early part of the dormant period.

Nitrogen remobilized from leaf tissue of apple and peach was stored primarily in roots and bark (Tromp 1970), and was comprised of soluble nitrogen in the form of arginine and asparagine and insoluble nitrogen in the form of protein (Taylor and May 1967). These authors debated as to whether soluble or insoluble nitrogen reserves predominated, and it appeared that nitrogen fertilization influenced the ratio. Fall-applied nitrogen increased the soluble nitrogen pool much more than the insoluble pool, suggesting that the amount of soluble nitrogen was determined by the rate of protein synthesis (Tromp 1970). In bilberry (*V. myrtillus*), nitrogen was stored primarily as arginine and ammonium compounds in the aerial stem and underground rhizomes (Lahdesmaki et al. 1990a, 1990b). There is no information on the location or composition of nitrogen reserves in rabbiteye blueberry.

IV. DORMANCY

A. General Literature Review

1. Reproductive apex morphology. In highbush blueberry, little or no floral bud development occurred during the dormant season (Gough et al. 1978). At the end of dormancy, ovules began their final phase of development, followed by pollen grain development. Embryo sac formation occurred just prior to bloom. However, differentiation of lowbush blueberry floral buds continued throughout the winter (Bell and Burchill 1955) as cell division in the embryo sac continued. Meiosis began towards the end of the dormant period and was completed first in the anthers, followed by the ovules. Differentiation and maturation was complete by the end of May, just prior to bloom. There is no information on floral bud differentiation in rabbiteye blueberry.

2. Dormancy avoidance.
    a. *Long days.* Long days can substitute for the dormancy requirement in blueberries, thus avoiding the need for chilling. Flowering occurred in highbush blueberries exposed to 16-h photoperiods after flower bud initiation occurred, although floral budbreak was not as uniform as when plants went through the normal dormancy cycle (Hall et al. 1963). Hall and Ludwig (1961) reported that lowbush blueberries will also break floral buds when placed under long day conditions, although they gave no indication as to whether flowering was normal. Rabbiteye blueberries flowered and fruited normally when long days (16 h) were substituted for the normal dormancy cycle as long as leaves remained on the plants and the plants were in "vigorous condition" (Sharpe and Sherman 1971).
    b. *Mineral nutrition.* Preliminary evidence suggests that maintaining vigorous growth of rabbiteye blueberries by application of nitrogen through late fall will avoid the dormancy cycle (if temperatures are maintained above freezing) and promote normal floral budbreak and fruit set even under conditions of short (<12 h) daylengths (R. K. Reeder and R. L. Darnell, unpublished data).

3. Dormancy induction and winter hardiness. Dormancy induction in blueberry, as in many other temperate fruit crops, probably involves the phytochrome-mediated perception of shortening daylengths, followed by decreasing temperatures, which result in gradual growth cessation (Wareing 1956). This dormant state allows the plant to withstand environmental extremes (mainly low temperature) that would be lethal during active growth. In the northeastern United States, highbush blueberry shoot tissue began acclimating in October and continued at a decreasing rate through January, when maximum hardiness developed (Costante and Boyce 1968). Bittenbender and Howell (1976) found that maximum hardiness of highbush blueberry cultivars occurred at the end of November under Michigan conditions. Dehardening began between mid-January and early March. Brierley and Hildreth (1928) reported that highbush blueberries in Minnesota were killed at −20°C, and concluded that winter survival in northern areas was more a function of the insulating properties of snow cover rather than the ability to physiologically withstand sub-freezing temperatures. Kender and Brightwell (1966) found that fully acclimated highbush blueberries can withstand temperatures of −29°C. However, they also observed that the short growing season in northern climates often prevented full acclimation, resulting in decreased hardiness and the reliance on snow cover as an insulator. More recently, however, half-high blueberry genotypes have been introduced from the University of Minnesota breeding program which exhibit resistance to winter injury without the protection of snow cover (Luby et al. 1986).

Flower bud hardiness has been examined in both highbush and rabbiteye blueberry cultivars. Terminal flower buds of highbush blueberry are less hardy than median or basal buds (Biermann et al. 1979; Hancock et al. 1987). Within the floral buds, Biermann et al. (1979) found that apical florets were less hardy than median or basal florets. They concluded that the degree of bud dehydration was positively correlated with the degree of winter hardiness. Bittenbender and Howell (1976) reported that the ovaries appeared to be the critical cold sensitive tissue.

Bud hardiness is an important consideration in the southern United States, where sudden, severe temperature changes can occur during the winter. Many low chill southern highbush and rabbiteye blueberry cultivars bloom early and thus are susceptible to late winter/early spring freezes. Flower buds from rabbiteye cultivars were less tolerant to freeze damage than were southern highbush flower buds at most stages of development (Patten et al. 1989). The degree of bud damage was dependent on cultivar and location on the bush. Spiers (1978b) reported that closed floral buds of the rabbiteye cultivar 'Tifblue' survived to −15°C, whereas 'Woodard' buds survived to −10°. Patten et al. (1989) found that, of the cultivars tested, 'Climax' rabbiteye blueberry flower buds were the most sensitive to freeze damage. They also reported that flower buds on previous spring growth were more susceptible to spring freeze injury than flower buds on fall growth. This was attributed to delayed bud development on fall growth.

4. Dormancy release and budbreak. Dormancy release in blueberry, as in other temperate fruit crops, is promoted by chilling temperatures. Cultural and/or environmental factors prior to chilling may influence chilling accumulation. During the chilling process, temperature and light conditions also influence chilling accumulation. After chilling completion, temperature appears to be the major factor influencing budbreak.

a. *Factors prior to chilling.* There is no information on the effects of environmental or cultural factors prior to dormancy onset on chilling accumulation in blueberry. In general, young, vigorously growing blueberries are delayed in the onset of dormancy and grow into late fall and early winter. There is no evidence to suggest, however, that this directly affects the chilling requirement. Spiers and Draper (1974) found that 'Tifblue' rabbiteye blueberries hand-defoliated in November broke vegetative bud faster the following spring than did non-defoliated plants, suggesting that defoliation may reduce the chilling requirement for vegetative budbreak. Alternatively, defoliation may hasten the development of vegetative bud endodormancy, resulting in earlier accumulation of chill units and therefore earlier budbreak.

'Tifblue' inflorescence buds formed on wood of the previous spring flush usually bloomed before inflorescences that formed on wood of fall growth flushes (Patten and Neuendorff 1989). In a study of several rabbiteye cultivars, Davies (1986) reported that lateral flowers, which initiated later in the season than terminal flowers, also opened later in the spring. This suggests that flower buds formed later in the fall may begin chilling accumulation at a later date than do flower buds initiated earlier. However, this may simply be due to later-initiated buds being less developed upon entering dormancy and therefore requiring further development after the chilling requirement is satisfied. There have been no studies in which precise bud developmental events have been correlated to chilling accumulation.

b. *Factors during chilling.* A substantial amount of research has focused on the effects of different chilling regimes on vegetative and/or floral budbreak in blueberries. In general, insufficient chilling results in delayed and erratic budbreak (Spiers 1976; Gilreath and Buchanan 1981). However, different temperature and light regimes during chilling may markedly influence the response.

Temperature. Interpretation of temperature effects on chilling accumulation is complicated by the use of different models to estimate chill unit accumulation, as well as the use of constant vs. fluctuating temperatures, and cold chamber vs. outdoor experiments. In general, the rate and amount of floral and vegetative budbreak in blueberry increases as constant chilling in the range of 0.6° to 15°C increases (Spiers and Draper 1974; Gilreath and Buchanan 1981; Norvell and Moore 1982; Shine and Buchanan 1982; Darnell and Davies 1990). Blueberry floral buds generally break before or concomitant with vegetative buds (depending on the chilling regime), implying that floral buds have a lower chilling requirement than vegetative buds (Spiers and Draper 1974; Darnell and Davies 1990). Cultivars vary with respect to temperature ranges and optima, as well as the number of chill units required. Northern highbush blueberries have a greater chilling requirement (800–1200 h below 7.2°C) than do rabbiteye blueberries (300–600 h below 7.2°). Within rabbiteye cultivars, there is also variability in chilling requirement. Shine and Buchanan (1982) reported that 'Woodard' and 'Aliceblue' had higher chilling temperature optima (7.2°C and 11.0°, respectively) and wider effective temperature ranges (−2.5° to 15.9° and −2.5° to 13.8°, respectively) than did 'Tifblue' (6.7° optima, −1.2° to 12.9° range). Furthermore, 'Tifblue' has a longer chilling requirement than many other rabbiteyes, even at optimum temperatures, requiring in excess of 500 h below 7°C for any floral budbreak to occur (Spiers and Draper 1974; Darnell and Davies 1990). 'Woodard' and 'Climax' require less chilling; 400–500 h at 7°C was sufficient for normal floral budbreak, although some floral budbreak occurred with less chilling (Spiers and Draper 1974; Darnell and Davies 1990). Normal budbreak occurred in 'Aliceblue' after 300 h chilling at 7.2°C (Shine and Buchanan 1982).

Temperature fluctuations during chilling may influence the chilling response in deciduous fruit crops (Erez and Couvillon 1987). Spiers (1976) reported that 'Tifblue' rabbiteye blueberries exposed to day/night temperatures of 7°/18°C or 7°/23° responded similarly in terms of vegetative budbreak to plants that received constant 7° chilling treatments. Floral budbreak was delayed by the 7°/23°C regime compared to the 7°/18° or the constant 7° regime. This suggests that floral buds of blueberry may be more sensitive than vegetative buds to negation of chilling by high temperatures. Gilreath and Buchanan (1981) found that budbreak in 'Woodard' under diurnal regimes of 0°/7°C, 7°/15°, and 7°/30° was similar to that obtained under constant conditions. Surprisingly, interruption of the chilling regimes midway through by 14 days at 30°C increased the rate of floral and vegetative budbreak compared to non-interrupted constant or diurnal chilling. In this study, high temperatures (30°C) during chilling had either no effect or a promoting effect on budbreak, in contrast to Spiers' findings, and may reflect cultivar differences in response to warm temperatures.

Light. Light intensity and photoperiod appear to influence chilling accumulation in some deciduous fruit crops (Erez et al. 1968); however, the effect of light intensity or photoperiod during chilling has not yet been tested in blueberries.

c. Factors after chilling. After chilling is completed, accumulation of heat units is required before floral and vegetative budbreak occurs in temperate fruit crops (Couvillon and Erez 1985). The heat unit requirement for budbreak in blueberry has not yet been determined.

B. Model Development: December 1 to March 15

1. Climatology. During the dormant period in Gainesville, Florida, the average maximum and minimum air temperatures are 21.2° and 7.7°C, respectively, in mid-December, decreasing to 20.0° and 6.4° by mid-

January, before increasing to 23.8° and 8.9°, respectively, by mid-March (Fig. 9.6). Average soil temperatures (at 10 cm) range from 16.8° to 14.8°C in mid-December, increasing to 20.3° to 17.3° in mid-March. The daylength decreases from 10:25 h to a minimum of 10:15 h, before increasing to 12:00 h by mid-March.

2. Reproductive development. Floral bud differentiation in rabbiteye blueberry probably continues throughout the dormant period in the north-central Florida area since average minimum air temperatures are above freezing. Environmental effects on chilling accumulation and floral/vegetative budbreak in rabbiteye blueberry are discussed in the vegetative growth section.

3. Vegetative growth. Although there are no data to support the idea that environmental factors or cultural practices prior to the onset of dormancy directly affect the ability to accumulate chilling, the degree of dormancy may be affected and thus indirectly affect the amount of chilling required for dormancy release. Fuchigami et al. (1977) found that defoliating dogwood between vegetative maturity, the point at which defoliation did not cause vegetative budbreak—i.e., onset of endodormancy (Lang 1989)—and natural defoliation in the fall did not injure plants, but resulted in delayed vegetative budbreak in the spring. There was no speculation as to the cause of the delay in budbreak, nor was there any indication as to whether the resulting growth was normal with respect to naturally defoliated plants. This suggests that premature defoliation after the onset of endodormancy may decrease the rate or amount of chilling accumulation, thus delaying budbreak. Significant bloom delay was observed on peach trees defoliated in mid-August, prior to vegetative maturity, but trees defoliated after vegetative maturity did not exhibit bloom delay (Couvillon and Lloyd 1978). However, Walser et al. (1981) found that peach vegetative bud dormancy was longer and more intense when leaves remained on trees later in autumn, suggesting that prolonged leaf retention in autumn delays dormancy release in the spring. This concurs with the observation of Spiers and Draper (1974) that autumn leaf retention on rabbiteye blueberry delayed vegetative budbreak the following spring. Jennings (1988) has speculated that dormancy and the amount of chilling required for breaking dormancy in raspberry canes are influenced by environmental conditions during the growing season, especially daylength and temperatures. He suggested that high summer temperatures increase dormancy intensity; however, he offered no data to support his contentions.

Mature, field-grown rabbiteye blueberries in the north-central Florida area may continue active, terminal growth through November in response to relatively high air temperatures (Fig. 9.5); however, shoot growth generally ceases by early to mid-December (Fig. 9.6). In the

absence of a hard freeze, plants may not defoliate fully throughout the entire winter. The effects of relatively high air temperatures and incomplete defoliation on the development of dormancy and subsequent chilling requirement are unknown. Dormancy studies in dogwood (Fuchigami et al. 1977) and peach (Couvillon and Lloyd 1978; Walser et al. 1981) were carried out in areas where winter temperatures presumably ensured complete defoliation and development of endodormancy. Under central Florida conditions, it is unclear to what extent rabbiteye blueberries develop endodormancy.

It is clear, however, that rabbiteye blueberries in the north-central Florida area have a chilling requirement that must be met for normal budbreak to occur. It is equally clear that this area often experiences mild winters when the chilling requirement is not met (Lyrene and Crocker 1983). In addition to the effects of different temperature regimes on dormancy release and budbreak discussed above, dormancy release in rabbiteye blueberry may also depend on light intensity and/or photoperiod during the chilling period. Erez et al. (1968) found the amount of vegetative peach budbreak was greater when plants were chilled in darkness or under reduced photoperiods than when plants were chilled under natural daylight conditions. Floral budbreak in peach was promoted when plants were chilled at constant bud temperatures under low light conditions (60–120 μmol m^{-2} s^{-1}) compared to high light conditions (650–1050 μmol m^{-2} s^{-1}) (Freeman and Martin 1981). The fact that both photoperiod and total irradiance influence chilling accumulation suggests that phytochrome may be involved, since phytochrome is the major photoreceptor involved in both photoperiodicity and high-irradiance responses (Mancinelli and Schwartz 1984). If photoperiod or light intensity does influence chilling perception in blueberries, then the relatively long days experienced in central Florida during the winter may exacerbate the problem of insufficient chilling during mild winters.

After the chilling requirement has been met, heat unit accumulation begins and buds begin to swell over a two- to three-week period, leading to budbreak (Figs. 9.6 and 9.7). Bloom time is determined by the interaction between the chilling and heat unit requirements (Richardson et al. 1975). Couvillon and Erez (1985) found that the heat unit requirement decreased as chilling increased beyond that required for 50% budbreak in four tree fruit species. Additionally, the base temperature for heat unit accumulation varies with species, and may partially explain differences in bloom time among species and cultivars (Werner et al. 1988). Heat unit requirements (both amount and base temperatures) of rabbiteye blueberry vegetative and reproductive buds are unknown.

Root growth of rabbiteye blueberry probably continues during the dormant period, although at a slower rate than at any other time during the season (Fig. 9.6). The relatively high soil temperatures, combined with the average maximum air temperatures, suggest that general plant

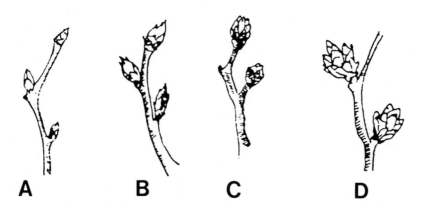

Figure 9.7.   Floral bud swell of rabbiteye blueberry in north-central Florida. A. Dormant bud. B. Visible swelling, bud scales separating (23 Feb.). C. Bud scales separated, floral apices visible (1 March). D. Bud scales abscised, florets visible (6 March). (After Spiers 1978b.)

metabolism is higher in the north-central Florida area during the dormant period than in more northerly areas. Root growth rate probably increases as buds begin to swell (Fig. 9.6).

4. Carbon exchange and partitioning. As in other temperate fruit crops, little carbon assimilation occurs during the dormant period in rabbiteye blueberry (Fig. 9.6). Even though some shoots remain green and the average day temperatures are high enough in the north-central Florida area to support photosynthesis during this period, the low surface area:volume ratio of the stems probably precludes significant contribution to the net carbon balance. This is similar to the situation in green fruit, where the low surface area:volume ratio often limits photosynthesis (Blanke and Lenz 1989).

Although carbon assimilation is limited during this period, carbon translocation and utilization continue. In general, carbohydrate reserves slowly decrease during dormancy of fruit trees due to maintenance respiration and continued bud development (Oliveira and Priestley 1988; Loescher et al. 1990). Since both of these phenomena are temperature related, the warm average winter temperatures in the north-central Florida area suggest that the rate of carbohydrate reserve depletion in rabbiteye blueberry may be more rapid than in northern areas. On the other hand, the brevity of the dormant season probably ensures that reserves are not exhausted prior to budbreak (Fig. 9.6).

Prior to budbreak, cambial activity in rabbiteye blueberry probably increases and carbohydrate reserves are depleted more rapidly as new phloem production begins. This appears to be true for many fruit crops

(Oliveira and Priestley 1988). As buds swell and begin breaking, carbohydrates (usually in the form of free sugars) move up through phloem (Oliveira and Priestley 1988) and/or xylem (Loescher et al. 1990) to these meristematic regions.

Carbohydrate reserve depletion may occur in different tissues at different times throughout the dormant season. Keller and Loescher (1989) found that total non-structural carbohydrates (TNSC) in wood of 10-year-old sweet cherry shoots began to decrease in late winter. Root wood TNSC levels decreased transiently in early winter, but did not decrease significantly until budbreak. This may reflect the effect of soil temperatures on root metabolism, since little or no root reserve depletion occurred in sweet cherry at soil temperatures < 10°C (Loescher et al. 1990). However, Young and Werner (1985) reported that starch reserves in one-year-old apple and peach roots decreased significantly throughout the dormancy/chilling period, with little decrease in shoot starch concentration. The contrasting results may be due to the low level of reserves in one-year-old vs. 10-year-old trees. Since average minimum soil temperatures in the north-central Florida area are well above 10°C during the dormant season (Fig. 9.6), root and shoot reserves of mature rabbiteye blueberry may be utilized to a similar extent throughout the dormant period.

5. Nitrogen partitioning. Nitrogen reserves in apple and peach are used for spring growth, regardless of the current, external nitrogen supply (Taylor and May 1967; Tromp 1970; Millard and Thomson 1989). Assuming an analagous situation occurs in rabbiteye blueberry, total nitrogen reserves probably remain constant throughout most of the dormant period, decreasing in late winter and early spring as cambial and bud activity commence (Fig. 9.6). The decrease may be quite significant, comprising 50% of the total reserve nitrogen in apple (Tromp 1970). There appears to be little difference among different apple tissues in the timing of nitrogen reserve utilization (Tromp 1970). Total nitrogen reserve concentrations declined concurrently in shoots, stems, and roots of apple, concomitant with the beginning of budbreak. Since average air and soil temperatures in the north-central Florida area are above freezing during the dormant period, nitrogen metabolism probably occurs in all plant parts of the rabbiteye blueberry.

In apple, about half of the total nitrogen utilized from reserves during the dormant season came from protein nitrogen (Tromp 1970). Soluble nitrogen levels increased during dormancy, indicating that protein hydrolysis occurred in woody tissues prior to budbreak. The contribution of protein vs. soluble nitrogen to the total nitrogen reserve pool in rabbiteye blueberry is unknown, as is the extent of dormant season utilization from protein vs. soluble nitrogen pools.

V. ANTHESIS

A. General Literature Review

1. Flower morphology. Blueberry flowers are perfect and are generally borne with the corolla opening inverted. Individual flowers consist of a five-lobed white or whitish pink urn-shaped corolla (Fig. 9.8A) that is 0.6–1.2 cm long and up to 8 mm wide (Eck and Mainland 1971). Rabbiteye flowers generally have narrower corollas than highbush flowers. The style is multi-lobed and hollow, with at least five stylar canals that are lined with transmitting tissue (Gough and Shutak 1977). The style is surmounted by a small stigma and extends near to, and eventually beyond (if unpollinated), the corolla opening. Ten stamens arise from the base of the corolla, with the bi-awned anthers borne at a position about halfway up the style on flattened, pubescent filaments. The ovary is inferior, with five locules containing numerous ovules (Mainland 1985) (Fig. 9.8B).

During stigmatic receptivity, nectar is produced at the base of the corolla and pollen is shed through pores at the end of each anther (USDA 1976). Pollen is shed in tetrads of four attached grains. Consequently, each tetrad may be capable of germinating up to four pollen tubes (Parrie 1990).

2. Anthesis. Budbreak in blueberry is followed by anthesis of individual flowers, culminating in corolla abscission after about a week (Fig. 9.9A–D). The overall period of anthesis, which can vary by cultivar as well as by year, depends on the amount of chilling and accumulation of heat units

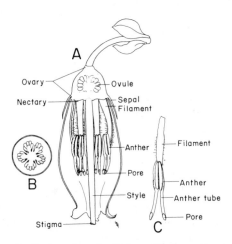

Figure 9.8.   A. Longitudinal section of rabbiteye blueberry flower. B. Cross-section of ovary. C. Longitudinal section of stamen. (After USDA 1976.)

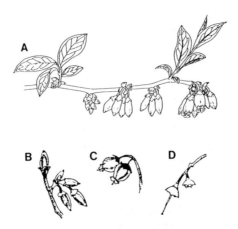

Figure 9.9.   Reproductive bud phenology during anthesis in rabbiteye blueberry in north-central Florida. A. Sequence of anthesis along a shoot. B. Individual flowers separated, corollas unexpanded and closed (11 March). C. Corollas expanded and open, stigma receptive and anthers dehiscent (15 March). D. Corolla abscised, followed by style desiccation and abscission (22 March). (After Spiers 1978b; Galletta and Himelrick 1990.)

in late winter. The greater the accumulation of heat units after the chilling requirement has been met, the shorter and more uniform the bloom period will be. As noted in section III, some rabbiteye cultivars exhibited shorter, more uniform bloom periods when short daylengths (8–10 h) preceded chilling. Insufficient chilling, regardless of late winter temperatures, may cause a protracted bloom period.

3. Pollination. Blueberry flower pollen is heavy and sticky. The pendant habit of open flowers, the small stigmatic surface shaped somewhat like an inverted funnel, and the location of the anthers (Fig. 9.8A) function to discourage self-pollination and facilitate cross-pollination (USDA 1976). Although pollen is shed copiously in most cultivars, it is unlikely to land on the stigma of its own or any other flower without transfer by an insect. Blueberry nectar and pollen are attractive to honeybees, although nectar of some cultivars is more attractive than that of others (Marucci 1966; Brewer and Dobson 1969a). Also, the narrow corolla of rabbiteye flowers makes it difficult for honeybees to obtain nectar through the corolla opening. Consequently, if carpenter bees are also present in a field, honeybees are likely to obtain nectar through the side cuts in the corolla made by carpenter bees, preventing pollination (Cane and Payne 1988). Thus, several native bees are more effective pollinators than honeybees, such as bumblebees (*Bombus* spp.) (Beckman and Tannenbaum 1939; Davies and Buchanan 1979) and the southeastern blueberry bee, *Habropoda laboriosa* (Cane and Payne 1988).

    Stigmatic receptivity (the ability to absorb pollen tetrads) is reported to range from five to eight days in highbush (Moore 1964), four to six days in lowbush (Wood 1962), and about six days in rabbiteye blueberries (Young and Sherman 1978), although there is cultivar variability within each species. Receptivity appeared to be better in experimentally-emasculated flowers if emasculation preceded pollination by several days (Moore 1964).

Pollen deposited on the stigmatic surface is rapidly absorbed into the liquid surface of the stigma (Parrie 1990). Parrie (1990) found that if the surface of the stigma became saturated with pollen, receptivity to further pollen was lost. Although it has been suggested that flowers remain "attractive" to bees until fertilization occurs (USDA 1976), the effect of stigmatic pollen saturation on the nectaries has not been examined. Parrie (1990) found that the stigma of southern and northern highbush blueberries permanently lost receptivity within minutes of pollen saturation. Interestingly, stigmatic pollen saturation occurred with fewer pollen tetrads when stigmas were cross-pollinated than when self-pollinated.

Pollen viability, germination, and pollen tube growth are affected by the temperature and relative humidity at anther dehiscence and pollination. These pollination factors, as well as pollen quantity, vary among different cultivars. Eaton (1966) noted that highbush cultivars that shed less pollen had higher levels of tetrad abortion. Stushnoff and Hough (1968) reported that much of the pollen produced by 'Coville' (a highbush blueberry cultivar that has chronic problems with poor fruit set) underwent irregular meiosis, resulting in pollen with collapsed walls or formation of a dyad (rather than a tetrad), poor dehiscence from anther sacs, and poor viability. Eck (1986) has reported further that 'Coville' required a longer period for pollen tube growth to the ovary compared to better setting cultivars. He suggests that pollen viability (and vigor) may influence fruit set and development in blueberries. Brewer and Dobson (1969c) found that 'Rubel', a highbush cultivar that generally has a heavy fruit set, released twice as much pollen, had twice the pollen germination rate, and exhibited a longer optimum period of pollen germination than 'Jersey', which sets a much lower percentage of fruits.

Conversely, Vander Kloet (1983) found that highbush pollen viability was not related directly to fruit set or to the length of the fruit development period. Rather, pollen viability was related to viable seed production and to fruit size, which may influence fruit set and the fruit developmental period indirectly. Plants with a mean pollen viability of 81% produced an average of 23 seeds; plants with a mean viability of 40% produced an average of 15 seeds. Pollen with different viabilities did not produce seeds of different weights. In a study of 19 native rabbiteye clones, Garvey and Lyrene (1987) found that no strong relationship existed between pollen viability and fruit set resulting from cross-pollination with 'Bluegem' or 'Climax'.

Of unknown importance to blueberry fertilization and fruit development is the germination of multiple pollen tubes from pollen tetrads deposited on the stigma. Although Eck (1986) and Brewer and Dobson (1969c) noted that less than 2% of highbush pollen tetrads germinated multiple pollen tubes in vitro, Parrie (1990) found a high incidence (up to 100%) of in vitro-germinated southern highbush tetrads had multiple pollen tubes. Consequently, cultivars with similar pollen tetrad viabilities (such as 'Sharpblue' and 'Flordablue' at 94% tetrad viability) may differ significantly in the total number of actual pollen tubes germinated (325 vs. 135 tubes per 100 germinated tetrads, respectively) (Lang et al. 1990). Reports of pollen viability percentages in the literature often do not imply clearly whether pollen grain viability or pollen tetrad viability was determined.

4. Fertilization. In highbush blueberries, pollen tube growth occurred over two to four days, depending on temperature (Knight and Scott 1964). Warmer temperatures (16°–27°C) decreased the time between pollen germination and fertilization. Wood (1962) reported that four to six days of warm temperatures were required for fertilization in lowbush blueberries. Eaton and Jamont (1966) reported that embryo sacs deteriorated with time in 'Rancocas', a highbush cultivar. At three days after anthesis, one-third of the embryo sacs were degenerated. Fertilization did not occur until one week after pollination.

Highbush blueberries are considered to be self-fertile (Eck 1988). Rabbiteye (Lyrene and Crocker 1983) and lowbush (Aalders and Hall 1961; Wood 1965, 1968) blueberries exhibit self- and, to some extent, cross-sterility. (Self- and cross-pollination effects on fruit set and developement are discussed further in section VI.) Knight and Scott (1964) noted, using style removal at various intervals, that pollen tube growth in self-pollinated highbush blueberry flowers appeared to be slower than that on cross-pollinated flowers. However, Krebs and Hancock (1988) found that pollen tube growth rates were similar for self- and cross-pollinated highbush flowers, and suggested that self-fertilization resulted in greater seed abortion than did cross-fertilization (Krebs and Hancock 1990). Microscopic examination of rabbiteye blueberry styles after pollination showed similar kinetics of pollen germination and pollen tube growth, regardless of whether flowers were self- or cross-pollinated (Garvey and Lyrene 1987)

B. Model Development: March 15 to April 1

1. Climatology. During the period of anthesis in Gainesville, Florida, the average maximum and minimum air temperatures are 25.3° to 10.7°C, respectively (Fig. 9.10). The average soil temperatures (at 10 cm) range

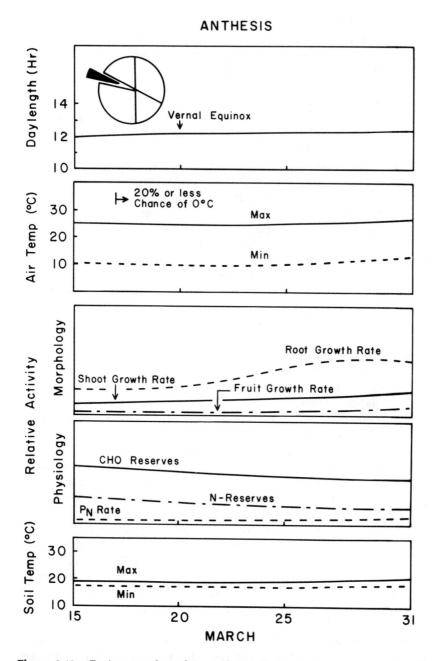

Figure 9.10. Environmental conditions (daylength, air and soil temperatures) and presumed physiological and morphological events during anthesis of rabbiteye blueberry in north-central Florida. Climatological data based on 10-year average (1979–1989). P_N rate = whole plant net photosynthetic rate; CHO reserves, N-reserves = whole plant carbohydrate and nitrogen reserves, respectively.

from 17.3° to 20.2°C, and the daylength increases from 12:00 to 12:30 h. March 2 and March 26 are the dates of 50% and 10% probabilities for 0°C (air temperature) freezes; Feb. 2 and March 4 are the dates of 50% and 10% probabilities of −2° (air temperature) freezes.

2. Reproductive development. For development of the rabbiteye blueberry model, anthesis is considered to be synonymous with 50% bloom, which is estimated to occur around March 15 in north-central Florida. Anthesis will occur earlier in years when winter chilling is sufficient and late winter temperatures are warm. Anthesis will be later in years when winter chilling is sufficient but late winter temperatures are cool. Over a four-year period (1981–84), the average date of 50% anthesis for 'Beckyblue' and 'Climax' rabbiteye blueberries in north-central Florida was March 16, with a range from March 1–29 (Lyrene 1985).

Floral budbreak in rabbiteye blueberry is followed by anthesis of individual flowers, which may last up to two weeks in a single inflorescence. The overall period of anthesis varies by cultivar as well as by the accumulation of heat units, assuming that chilling has been sufficent. Vasilakakis and Porliggis (1984) reported that pear pollen tube growth rates increased with increasing temperature between 5° and 25°C, while stigmatic receptivity and ovule longevity decreased. It is likely that the relatively high air temperatures in the north-central Florida area during anthesis accelerate the rate of budbreak, increase pollen germination and pollen tube growth rate, and yet decrease ovule longevity and the effective pollination period.

Pollen source is a critical factor for cropping of rabbiteye blueberry cultivars. Little information is available on cultivar differences in pollen viability, vigor, or post-pollination competition effects in blueberry pistils. Also, there is no information available on potential differences in attractiveness of rabbiteye cultivar nectaries. Darnell and Lyrene (1989) demonstrated that some closely-related cultivars, such as 'Beckyblue' and 'Aliceblue', were cross-incompatible. Some progeny of self-incompatible parents appear to be self-compatible, such as 'Premier' (J. R. Ballington, personal communication). Until breeders and geneticists can confer self-compatibility to new rabbiteye blueberry cultivars, management strategies must be developed to maximize the deposition of vigorous, viable, cross-compatible pollen on well-developed flowers during warm temperatures.

3. Vegetative growth. The first flush of root growth in highbush blueberry occurred during leaf bud swell, bloom, and fruit set (Abbott and Gough 1987a). It is likely that root growth rates of rabbiteye blueberries increase during anthesis, probably reaching a peak towards the end of the period (Fig. 9.10). At this time, soil temperatures in north-central Florida are probably near optimal for root growth, and competition with other sink organs for reserves is very low. If chilling is sufficient, vegetative budbreak in most cultivars of rabbiteye blueberry is concomitant with floral budbreak (Darnell and Davies 1990); however, little shoot extension occurs before the end of the anthesis period (Fig. 9.10).

4. Carbon exchange and partitioning. Although leaf photosynthesis begins as vegetative buds start to break, the high respiration rates associated with budbreak result in only a minor increase in the rate of net photosynthesis during anthesis (Fig. 9.10). This photosynthetic supply is usually insufficient to meet the carbon demands of developing leaves, resulting in no net export of currently assimilated carbon to other sinks. The timing of net carbohydrate translocation from developing leaves depends on the stage of leaf development and the carbohydrate status of the plant. Net export of carbohydrate from apple leaves did not begin until about 15 days after budbreak (Johnson and Lakso 1986). Net export from the earliest formed sweet cherry leaves (when carbohydrate levels are presumably at their lowest) began when leaves reach about 30% full expansion (about five days after budbreak), net export from later formed leaves did not begin until 50–75% full expansion (Kappes and Flore 1989).

During anthesis, current photosynthesis in rabbiteye blueberry probably supplies carbon only to the developing leaves, with no net export of carbohydrates. Thus, early floral and vegetative growth in rabbiteye blueberry depends on root and/or shoot carbohydrate reserves. Hansen and Grauslund (1973) reported that growth of the first five to six leaves of apple depended on reserves translocated from the roots. Root TNSC levels in sweet cherry decreased dramatically during full bloom, due to translocation to the flowers (Keller and Loescher 1989). In apple and peach, whole plant starch concentrations decreased until eight weeks after budbreak (Oliveira and Priestley 1988). Consequently, total carbohydrate reserves in rabbiteye blueberry probably decrease during anthesis (Fig. 9.10).

5. Nitrogen partitioning. In the early spring, even where soil nitrogen availability is high, nitrogen is not readily absorbed by plant roots, possibly due to insufficient carbohydrate availability (Taylor et al. 1975; Beevers 1976; Weinbaum et al. 1978). Consequently, budbreak and early reproductive and vegetative growth are dependent upon storage forms of nitrogen. During the spring flush of growth in citrus, the nitrogen compounds present in xylem sap originated primarily from root reserves and reserves in parenchymatous tissues (Moreno and Garcia-Martinez 1983). The primary transportable forms of nitrogen in citrus xylem sap included arginine, which peaked just after full bloom, and proline, which declined concomitantly. Levels of total nitrogen, amino acids, and nitrates were at a maximum during budbreak, decreasing from budbreak to full bloom. In

'Golden Delicious' apple, the xylem sap amino acid content increased prior to budbreak and flowering, remaining constant during fruit growth and shoot extension (Tromp and Ovaa 1967). Although the total amino acid content of the xylem did not change, amounts of individual amino acids varied during development. Prior to budbreak, glutamine predominated, while asparagine and aspartic acid predominated during anthesis and fruit development. The increase in soluble nitrogen compounds during spring growth of apple was concomitant with decreases in protein levels in woody tissues (Tromp and Ovaa 1971, 1973).

It is likely that nitrogen reserves in rabbiteye blueberry decrease during anthesis as soluble nitrogenous compounds are translocated from roots and bark to developing flowers and leaves (Fig. 9.10). The extent of remobilization and the translocated forms of nitrogen in rabbiteye blueberry are unknown.

VI. FRUIT DEVELOPMENT AND RIPENING

A. General Literature Review

1. Fruit morphology. The blueberry fruit is a true berry, consisting of a multi-carpellate ovary with 100 or more ovules. The endocarp consists of a single layer of stone cells which constitute the lining of the five compound locules (Yarbrough and Morrow 1947). Stone cells are also found throughout the mesocarp, which is the largest tissue in the fruit and accounts for the major changes in fruit size (Edwards et al. 1970). Rabbiteye blueberries contain a substantial number of stone cells, as well as enlarged, lignified placental tissue, both of which contribute to fruit grittiness. The epicarp consists of a layer of epidermis and a layer of hypodermis. During ripening, the epicarp cells elongate and become filled with pigment (Yarbrough and Morrow 1947). Some cultivars, such as 'Rubel' and 'Dixi' highbush blueberries, contain pigment throughout the fleshy pericarp. Ripe fruits are generally dark purple with an epicuticular wax fine structure that gives the overall impression of a light blue "bloom" on the berry. Specific morphological and anatomical descriptions have been made for lowbush (Stevens 1919), rabbiteye (Yarbrough and Morrow 1947), and highbush (Bell 1957) blueberry fruit.

2. Fruit and seed growth. In blueberry, up to 100% fruit set and development is possible (Eck 1986). Eck and Mainland (1971) found fruit set in highbush cultivars ranged from less than 56% to as much as 100%. Flowers with short, wide corollas that narrow at the base and have a short distance between the anthers and the stigma were associated with the highest level of fruit set among 35 highbush cultivars. Davies (1986) reported that percent fruit set varied from 17% to 60% among rabbiteye cultivars in Florida.

Cross-pollination results in higher fruit set and larger fruit size in high-bush (Bailey 1937; Shaw and Bailey 1937; Shaw et al. 1939; Morrow 1943; Meader and Darrow 1944, 1947), lowbush (Lee 1958; Wood 1965), and rabbiteye (Eck and Childers 1966) blueberries, although Merrill and Johnston (1939) considered highbush blueberries to be self-fruitful. It is now generally recognized that most highbush cultivars are self-fruitful, but will have larger, earlier ripening fruit with more seeds if cross-pollinated (Knight and Scott 1964; Moore 1964; Krebs and Hancock 1988).

Most rabbiteye cultivars are less self-fertile than highbush (Meader and Darrow 1944; El-Agamy et al. 1981; Garvey and Lyrene 1987; Payne et al. 1989) and generally exhibit lower fruit set than highbush blueberry. In southern climates that have relatively warm winters, insufficient chilling has been implicated in poor rabbiteye fruit set (Lyrene and Crocker 1983; Lyrene and Sherman 1985). More recent work, however, suggests that low yield in rabbiteyes after mild winters was not due to a decrease in fruit set, but rather to a decrease in the number of flower buds that open (Darnell and Davies 1990). Southern highbush plants exhibited delayed bloom when chilled insufficientlly, but still set fruit well.

Rabbiteye blueberries undergo a period of fruit abscission during the three to four weeks after full bloom, with lesser amounts five to seven weeks after full bloom (Davies 1986). Fruit abscission was not related to time of flower opening (provided that suitable temperatures and bee activity occurred), shoot location within the canopy, or flower position. In 'Woodard', terminal fruit set was generally higher than that for laterally-borne fruit, whereas the lower laterals on 'Tifblue' had the highest fruit set. Davies suggested that flowers on shoots of moderate vigor may have the highest levels of fruit set. Other factors that influence fruit set include cross-compatibility (El-Agamy et al. 1981), presence and activity of pollinators (Davies and Buchanan 1979), water stress (Buchanan et al. 1978; Lyrene and Crocker 1983), and plant nutrition (Lyrene and Crocker 1983).

Highbush, lowbush, and rabbiteye blueberry fruit exhibit a double sigmoid growth curve (Fig. 9.11A) with three stages of growth (Young 1952; Edwards et al. 1970). Stage I is characterized by a high, but declining, respiration rate, rapid growth, and (at least in lowbush blueberry) a high rate of ethylene production (Hall and Forsyth 1967b; Forsyth and Hall 1969). It begins with fertilization, after which the corolla and stamens abscise, followed by browning and abscission of the style. Eaton and Jamont (1966) reported that the first developmental events following fertilization in highbush blueberry were two cellular divisions of the endosperm followed by enlargement of the ovary by cell division, which accounted for the majority of size increase during this stage (Edwards et al. 1970). Embryo and endosperm development are

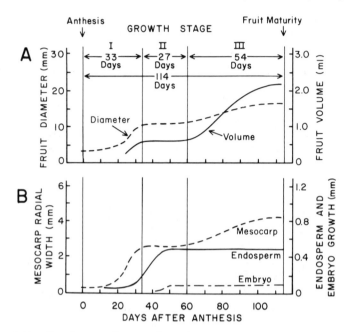

Figure 9.11.   Developmental stages of A. whole fruit, and B. mesocarp, endosperm, and embryo growth of 'Bluegem' rabbiteye blueberry in north-central Florida. (After Edwards et al. 1970.)

arrested during this stage. Empirical evidence suggests that the length of stage I is tied closely to temperature, as late blooms followed by warm temperatures may exhibit shorter stage I periods than early blooms followed by cool temperatures (Edwards et al. 1970; Davies 1986).

Variation in the length of stage I among highbush and lowbush fruits is related to harvest date (Young 1952). The stage I developmental period for fruit from the first and second harvests was 25 days, whereas that for fruit from the third or later harvests was 35 days. Spiers (1981) reported that the average length of stage I among various rabbiteye cultivars was about 35 days. He found that the transition between stages I and II was not as clear in some rabbiteye cultivars as in highbush or lowbush blueberries.

Pericarp growth and metabolic activity slow during stage II, as embryo and endosperm development increase (Edwards et al. 1970) (Fig. 9.11B). Young (1952) found that the length of the stage II developmental period in highbush and lowbush blueberries in Michigan was 30–40 days for fruit from the first and second harvests, regardless of species or cultivar. During the third or later harvests, the stage II developmental period was about 60 days, with lowbush cultivars exhibiting slightly shorter periods than highbush. In the highbush cultivar 'Jersey', stage II varied from 14 days for early-ripening fruit to 56 days for late-ripening fruit (Eck

1986). The variability in length of stage II within cultivars is thought to be strongly influenced by the number of viable seeds present. The length of stage II is also genetically linked. Edwards et al. (1970) found that stage II in developing rabbiteye fruit was 50–90% longer than in southern highbush blueberry selections.

The pericarp increases rapidly in size during stage III, as fruit grow from about 40% final size to final size (Eck 1986). As in most fruits, the final increase in size is primarily by cell enlargement. Edwards et al. (1970) found that the length of stage III was more than twice as long in rabbiteye fruit as in southern highbush blueberry.

Darrow (1958) surveyed several species of blueberry for seed production and reported a range of 16–74 seeds per highbush fruit, and 38–82 seeds per rabbiteye fruit. Highbush fruits averaged about 64 seeds, many of which were not fully developed (Eck 1986). Developed seed tended to be located around the top of the locule, suggesting that the number of pollen tubes was insufficient to fertilize the lower, basal portion of the locules. Based on ovule counts, Parrie (1990) found that about 82–112 potential seed existed in several southern and northern highbush cultivars. Fully developed seed from hand pollinations (that saturated the stigmatic surface) of these cultivars numbered less than 20 per fruit.

Bell (1957) found embryo abortion rates in lowbush blueberry as high as 80%. Edwards et al. (1972), in a study of 'Bluegem' rabbiteye blueberry, reported 80% ovule abortion during fruit development. They identified three periods during which embryo abortion occurred: (A) immediately following fertilization, (B) at the transition from fruit development stage I to stage II, and (C) during early stage II fruit development. Embryos "emerge from the micropylar haustorium" at the end of stage I, entering a period of rapid development until the end of stage II. Embryo size remained constant from the end of stage II until maturity. The authors suggested that abortion during period A probably occurred due to insufficient fertilization of the egg, although it may occasionally be due to degeneration of the zygote and endosperm nuclei; it accounted for about half of the ovule abortions (40% of total ovules). Abortion during period B was "characterized by failure of the embryo in the micropylar haustorium to penetrate the endospermic plug" and develop normally into the endosperm. Such aborted embryos produced nearly normal-sized seeds, accounting for about half of the aborted seeds present. Abortion during period C accounted for only about 5% of the total abortions. It was associated with "the normal rapid filling of endosperm cells with food storage bodies" and a degenerated embryo in a normal-sized seed.

3. Fruit maturation and ripening. During stage III, sugars accumulate in the flesh, the calyx end of the fruit turns purple, and the dark green berry becomes translucent in highbush (Hindle et al. 1957a) and lowbush species or pink in rabbiteye species (Eck 1986). The phase of color change from light purple to the characteristic dark blue-black generally takes about four days (Young 1952) and is accompanied by a substantial increase in respiration and softening of the fruit flesh. Similar data by numerous researchers has led to the classification of lowbush, highbush, and rabbiteye blueberry fruits as climacteric (Bergman 1929; Ismail and Kender 1969; Windus et al. 1976; Lipe 1978), although a few researchers have been unable to corroborate such data (Hall and Forsyth 1967a; Frenkel 1972; El-Agamy et al. 1982).

According to Eck (1986), fully ripe blueberries continued to increase in soluble solids and other flavor components after full coloration occurred, unless they were removed from the plant. Spiers (1981), however, reported that soluble solids content in several rabbiteye cultivars peaked and remained constant as maximum fruit size was attained. Woodruff et al. (1960) characterized the changes that occurred in highbush fruit flavor and textural components during ripening. Pigmentation intensity increased, followed by an increase in sugars from 50% to 70% on a dry weight basis and a decrease in titratable acidity from 9% to 1%. Soluble pectin decreased, concomitant with increasing pectin methylesterase activity. Starch and acid hydrolyzable polysaccharides increased slightly during ripening, while ether-soluble material, lignin, and cellulose levels decreased. Glucose and fructose were reported to be the main sugars in ripe fruits and were present in equal amounts; the major organic acid was citric, followed by malic and quinic (Kushman and Ballinger 1967).

Fruit coloration depends on temperature and occurs in either light or dark (Shutak et al. 1956). Below 10°C, coloration changes did not occur. At 10°C, only red pigmentation developed. At 16°C and above, blue coloration occurred, with the rate of coloration dependent upon temperature. Ripening was delayed in total darkness. Fruit pigment is primarily composed of anthocyanins in the skin of the fruit, although purple streaks may be seen in the white flesh of some cultivars (Eck 1986). Gough and Litke (1980) reported that the pedicel disk also turns red or blue. The 3-glucose, 3-galactose, and 3-arabinose forms of malvidin, peonidin, petunidin, cyanidin, and delphinidin were the main anthocyanins found (Ballinger et al. 1970; Makus and Ballinger 1973). The 3-glucose and 3-galactose forms of cyanidin predominated in ripe fruit while those same forms of malvidin predominated in overripe fruit.

4. Factors affecting fruit development. The fruit development period ranges from 42 to 90 days for highbush, 70 to 90 days for lowbush, and 60 to 135 days for rabbiteye blueberries (Meader and Darrow 1944; Forsyth and Hall 1969; Eck 1986). Some native blueberries in Florida have fruit development periods up to 180 days (Lyrene and Sherman 1984). Bailey (1947) found the fruit development period differed by as much as 25 days between years in Massachusetts. The genetic influence on fruit development period between cultivars appears to be manifested primarily during stage II of development; the genetic influence on the fruit development period between species appears to be manifested during both stages II and III (Edwards et al. 1970). Unlike the rabbiteye fruit development period of 114 days depicted in Figure 9.11, most rabbiteye cultivars currently grown around the north-central Florida area have fruit development periods of 70 to 90 days.

The largest fruits are generally those from the first or second harvest (Darrow 1958). There is an inverse linear relationship between fruit size and harvest date in rabbiteye blueberry (Spiers 1981). Hindle et al. (1957b) found that the largest fruits within a cluster were from basal flowers (which bloomed first) and from inflorescences on thick wood (which generally did not bloom first). The effect of wood thickness on harvest date varies among cultivars. Medial flowers in the inflorescence tend to ripen first. Davies (1986) has also noted that the earliest-ripening fruits were not necessarily the earliest to bloom. Late blooming inflorescences tended to bear uniformly ripening fruit clusters (Young 1952).

The number of flowers per bud affects total yield (Lyrene 1984). Davies (1986) noted that flower number per bush was a key determinant of potential yield, since fruit set percentage was correlated with yield in 'Tifblue', but not in 'Woodard' and 'Bluegem' rabbiteye blueberries. Other factors, such as individual berry sink strength, the availability of water and nutrients, and conditions favorable for maximum photosynthesis, combine to affect final fruit size.

a. *Seed factors.* Pollen source, and the effectiveness of pollination and fertilization, significantly affect fruit growth of most blueberry cultivars, since viable seed number is generally associated with larger and earlier-ripening fruit. Knight and Scott (1964) reduced the fruit development period in highbush blueberry cultivars from an average of 77 days to 65 days by cross-pollination. Garvey and Lyrene (1987) reduced the fruit development period within rabbiteye blueberry selections from 106 days to 92 days by cross-pollination.

There are numerous reports of correlations between seed number and fruit weight in self-and cross-pollinated fruits of highbush and rabbiteye blueberries (Meader and Darrow 1944; Darrow 1958; Brewer and Dobson 1969b; Moore et al. 1972; Hellman and Moore 1983; Krebs and Hancock 1988; Darnell and Lyrene 1989). The correlation coefficients for rabbiteye fruits indicate that less than 50% of the variability in fruit size was accounted for by differences in seed number. The level of correlation was higher for rabbiteye fruits than for northern highbush fruits (Moore et al. 1972). In southern highbush, r^2 values of 0.61–0.76 were found for developed seed number to fruit size within a cultivar, but absolute numbers of seeds for a given berry size varied greatly among cultivars (G. A. Lang, unpublished data).

Others have found low correlations of seed number with fruit size, often depending on cultivar (Darrow 1958; Rabaey and Luby 1988). Eaton

(1967), in a study of 13 highbush cultivars, found that fruit from early-ripening cultivars generally had fewer seeds than later-ripening cultivars. Although seed number and fruit size were correlated, the r^2 values were below 0.40. He concluded that seed number accounted for a relatively small amount of the total fruit fresh weight variability. Kushima and Austin (1979) noted that correlations were especially poor for rabbiteye fruits of low weight (<1.5 g). A major factor in this phenomena is probably pollen source (Morrow 1943; Meader and Darrow 1944, 1947; Gupton 1984).

Below a minimum or threshold seed number, fruit size usually correlates well to seed number; however, this relationship is cultivar dependent. Meader and Darrow (1947) reported large fruit from a pentaploid hybrid (rabbiteye × highbush blueberry) with only two seeds/fruit, compared to normal rabbiteye fruits of similar size with 30 seeds/fruit. Likewise, 'Bluechip' northern highbush has been observed to develop full-sized, non-seeded fruit (G. A. Lang, unpublished data). Seed number is generally more highly correlated with earliness of harvest within a cultivar than with fruit size; that is, the earliest harvested fruits from an individual cultivar generally have the most seeds (Darrow 1958; Lang et al. 1990). The date of individual inflorescence bloom during the overall period of anthesis had little effect on seed number in rabbiteye blueberries (Davies 1986). Seed number tended to be consistent, regardless of whether fruit were from early or late blooming flowers, providing there was sufficient bloom overlap of pollenizers.

b. Water relations. Davies (1986) noted that rabbiteye fruit size was consistent, regardless of the cropping level (by total yield or percent fruit set) as long as water was not limiting. Andersen et al. (1979b) and Buchanan et al. (1978) have shown that drip irrigation increased rabbiteye yield and berry fresh weight by 20–25% over non-irrigated plants in Florida. The use of antitranspirants at moderate rates on rabbiteye cultivars increased fruit weight, but delayed maturation and decreased soluble solids content (Andersen et al. 1979a). Antitranspirants did not increase midday xylem potential, but did decrease stomatal conductance and transpiration.

Davies et al. (1979) found that yield was correlated with feeder root density in some rabbiteye cultivars. This was particularly pronounced in 'Tifblue', which develops a large canopy volume and limited feeder root density relative to the other cultivars examined. Thus, 'Tifblue' feeder roots are unable to supply sufficient water to the canopy during periods of high demand, resulting in decreased yields.

Excessive water can also influence reproductive development in blueberry. Abbott and Gough (1987b) found that flooding reduced the number of flower buds per shoot and florets per inflorescence; delayed anthesis; reduced fruit set; increased fruit abscission; and caused significant reductions in fruit weight, size, and soluble solids in container-grown highbush blueberries. Adverse effects generally were severe regardless of the time of year. It was suggested that flooding may decrease root growth, leading to a decreased production of hormones (e.g., cytokinins and gibberellins) important to bud formation and development. For rabbiteye blueberries, Crane and Davies (1988a) reported that two weeks of spring flooding delayed fruit development and ripening in field-grown 'Woodard'. Three weeks of spring flooding decreased fruit set percentage and increased fruit abscission. In further experiments with field-grown 'Tifblue', Crane and Davies (1988b) found that two 7-day periods of spring flooding reduced fruit set percentage and yields. Fruit shriveling and abscission were suggested to be due to flooding-induced restriction of water uptake.

   c.Photosynthetic factors. In a study of lifetime biomass partitioning in native populations of V. corymbosum, Pritts and Hancock (1985) found that fruit accounted for more than 50% of the annual dry matter produced during years 3 to 11. Forsyth and Hall (1965) reported that leaf photosynthetic light saturation occurred at a laboratory light intensity of 0.01 lux in lowbush blueberry. However, Aalders et al. (1969) found a delay in fruit maturation and a decrease in reducing sugar content of lowbush blueberry fruit when light intensities were less than 0.02 lux.

   There is no published information on photosynthetic levels in blueberry fruits. Young (1952) observed that fruit that receive the most sunlight are usually the largest. Preliminary data indicate that net photosynthesis occurs in rabbiteye fruit from petal fall to color change (K. T. Birkhold and R. L. Darnell, unpublished data). During this period, net fruit photosynthesis decreased from 14 μmol CO_2 g $fw^{-1}h^{-1}$ to 1 μmol CO_2 g fw^{-1} h^{-1}; and dark respiration decreased from 14 μmol CO_2 g fw^{-1} h^{-1} to 3 μmol CO_2 g fw^{-1} h^{-1}. After coloration, the photosynthetic rate fell to zero, and the respiration rate increased transiently to 8 μmol CO_2 g fw^{-1} h^{-1}, before decreasing. Fruit composition analysis indicated that the cost for dry matter development of a 1.0 g fw blueberry fruit was 76 mg carbon. Respiratory loss over the course of development averaged 45 mg carbon g fw^{-1} (calculated at 22°C), while fruit photosynthetic gain averaged 17 mg carbon g fw^{-1} (calculated at 25°, 12 h day^{-1}). Thus, a total of 121 mg carbon was required for development of a 1.0 g fw berry. Berry photosynthesis itself supplied 17 mg carbon, requiring the import of 104 mg carbon from other plant sources.

   Patten and Neuendorff (1989) examined light penetration effects on flower and fruit development within the canopy of rabbiteye blueberry plants. 'Delite' inflorescences in the top of the canopy (formed the previous fall) flowered after inflorescences located within the interior or the bottom exterior of the canopy (which formed the previous spring). However, fruits from the top of the canopy were equally large, with similar soluble solids content. Fruit in the outer 25 cm of canopy received 50–99% full sun and had significantly higher soluble solids than fruit from 25 to 75 cm within the canopy (which received 14–25% full sun).

Fruit from the center of the canopy received 11% full sun and had the lowest soluble solids content. Overall, light exposure accounted for 64% of the variability in maturity and 30% of the variability in soluble solids. Three hours of photosynthetic light saturation (25–30% full sun) were considered sufficient to ensure that light was not limiting. However, the importance of light distribution over the course of the day was not addressed. In the case of non-limiting light, these authors found that the leaf:fruit ratio became the limiting factor, accounting for 76% of the variability in maturity. The critical leaf:fruit ratio value for 'Tifblue' was 20 cm^2 leaf area/fruit (approximately one large leaf per fruit). Below this value, leaf:fruit area is probably a critical limiting factor to maturity. However, translocation of photosynthates from reserves or sources distant from the fruit cluster may be significant as well.

B. Model Development: April 1 to June 1

1. Climatology. At the beginning of the fruit development period in Gainesville, Florida, the average maximum and minimum air temperatures are 27.9° and 13.6°C, respectively, increasing to 31.0° and 17.3° during fruit ripening (Fig. 9.12). The average soil tempertures (at 10 cm) range from 24.5° to 20.5°C in mid-April, increasing to 27.6° to 24.4° by mid-May; daylength increases from 12:30 to 14:00 h.

2. Reproductive development. For development of the rabbiteye blueberry model, harvest is considered to be synonymous with 50% ripening, which is estimated to be June 1 in north-central Florida. This date may vary by as much as two full weeks before or after the model date, depending on flowering date and post-anthesis temperatures. Temperature is a major determinant of the total fruit development period length, as it exerts a strong influence on all three growth stages. The average fruit development period is considered to be 75 days. However, a brief review of several consecutive years' crops of rabbiteye blueberries from 1981 to 1984 revealed harvest dates that differed by as much as 17 days for 'Climax' and fruit development periods that differed by as much as 19 days for 'Beckyblue' (Lyrene 1985).

After syngamy, rapid cell-division leads to the initial fruit size increase. In peach, there was rapid cell division for two to four weeks after anthesis with little increase in cell volume (Addoms et al. 1930; Ragland 1934). It is difficult to measure increases in fruit size until two to three weeks following anthesis for most fruit, especially small fruited species such as blueberry, in which the diameter of the ovary after corolla drop, but before rapid growth, is about 4 mm (G. A. Lang, unpublished data). An indication of initial fruit set can be determined by comparison of slightly swollen, green persisting ovaries with those that may be red-tinged and

FRUIT DEVELOPMENT AND RIPENING

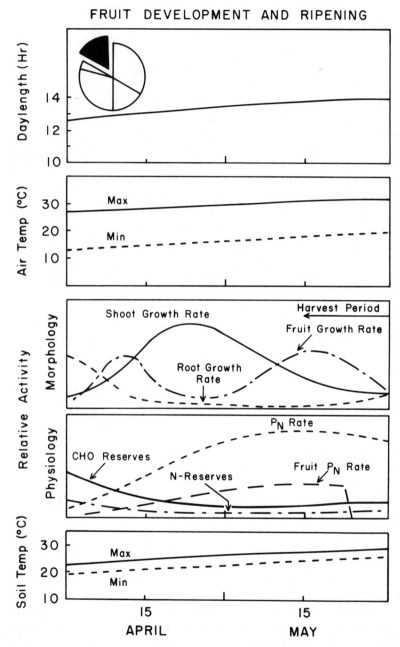

Figure 9.12. Environmental conditions (daylength, air and soil temperatures) and presumed physiological and morphological events during fruit development of rabbiteye blueberry in north-central Florida. Climatological data based on 10-year average (1979–1989). P_N rate = whole plant net photosynthetic rate; CHO reserves, N-reserves = whole plant carbohydrate and nitrogen reserves, respectively.

slightly shriveled. The latter abscise, generally within four weeks of anthesis.

Rabbiteye blueberry fruit in the north-central Florida area probably enter stage II of fruit growth in mid-April (Fig. 9.12). During this stage of development, as in other fruit crops exhibiting double sigmoidal growth curves, fruit growth rate declines and there is little or no measurable increase in fruit volume, as the endosperm and embryo tissues develop.

Fruit of rabbiteye blueberry in the north-central Florida area enter stage III by the beginning of May, and fruit growth rate and size begins to increase again (Fig. 9.12). The remaining four-week period of fruit growth is characterized by a logarithmic increase in fruit volume, accounting for as much as 60% of total fruit size in rabbiteye blueberry (Eck 1986). This logarithmic increase makes the timing of fruit harvest critical, since harvesting even one or two days before maximum fruit size is obtained may result in significant reductions of total yield.

Characteristic flavor components begin accumulating during the latter part of May, as ripening commences. Pigmentation occurs first in fruit that are well exposed to light. Anthocyanin accumulation begins as chlorophyll content decreases; this accumulation is possibly associated with the induction of phenylalanine ammonia-lyase activity and increased ethylene biosynthesis, as has been demonstrated in apple (Faragher and Brohier 1984), grape (Hrazdina et al. 1984), and strawberry (Given et al. 1988). Little is known about the degree-day heat unit accumlation necessary for blueberry fruit maturation, although cultivar variation almost certainly exists, as has been reported for other deciduous fruit crops (Baker and Brooks 1944; Fisher 1962; Munoz et al. 1986).

3. Vegetative growth. At the beginning of the fruit development period, leaf and shoot expansion continues, coincident with fruit set and early fruit development. As the photosynthetic canopy fills in over the next two to three weeks, the steadily increasing temperatures and daylengths during this period promote high photosynthetic rates in support of continued shoot, fruit, and root growth (Fig. 9.12). Shoot growth in highbush blueberry began to increase rapidly during the early stages of fruit growth, peaking during stage II of fruit growth (Abbott and Gough 1987a). Shoot growth of rabbiteye blueberry in north-central Florida also begins to increase with early fruit development (Fig. 9.12). The growth rate of this first shoot flush probably peaks during stage II of fruit growth, then decreases as increased competition for resources occurs during stage III of fruit development.

Little is known about root growth during the fruit developoment period. In highbush blueberry, after the initial increase in root growth rate at the end of anthesis, the rate declined and continued at a relatively low level until after harvest (Abbott and Gough 1987a). A similar situation likely exists in rabbiteye blueberry in the north-central Florida area. Root growth rate probably declines coincident with the increasing rates of fruit and shoot growth, continuing at a much reduced rate throughout the fruit development period (Fig. 9.12).

4. Carbon exchange and partitioning. Although leaf and shoot expansion, and therefore photosynthesis, increases at the beginning of fruit development, the carbon supply from leaves is insufficient to support the growth of the developing fruit. Maximum photosynthetic capacity was not reached in lingonberry until two weeks after budbreak, while leaves of *Vaccinium uliginosum* required up to four weeks to reach full photosynthetic capacity (Karlsson 1985a, 1985b). As a consequence, much of the carbon required for early stage I fruit development is drawn from carbon reserves in the roots and bark, resulting in a rapid depletion of carbohydrate reserve pools. Little information is available on the contribution of current vs. reserve carbohydrate to fruit development in most deciduous crops. Quinlan and Preston (1969) reported that early fruit development in apple depended partially on carbon reserves, although spur leaf photosynthesis supplied some of the required carbon. In crops which flower prior to or concomitantly with leafing, such as blueberry, early fruit development may depend both on carbon reserves and fruit photosynthesis. Kappes (1985) reported that the first peak of sour cherry fruit growth competed with early vegetative growth for carbohydrate reserves. He found that fruit photosynthesis was very important for supplying carbon during stage I of fruit development, contributing 19% of the total carbon required for dry weight gain and respiratory loss. Preliminary evidence obtained from two cultivars of rabbiteye blueberry suggested that 15% of the total carbon required for dry weight gain and respiration during stage I was supplied by fruit photosynthesis (K. T. Birkhold and R. L. Darnell, unpublished data). The contribution of leaf photosynthesis and carbon reserves to stage I of fruit development in rabbiteye blueberry is unkown.

Stage II of blueberry fruit growth is marked by arrested pericarp development and increased development of embryo and endosperm tissues. Kappes (1985) reported that sour cherry fruit respiration rate was high during stage II due to lignin and lipid synthesis in the endosperm and embryo. During this stage, sour cherry fruit photosynthesis supplied 30% of the carbon required. Stage II of fruit development in rabbiteye blueberry probably draws very little on carbohydrate supplies from the plant, since the cost of endosperm development in blueberry is probably much less than that of sour cherry. Thus, carbohydrates would be more available for shoot growth, resulting in an increased rate of shoot growth during stage II of fruit development (Fig. 9.12). This idea of carbohydrate reallocation between vegetative and reproductive sinks during development is supported by work of Chalmers et al. (1986), in which regulated deficit irrigation prior to the commencement of rapid fruit growth in pear decreased shoot growth (and therefore the assimilate demand of the vegetative sink), resulting in increased fruit size at harvest.

During stage III of fruit development in rabbiteye blueberry, fruit growth rate is high (Eck 1986). The leaf canopy is well developed, and the leaves are the primary source of carbon for fruit development. Although fruiting effects on leaf photosynthetic rates are inconsistent, and appear to be species and/or stage of fruit development related (Chalmers et al. 1975; Sams and Flore 1983; Downton et al. 1987), the high demand for carbon by the developing blueberry fruit during this stage probably results in increased leaf photosynthetic rates. Furthermore, little carbon is likely to be allocated to reserve pools during stage III, resulting in continued low levels of carbohydrate reserves until after harvest (Fig. 9.12)

5. Nitrogen partitioning. As is the case with carbon, nitrogen mobilized to developing fruit during early spring growth is derived primarily from nitrogen reserves (Weinbaum et al. 1978, 1984). Prune trees did not take up significant amounts of soil nitrogen until the period of rapid shoot growth, when leaf area was sufficient to supply carbohydrates for root uptake (Weinbaum et al. 1978). Even then, soil nitrogen was partitioned preferentially into vegetative growth, then recycled into reproductive growth when endogenous nitrogen pools were depleted. A similar situation may exist in rabbiteye blueberry, where fruit growth is dependent on stored nitrogen reserves until enough shoot growth occurs to supply needed carbohydrates for root uptake of soil nitrogen (Fig. 9.12) Over the remainder of the fruit development period, soil nitrogen uptake probably remains high (assuming adequate availability) even when shoot growth slows, as has been reported for prune (Weinbaum et al. 1978). Thus, leaf and fruit nitrogen demands later in the season are probably met by current uptake of soil nitrogen. The nitrogen requirements of leaves and fruit probably prevent the replenishment of nitrogen reserves in rabbiteye blueberry until late in the season (Figs. 9.4 and 9.5).

VII. MODEL INTEGRATION

A large informational data base now exists to describe individual plant growth components. This has resulted in research, and reviews of research, which cover small components of plant growth in great detail. The outcome of continued detailed research, without synthesis, leads to fragmented understanding of the whole plant. Although this reductionist approach is necessary for continued elucidation of plant growth processes, it is important to integrate these pieces of information in order to understand the whole plant and the complex relationship between the whole plant and its environment. The goal of this review was to integrate physiological, morphological, and climatological information from the blueberry, as well as other crops, in order to develop a model describing the interactions that influence reproductive development of rabbiteye blueberry. Environmental data from Gainesville, Florida, were used to explore yield limitations in the increasingly important low-chilling Gulf coast production area.

The annual model of rabbiteye blueberry reproduction was divided into five periods that correspond to well-established stages of development (Fig. 9.1): vegetative development following harvest (June 1 to Oct. 1), flower bud initiation (Oct. 1 to Dec. 1), dormancy (Dec. 1 to Mar. 15), anthesis (Mar. 15 to Apr.1), and finally fruit development and ripening (Apr. 1 to June 1). The models, represented schematically in Figures 9.4–9.6, 9.10, and 9.12, are designed around these five stages. Evaluated together, the model describes the annual cycle of morphological and physiological development of rabbiteye blueberry in response to major environmental inputs. Although portions of the model are speculative, the purpose of research is not to prove what is known, but test what is hypothesized based on insufficient previous knowledge. Towards that end, this model demands calculated speculation until more precise information is available, and also identifies starting points for further problem description and experimentation.

Reproductive development. Our model of rabbiteye blueberry reproduction begins not with budbreak, but with fruit harvest (Fig. 9.4). With this perspective, the reproductive year requires several months of vegetative growth and development to produce the vegetative canopy and carbon and nitrogen resources necessary to support reproductive growth. Although some flower bud initiation may occur towards the end of summer, the major period of initiation occurs during the autumn, after daylength falls below 12 h (Fig. 9.5). The naturally-occurring daylengths along the Gulf coast during this period may provide a limitation to yield in some rabbiteye cultivars, such as 'Beckyblue', which initiates significantly fewer flower buds under natural daylengths than under 8- to 10-h daylengths.

A better understanding of the influence of mild winter climates, such as those along the Gulf coast, on reproductive development is clearly important for rabbiteye blueberry production. The interaction of temperature and photoperiod may influence not only the extent and/or timing of flower bud initiation, but also the onset of dormancy. During the dormant period, which may be avoided by exposure to 16-h days or alleviated by a specific period of chilling temperatures, flower bud differentiation occurs (Fig. 9.6). Chilling requirements and temperature optima for chill unit accumulation differ among cultivars. A period of warm temperatures is required for budbreak following dormancy. The base temperature and heat unit accumulation requirement for growth have not yet been determined for any rabbiteye blueberry cultivar.

Anthesis generally precedes or occurs concomitantly with vegetative budbreak, depending on chilling sufficiency and heat unit accumulation. The pollination and fertilization periods are directly related to temperature effects on pollen tube growth and ovule longevity. With average maximum temperatures of 25°C during anthesis in north-central Florida (Fig. 9.10), pollen germination and growth is probably rapid, whereas ovule longevity is decreased.

Blueberry fruit growth is double sigmoidal (Fig. 9.11). Temperature has a significant influence on stage I of fruit growth, as warm or cool temperatures can either decrease or increase, respectively, the length of this stage. The length of stage II appears to be more closely linked to the number of viable seeds. Fruit growth rate is greatest during stage III, and is influenced significantly by temperature and water status. If water is limiting during this stage, fruit size would be expected to decrease accordingly. The roles of seed development and genetic expression of fruit size on subsequent fruit development and ripening have yet to be thoroughly documented and explained. The frequent poor fruit sets experienced in some parts of the Gulf region may be due to such environmental influences as insufficient chilling temperatures during the dormant period, high temperature during the pollination period or early stage I of fruit growth, or unfavorable conditions for pollinator activity.

Vegetative growth. There are three main flushes of shoot growth for rabbiteye blueberry in north-central Florida. In the context of the model, the first flush follows harvest (Fig. 9.4), the second flush precedes flower bud initiation (Fig. 9.4), and the third flush follows anthesis (Fig. 9.12). The first flush, which provides the canopy for subsequent development of flower buds, may be initiated by decreased reproductive demand for carbon and nutrients, in concert with long days and high temperatures. In fact, it may be supraoptimal temperatures, possibly coupled with high evapotranspirative demands, that bring this growth flush to a close. The second flush of shoot growth occurs just prior to flower bud initiation, as air temperatures decrease and presumably the water status of the plant improves. The third flush of shoot growth occurs after anthesis and may be related to favorable environmental conditions such as optimal air and soil temperatures and increasing daylength. Vegetative to reproductive ratios, such as number of leaves per fruit and/or number of inflorescences per shoot, need to be characterized as part of a strategy to optimize the vegetative support system for fruit development.

Our model suggests that roots also have three flushes of growth. In the context of the model, the first flush occurs near harvest and precedes the shoot growth that occurs after harvest (Fig. 9.4). Although direct evidence is lacking, the lack of competition from either vegetative or reproductive sinks and moderate soil temperatures appear to trigger this period of root development. The rate of root growth probably declines as shoot growth is initiated and soil temperatures increase (Fig. 9.4). A second flush of root growth occurs in the autumn, following shoot growth, and may be a response to reduced sink activity and lower soil temperatures (Fig. 9.4). There is also a flush of root growth prior to anthesis (Fig. 9.6). It is assumed that soil moisture and temperatures are near optimum for growth during this period. The root growth rate declines following fruit set and the initiation of vegetative growth.

The relationships between root and shoot growth with respect to soil temperature, intra-plant competition, nutrient uptake, and carbon and nitrogen partitioning are complex and require characterization in order to successfully balance vegetative and reproductive development.

Carbon exchange and partitioning. Following harvest, a transient decrease in canopy photosynthetic rate may occur (Fig. 9.4), although evidence to confirm this is lacking. Although photosynthetic rate is presumed to recover, climatic conditions in north-central Florida may limit the full photosynthetic potential during the period following fruit harvest. There is very little known in rabbiteye blueberry about photoinhibition, optimum photosynthetic temperatures, or the influence of light intensity and vapor pressure deficit on optimum temperatures for photosynthesis. Studies of photosynthetic light saturation levels indicate that light levels in the Gulf coast states are rarely limiting for photosynthesis. On the other hand, optimal temperatures for photosynthesis have only been examined for a single rabbiteye cultivar (Davies and Flore 1986a), and cultivar variation for photosynthetic temperature optima is known to be significant in highbush blueberry (Moon et al. 1987). The combination of high light intensity and supraoptimal temperatures may limit photosynthetic potential during this period. A significant reduction in the photosynthetic potential has the dual effect of reducing current photosynthates for canopy and root development and reducing or delaying the accumulation of storage carbohydrates. The reduction in shoot growth rate observed during this period may, in part, be a consequence of this reduction in photosynthetic potential. However, observational evidence suggests that sufficient carbohydrates are available for renewed shoot growth and flower bud initiation later in the season.

Flower bud initiation is likely marked by a reduction in total photosynthesis, as daylength decreases and shoot growth rate declines (Fig. 9.5). Maximum accumulation of carbon reserves probably occurs during this period. Light levels are presumably not limiting, although light intensity and carbohydrate requirements for initiation in rabbiteye blueberry are unknown.

Mild winter temperatures during the dormant period may allow rapid depletion of carbohydrate reserves due to high respiration and continued flower bud development (Fig. 9.6). This, coupled with the opening of floral buds prior to or concomitant with vegetative buds, suggests that accumulation of sufficient carbohydrate reserves is crucial to maximizing reproductive development. There is no information on the extent or composition of carbohydrate reserves, or how reserve levels influence budbreak, anthesis, and fruit set in rabbiteye blueberry.

The carbon requirements for early fruit growth in rabbiteye blueberry depend on carbohydrate reserves and fruit photosynthesis, since maximum photosynthetic rates have not been reached. As shoots and leaves expand, and daylength and air temperatures increase, the total canopy photosynthetic rate increases, and net export of carbohydrates from leaves occurs. Stage II of fruit growth probably draws little on carbohydrate sources, making carbohydrates more available for shoot growth (Fig. 9.12). By stage III of the fruit development period, canopy photosynthetic rate is at its highest, and leaves are the primary source of carbon for fruit development, as temperatures are near the reported optimum for rabbiteye blueberry leaf photosynthesis. The competitive partitioning of carbon between the major root, shoot, fruit, and storage pool sinks needs to examined to characterize temporal changes in sink strength among these organs, as well as between fruits within and among clusters, and between fruit clusters and leaves.

Nitrogen partitioning. Although optimal plant development depends on the proper balance among all essential elements, nitrogen demands special attention since extremes in nitrogen levels are likely to significantly affect vegetative and reproductive development. During the vegetative period following harvest, supraoptimal soil temperatures may limit nitrogen uptake, which would result in limited reserve accumulation (Fig. 9.4) Towards the end of this period, soil nitrogen uptake probably begins to increase, as root growth increases in response to more favorable soil temperatures and decreased competition from other sink organs. As shoot and root growth rates slow at the end of this period, and demand for nitrogen decreases, nitrogen reserve levels begin to increase.

Nitrogen reserve levels probably continue to increase throughout the flower bud initiation phase (Fig. 9.5), as shoot and root growth rates continue to decrease, and the metabolic requirements for nitrogen decrease. Although nitrogen is required for flower bud initiation, the demand is small compared to the nitrogen demand during active shoot and root growth. Furthermore, continued nitrogen uptake from the soil probably occurs throughout this period in response to favorable soil temperatures. Information regarding the levels, composition, and location of nitrogen reserves in rabbiteye blueberry, as well as the influence of reserves on budbreak and early fruit development, is unknown.

During the latter stages of dormancy (Fig. 9.6) and throughout the anthesis period (Fig. 9.10), nitrogen levels continue to decrease as reserves are utilized by developing organs. Nitrogen uptake during these periods is low, due to unfavorably low soil temperatures during dormancy and insufficient carbohydrate availability for nitrogen assimilation during anthesis.

Early fruit growth of rabbiteye blueberry is probably also dependent on stored nitrogen reserves (Fig 9.12), as net export of current photosynthates to roots for nitrogen assimilation would be limited. However, as net export increases, and carbohydrate supply to roots increases, soil nitrogen uptake is expected to increase. This uptake rate would be enhanced by the favorable soil temperatures during this period. During the remainder of the fruit development period, nitrogen demands for shoot, root, and fruit growth are probably met by current uptake of soil nitrogen. As with carbohydrates, there is little information on the competitive partitioning of nitrogen among sink organs.

We have presented this review as the foundation for a comprehensive, integrative model of rabbiteye blueberry reproductive development in one of its cropping environments. We recognize the speculative nature of much of the model, and urge experimentation under a range of environmental and cultural conditions in order to elucidate the speculative areas. Interpretation of experimental results with respect to the complex relationships between the whole plant and the environment will lead to a greater understanding of those relationships and allow ongoing refinement of the model.

LITERATURE CITED

Aalders, L. E., and I. V. Hall. 1961. Pollen incompatibility and fruit set in lowbush blueberries. Can. J. Genet. Cytol. 3:300–307.

_____. 1964. A comparison of flower bud development in the lowbush blueberry, V. angustifolium Ait., under greenhouse and field conditions. Proc. Am. Soc. Hort. Sci. 85:281–284.

Aalders, L. E., I. V. Hall, and F. R. Forsyth. 1969. Effects of partial defoliation and light intensity on fruit set and berry development in the lowbush blueberry. Hort. Res. 9:124–129.

Abbott, J. D., and R. E. Gough. 1987a. Seasonal development of highbush blueberry under sawdust mulch. J. Am. Soc. Hort. Sci. 112:60–62.

_____. 1987b. Reproductive response of the highbush blueberry to root-zone flooding. HortScience 22:40–42.

Addoms, R. H., G. T. Nightingale, and M. A. Blake. 1930. Development and ripening of peaches as correlated with physical characteristics, chemical composition, and histological structure of the fruit flesh: II. Histology and Microchemistry, N.J. Agr. Expt. Sta. Bull. 507:3–19.

Andersen, P. C., D. W. Buchanan, and L. G. Albrigo. 1979a. Antitranspirant effects on water relations and fruit growth of rabbiteye blueberry. J. Am. Soc. Hort. Sci. 104:378–393.

_____. 1979b. Water relations and yields of three rabbiteye blueberry cultivars with and without drip irrigation. J. Am. Soc. Hort. Sci. 104:731–736.

Atkinson, D. 1980. The distribution and effectiveness of the roots of tree crops. Hort. Rev. 2:424–490.

Austin, M. E. 1982. An observation on rooting depth of rabbiteye blueberry plants. Ga. Agr. Expt. Sta. Res. Rpt. 394.

Austin, M. E., and K. Bondari. 1986. Effect of fertilizer applications, plant age, and rainfall on yield and fruit size of rabbiteye blueberry clone T-110. J. Am. Soc. Hort. Sci. 111:840–844.

Bailey, J. S. 1937. The pollination of the cultivated blueberry. Proc. Am. Soc. Hort. Sci. 35:71–72.

_____. 1947. Development time from bloom to maturity in cultivated blueberries. Proc. Am. Soc. Hort. Sci. 49:193–195.

Bailey, J. S., and L. H. Jones. 1941. The effect of soil temperature on the growth of cultivated blueberry bushes.Proc. Am. Soc. Hort. Sci. 38:462–464.

Bailey, J. S., A. F. Spelman, and B. Gersten. 1962. Seasonal changes in the nutrients in the leaves of 'Rubel' blueberry bushes. Proc. Am. Soc. Hort. Sci. 80:327–330.

Baker, G. A., and R. M. Brooks. 1944. Climate in relation to deciduous fruit production in California. III. Effect of temperature on number of days from full bloom to harvest of apricot and prune fruits. Proc. Am. Soc. Hort. Sci. 45:95–104.

Ballinger, W. E. 1966. Seasonal trends in 'Wolcott' blueberry (Vaccinium corymbosum L.) leaf and berry composition. N.C. Agr. Expt. Sta. Tech. Bull. 173.

Ballinger, W. E., E. P. Maness, and L. J. Kushman. 1970. Anthocyanins in ripe fruit of the highbush blueberry, Vaccinium corymbosum L. J. Am. Soc. Hort. Sci. 95:283–285.

Barden, J. A. 1978. Apple leaves, their morphology and photosynthetic potential. HortScience 13:644–646.

Batjer, L. P., and M. N. Westwood. 1958. Seasonal trend of several nutrient elements in leaves and fruits of 'Elberta' peach. Proc. Am. Soc. Hort. Sci. 71:116–126.

Beckman, W., and L. Tannenbaum. 1939. Insects pollinating cultivated blueberries. Am. Bee J. 79:436–437.

Beevers, L. 1976. Nitrogen Metabolism in Plants. Edward Arnold Ltd., London.

Bell, H. P. 1957. The development of the blueberry seed. Can. J. Bot. 35:139–153.

Bell, H. P., and J. Burchill. 1955. Flower development in the lowbush blueberry. Can. J. Bot. 33:251–258.

Bergman, H. F. 1929. Changes in the rate of respiration of the fruits of cultivated blueberry during ripening. Science 70:15.

Biermann, J., C. Stushnoff, and M. J. Burke. 1979. Differential thermal analysis and freezing injury in cold hardy blueberry flower buds. J. Am. Soc. Hort. Sci. 104:444–449.

Bittenbender, B. C., and G. S. Howell. 1976. Cold hardiness of flower buds from selected highbush blueberry cultivars. J. Am. Soc. Hort. Sci. 101:135–139.

Bjorkman, O. 1981. Responses to different quantum flux densities. p. 57–107. In: Lange, O. L., P. S. Nobel, C. B. Osmond, and H. Ziegler (eds.), Encyclopedia of Plant Physiology, new series vol. 12A, Physiological plant ecology. Springer-Verlag, Berlin.

Blanke, M. M., and F. Lenz. 1989. Fruit photosynthesis. Plant, Cell Environ. 12:31–46.

Borchers, E. A. 1979. Blueberries for Virginia. The Veg. Growers News 34(4):1–4.

Boyle, T. H. 1990. Flowering of Rhipsalidopsis rosea in response to temperature and photoperiod. HortScience 25:217–219.

Brewer, J. W., and R. C. Dobson. 1969a. Varietal attractiveness of blueberry blossoms to honeybees. Am. Bee J. 109:423–425.

_____. 1969b. Seed count and berry size in relation to pollinator level and harvest date for the highbush blueberry, Vaccinium corymbosum. J. Econ. Entomol. 62:1353–1356.

_____. 1969c. Pollen analysis of two highbush blueberry varieties, Vaccinium corymbosum. J. Am. Soc. Hort. Sci. 94:251–252.

Brierley, W. G., and A. C. Hildreth. 1928. Some studies on the hardiness of certain species of Vaccinium. Plant Physiol. 3:303–308.

Buchanan, D. W., F. S. Davies, and A. H. Teramura. 1978. Yield responses of three rabbiteye blueberry cultivars to drip irrigation and Vapor Guard. Proc. Fla. State Hort. Soc. 91:162–163.

Cain, J. C. 1952. A comparison of ammonium and nitrate nitrogen for blueberries. *Proc. Am. Soc. Hort. Sci.* 59:161–166.

Cain, J. C., and P. Eck. 1966. Blueberry and cranberry. p. 101–129. In: N. F. Childers (ed.), *Temperate to Tropical Fruit Nutrition.* Horticultural Publications, New Brunswick, NJ.

Camp, W. H. 1945. The North American blueberries with notes on other groups of *Vacciniaceae. Brittonia* 5:203–275.

Cane, J. H., and J. A. Payne. 1988. Foraging ecology of the bee *Habropoda laboriosa* (Hymenoptera: Anthophoridae), an oligolege of blueberries (Ericaceae: Vaccinium) in the southeastern United States. *Ann. Entomol. Soc. Am.* 81:419–427.

Castagnoli, S. P., T. M. DeJong, S. A. Weinbaum, and R. S. Johnson. 1990. Autumn foliage applications of $ZnSO_4$ reduced leaf nitrogen remobilization in peach and nectarine. *J. Am. Soc. Hort. Sci.* 115:79–83.

Chalmers, D. J., G. Burge, P. H. Jerie, and P. D. Mitchell. 1986. The mechanism of regulation of 'Bartlett' pear fruit and vegetative growth by irrigation withholding and regulated deficit irrigation. *J. Am. Soc. Hort. Sci.* 111:904–907.

Chalmers, D. J., R. L. Canterford, P. H. Jerie, T. R. Jones, and T. D. Ugalde. 1975. Photosynthesis in relation to growth and distribution of fruits in peach trees. *Austral. J. Plant Physiol.* 2:635–645.

Clark, J. R., D. B. Marx, and D. G. Sombek. 1989. Seasonal variation in elemental content of 'Bluecrop' blueberry leaves. *Ark. Agr. Expt. Sta. Bull.* 920.

Costante, J. F., and B. R. Boyce. 1968. Low temperature injury of highbush blueberry shoots at various times of the year. *Proc. Am. Soc. Hort. Sci.* 93:267–272.

Couvillon, G. A., and A. Erez. 1985. Effect of level and duration of high temperatures on rest in the peach. *J. Am. Soc. Hort. Sci.* 110:579–581.

Couvillon, G. A., and D. A. Lloyd. 1978. Summer defoliation effects on peach spring bud development. *HortScience* 13:53–54.

Coville, F. V. 1910. Experiments in blueberry culture. U.S. Dept. Agr. Bull. 193.

Crane, J. H., and F. S. Davies. 1988a. Flooding duration and seasonal effects on growth and development of young rabbiteye blueberry plants. *J. Am. Soc. Hort. Sci.* 113:180–184.

———. 1988b. Periodic and seasonal flooding effects on survival, growth, and stomatal conductance of young rabbiteye blueberry plants. *J. Am. Soc. Hort. Sci.* 113:488–493.

Creech, D. 1988. Influence of three above-ground mulch treatments and four in-ground amendment treatments on growth and yield of 'Brightwell' and 'Climax' rabbiteye blueberries. p. 66–77. In: Proc. 1988 Texas Blueberry Growers Assoc. Tyler.

Darnell, R. L., and F. S. Davies. 1990. Chilling accumulation, budbreak, and fruit set of young rabbiteye blueberry plants. *HortScience* 25:635–638.

Darnell, R. L., and P. M. Lyrene. 1989. Cross-incompatibility of two related rabbiteye blueberry cultivars. *HortScience* 24:1017–1018.

Darrow, G. M. 1958. Seed number in blueberry fruits. *Proc. Am. Soc. Hort. Sci.* 72:212–215.

Daughty, C. C., E. B. Adams, and L. W. Martin. 1984. Highbush blueberry production. Washington, Oregon and Idaho Coop. Ext. Serv. Bull. 215.

Davies, F. S. 1986. Flower position, growth regulators, and fruit set of rabbiteye blueberries. *J. Am. Soc. Hort. Sci.* 111:338–341.

Davies, F. S., and D. W. Buchanan. 1979. Fruit set and bee activity in four rabbiteye blueberry cultivars. *Proc. Fla. State Hort. Soc.* 92:246–247.

Davies, F. S., and J. A. Flore. 1986a. Gas exchange and flooding stress of highbush and rabbiteye blueberries. *J. Am. Soc. Hort. Sci.* 111:565–571.

———. 1986b. Short-term flooding effects on gas exchange and quantum yield of rabbiteye blueberry (*Vaccinium ashei* Reade). *Plant Physiol.* 81:289–292.

Davies, F. S., and C. R. Johnson. 1982. Water stress, growth, and critical water potentials of rabbiteye blueberry (*Vaccinium ashei* Reade). *J. Am. Soc. Hort. Sci.* 107:6–8.

Davies, F. S., A. H. Teramura, and D. W. Buchanan. 1979. Yield, stomatal resistance, xylem pressure potential, and feeder root density in three rabbiteye blueberry cultivars. *HortScience* 14:725–726.

DeJong, T. M. 1986. Fruit effects on photosynthesis in *Prunus persica*. *Physiol. Plant.* 66:149–153.

Doehlert, C. A. 1937. Blueberry tillage problems and a new harrow. N.J. Agr. Expt. Sta. Bull. 625.

Downton, W. J. S., W. J. R. Grant, and B. R. Loveys. 1987. Diurnal changes in the photosynthesis of field-grown grape vines. *New Phytol.* 105:71–80.

Durner, E. F., J. A. Barden, D. G. Himelrick, and E. B. Poling. 1984. Photoperiod and temperature effects on flower and runner development in day-neutral, Junebearing, and everbearing strawberries. *J. Am. Soc. Hort. Sci.* 109:396–400.

Eaton, G. W. 1966. Production of highbush blueberry pollen and its germination *in vitro* as affected by pH and sucrose concentration. *Can. J. Plant Sci.* 46:207–209.

_____. 1967. The relationship between seed number and berry weight in open-pollinated highbush blueberries. *HortScience* 2:14–15.

Eaton, G. W., and A. M. Jamont. 1966. Megagametogenesis in *V. corymbosum*. *Can. J. Bot.* 44:712–714.

Eaton, G. W., and C. N. Meehan. 1971. Effects of leaf position and sampling date on leaf nutrient composition of eleven highbush blueberry cultivars. *J. Am. Soc. Hort. Sci.* 96:378–380.

Eck, P. 1966. Botany. p. 14–44. In: Eck, P., and N. Childers (eds.), *Blueberry Culture*. Rutgers Univ. Press, New Brunswick, NJ.

_____. 1977. Nitrogen requirement of the highbush blueberry, *Vaccinium corymbosum* L. *J. Am. Soc. Hort. Sci.* 102:816–818.

_____. 1986. Blueberry. p. 75–85. In: S. P. Monselise (ed.), *Handbook of Fruit Set and Development*. CRC Press, Boca Raton, FL.

_____. 1988. *Blueberry Science*. Rutgers Univ. Press, New Brunswick, NJ.

Eck, P., and N. F. Childers. 1966. *Blueberry Culture*. Rutgers Univ. Press, New Brunswick, NJ.

Eck, P., and C. M. Mainland. 1971. Highbush blueberry fruit set in relation to flower morphology. *HortScience* 6:494–495.

Edwards, T. W., W. B. Sherman, and R. H. Sharpe. 1970. Fruit development in short and long cycle blueberries. *HortScience* 5:274–275.

_____. 1972. Seed development in certain Florida tetraploid and hexaploid blueberries. *HortScience* 7:127–128.

El-Agamy, S. Z. A., M. M. Aly, and R. H. Biggs. 1982. Fruit maturity as related to ethylene in 'Delite' blueberry. *Proc. Fla. State Hort. Soc.* 95:245–246.

El-Agamy, S. Z. A., W. B. Sherman, and P. M. Lyrene. 1981. Fruit set and seed number from self- and cross-pollinated highbush (4X) and rabbiteye (6X) blueberries. *J. Am. Soc. Hort. Sci.* 106:443–445.

Erez, A., and G. A. Couvillon. 1987. Characterization of the influence of moderate temperatures on rest completion in peach. *J. Am. Soc. Hort. Sci.* 112:677–680.

Erez, A., S. Lavee, and R. M. Samish. 1968. The effect of limitation in light during the rest period on leaf bud break of the peach. *Physiol. Plant.* 21:759–764.

Faragher, J. D., and R. L. Brohier. 1984. Anthocyanin accumulation in apple skin during ripening: Regulation by ethylene and phenylalanine ammonia-lyase. *Scientia Hort.* 22:89–96.

Fisher, D. V. 1962. Heat units and number of days required to mature some pome and stone fruits in various areas of North America. *Proc. Am. Soc. Hort. Sci.* 80:114–124.

Forsyth, F. R., and I. V. Hall. 1965. Effect of leaf maturity, temperature, carbon dioxide concentration and light intensity on rate of photosynthesis in clonal lines of the lowbush blueberry, *Vaccinium angustifolium* Ait. under laboratory conditions. *Can. J. Bot.* 43:893–900.

_____. 1969. Ethylene production with accompanying respiration rates from the time of blossoming to fruit maturity in three *Vaccinium* species. *Nat. Can.* 96:257–259.

Freeman, M. W., and G. C. Martin. 1981. Peach floral bud break and abscisic acid content as affected by mist, light, and temperature treatments during rest. *J. Am. Soc. Hort. Sci.* 106:333–336.

Frenkel, C. 1972. Involvement of peroxidase and indole-3-acetic acid oxidase isozymes from pear, tomato, and blueberry fruit in ripening. *Plant Physiol.* 49:757–763.

Fuchigami, L. H., M. Hotze, and C. J. Weiser. 1977. The relationship of vegetative maturity to rest development and spring bud-break. *J. Am. Soc. Hort. Sci.* 102:450–452.

Fujii, J. A., and R. A. Kennedy. 1985. Seasonal changes in the photosynthetic rate in apple trees: A comparison between fruiting and non-fruiting trees. *Plant Physiol.* 78:519–524.

Galletta, G. J., and D. G. Himelrick. 1990. *Small Fruit Crop Management.* Prentice Hall, Englewood Cliffs, NJ.

Garvey, E. J., and P. M. Lyrene. 1987. Self-incompatibility in 19 native blueberry selections. *J. Am. Soc. Hort. Sci.* 112:856–858.

Gilreath, P. R., and D. W. Buchanan. 1981. Temperature and cultivar influences on the chilling period of rabbiteye blueberry. *J. Am. Soc. Hort. Sci.* 106:625–628.

Given, N. K., M. A. Venis, and D. Grierson. 1988. Phenylalanine ammonia-lyase activity and anthocyanin synthesis in ripening strawberry fruit. *J. Plant Physiol.* 133:25–30.

Gough, R. E. 1980. Root distribution of 'Coville' and 'Lateblue' highbush blueberry under sawdust mulch. *J. Am. Soc. Hort. Sci.* 105:576–578.

Gough, R. E., and W. Litke. 1980. An anatomical and morphological study of abscission in highbush blueberry fruit. *J. Am. Soc. Hort. Sci.* 105:335–341.

Gough, R. E., and V. G. Shutak. 1977. Anatomy and morphology of cultivated highbush blueberry. Rhode Island Agr. Expt. Sta. Bull. 423.

Gough, R. E., V. G. Shutak, and R. L. Hauke. 1978. Growth and development of highbush blueberry. II. Reproductive growth, histological studies. *J. Am. Soc. Hort. Sci.* 103:476–479.

Gough, R. E., V. G. Shutak, and N. D. Windus. 1976. Observations on vegetative and reproductive growth in blueberry. *HortScience* 11:260–261.

Gucci, R. 1988. The effect of fruit removal on leaf photosynthesis, water relations, and carbon partitioning in sour cherry and plum. Ph.D. disseration. Mich. State University, East Lansing.

Gupton, C. L. 1984. Effect of pollen source on fruit characteristics of low-chilling highbush type blueberries. *HortScience* 19:531–532.

Hall, A. J., and C. J. Brady. 1977. Assimilate source-sink relationships in *Capsicum annuum* L. II. Effects of fruiting and defloration on the photosynthetic capacity and senescence of leaves. *Austral. J. Plant Physiol.* 4:771–783.

Hall, I. V. 1958. Some effects of light on native lowbush blueberries. *Proc. Am. Soc. Hort. Sci.* 72:216–218.

Hall, I. V., D. L. Craig, and L. E. Aalders. 1963. The effect of photoperiod on the growth and flowering of highbush blueberry. *Proc. Am. Soc. Hort. Sci.* 82:260–263.

Hall, I. V., and F. R, Forsyth. 1967a. Respiration rates of developing fruits of the lowbush blueberry. *Can. J. Plant Sci.* 47:157–159.

——. 1967b. Production of ethylene by flowers following pollination and treatments with water and auxin. *Can. J. Bot.* 45:1163–1166.

Hall, I. V., and R. A. Ludwig. 1961. Effects of photoperiod, temperature, and light intensity on the growth of the lowbush blueberry. *Can. J. Bot.* 39:1733–1739.

Hammett, L. K., and W. E. Ballinger. 1972. A nutrient solution-sand culture system for studying the influence of N form on highbush blueberries. *HortScience* 7:498–500.

Hancock, J. F., and A. D. Draper. 1989. Blueberry culture in North America. *HortScience* 24:551–556.

Hancock, J. F., J. W. Nelson, H. C. Bittenbender, P. W. Callow, J. S. Cameron, S. L. Krebs, M. P. Pritts, and C. M. Schumann. 1987. Variation among highbush blueberry cultivars in susceptibility to spring frost. *J. Am. Soc. Hort. Sci.* 112:702–706.

Hansen, P., and J. Grauslund. 1973. ^{14}C-Studies on apple trees. VIII. The seasonal variation and nature of reserves. *Physiol. Plant.* 28:24–32.

Hellman, E. W., and J. N. Moore. 1983. Effect of genetic relationship to pollinizer on fruit, seed, and seedling parameters in highbush and rabbiteye blueberries. *J. Am. Soc. Hort. Sci.* 108:401–405.

Hindle, R., V. G. Shutak, and E. P. Christopher. 1957a. Growth studies of the highbush blueberry fruit. *Proc. Am. Soc. Hort. Sci.* 69:282–287.

_____ . 1957b. Relationship of wood thickness to blossoming, rate of ripening, and size of fruit on highbush blueberry. *Proc. Am. Soc. Hort. Sci.* 70:150–155.

Hrazdina, G., G. F. Parsons, and L. R. Mattick. 1984. Physiological and biochemical events during development and maturation of grape berries. *Am. J. Enol. Vit.* 35:220–227.

Ismail, A. A., and W. J. Kender. 1969. Evidence of a respiratory climacteric in highbush and lowbush blueberry fruit. *HortScience* 4:342–344.

Jennings, D. L. 1988. *Raspberries and Blackberries: Their Breeding, Diseases, and Growth.* Academic Press, New York.

Johnson, R. S., and A. N. Lakso. 1986. Carbon balance model of a growing apple shoot. I. Development of the model. *J. Am. Soc. Hort. Sci.* 111:160–164.

Kappes, E. M. 1985. Carbohydrate production, balance and translocation in leaves, shoots and fruits of 'Montmorency' sour cherry. Ph.D. Dissertation. Michigan State University, East Lansing.

Kappes, E. M., and J. A. Flore. 1989. Phyllotaxy and stage of leaf and fruit development influence initiation and direction of carbohydrate export from sour cherry leaves. *J. Am. Soc. Hort. Sci.* 114:642–648.

Karlsson, P. S. 1985a. Patterns of carbon allocation above ground in a deciduous (*Vaccinium uliginosum*) and an evergreen (*Vaccinium vitis-idaea*) dwarf shrub. *Physiol. Plant.* 63:1–7.

_____ . 1985b. Photosynthetic characteristics and leaf carbon economy of a deciduous and an evergreen dwarf shrub: *Vaccinium uliginosum* L. and *V. vitis-idaea* L. *Holarct. Ecol.* 8:9–17.

Keller, J. D., and W. H. Loescher. 1989. Nonstructural carbohydrate partitioning in perennial parts of sweet cherry. *J. Am. Soc. Hort. Sci.* 114:969–975.

Kender, W. J., and W. T. Brightwell. 1966. Environmental relationships. p. 75–93. In: Eck, P., and N. Childers (eds.), *Blueberry Culture.* Rutgers Univ. Press, New Brunswick, NJ.

Knight, R. J., Jr. and D. H. Scott. 1964. Effect of temperatures on self- and cross-pollination and fruiting of four highbush blueberry varieties. *Proc. Am. Soc. Hort. Sci.* 85:302–306.

Korcak, R. F. 1986. Adaptability of blueberry species to various soil types. II. Leaf and soil analysis. *J. Am. Soc. Hort. Sci.* 111:822–828.

_____ . 1988. Nutrition of blueberry and other calcifuges. *Hort. Rev.* 10:183–227.

Krebs, S. L., and J. F. Hancock. 1988. The consequences of inbreeding on fertility in *Vaccinium corymbosum* L. *J. Am. Soc. Hort. Sci.* 113:914–918.

_____ . 1990. Early-acting inbreeding depression and reproductive success in the highbush blueberry, *Vaccinium corymbosum* L. *Theor. Appl. Genet.* 79:825–832.

Krewer, G., S. Myers, P. Bertrand, D. Horton, T. Murphy, and M. Austin. 1986. Commercial blueberry culture. Ga. Coop. Ext. Circ. 713.

Kushima, T., and M. E. Austin. 1979. Seed number and size in rabbiteye blueberry fruit. *HortScience* 14:721–723.

Kushman, L. J., and W. E. Ballinger. 1967. Acid and sugar changes during ripening in 'Wolcott' blueberries. *Proc. Am. Soc. Hort. Sci.* 92:290–295.

Lahdesmaki, P., T. Pakonen, E. Saari, K. Laine, L. Tasanen and P. Havas. 1990a. Environmental factors affecting basic nitrogen metabolism and seasonal levels of various nitrogen fractions in tissues of bilberry, *Vaccinium myrtillus*. *Holarct. Ecol.* 13:19–30.

_____ . 1990b. Changes in total nitrogen, protein, amino acids and NH$_4^+$ in tissues of bilberry, *Vaccinium myrtillus*, during the growing season. *Holarct. Ecol.* 13:31–38.

Landsberg, J. J., and H. G. Jones. 1981. Apple orchards. p. 419–469. In: T. T. Kozlowski (ed.), *Water Deficits and Plant Growth*. Academic Press, New York.

Lang, G. A. 1989. Dormancy-models and manipulations of environmental/physiological regulation. p. 79–98. In: C. J. Wright (ed.), *Manipulation of Fruiting.* Butterworth and Co., London.

Lang, G. A., R. G. Danka, and E. J. Parrie. 1990. Pollen-stigma interactions and relationship to fruit development in low-chill southern highbush blueberry. In: C. A. Brun (ed.), Proc. 6th North American Bluebuerry Research Workers Conference, Portland, OR. (In Press).

Lange, O. L., E. D. Schulze, M. Evenari, L. Kappen, and U. Buschbom. 1974. The temperature related photosynthetic capacity of plants under desert conditions. I. Seasonal changes of the photosynthetic response to temperature. *Oecologia* 17:97–110.

Lee, W. R. 1958. Pollination studies on lowbush blueberries. *J. Econ. Entomol.* 51:544–545.

Lenz, F. 1979. Fruit effects on photosynthesis, light- and dark-respiration. p. 271–281. In: Marcelle, R., H. Clijsters, and M. van Poucke (eds.), *Photosynthesis and Plant Development.* W. Junk, The Hague.

Lipe, J. A. 1978. Ethylene in fruits of blackberry and rabbiteye blueberry. *J. Am. Soc. Hort. Sci.* 103:76–77.

Loescher, W. H., T. McCamant, and J. D. Keller. 1990. Carbohydrate reserves, translocation, and storage in woody plant roots. *HortScience* 25:274–281.

Luby, J. J., D. K. Wildung, C. Stushnoff, S. T. Munson, P. E. Read, and E. E. Hoover. 1986. 'Northblue', 'Northsky', and 'Northcountry' blueberries. *HortScience* 21:1240–1242.

Lyrene, P. M. 1984. Late pruning, twig orientation, and flower bud formation in rabbiteye blueberry. *HortScience* 19:98–99.

_____ . 1985. Effects of year and genotype on flowering and ripening dates in rabbiteye blueberry. *HortScience* 20:407–409.

_____ . 1989a. Variation in flower bud numbers on rabbiteye blueberry. *Acta Hort.* 241:201–203.

_____ . 1989b. Blueberry news. Fla. Blueberry Growers Assoc. Newsletter, Summer.

Lyrene, P. M., and T. E. Crocker. 1983. Poor fruit set on rabbiteye blueberries after mild winters: Possible causes and remedies. *Proc. Fla. State Hort. Soc.* 96:195–197.

Lyrene, P. M., and W. B. Sherman. 1984. Breeding early-ripening blueberries for Florida. *Proc. Fla. State Hort. Soc.* 97:322–325.

_____ . 1985. Breeding blueberry cultivars for the central Florida peninsula. *Proc. Fla. State Hort. Soc.* 98:158–162.

Mainland, C. M. 1985. Vaccinium. p. 451–455. In: A. H. Halevy (ed.), *Handbook of Flowering,* Vol. IV. CRC Press, Boca Raton, FL.

_____ . 1986. Commercial blueberry production guide for North Carolina. N.C. Agr. Ext. Bull. AG-115.

_____ . 1987. N.C. blueberry industry got its start in '20s, was thretened in '30s. *Fruit South* 8(5):14–16.

Makus, D. J., and W. E. Ballinger. 1973. Characterization of anthocyanins during ripening of fruit of *Vaccinium corymbosum* L., cv. Wolcott. *J. Am. Soc. Hort. Sci.* 98:99–101.

Mancinelli, A. L., and O. M. Schwartz. 1984. The photoregulation of anthocyanin synthesis. IX. The photosensitivity of the response in dark and light grown tomato seedlings. *Plant and Cell Physiol.* 25:93–105.

Marucci, P. E. 1966. Blueberry pollination. *Am. Bee J.* 106:250–251.

Meader, E. M., and G. M. Darrow. 1944. Pollination of the rabbiteye blueberry and related species. *Proc. Am. Soc. Hort. Sci.* 45:267–274.

_____ . 1947. Highbush blueberry pollination experiments. *Proc. Am. Soc. Hort. Sci.* 49:196–204.

Merrill, T. A., and S. Johnston. 1939. Further observations on the pollination of the highbush blueberry. *Proc. Am. Soc. Hort. Sci.* 37:617–619.

Millard, P., and C. M. Thomson. 1989. The effect of the autumn senescence of leaves on the internal cycling of nitrogen for the spring growth of apple trees. *J. Expt. Bot.* 40:1285–1289.

Mondal, M. H., W. A. Brun, and M. L. Brenner. 1978. Effects of sink removal on photosynthesis and senescence in leaves of soybean (*Glycine max* L.) plants. *Plant Physiol.* 61:394–397.

Moon, J. W., J. F. Hancock, Jr., A. D. Draper, and J. A. Flore. 1987. Genotypic differences in the effect of temperature on CO_2 assimilation and water use efficiency in blueberry. *J. Am. Soc. Hort. Sci.* 112:170–173.

Moon, J. W., Jr., J. A. Flore, and J. F. Hancock, Jr. 1987. A comparison of carbon and water vapor gas exchange characteristics between a diploid and highbush blueberry. *J. Am. Soc. Hort. Sci.* 112:134–138.

Moore, J. N. 1964. Duration of receptivity to pollination of flowers of the highbush blueberry and the cultivated strawberry. *Proc. Am. Soc. Hort. Sci.* 85:295–301.

Moore, J. N., B. D. Reynolds, and G. R. Brown. 1972. Effects of seed number, size and development on fruit size of cultivated blueberries. *HortScience* 7:268–269.

Moreno, J., and J. L. Garcia-Martinez. 1983. Seasonal variation of nitrogenous compounds in the xylem sap of *Citrus*. *Physiol. Plant.* 59:669–675.

Morrow, E. B. 1943. Some effects of cross-pollination versus self-pollination in the cultivated blueberry. *Proc. Am. Soc. Hort. Sci.* 42:469–472.

Mowry, H., and A. F. Camp. 1928. Blueberry culture in Florida. *Fla. Agr. Expt. Sta. Bull.* 194.

Munoz, C., G. Sepulveda, J. Garcia-Huidobro, and W. B. Sherman. 1986. Determining thermal time and base temperature required for fruit development in low chilling peaches. *HortScience* 21:520–522.

Nelson, J. W., and J. P. Holbein. 1986. North American blueberry acreage and production. North Amer. Blueberry Council Ann. Report, p. 27.

Norvell, D. J., and J. N. Moore. 1982. An evaluation of chilling models for estimating rest requirements of highbush blueberries. *J. Am. Soc. Hort. Sci.* 107:54–56.

Oland, K. 1963. Changes in the content of dry matter and major nutrient elements of apple foliage during senescence and abscission. *Physiol. Plant.* 16:682–694.

Oliveira, C. M., and C. A. Priestley. 1988. Carbohydrate reserves in deciduous fruit trees. *Hort. Rev.* 10:403–430.

Parrie, E. J. 1990. Pollination of hybrid southern highbush blueberries (*Vaccinium corymbosum* L.). MS Thesis, Louisiana State University, Baton Rouge.

Patten, K. D., E. W. Neuendorff, and S. C. Peters. 1988. Root distribution of 'Climax' rabbiteye blueberry as affected by mulch and irrigation geometry. *J. Am. Soc. Hort. Sci.* 113:657–661.

Patten, K. D., and E. W. Neuendorff. 1989. Influence of light and other parameters on the development and quality of rabbiteye blueberry fruit. p. 109–117. In: Proc. 1989 Texas Blueberry Growers Assoc. Beaumont.

Patten, K., G. Nimr, E. Neuendorff, J. Clark, and G. Fernandez. 1989. Freeze and frost tolerance of southern highbush and rabbiteye blueberries. p. 127–131. In: Proc. 1989 Texas Blueberry Growers Assoc. Beaumont.

Payne, J. A., J. H. Cane, A. A. Amis, and P. M. Lyrene. 1989. Fruit size, seed size, seed viability, and pollination of rabbiteye blueberries (*Vaccinium ashei* Reade). *Acta Hort.* 241:38–43.

Peterson, L. A., E. J. Stang, and M. N. Dana. 1988. Blueberry response to NH_4-N and NO_3-N. *J. Am. Soc. Hort. Sci.* 113:9–12.

Pettersen, H. 1972. The effect of temperature and daylength on shoot growth and bud formation in azaleas. *J. Am. Soc. Hort. Sci.* 97:17–24.

Pritts, M. P., and J. F. Hancock. 1985. Lifetime biomass partitioning and yield component relationships in the highbush blueberry, *Vaccinium corymbosum* L. (Ericaceae). *Am. J. Bot.* 72:446–452.

Quinlan, J. D., and A. P. Preston. 1968. Effects of thinning blossom and fruitlets on growth and cropping of 'Sunset' apple. *J. Hort. Sci.* 43:373–381.

Rabaey, A., and J. Luby. 1988. Fruit set in half-high blueberry genotypes following self and cross pollination. *Fruit Var. J.* 42:126–129.

Ragland, C. H. 1934. The development of the peach fruit with special reference to split pit and gumming. *Proc. Am. Soc. Hort. Sci.* 31:1–21.

Retamales, J. B., and E. J. Hanson. 1989. Fate of ^{15}N-labeled urea applied to mature highbush blueberries. *J. Am. Soc. Hort. Sci.* 114:920–923.

_____. 1990. Effect of nitrogen fertilizers on leaf and soil nitrogen levels in high bush blueberry. *Comm. Soil Sci. Plant Anal.* (in press).

Richardson, E. A., S. D. Seeley, D. R. Walker, J. L. Anderson, and G. L. Ashcroft. 1975. Phenoclimatography of spring peach bud development. *HortScience* 10:236–237.

Rogers, B. L., L. P. Batjer, and A. H. Thompson. 1953. Seasonal trend of several nutrient elements in 'Delicious' apple leaves expressed on a percent and unit area basis. *Proc. Am. Soc. Hort. Sci.* 61:1–5.

Roper, T. R., J. D. Keller, W. H. Loescher, and C. R. Rom. 1988. Photosynthesis and carbohydrate partitioning in sweet cherry: Fruiting effects. *Physiol. Plant.* 72:42–47.

Rosen, C. J., D. L. Allan, and J. J. Luby. 1990. Nitrogen form and solution pH influence growth and nutrition of two *Vaccinium* clones. *J. Am. Soc. Hort. Sci.* 115:83–89.

Ryugo, K. 1988. *Fruit Culture: Its Science and Art.* Wiley, New York.

Sams, C. E., and J. A. Flore. 1982. The influence of age, position and environmental variables on net photosynthetic rate of sour cherry leaves. *J. Am. Soc. Hort. Sci.* 107:339–344.

_____. 1983. Net photosynthetic rate of sour cherry (*Prunus cerasus* L. 'Montmorency') during the growing season with particular reference to fruiting. *Photosyn. Res.* 4:307–316.

Schaller, C. C. 1983. Growing blueberries in Arkansas. *Univ. Arkansas Coop. Ext. Bull.* EL-600.

Schulze, E. D., O. L. Lange, M. Evenari, L. Kappen, and U. Buschbom. 1974. The role of air humidity and leaf temperature in controlling stomatal resistance of *Prunus armeniaca* L. under desert conditions. I. A simulation of the daily course of stomatal resistance. *Oecologia* 17:159–170.

Schulze, E. D., O. L. Lange, L. Kappen, M. Evenari, and U. Buschbom. 1975. The role of air humidity and leaf temperature in controlling stomatal resistance of *Prunus armeniaca* L. under desert conditions. II. The significance of leaf water status and internal carbon dioxide concentration. *Oecologia* 18:219–233.

Sharpe, R. H., and W. B. Sherman. 1971. Breeding blueberries for low-chilling requirement. *HortScience* 6:145–147.

Shaw, F. R., and J. S. Bailey. 1937. The honeybee as pollinator of cultivated blueberries. *Am. Bee J.* 77:30.

Shaw, F. R., J. S. Bailey, and A. I. Bourne. 1939. The comparative value of honeybees in the pollination of cultivated blueberries. *J. Econ. Entomol.* 33:872–876.

Shine, J., and D. W. Buchanan. 1982. Chilling requirements of 3 Florida blueberry cultivars. *Proc. Fla. State Hort. Soc.* 95:85–87.

Shutak, V. G. 1957. Growth studies of the cultivated blueberry. *Rhode Island Agr. Expt. Sta. Bull.* 339.

Shutak, V. G., R. Hindle, and E. P. Christopher. 1956. Factors associated with ripening of highbush blueberry fruit. *Proc. Am. Soc. Hort. Sci.* 68:178–183.

Spiers, J. M. 1976. Chilling regimes affect bud break in 'Tifblue' rabbiteye blueberry. *J. Am. Soc. Hort. Sci.* 101:88–90.

_____. 1978a. Effects of pH level and nitrogen source on elemental leaf content of 'Tifblue' rabbiteye blueberry. *J. Am. Soc. Hort. Sci.* 103:705–708.

_____. 1978b. Effect of stage of bud development on cold injury in rabbiteye blueberry. *J. Am. Soc. Hort. Sci.* 103:452–455.

_____. 1979. Calcium and nitrogen nutrition of 'Tifblue' rabbiteye blueberry in sand culture. *HortScience* 14:523–525.

_____. 1981. Fruit development in rabbiteye blueberry *Vaccinium ashei* cultivars. *HortScience* 16:175–176.

_____ . 1986. Root distribution of 'Tifblue' rabbiteye blueberry as influenced by irrigation, incorporated peatmoss and mulch. *J. Am. Soc. Hort. Sci.* 111:877–880.

Spiers, J. M., and A. D. Draper. 1974. Effect of chilling on bud break in rabbiteye blueberry. *J. Am. Soc. Hort. Sci.* 99:398–399.

Starck, Z., M. Kozinska, and R. Szaniawski. 1979. Photosynthesis in tomato plants with modified source-sink relationship. p. 233–241. In: Marcelle, R., H. Clijsters, and M. van Poucke (eds.), *Photosynthesis and Plant Development.* W. Junk, The Hague.

Stephenson, A. G., and R. I. Bertin. 1983. Male competition, female choice, and sexual selection in plants. p. 110–149. In: L. Real (ed.), *Pollination Biology.* Academic Press, New York.

Stevens, N. E. 1919. The development of the endosperm in *Vaccinium corymbosum. Torrey Bot. Club Bull.* 46:465–468.

Stushnoff, C., and L. F. Hough. 1968. Sporogenesis and gametophyte development in 'Bluecrop' and 'Coville' highbush blueberries. *Proc. Am. Soc. Hort. Sci.* 93:242–247.

Taylor, B. K., and L. H. May. 1967. The nitrogen nutrition of the peach tree. II. Storage and mobilization of nitrogen in young trees. *Austral. J. Biol. Sci.* 20:389–411.

Taylor, B. K., and B. van den Ende. 1969. The nitrogen nutrition of the peach tree. IV. Storage and mobilization of nitrogen in mature trees. *Austral. J. Agr. Res.* 20:869–881.

Taylor, B. K., B. van den Ende, and R. L. Canterford. 1975. Effects of rate and timing of nitrogen applications on the performance and chemical composition of young pear trees, cv. Williams' Bon Chretien. *J. Hort. Sci.* 50:29–40.

Teramura, A. H., F. S. Davies, and D. W. Buchanan. 1979. Comparative photosynthesis and transpiration in excised shoots of rabbiteye blueberry. *HortScience* 14:723–724.

Titus, J. S., and S. M. Kang. 1982. Nitrogen metabolism, translocation, and recycling in apples. *Hort. Rev.* 4:204–246.

Townsend, L. R. 1967. Effect of ammonium nitrogen and nitrate nitrogen separately and in combination, on the growth of the highbush blueberry. *Can. J. Plant Sci.* 47:555–562.

_____ . 1969. Influence of form of nitrogen and pH on growth and nutrient levels in the leaves and roots of the lowbush blueberry. *Can. J. Plant Sci.* 49:333–338.

Townsend, L. R., and I. V. Hall. 1970. Trends in nutrient levels of lowbush blueberry leaves during four consecutive years of sampling. *Nat. Can.* 97:461–466.

Trevett, M. F., P. N. Carpenter, and R. E. Durgin. 1968. Seasonal trend and inter-relation of mineral nutrients in lowbush blueberry leaves. *Maine Agr. Expt. Sta. Bull.* 665.

Tromp, J. 1970. Storage and mobilization of nitrogenous compounds in apple trees with special reference to arginine. p. 143–159. In: Luckwill, L. C., and C. V. Cutting (eds.), *Physiology of Tree Crops.* Academic Press, New York.

_____ . 1983. Nutrient reserves in roots of fruit trees, in particular carbohydrates and nitrogen. *Plant & Soil* 71:401–413.

Tromp, J., and J. C. Ovaa. 1967. Seasonal variations in the amino acid composition of xylem sap of apple. *Z. Pflanzenphysiol.* 57:11–21.

_____ . 1971. Phloem translocation of storage nitrogen in apple. *Physiol. Plant.* 25:407–413.

_____ . 1973. Spring mobilization of protein nitrogen in apple bark. *Physiol. Plant.* 29:1–5.

U. S. Department of Agriculture. 1976. Insect pollination of cultivated crop plants. USDA/ARS Agricultural Handbk. 496, Washington, D.C., p. 110–116.

Vander Kloet, S. P. 1983. The relationship between seed number and pollen viability in *Vaccinium corymbosum* L. *HortScience* 18:225–226.

Vasilakakis, M., and I. C. Porliggis. 1984. Effect of temperature on pollen germination, pollen tube growth, stigma receptivity and fruit set of 'Tsakoniki' or 'Spada' pear. *HortScience* 19:528 (Abstr.).

Walser, R. H., D. R. Walker, and S. D. Seeley. 1981. Effect of temperature, fall defoliation, and gibberellic acid on the rest period of peach leaf buds. *J. Am. Soc. Hort. Sci.* 106:91–94.

Wareing, P. F. 1956. Photoperiodism in woody plants. *Ann. Rev. Plant Physiol.* 7:191–214.

Weinbaum, S. A., I. Klein, F. E. Broadbent, W. C. Micke, and T. T. Muraoka. 1984. Effects of time of nitrogen application and soil texture on the availability of isotopically labeled fertilizer nitrogen to reproductive and vegetative tissue of mature almond trees. *J. Am. Soc. Hort. Sci.* 109:339–343.

Weinbaum, S. A., M. L. Merwin, and T. T. Muraoka. 1978. Seasonal variation in nitrate uptake efficiency and distribution of absorbed nitrogen in non-bearing prune trees. *J. Am. Soc. Hort. Sci.* 103:516–519.

Werner, D. J., B. D. Mowrey, and E. Young. 1988. Chilling requirement and post-rest heat accumulation as related to difference in time of bloom between peach and western sand cherry. *J. Am. Soc. Hort. Sci.* 113:775–778.

Williams, R. R. 1965. The effect of summer nitrogen applications on the quality of apple blossoms. *J. Hort. Sci.* 40:31–41.

Windus, N. D., V. G. Shutak, and R. E. Gough. 1976. CO_2 and C_2H_4 evolution by highbush blueberry fruit. *HortScience* 11:515–517.

Wood, G. W. 1962. The period of receptivity in flowers of the lowbush blueberry. *Can. J. Bot.* 40:685–686.

———. 1965. Evidence in support of reduced fruit set in lowbush blueberry by pollen incompatibility. *Can. J. Plant Sci.* 45:601–602.

———. 1968. Self-fertility in the lowbush blueberry. *Can. J. Plant Sci.* 48:431–433.

Woodruff, R. E., D. H. Dewey, and H. M. Sell. 1960. Chemical changes of 'Jersey' and 'Rubel' blueberry fruit associated with ripening and deterioration. *Proc. Am. Soc. Hort. Sci.* 75:387–401.

Worley, R. E. 1979. Fall defoliation date and seasonal carbohydrate concentration of pecan wood tissue. *J. Am. Soc. Hort. Sci.* 104:195–199.

Yarbrough, J. A., and E. B. Morrow. 1947. Stone cells in *Vaccinium. Proc. Am. Soc. Hort. Sci.* 50:224–228.

Young, E., and D. J. Werner. 1985. Effects of shoot, root, and shank chilling during rest in apple and peach on growth resumption and carbohydrates. *J. Am. Soc. Hort. Sci.* 110:769–774.

Young, M. J., and W. B. Sherman. 1978. Duration of pistil receptivity, fruit set, and seed production in rabbiteye and tetraploid blueberries. *HortScience* 13:278–279.

Young, R. S. 1952. Growth and development of the blueberry fruit (*Vaccinium corymbosum* L. and *V. angustifolium* Ait.). *Proc. Am. Soc. Hort. Sci* 59:167–172.

10

Maturity Indices for Apple and Pear

*C. M. Kingston**
Ministry of Agriculture and Fisheries
Levin Horticultural Research Centre
Private Bag, Levin, New Zealand

I. INTRODUCTION

Every consumer has a slightly different idea about what constitutes a high-quality apple or pear. This concept results from the assessment of several factors including size, color, flavor, aroma, and texture (Vangdal 1985). Fruit that the consumer finds acceptable are often said to be "mature" or "ripe." However, the stage of development at which fruit has maximum appeal to the consumer does not necessarily coincide with the point at which fruit is "physiologically" mature, i.e., able to continue ontogeny even if detached from the plant (Watada et. al. 1984). For this reason the terms "commercial" and "horticultural" maturity have been developed to describe the stage of development at which the fruit has the

*I thank my colleagues, especially Graeme King and Nigel Given, for suggestions made during both preparation and review of this manuscript. Thanks also to Jo Mulhane for help in preparing the typed version.

desired characteristics for a particular use by consumers (Watada et al. 1984; Wills et al. 1989). In this review the term "maturity" implies the commercial rather than the physiological state.

External factors such as size and color strongly influence our initial reaction to fresh fruit, but a final judgement cannot be made until the fruit is bitten into and tasted. Customers purchasing fruit must choose the fruit on the basis of its external appearance. They rely on the reputation, experience, and knowledge of the seller to ensure that the fruit not only looks good but also tastes good. If their expectations are realized, consumers maintain their faith in the seller and continue to purchase and pay maximum prices for the product. The same principle applies on world markets: fruit of consistently high quality wins consumer favor and realizes top prices. Unfortunately, assessing fruit quality is not easy. External characteristics such as skin color and fruit size are not always good indicators of internal composition. Environmental factors, cultivar, and tree management have diverse effects on external and internal characteristics. Excessive nitrogen applications, for example, increase green coloration yet decrease soluble solids concentration, titratable acidity, and fruit firmness (Hikasa et al. 1986; Olsen et al. 1986). In contrast, increasing exposure to sunlight hastens development of red coloration and increases fruit size, soluble solids concentration, and fruit firmness (Shaw and Rowe 1982; Barritt et al. 1987).

Apples and pears are not always marketed directly after harvest. To extend the period over which fruit is available, and moderate fluctuations in supply, fruit may be stored at cool temperatures. These temperatures slow down the biochemical processes occurring in fruit but do not halt them completely. This introduces a further complication—to allow for changes occurring during storage, fruit to be stored necessarily has a different "optimum" maturity at harvest than fruit destined for immediate consumption. With the advent of increasingly longer-term storage, with controlled atmosphere (CA) techniques, picking fruit before or after it reaches optimum maturity for storage greatly increases the risk of physiological disorders developing during the storage period (Lau 1985). To minimize this risk it is essential to clearly define what optimum maturity is.

Many pome fruit can ripen over an extended period, requiring several harvests to ensure the fruit is picked as near to optimum maturity as possible. This variation in the time taken to reach maturity arises from a number of factors. Often it reflects the time period over which flowers reached anthesis, and hence fruit set. Fruit are produced at many places within the canopy, and this affects the amount of light, ambient temperatures, and endogenous hormone supply received by fruit. Growth regulators produced by growing shoots may delay physiological maturity of fruit in close proximity. Apple fruit at the ends of branches produce less ethylene than fruit from within the canopy at any sampling date after full bloom (Farhoomand et al. 1977) Despite the apparent delay in reaching physiological maturity, the increased color, firmness and soluble solids concentration of fruit in exposed positions (Shaw and Rowe 1982; Barritt et al. 1987) may mean they reach commercial maturity earlier than fruit within the canopy. These variations need to be taken into account if fruit with maximum quality is to be harvested.

With the approach of maturity, many physical and chemical changes occur which enable the fruit to ripen after harvest. Usually fruit firmness and titratable acidity decrease while total sugars and the rate of starch disappearance increase (Divikar et al. 1981; Mann and Singh 1985). The rate at which these parameters change and the absolute value of the change depend on a number of factors, including climate and cultivar. The extent of any change is rarely duplicated between seasons, so the rate at which these changes occur can produce a more meaningful assessment of how close to maturity fruit are (Lau 1985; Truter and Hurndall 1988a, 1988b). The change occurring must necessarily be greater than the sampling error. There is a poor correlation between some maturity indices, for example fruit firmness and starch content (Knee et al. 1989). This, coupled with the variability between seasons and individual fruit, suggests that several parameters should be used to define maturity in any one cultivar (Shaw and Rowe 1982; Lau 1985; Blankenship and Unrath 1988; Truter and Hurndall 1988a, 1988b). In this review, the methods most commonly used to assess maturity are examined, and their success either alone or in combination with other indices is discussed.

II. METHODS FOR ASSESSING MATURITY

A. Titratable Acidity

Total organic acid content gradually declines in fruit during maturation, ripening, and storage (Olsen and Martin 1980; Divikar et al. 1981; Lau 1985; Mann and Singh 1986; Truter and Hurndall 1988a, 1988b). Fruit flavor results from the combination of sugars, acids, and astringent and aromatic materials within the fruit. Apples have a relatively high acid content compared with other fruit (Vangdal 1985), and consumer acceptability in European countries is closely correlated to acid content (Gorin et al. 1975; Blanpied and Blak 1977; Wills et al. 1980).

The titratable acidity (TA) at which fruit should be harvested varies with the cultivar: 'Delicious' apples, for example, have a much lower TA at maturity than most other apple cultivars (Blanpied and Blak 1977; Bartram 1986; Lau 1988). TA can also vary with season (Divikar et al. 1981; Lau 1988), and with the amount of nitrogen fertilizer applied, increased applications decreasing TA (Hikasa et al. 1986; Olsen et al. 1986). TA often varies more between seasons and orchards than between harvest date, so setting a specific value for optimum harvest quality is inappropriate (Olsen 1982; Truter and Hurndall 1988b; Knee and Smith 1989). TA at harvest cannot be used to predict TA after storage (Blanpied and Blak 1977; Olsen and Martin 1980; Lau 1988). However, if TA is "low" when fruit is harvested, eating quality after storage may be unacceptable because TA falls markedly during storage. A minimum level for TA, below which fruit should not be stored, may therefore be specified for each cultivar (Bartram 1986).

The rate of change of titratable acidity, rather than an absolute value, has been suggested as one way to determine optimum maturity (Olsen 1982; Lau 1985; Truter and Hurndall 1988a, 1988b). Not all apple cultivars exhibit a significant change in the rate of change of titratable acidity (Truter and Hurndall 1988a), so titratable acidity should be used in conjunction with other maturity indices to assess optimum harvest date. Hammett et al. (1977) found a soluble solids:(% acid of juice, titrated to pH 8.1) ratio had the best correlation with days from full bloom, a commonly used guide to assist in determining picking date. The percent acid content alone was not as reliable in predicting harvest date.

Increasing exposure to light decreased TA in 'Delicious' apple (Robinson et al. 1983). If fruit from more exposed positions is not considered in a different group to that for fruit taken from within the tree canopy, this may increase sampling error associated with any measure of TA. Within each group a standardized sampling pattern should be used, again to reduce sampling error at each test date. Given the close correlation between TA and eating quality, TA can be useful to help determine acceptability of fruit for immediate consumption or storage if precautions are taken to reduce sampling error.

The acid content of pears is low and has much less impact on flavor quality (Vangdal 1982), and soluble solids content may be a more appropriate quality measurement (Vangdal 1985).

B. Soluble Solids

As apples and pears mature, starches are converted to sugars, increasing fruit sweetness and changing fruit taste, particularly in pears (Vangdal 1982; Wang 1982). Sugar content can be measured directly by chemical means, but as sugar is the major component of soluble solids it is easier to measure soluble solids in extracted juice using a refractometer (Wills et al. 1989). Soluble solids concentration tends to increase as apples and pears ripen (Smock 1948; Kajiura et al. 1979; Divakar et al. 1981; Olsen and Martin 1982; Reid et al. 1982; Vangdal 1982; Wang 1982; Mann and Singh 1985,1986; Lallu and Searle 1987; Blankenship and Unrath 1988), and so it could be a useful maturity index for these fruit. Position of fruit on the tree can influence soluble solids concentration in apples (Shaw and Rowe 1982). This is largely due to the exposure of fruit and attendant leaves to light, increased exposure to light increasing soluble solids concentration (Heinicke 1966; Robinson et al. 1983; Kupferman 1986; Barritt et al. 1987; Tustin et al. 1988). Therefore, the method of sampling fruit from the tree needs to be standardized to minimize variation. Kupferman (1986) suggests average fruit soluble solids concentration will be obtained by sampling fruit growing midway between the trunk and canopy perimeter in each of the four tree quadrants. The same trees should be used at each sampling date, and results applied only to the block in which the trees are situated. Absolute values of soluble solid concentrations at commercial maturity, in any season, are characteristic of each individual orchard (Hammett et al. 1977; Reid et al. 1982). Some variation may be due to differences in chemical applications; fungicide applications can affect sugar content within apples, usually increasing it (Rouchaud et al. 1983).

The value of the soluble solids concentration for one season will not necessarily be the same in successive seasons (Ingle and D'Souza 1989). Truter and Hurndall (1988b) found soluble solids content of 'Starking Delicious' apples varied between 10.8 and 12.2% at the optimum picking stage. In 'Granny Smith' apples the range of values between orchards was similar to the magnitude of change over the season studied (Reid et al. 1982). Similarly, when sugar concentration was measured by chemical methods, variation between seasons and orchards was greater than between harvest dates for 'Cox's Orange Pippin' apples (Knee and Smith 1989). Because of the wide range in values between orchards and seasons, an absolute value for soluble solids cannot be set as a guide to maturity (Smock 1948; Reid et al. 1982; Lallu and Searle 1987; Truter and Hurndall 1988b). The rate of change in soluble solids concentration may be a more useful maturity index for apple (Olsen 1982; Truter and Hurndall 1988a). However, in some cultivars, 'McIntosh' for example, the rate of change during the season is slow (Smock 1948; Shaw and Rowe 1982). In these cultivars sampling error on any one day may be greater than differences in soluble solids concentration between successive weekly sampling dates (Smock 1948).

Soluble solids concentration incorporates not only soluble sugars but also organic acids and inorganic salts, some of which rise and some fall during maturation (N. K. Given, unpublished data). The effects of these compounds do not always cancel each other out (Harman and Watkins 1981), so it is likely to be a less precise indicator of changes occurring within the fruit than a measure of maturity which assesses only one variable. Given this and the large influence of season and location on, plus the variation in, rate of change in soluble solids concentration, it seems inadvisable to use it as the sole guide to picking maturity. Most researchers have found soluble solids concentration gives best prediction of harvest date when used in conjunction with at least one other index such as firmness (Blanpied 1974) or acidity (Hammett et al. 1977).

Position of fruit on the tree can also influence soluble solids content in

European pear (Wang 1982). This, plus the influence of weather conditions, causes too great a variation between and within trees to make soluble solids concentration a reliable index when used alone. A minimum value may, however, be specified to avoid freezing injury occurring during storage (Wang 1982).

In Asian pears, soluble solids concentration increases from the stem end towards the calyx and from the inside towards the outside of the fruit; the more exposed side of the fruit also tends to have a higher soluble solids level (Kajiura et al. 1979). It is important to standardize sampling techniques within the fruit, to ensure accurate estimation of soluble solids. Soluble solids concentrations varied up to 4° Brix[1] between seasons for the same color grade of Asian pear (Lallu and Searle 1987). This suggests that other maturity indices need to be used in conjunction with soluble solids concentration to improve estimation of harvest date.

C. Fruit Firmness

As pome fruits ripen, the middle lamella, the cementing material between cells, dissolves. This, and changes in the cell sap, result in fruit starting to soften. This softening can be measured using a penetrometer which records the resistance of peeled fruit flesh to the insertion of a plunger of known diameter, usually 8 mm for pear and 11 mm for apple. Decreasing fruit firmness as maturity approaches is well documented in apple (Hesse and Hitz 1938; Smock 1948; Olsen and Martin 1980; Divikar et al. 1981; Reid et al. 1982; Shaw and Rowe 1982; Blankenship and Unrath 1988; Lau 1988; Ingle and D'Souza 1989), as are the effects of environmental and cultural factors on penetrometer values. There is an inverse relationship between fruit size and penetrometer readings (Hesse and Hitz 1938; Blanpied et al. 1978). Fruit firmness decreases as the amount of nitrogen fertilizer applied increases (Smock and Boynton 1944; Blanpied et al. 1978). Increased exposure to light is reported to increase fruit firmness especially on the most exposed side of the fruit (Blanpied et al. 1978; Shaw and Rowe 1982). As fruit shading increased, Heinicke (1966) and Robinson et al. (1983) found fruit firmness to increase as fruit size decreased with light exposure. The most important factors influencing penetrometer readings (nitrogen status, fruit position in canopy, watercore, fruit size, and fruit temperature) were considered by Blanpied et al. (1978) when developing a standardized sampling procedure to minimize variation.

Like soluble solids, optimum values for fruit firmness at commercial harvest date can vary markedly between orchards and seasons (Ryall et al. 1941; Haller 1942; Smock 1948; Blanpied and Blak 1971; Truter and

---

[1]°Brix = % pure sucrose by weight at 17.5°C (assumes all actively refractive soluble solids are sucrose).

Hurndall 1988b; Ingle and D'Souza 1989); often the variation between orchards and between seasons is similar to the size of change in flesh firmness over the period studied (Reid et al. 1982; Knee and Smith 1989). Thus absolute values of fruit firmness cannot be used as the sole indicator of maturity. Again, rate of change in fruit firmness values may be used, in conjunction with other indices, to determine the optimum time for commercial harvest (Truter and Hurndall 1988b). In 'Starking Delicious' and 'Starkrimson' apples the rate of change in flesh firmness changed significantly in the week preceding optimum maturity (Truter and Hurndall 1988a). In some cultivars the rate of change in flesh firmness as they ripen is slow (Hesse and Hitz 1938; Blanpied 1969; Reid et al. 1982) and not great enough to enable differentiation between commercially immature fruit and those at commercial maturity (Shaw and Rowe 1982). The rate of change may also vary between seasons and in relation to other maturity indices (Blankenship and Unrath 1988). These inconsistences, and the large influence of outside factors on the level of fruit firmness, indicate that this measure of maturity is not reliable when used alone. However, fruit firmness is very highly correlated with overall quality and texture (Wills et al. 1980; Vangdal 1982), and is a good indication of fruit crispness and juiciness (Lovelidge 1984; Vangdal 1985). High fruit firmness is necessary if fruit are to have a long shelf life and are to transport or store well (Lau 1985; Vangdal 1985). Recommended levels of firmness are often specified for apple cultivars at harvest (Lau 1985; Vangdal 1985) to ensure fruit will be acceptable to the consumer. These levels are cultivar dependent and usually expressed as a range to accommodate fluctuations between seasons (Lau 1985). Levels may be linked with other maturity indices to improve maturity prediction. Blanpied (1974) developed a sliding scale incorporating fruit firmness and soluble solids concentration which could separate apples, one day after harvest, into immature and mature categories that agreed with taste panel evaluations. As soluble solids concentration increased the firmness at which apples had acceptable eating quality also increased. The slope of the line varied between geographic regions but was identical for any one region over three seasons. This sliding scale is considered one of the easiest and best maturity indices for apples intended for prompt consumption (S. R. Blankenship, personal communication).

Fruit firmness decreases continuously as pear fruit develop (El-Azzouni et al. 1977; Wang 1982; Mann and Singh 1985, 1986). Mann and Singh (1986) suggest firmness should be used in conjunction with other indices such as color and chemical constituents to predict maturity, whereas Wang (1982) found flesh firmness alone one of the most satisfactory ways to determine pear maturity. Recommended levels of firmness have been developed for each pear cultivar to ensure acceptability to the consumer after storage. A range of acceptable firmness levels allows for fluctuations between seasons (Wang 1982).

D. Starch-Iodine Test

Starch accumulates in apple and in European and Asian pear during fruit growth and is hydrolyzed to sugar as fruit mature (North 1971: Smith et al. 1979; Yamaki et al. 1979; Lau 1985; Mann and Singh 1985; Stow 1988). Starch hydrolysis occurs in the core area first, and tends to progress outwards. The amount of starch present and its distribution can be assessed by exposing a cut surface of the fruit for about 60 seconds to an iodine solution which stains starch blue-black (Phillips and Poapst 1952; North 1971; Smith et al. 1979; Lau 1985). Quantification is provided by allocating a value to each distinct stage of starch disappearance (value increasing as starch disappears). This produces a starch index chart. Most cultivars have a characteristic pattern of disappearance, so starch charts are often cultivar specific; hence the starch index value (SIV) at any optimum harvest date can vary between cultivars (Smith et al. 1979; Lau 1985). Starch hydrolysis accelerates once fruit is harvested (North 1971), so the test should be performed as quickly as possible after harvest to get a true indication of starch content in fruit remaining on the tree. Weather conditions, particularly temperature, may also influence starch pattern (Phillips and Poapst 1952), so the time of day at which fruit is taken for testing should also be standardized.

A distinct change in starch pattern and stain intensity usually occurs two to three weeks before the onset of ethylene production (climacteric) in apples (Lau 1985) and about 10 days before the climacteric in European pears (North 1971). The rate of change in the starch index is influenced by the cultivar. Some apples, particularly those ripening late in the season, show little starch disappearance by the earliest suitable picking date for short-term cold storage (Hesse and Hitz 1938; Haller 1942; Phillips and Poapst 1952). Other maturity indices are needed to establish the optimum harvest date for these cultivars. In most of the cultivars forming the basis of the current apple industry, change in the SIV is clearly defined as ripening progresses, so can be a suitable guide to picking maturity (Reid et al. 1982; Shaw and Rowe 1982; Lau 1985; Truter and Hurndall 1988a, 1988b). Changes in percentage starch in fruit are well correlated with changes in SIVs in any one season (Poapst et al. 1959; Knee 1987; Lau 1988) but fluctuate widely between seasons (Poapst et al. 1959). Absolute values for percentage starch at optimum picking date are unsuitable as a guide to maturity (Truter and Hurndall 1988b); a SIV is more reliable, and determining it a simpler procedure than estimating starch content.

Climate, particularly temperature, influences the accumulation of starch and its rate of hydrolysis as maturity approaches. Increased temperatures during summer favor the conversion of sugar to starch and can increase the measurable starch content of fruit (Smith et al. 1979; Quast 1987) as well as delay the onset of starch hydrolysis (Beattie and Wild 1973; Watkins et al. 1982). As a result, the time at which a specific

SIV is reached can vary between years, districts, and orchards within districts. Starch hydrolysis proceeds more rapidly in warmer districts but starts later, so fruit still tend to reach specific SIVs after those fruit from cooler districts (Beattie and Wild 1973; Watkins et al. 1982). The optimum SIV for fruit harvest will depend on the intended use for the fruit—the longer the product is to be stored the lower the SIV should be at harvest (Lau 1985). Smith et al. (1979) report that SIVs at first acceptable harvest (determined by quality assessment after storage) were higher as latitude of sampling location increased. However Watkins et al. (1982) found a SIV of 3 (range 0–6) gave best-quality 'Granny Smith' apples after storage whether fruit was harvested from a warmer or a cooler climate. If fruit was harvested before SIV 3, the incidence of physiological disorders, such as scald, in storage was unacceptably high. Beattie and Wild (1973), using 'Granny Smith' and the starch test pattern later used by Watkins et al. (1982), found a value of 1 to be best if apples were to be stored before consumption. They noted that in some seasons physiological disorders could occur in early picked fruit. Physiological disorders in early picked fruit are a greater problem in some seasons and climates than in others, and the most suitable SIV for best fruit quality after storage will vary accordingly. A range of SIVs may ensure an acceptable product after storage if physiological diseases are not a problem and other factors affecting fruit quality such as titratable acidity or fruit firmness do not vary widely.

Starch disappearance does not appear to be closely related to organoleptic changes in fruit (Wills et al. 1980). Although fruit quality often increases as SIVs increase, (Beattie et al. 1972; Blankenship and Unrath 1988), fruit can also be judged mature in response to a decrease in flesh firmness or increase in soluble solids in the absence of any increase in SIV. Thus the SIV for fruit found acceptable by a taste panel may cover a wide range of the scale (Blankenship and Unrath 1988), and the average SIV at optimum maturity can differ significantly between orchard and districts (Beattie et al. 1972). Fruit abscission in some apple cultivars always occurs at a fixed starch concentration (Poapst et al. 1959). SIVs are positively correlated with internal ethylene concentration (Lau 1988), and are relatively uniform at the preclimacteric minimum (Smith et al. 1979). So, starch hydrolysis may have some relationship to physiological maturity. The variation in fruit firmness, acidity, and soluble solids content may alter the stage of physiological development at which fruit is perceived to have attained maximum eating quality. Because these three fruit characteristics can vary widely and independently between seasons and also between fruit on any one tree, use of SIVs alone may be inadequate to ensure optimum eating quality, accounting for the reported inadequacy (Haller 1942; Smock 1948; Blanpied 1974; Stow 1988) of the test. Variation of individual SIVs around the mean (SD = 0.92) is acceptable if the rate of change of the SIV is relatively large and uniform between seasons, as it is for 'Granny Smith' apple (Reid et al. 1982; Shaw and Rowe 1982). A standardized sampling technique may help minimize variation: percent full sunlight received by the fruit was positively correlated with starch content (Robinson et al. 1983). Perhaps this was in part due to some temperature effect, temperature influencing both initiation and rate of starch hydrolysis (Beattie and Wild 1973; Watkins et al. 1982).

Starch content has been used as a maturity index for European pears. North (1971) suggests pears should be harvested for CA storage when the blue-black color has decreased to two-thirds of its maximum coverage, irrespective of the pattern of disappearance. However, variation from year to year in particular is often too great to make this index reliable (Wang 1982; Stow 1988).

E. Skin and Flesh Color

As apples and pears mature they characteristically lose their green skin and flesh color as chlorophyll is lost. Until maturation, chlorophyll is continuously regenerated; once maturation begins, the rate of chlorophyll production slows, initially causing a loss in intensity of green coloration. As chlorophyll disappears further, other pigmentation, often yellow, appears in the skin (Fidler 1973; Olsen 1982; Wills et al. 1989). The changes in skin and flesh color can be measured analytically by assessing the concentration of chlorophyll or other colored pigments. Skin ground color may be assessed subjectively using color cards, where scores are allocated to each specified stage of color development (Kajiura et al. 1975; Olsen et al. 1986). A more objective nondestructive measure of fruit color can be made using a tristimulus colorimeter which assigns a value to any color in terms of its red, green, or blue components (Francis 1980). As some pome fruit (particularly apples) mature, a red blush may develop masking some or all of the green fruit ground color. Development of this red blush depends on light exposure received by the fruit. Those fruit at the outer edge and in the upper portion of the canopy develop more red color than fruit from the inner and bottom portions (Heinicke 1966; Warrington et al. 1984; Barritt et al. 1987), although differences between fruit may be less noticeable in highly colored cultivars or strains (Robinson et al. 1983; Barritt et al. 1987). Red color development is also encouraged by cool nights and warm, sunny days before harvest (Olsen and Martin 1980; Lau 1985). The degree of red blush development on apples is not a good index of fruit maturity because it can vary markedly between fruit positions in the tree, orchards, and seasons. Ground color, or the color of the unblushed portion of the skin, may be a more reliable indicator. In many pome fruit, ground color changes from green to yellow as fruit mature (Kajiura et al. 1975; El-Azzouni et al. 1977;

Mann and Singh 1985; Lallu and Searle 1987; Lau 1988). For some apple cultivars the rate of change in ground color is slow, limiting its value as a maturity index (Hesse and Hitz 1938; Haller 1942; Smock 1948; Shaw and Rowe 1982). Nitrogen level also influences ground color development. High nitrogen levels enhance chlorophyll retention and retard yellow ground color development (Smock and Boynton 1944; Delver 1986; Hikasa et al. 1986; Olsen et al. 1986). Warmer night temperatures have similar effects (Olsen and Martin 1980). Light exposure can also influence ground color; fruit in the outer regions and mid and upper tiers of apple trees are paler green than those in inner regions and bottom tiers (Jackson et al. 1971; Shaw and Rowe 1982; Tustin et al. 1988). This seems to be due to the more rapid degradation of chlorophyll in fruit in these exposed positions (Olsen 1982), either by photodegradation or a higher rate of chlorophyll cycling (Tustin et al. 1988), but may also be a response to the lower nitrogen concentration found in these fruit (Jackson et al. 1971). The large influence of light, temperature, and nitrogen on chlorophyll content suggests color indices for flesh or skin ground color cannot be relied upon to determine optimum harvest dates.

The color changes occurring as European pears develop can be used as a guide for picking, but are not considered a reliable enough index to be used alone (Wang 1982; Mann and Singh 1985, 1986). In Asian pears ground color is often used as a picking guide. There is a close relationship between skin color at harvest and development of disorders during storage (Lallu and Searle 1987). Ground color is correlated positively with soluble solids concentration and negatively with fruit firmness (Kajiura et al. 1975), but the absolute value of soluble solids concentration associated with any color grade varies considerably between seasons (Lallu and Searle 1987).

F. Seed Color

The seeds in pome fruit become brown (apple) or black (pear) as maturity advances. In some apple cultivars the development of seed color parallels the disappearance of starch (Lau 1985; Truter and Hurndall 1988a, 1988b). In others seed color changes are less reliable, starting at varying times before optimum harvest date so, at any one sampling time, the range of seed colors present can be quite large (Haller 1942; Truter and Hurndall 1988a). There is considerable seasonal variation in color development at harvest maturity (Tripathi et al. 1986; Truter and Hundall 1988a, 1988b), so an optimum value for seed color has no use as a maturity index. Seed color may be useful as an additional maturity index in those cultivars where it changes significantly and with little variation prior to harvest, in conjunction with another destructive sampling technique.

G. Calendar Date and Days from Full Bloom

In climates with little year-to-year variation, the calendar date may be used as a reliable guide to the optimum harvest date. It will only be useful when flowering occurs at a similar time each season and climate follows a consistent pattern throughout the season (Lau 1985; Wills et al. 1989). It will be most appropriate for late-maturing cultivars where temperature fluctuations tend to balance out in the longer growing season (Tukey 1942; Fidler 1973; Luton and Hamer 1983). When temperatures fluctuate widely within and between seasons, time of flowering and time taken for fruit to develop will also vary, so calendar date will not accurately predict the date of harvest maturity (Lau 1985). Calendar date is, however, a useful reference point for researchers and growers alike, providing some idea when harvest is likely to occur. Days from full bloom may provide a better index of maturity than calendar date (Haller 1942; Tukey 1942; Truter and Hurndall 1988b), provided the number of days used has been derived in the district where it is being used as an index (Ryall et al. 1941; Fidler 1973). In some districts temperature fluctuations are too great to allow days from full bloom to be used with any accuracy as a maturity index (Smock 1948; Wang 1982). In other areas prolonged blossoming makes definition of full bloom date difficult (Tukey 1942; Reid et al. 1982), leading to errors in predicting harvest maturity. The range of days from full bloom over which pear quality can be acceptable (Ryall et al. 1941) suggests other maturity indices are needed to define the beginning and end of harvest.

The apparent influence of temperature on flowering and fruit development suggests the use of degree days or some other form of heat summation might enable more accurate prediction of harvest date. These methods use a base or minimum temperature, mean daily temperatures above or below which are summed positively or negatively respectively. Fruit development in orchards with different bloom dates varies markedly, but plotting morphological changes against accumulated heat units, rather than days from full bloom, provides a much more consistent pattern (Olsen 1982). Although for some apple and pear cultivars a characteristic number of heat units are required to develop fruit to maturity, for others heat units accumulated between full bloom and maturity do not always explain the variation in number of days elapsed during this period (Ryall et al. 1941; Eggert 1960; Fisher 1962; Blanpied 1964; Lombard et al. 1971). Heat units accumulated early in the season, usually in the 6–9 weeks following full bloom for many pear cultivars (Mellenthin 1966; Lombard et al. 1971; Wang 1982) and the first 4–12 weeks following full bloom in apples (Eggert 1960; Luton and Hamer 1983), seemed to have the best correlation with harvest date. Temperatures later in the season have much less effect (Mathee 1988). The base temperature and period for accumulating the heat units are specific for each cultivar and need to be developed in each growing region (Wang 1982). This may explain why heat summation has been found to be of limited importance in some research studies. Other reasons put forward when heat summation techniques failed to account for variation in days from full bloom to maturity include Fisher's (1962) suggestion that hotter-than-average climates might increase respiration losses counteracting any increased carbohydrate accumulation. In colder-than-average climates the reverse would hold true so there would be minimal effect of changes in heat unit summation between seasons. The influence, on days from full bloom to maturity, of factors such as number of cold units accumulated in winter and soil type cannot be discounted either (Ryall et al. 1941; Tukey 1942; Mathee 1988). Temperature, rain, and snowfall in the pre-bloom period, as well as temperature and rainfall in the post-bloom period, have to be considered to predict optimum harvest date accurately. Using environmental conditions post-bloom, or temperature alone, cannot reproduce such accuracy of harvest date prediction (Narasimham et al. 1988). Length of the growing season and time of blossoming also seem to affect the effectiveness with which days from full bloom can be used as an index of maturity. Apple cultivars reaching commercial maturity early in the season exhibit greatest variation in time elapsed between bloom and harvest, possibly because temperature fluctuations can have greater impact in a shorter season (Tukey 1942). Early bloom can also affect length of the fruit growing season: the more advanced the bloom period, the longer the growing season (Blanpied 1964; Olsen and Martin 1980; Luton and Hamer 1983). For 'Delicious' and 'McIntosh' apples, Blanpied (1964) proposed the following "correction" factor: one day of increase or decrease from the average length of growing season for every two days of deviation from the average date of full bloom. This may be a result of cooler temperatures in the early part of the season, hence a longer time required to accumulate sufficient heat units to ripen satisfactorily.

Days from full bloom or calendar date do not seem to be sufficiently reliable in themselves to do much more than predict an approximate date for commercial harvest. Modifying number of days from bloom to maturity using some measure of temperature experienced over varying lengths of the growing period may improve the accuracy of prediction. Unfortunately such modifications are cultivar- and often district-specific, thus limiting their general applicability. The inclusion of modifying factors (such as heat units accumulated during a specified time interval) which have been established and proven to work satisfactorily over a number of years can make days from full bloom a useful maturity index for some cultivars.

H. Fruit Retention Strength

The strength with which an apple or pear fruit is attached to the spur at the natural abscission site decreases as the fruit matures, and eventually fruit will drop. The force required to pull fruit from the tree may indicate how mature fruit is: fruit drop from the tree tends to occur at a fixed starch index value in any one season (Poapst et al. 1959); Smock (1948) found that fruit with the best storage quality could be removed from spurs using a slight rolling action. However, ease of fruit separation is influenced by environmental conditions, particularly temperature (Poapst et al. 1959). This, and the subjective nature of the test, make it of limited value as a picking guide.

I. Respiration Rate

Apples and European pears are climacteric: when they reach physiological maturity and ripening processes are initiated, there is a marked increase in respiratory activity resulting in the increased evolution of CO_2 (Kidd 1934; Kidd et al. 1940; Kitamura et al. 1981). The preclimacteric minimum (PCM), when CO_2 production is at its lowest level, is therefore a useful indicator of the physiological state of the fruit. It was initially thought to provide a good method of determining the suitability of fruit for storage, the optimum time of harvest being at the PCM, before ripening commenced (Kidd and West 1973). However, factors other than rate of CO_2 evolution, such as fruit firmness, texture, and flavor, also have an influence on fruit quality (Blanpied 1969), so the time to harvest fruit for optimum quality often occurs at varying positions along the climacteric curve (Phillips 1939; Smock 1948; Blanpied 1969; Smith et al. 1969) rather than at the PCM. In addition, the pattern of CO_2 production varies, for any cultivar, between picking dates in any one season or between seasons, so it is often quite difficult to determine at what point the PCM actually occurs (Blanpied 1969; Meheuriuk 1977; Knee et al. 1989). However, for some cultivars, such as 'Cox's Orange', the time interval between the onset and peak of the climacteric rise is constant over several seasons (Hulme 1954). The respiratory rate and the extent of its rise during the climacteric also vary between seasons (Hulme 1954; Blanpied 1969; Knee et al. 1989), so setting an absolute minimum or threshold value for CO_2 evolution to correspond to PCM is inappropriate.

Variability among fruit samples (Hulme 1954) and in CO_2 evolution patterns (Blanpied 1969; Meheriuk 1977; Knee et al. 1989) means it may take several days to ascertain whether the climacteric has been initiated, by which time it is too late to harvest fruit at the optimum date. Thus, measurement of respiratory activity is a useful test "in retrospect" to determine, for example, what physiological state fruit with best storage was in when harvested, but not as a practical maturity index (Smock 1948; Fidler 1973).

J. Ethylene Production

During the climacteric, the temporary but marked increase in CO_2 production is not the only phenomenon occurring. Production of ethylene, a ripening hormone in climacteric fruit (Burg and Burg 1965), follows a similar pattern concurrently (Smith et al. 1969; Reid et al. 1973; Knee et al. 1983), and hence can be a useful indicator of the physiological maturity of the fruit. Measurement of ethylene concentration is just as quick, but the change at the climacteric much more pronounced than that for CO_2 (Reid et al. 1973), so it has largely replaced tests involving CO_2 to monitor progress through the climacteric.

Ethylene concentration can be measured in several ways. Extracting a gas sample from the internal core space of fruit, or from a sealed container in which fruit have been kept for a length of time and analyzing it using gas chromatography, are the most commonly used methods. Ethylene concentration may increase between 10- and 1000-fold during the short climacteric period (Reid et al. 1973; Sfakiotakis and Dilley 1973). The rate of change and the absolute concentration of ethylene reached can vary markedly between cultivars (Chu 1948, 1988; Blankenship and Unrath 1988), the same cultivars on different rootstocks (Fallahi et al. 1985), trees in the orchard (Farhoomand et al. 1977; Knee et al. 1989), blocks in the orchard (Blanpied 1984), as well as locations and seasons (Knee et al. 1983, 1989; Blanpied 1984; Blankenship and Unrath 1988). Ethylene is only one of the factors that influence ripening (Dilley 1969), and it is not unexpected that the same absolute value does not indicate the same physiological state each season (Knee et al. 1989). Fruit abscission occurs when ethylene concentration is maximal or declining, irrespective of absolute value attained. Blanpied (1972) found abscission occurred once an ethylene gradient had been created between pedicel and cluster base of 'McIntosh' apple. The time to establish this gradient is determined by temperature, higher temperature reducing the time taken to drop (Walsh 1977). So, it may be the length of time that fruit is exposed to above-threshold levels of ethylene and temperature during this time, rather than the concentration per se, that is most important for ripening (Walsh 1977). However, cultivars have a characteristic pattern of internal ethylene production which remains unchanged irrespective of cultural or environmental factors (Chu 1984; Fallahi et al. 1985). This pattern, rather than the rate of change or any absolute value can be used as a guide to physiological maturity of the fruit in any season. The optimum harvest date for many cultivars coincides with the beginning of the climacteric (onset of ethylene production) or within 1–2 weeks of its occurrence (Smith et al. 1969; Douglas 1983; Chu 1984, 1988). Hence,

measuring time of onset of ethylene generation may help to determine
optimum commercial harvest date for apples, at least for some end
uses.

A few apple cultivars, 'Northern Spy' for example, do not begin their
climacteric until well after optimum harvest date. Others, 'Idared' for
example, generate little ethylene at any time during the harvest season
(Chu 1984). In these instances, ethylene evolution is of little use as a
maturity index.

In apples there can be considerable variation in the ethylene concentration, either internally (IEC) or externally, of individual fruit picked on any
one date (Reid et al. 1973; Knee et al. 1983, 1989). To get best prediction of
harvest date, individual blocks even within one orchard should be
monitored separately and enough fruit taken from a selection of trees in
this block (30 fruit seems to be the optimum number) to reduce sampling
variation (Blanpied 1984; Blanpied and Pritts 1987; Knee et al. 1989). In
addition, fruit should be taken from specific locations within the tree, as
ethylene generation is lower in fruit growing nearer the apex of the tree or
the distal part of lower branches (Farhoomand et al. 1977). Position on the
tree may also modify temperature of the fruit flesh, and higher temperatures enhance ethylene generation (Reid et al. 1973; Fallahi et al. 1985).
Ethylene concentrations in harvested fruits should be measured as
quickly as possible (no more than 24 h later), as removal of fruit from the
tree reduces the time taken before the onset of climacteric ethylene
generation occurs (Sfakiotakis and Dilley 1973).

Ethylene production is not always closely linked with other maturity
indices. Ethylene concentration increases curvilinearly and can be difficult to correlate directly with the linear increases in other aspects of fruit
quality (Ingle and D'Souza 1989; Knee et al. 1989). A logarithmic conversion of ethylene concentration may enable better comparison with other
indices (Reid et al. 1973; Fallahi et al. 1985; Lau 1988). An increase in $\log_{10}$
IEC coincided with the initial rapid increase in starch index and yellow
ground color in 'Jonagold' apple (Lau 1988). Taste panelists can distinguish between apples divided into four categories on the basis of their
ethylene evolution (Saltveit 1983), indicating the influence of physiological maturity on eating quality. Many maturity indices such as firmness, soluble solids concentration, and titratable acidity change
markedly, however, both before and during the time at which changes are
recorded in IEC (Douglas 1983; Blankenship and Unrath 1988).
Undoubtedly these influence fruit quality to differing degrees each
season. Hence, no one position on the ethylene evolution curve can be
used to satisfactorily predict optimum harvest date for storage or immediate consumption every season (Douglas 1983; Blankenship and Unrath
1987).

Lau (1985) suggests fruit should be harvested just before the commencement of the climacteric, whilst IEC is still low, for best storage.

However, low ethylene production does not necessarily mean fruit will have good eating quality after storage. Ethylene measurements should be used in conjunction with other maturity indices to determine which fruit have best potential for long term storage, and which would be more suited for immediate consumption or other end uses.

Many European pears will only commence ripening after a preconditioning treatment, often cool storage (Wang 1982). A number of Asian pear cultivars are non-climacteric (Kitamura et al. 1981). For these fruit, ethylene measurements are of no use as a maturity index.

K. Other Indices

There is an continuing search for new indices which may improve harvest date prediction. Light-transmittance techniques, initially developed for tomatoes (Worthington 1974), have been modified to determine maturity in intact apples and pears. As fruit ripen internal optical density decreases. This is thought to occur as bound water is converted to free water in tissues, changing light scattering properties. Changes in optical density were larger than those in skin color as European pears ripened. Hence optical density was a better indicator of maturity (Worthington et al. 1977). Light transmittance has also been used to segregate intact apples ('Golden Delicious') on the basis of chlorophyll content (Olsen et al. 1967), and detect watercore and other internal disorders in 'Delicious' apple (Bramlage and Shipway 1967). Another nondestructive technique, magnetic resonance imaging, increases the accuracy of detection of watercore, and the amount of information able to be obtained about its distribution (Wang et al. 1988). It relies on the interaction of radio frequency waves with magnetic moments of hydrogen atoms to produce images. Similar techniques may be able to be developed to assess other aspects of fruit quality.

Changes in substrates or products involved in ripening processes have also been monitored. Insufficient changes were found in enzyme activities (Kumar et al. 1986; Knee et al. 1989) and substrates and products in carbohydrate metabolism (Gorin et al. 1976; Knee et al. 1989) to enable satisfactory assessment of maturity. The potential of NAD(P)H fluorescence as a nondestructive method of determining maturity is being investigated by Cavalieri et al. (1989). NAD(P)H concentration changes result during cell metabolism.

III. SOME EXISTING MATURITY ASSESSMENT PROGRAMS

Recognizing the difficulties inherent in assessing optimum maturity of fruit, some apple growing regions have developed large-scale maturity testing programs. These assist grower prediction of optimum harvest time for fruit destined for long-term CA storage. These programs rely on use of a standardized sampling technique and evaluation of several (at least five) maturity indices, often a combination of field observations and laboratory analyses, to maximize accuracy of prediction. With several years' experience, desired or target values for each index may be identified. These and the rate of change in the selected indices are considered, recognizing seasonal fluctuations make it unlikely that desired values of all indices will occur coincidentally. Usually when at least three indices have reached the desired level, harvest is started. The rate of change in the other indices is often a good indicator of how close fruit are to commercial maturity. For best long-term storage fruit should be harvested before autocatalytic ethylene production is initiated during the climacteric, as high levels of ethylene accelerate ripening and will result in fruit of inferior quality after storage.

In British Columbia (Lau 1985) measurements made include the starch-iodine test, internal ethylene concentration, flesh firmness, soluble solids concentration, and seed color. Days from full bloom, calendar date, skin color, and titratable acidity are considered less reliable maturity indices. Fruit being stored long-term must have, as a prerequisite, low ethylene production. SIVs are used to predict harvest date for best storage; once the SIV exceeds a specified value fruit is not considered suitable for long-term storage. Desirable values are set for flesh firmness and soluble solids concentration to ensure fruit are of good quality when removed from storage. Because of seasonal variations these levels may not always be achieved before harvest date. The preharvest rate of change of these indices is also determined to assist in harvest date prediction. For example, in a season when fruit softens rapidly, harvest is started when fruit has fallen to the desired firmness level, irrespective of the soluble solids concentration. Delaying harvest until soluble solids concentration increases to the desired level may leave fruit too soft to store satisfactorily for long periods. The apple cultivars included in this program are 'McIntosh', 'Golden Delicious', 'Spartan', 'Delicious', 'Newtown', and 'Granny Smith'.

The Apple Maturity Program based in Washington State, USA (Bartram 1986), uses flesh chlorophyll level, skin color, fruit firmness, soluble solids, acidity, starch level, and ethylene emission from fruit as maturity indices. Desired levels for each of these indices have been established, as have rates of change that may indicate the onset of commercial maturity. Usually when three or four of these indices reach desired levels, harvest is started. The decision when to harvest is also influenced by rates of change of these indices; if three or four exhibit a rate of change that may indicate the onset of maturity this suggests harvest should commence. The indices reaching desired levels or exhibiting marked rates of change usually vary between seasons. Acid levels and ethylene emission from fruit at harvest are considered to have the greatest effect on fruit quality after storage. Low acid levels and high ethylene emission produce poor quality fruit at the end of the storage period. The cultivars included in this program are 'Starking', spur and non-spur 'Delicious', and spur and non-spur 'Golden Delicious'.

A method to determine the eating quality of apples for immediate consumption has been developed by Alavoine et al. (1988). Concentrating on the organoleptic qualities of fruit, measurements of size, shape, color, firmness, soluble solids concentration, and acidity are made and compared to critical values. These indicate whether fruit eating quality has reached a "satisfactory" or "high" level.

The New Zealand Apple and Pear Marketing Board routinely specify calendar dates as an approximate indicator of harvest date. The maturity indices flesh firmness, soluble solids, starch index, background color, fruit size and ethylene production are monitored for fruit from several orchards within each growing region. Results are used to finalize opening and closing dates for each cultivar in each area. The cultivars monitored include 'Gala', 'Royal Gala', 'Braeburn', 'Granny Smith', 'Cox's Orange Pippin', 'Delicious' strains, and 'Fuji'. More apple cultivars are being added to this list each season.

A working party established by the European Economic Community (EEC) has produced recommended methods for testing apple quality (Smith 1985). A general discussion on reducing variation within fruit samples is also included. It is suggested that researchers within member countries should adopt the recommended procedures to enable comparison to be made more easily between data sets. Starch pattern and ethylene production are recommended to determine apple maturity. The recommended method for starch assessment is "production of a standard set of reference photographs." For ethylene, determination of both ethylene emission and IEC under specified conditions including temperature, airflow, and volume is recommended (Smith 1985).

Some European pear cultivars ripen only after a preconditioning treatment, often cool storage. Fruit should have attained full maturity prior to this treatment to ensure they ripe fully afterwards. The most commonly used maturity indices are flesh firmness, soluble solids concentration, starch index, and for some cultivars, heat unit summation. Desirable values, often cultivar specific, are well documented for each of these indices (Wang 1982), although there are few recognized maturity testing programs as for apples. The New Zealand Apple and Pear Marketing Board is currently developing a maturity testing program for European pears similar to that used for apples.

IV. CONCLUSIONS

Many factors influence the eating and storing quality of apple and pear; no one test can be used to predict fruit quality satisfactorily. With climatic and cultural factors having varying effects on time taken to attain given maturity, or the absolute level of an index attained in any one season, a number of tests are needed to improve success in assessing commercial maturity. There are two broad types of maturity tests; those that assess eating quality or visual appeal, such as acidity and skin color and those that indicate physiological maturity such as ethylene or CO_2 evolution. Ripening cannot proceed to its fullest extent unless fruit is physiologically mature, hence measures of physiological maturity can provide guidelines to fruit quality. Consumer perception of fruit quality is greatly influenced by visual and organoleptic appeal. Both types of index must therefore be considered when assessing fruit maturity. The most universally applicable indices are titratable acidity, soluble solids concentration, flesh firmness, starch-iodine level, and ethylene concentration within the fruit. Skin ground color, seed color, and some form of heat unit summation may be applicable in some circumstances for some cultivars.

The most suitable maturity indices, and desirable values for these indices, have been established after several seasons' evaluation of quality of fruit at its end use and fruit attributes at harvest. Desirable values for maturity indices are specific for each end use of fruit, and usually for each cultivar. When fruit are to be consumed immediately, two or three indices (firmness, soluble solids concentration, and titratable acidity) will probably suffice to determine if fruit are mature. In Blanpied's (1974) sliding scale firmness and soluble solids are the two indices used to determine suitability for immediate consumption. Skin color may be used to further separate fruit into grades. Fruit to be stored should have a higher starch content and lower ethylene production than fruit for immediate consumption. They should also have lower soluble solid concentrations and higher TA and fruit firmness to allow for the changes in these characteristics which occur, albeit slowly, during the storage period. Fruit will then have acceptable eating quality when removed from storage. In the maturity testing programs examined, desired levels of these indices were established, which in past years had resulted in high-quality fruit after removal from long-term storage. If too many desirable values were exceeded fruit was considered more suitable for short-term storage or immediate consumption.

Standardized sampling techniques are essential to improve harvest date prediction. It is undeniable that position within the fruit, of the fruit within the canopy, and of the canopy within the orchard can have a significant effect on the value reached by many maturity indices. Taking samples for analysis from a similar part of the fruit, fruit from similar positions within the canopy, and monitoring blocks within the orchard separately will all help to reduce variation in maturity index values in any one season. Although sampling techniques can be standardized, environmental and cultural conditions still fluctuate between seasons, so some variation in the absolute value reached each season must be expected. Optimum values for maturity indices are therefore only guidelines. The rate of change of an index during the period preceding harvest should be monitored to ascertain how prevailing conditions are influencing progression. The desired value may not be reached for each index every season; the progress of all indices monitored must be considered. Sometimes the rate of change of an index will itself change significantly just before commercial maturity. This can assist in defining optimum harvest date.

Phenological indices are already used to predict outbreaks of commercially important diseases and susceptibility to various maturity disorders. The potential of phenological indices to help improve harvest date prediction should also be taken into account. Computer analysis enables manipulation of data in ways that have not previously been practical. There is evidence that consideration of data accumulated over past seasons and prevailing weather conditions, particularly temperature, during the season in question can allow prediction of harvest date to within five days at best. The earliness in the season at which these predictions can be made allows for organization well in advance of picking, packing, and despatch of fruit.

Eating quality is largely dependent on factors that cannot be consistently related to physiological maturity. Ripening is not only the result of DNA expression and subsequent events but is modified by growth conditions, particularly temperature, which may affect maturity indices in different ways. It is improbable, therefore, that a perfect maturity assessment scheme can be evolved. In most fruit growing areas there are large fluctuations in weather conditions among seasons. So, no matter how carefully one documents the physiological changes preceding ripening or defines optimum values for maturity indices, it is highly improbable that the stage at which fruit reaches optimum quality for the desired end use will be similar each season. The programs currently used get around this problem by specifying a range for the optimum value of each index and monitoring a range of indices. All the information collected is used to determine when fruit are at optimum maturity for the desired end use. The decision when to harvest cannot be totally objective. Indices do not reach the desired range of values simultaneously, hence some subjective judgement must be used to determine when to start and also when to finish harvesting for each cultivar × end use combination. A consistent relationship between one of the fruit characteristics influencing eating quality (titratable acidity, firmness, soluble solids) and some attribute connected with physiological maturity would improve maturity assessment. Unless this can be found, the use of several indices plus standardized sampling and analytical methods for each cultivar every season is the best compromise.

LITERATURE CITED

Alavoine, F., M. Crochon, C. Fady, J. Fallot, P. Moras, and J. C. Pech. 1988. *Eating Quality of Fruits*. Practical methods of analysis (in French). Antony, France.

Barritt, B. H., C. R. Rom, and S. Drake. 1987. Management of apple fruiting spurs for fruit quality and profitability. V. Spur quality and fruit quality. *Goodfruit Grower* 38(21):102.

Bartram, D. 1986. Interpretation of laboratory test. p. 45–47. In: Apple maturity program. Wenatchee, Washington, USA.

Beattie, B. B., and B. L. Wild. 1973. Assessing harvest maturity of 'Granny Smith' apples for export. *Agr. Gaz. New S. Wales* 84:30–33.

Beattie, B. B., B. L. Wild, and G. C. Coote. 1972. Maturity and acceptability of early-picked 'Granny Smith' apples for export. *Austral. J. Expt. Agr. Anim. Hus.* 12:323–327.

Blankenship, S. M., and C. R. Unrath. 1987. Use of ethylene production for harvest date prediction of apples for immediate fresh market. *HortScience* 22:1298–1300.

_____. 1988. Internal ethylene levels and maturity of 'Delicious' and 'Golden Delicious' apples destined for prompt consumption. *J. Am. Soc. Hort. Sci.* 113:88–91.

Blanpied, G. D. 1964. The relationship between growing season temperatures, bloom dates and the length of the growing season of 'Red Delicious' apples in North America. *Proc. Am. Soc. Hort. Sci.* 84:72–81.

_____. 1969. A study of the relationship between optimum harvest dates for storage and the respiratory climacteric rise in apple fruits. *J. Am. Soc. Hort. Sci.* 94:177–179.

_____. 1972. A study of ethylene in apple, red raspberry and cherry. *Plant Physiol.* 49:627–630.

_____. 1974. A study of indices for earliest acceptable harvest of 'Delicious' apples. *J. Am. Soc. Hort. Sci.* 99:537–539.

_____. 1979. Predicting early harvest maturity dates for 'Delicious' apples in New York. *HortScience* 14:710–711.

_____. 1984. Observations of the variations in the initiation of the ethylene climacteric in apples attached to the tree. *Acta Hort.* 157:143–148.

Blanpied, G. D., and V. A. Blak. 1977. A comparison of pressure tests, acid levels and sensory evaluation of overripeness in apples. *HortScience* 12:73–74.

Blanpied, G. D., W. J. Bramlage, D. H. Dewey, R. L. Labelle, L. M. Massey, G. E. Mattus, W. C. Stiles, and A. E. Watada. 1978. A standardized method for collecting apple pressure test data. New York's Food and Life Sciences Bulletin No. 74. 8 p. New York State College of Agriculture and Life Sciences, Ithaca, NY.

Blanpied, G. D., and M. P. Pritts. 1987. Estimating ethylene climacteric initiation dates in apple orchards. *Acta Hort.* 201:61–68.

Burg, S. P., and E. A. Burg. 1965. Ethylene action and the ripening of fruits. *Science* 148:1190–1196.

Cavalieri, R. P., T. D. Strecker, and J. K. Fellman. 1988. Reduced pyridine nucleotides as a measure of fruit maturity. Paper, American Soc. Agric. Engineers, No. 88-6565. St. Joseph, MI.

Chu, C. L. 1985. Use of internal ethylene concentrations as a maturity index of eleven apple cultivars. *Acta Hort.* 157:129–134.

_____. 1988. Internal ethylene concentration of 'McIntosh', 'Northern Spy', 'Empire', 'Mutsu', and 'Idared' apples during the harvest season. *J. Am. Soc. Hort. Sci.* 113:226–229.

Delver, P. 1986. Effect of groundwater regime, nitrogen fertilization and chemical fruit thinning on yield and quality of apples (in Dutch). Rapport, Instituut voor Bodemvruchtbaarheid 5. 99 p. [Hort. Abstr. 59(6):4542; 1989.]

Dilley, D. R. 1969. Hormonal control of fruit ripening. HortScience 4: 111–114.

Divikar, B. L., P. D. Shukla, and K. S. Adhikari. 1981. Physical and biochemical changes during maturation in fruits of apple variety 'Early Shanburry'. Prog. Hort. 13:61–65.

Douglas, J. B. 1983. An evaluation of harvest indices for 'McIntosh' apples in two orchards. HortScience 18:216–218.

Eggert, F. D. 1960. The relation between heat unit accumulation and length of time required to mature 'McIntosh' apples in Maine. Proc. Am. Soc. Hort. Sci. 76:98–105.

El-Azzouni, M. M., F. I. A. El-Latief, and E. A. Kenawi. 1977. Determination of maturity in 'LeConte', 'Shoubra' and 'Pineapple' pear varieties. Acta Agron. Acad. Sci. Hung. 26(1/2):140–144.

Fallahi, E., D. G. Richardson, and M. N. Westwood. 1985. Influence of rootstocks and fertilizers on ethylene in apple fruit during maturation and storage. J. Am. Soc. Hort. Sci. 110:149–153.

Farhoomand, M. B., M. E. Patterson, and C. L. Chu. 1977. The ripening pattern of 'Delicious' apples in relation to position on the tree. J. Am. Soc. Hort. Sci. 102:771–774.

Fidler, J. C. 1973. The optimum date of harvest. p 10–16. In: F. C. Fidler, B. G. Wilkinson, K. L. Edney, and R. O. Sharples (eds), The Biology of Apple and Pear Storage. Commonwealth Agr. Bureaux, Farnham Royal, Slough, England.

Fisher, D. V. 1962. Heat units and number of days required to mature some pome and stone fruits in various areas of North America. Proc. Am. Soc. Hort. Sci. 80:114–124.

Francis, F. J. 1980. Color quality evaluation of horticultural crops. HortScience 15:58–59.

Haller, M. H. 1942. Days from full bloom as an index of maturity for apples. Proc. Am. Soc. Hort. Sci. 40:141–145.

Hammett, L. H., H. J. Kirk, H. G. Todd, and S. A. Hale. 1977. Association between soluble solids/acid content and days from full bloom of three red strains of 'Delicious' and 'Law Rome' apple fruits. J. Am. Soc. Hort. Sci. 102:733–738.

Harman, J. E., and C. B. Watkins. 1981. Fruit testing: Use of refractometers. AgLink HPP212, Media Services, MAF, Wellington, New Zealand.

Heinicke, D. R. 1966. Characteristics of 'McIntosh' and 'Red Delicious' apples as influenced by exposure to sunlight during the growing season. Proc. Am. Soc. Hort. Sci. 89:10–13.

Hesse, C. O., and C. W. Hitz. 1938. Maturity studies with 'Jonathan' and 'Grimes Golden' apples. Proc. Am. Soc. Hort. Sci. 36:351–357.

Hikasa, Y., H. Muramatsu, and T. Minegishi. 1986. Effects of nitrogen application on the growth, yield and fruit quality of dwarf apple trees (in Japanese). Bull. Hokkaido Prefectural Agr. Expt. Sta. 55:23–31 [Hort. Abstr. 58:63;1988.]

Hulme, A. C. 1954. Studies on the maturity of apples. Respiration progress curves for 'Cox's Orange Pippin' apples for a number of consecutive seasons. J. Hort. Sci. 29:142–149.

Ingle, M., and M. C. D'Souza. 1989. Fruit characteristics of 'Red Delicious' apple strains during maturation and storage. J. Am. Soc. Hort. Sci. 114:776–780.

Kajiura, I., Y. Sato, M. Omura, and Y. Machida. 1979. Local differences in the concentration of soluble solids with fruit of Japanese pear and the sampling method for the measuring (in Japanese, English summary). Bull. Fruit Tree Res. Sta. A 6:1–14.

Kajiura, I., K. Suzuki, and T. Yamazaki. 1975. Color chart for Japanese pear (Pyrus serotina var. culta Rehder). HortScience 10:257–258.

Kidd, F. 1934. The respiration of fruits. Paper, Royal Institute of Great Britain. [Hort. Abstr. 5:25; 1935.]

Kidd, F., and C. West. 1937. The keeping qualities of apples in relation to their maturity when gathered. Sci. Hort. 5:78–86. [Hort. Abstr. 7:487; 1937.]

Kidd, F., C. West, D. G. Griffiths, and N. A. Potter. 1940. An investigation of the changes in chemical composition and respiration during the ripening and storage of Conference pears. *Ann. Bot. Lond.* 4:1–30. [Hort. Abstr. 10:380; 1940.]

Kitamura, T., T. Iwata, T. Fukushima, Y. Furukawa, and T. Ishiguro. 1981. Studies on the maturation physiology and storage of fruits and vegetables. II. Respiration and ethylene production in reference to pear species and cultivars (in Japanese). *J. Japan. Soc. Hort. Sci.* 49:608–616. [Hort. Abstr. 41:9250; 1971.]

Knee, M., S. G. S. Hatfield, and S. M. Smith. 1989. Evaluation of various indicators of maturity for harvest of apple fruit intended for long-term storage. *J. Hort. Sci.* 64:403–411.

Knee, M., and S. M. Smith. 1989. Variation in quality of apple fruits stored after harvest on different dates. *J. Hort. Sci.* 64:413–419.

Knee, M., S. M. Smith, and D. S. Johnson. 1983. Comparison of methods for estimating the onset of the respiratory climacteric in unpicked apples. *J. Hort. Sci.* 58:521–526.

Kumar, S., T. R. Sharma, and A. K. Goswami. 1986. Changes in the total phenolic content and in the activities of some oxido-reductive and hydrolytic enzymes in relation to development and ripening of apple. p. 247–250. In: T. R. Chadha, V. P. Bhutani, and J. L. Kaul (eds), *Advances in Research on Temperate Fruits.* Dr. Y. S. Parmar Univ. of Horticulture and Forestry, Solan, India.

Kupferman, E. M. 1986. Sampling for accurate analysis. p. 17–18. In: Apple maturity program. Wenatchee, WA.

Lallu, N., and A. N. Searle. 1987. Postharvest research on nashi. *Nashi News* (New Zealand Nashi Growers Assoc. Inc.) 3(4):4–5.

Lau, O. L. 1985. Harvest indices for B. C. apples. *B. C. Orchardist* 7(7):1A–20A.

‎_____. 1988. Harvest indices, dessert quality, and storability of 'Jonagold' apples in air and controlled atmosphere storage. *J. Am. Soc. Hort. Sci.* 113:564–569.

Lovelidge, B. 1987. A penetrating assessment of fruit firmness. *Grower* 108(13):53–54.

Luton, M. T., and P. J. C. Hamer. 1983. Predicting the optimum harvest dates for apple using temperature and full bloom records. *J. Hort. Sci.* 58:37–44.

Mann, S. S., and B. Singh. 1985. Some aspects of developmental physiology of 'LeConte' pear. *Acta Hort.* 158:211–215.

‎_____. 1986. Studies on change during development and maturation of pear fruit. p. 251–254. In: T. R. Chadha, V. P. Bhutani, and J. L. Kaul (eds), *Advances in Research on Temperate Fruits.* Dr. Y. S. Parmar Univ. of Horticulture and Forestry, Solan, India.

Mathee, G. W. 1988. The relationship between weather factors and the optimum picking date of 'Starking' and Granny Smith' apples. *Decid. Fruit Grower* 38:370–372.

Mattus, G. E. 1966. Maturity standards for 'Red Delicious'. *Am. Fruit Grower* 86(8):16.

Meheuriuk, M. 1977. Maturity studies on red sports of 'Delicious' apples. *Can. J. Plant Sci.* 57:963–967.

Narashimham, P., S. Dhanaraj, M. S. Krishnaprakesh, B. Arvindaprasad, C. A. Krishnaprasad, S. Habibunnisa, and S. M. Ananthakrishna. 1988. Effect of meteorological factors on fruit maturation and the prediction of optimum harvest date for apples. *Scientia Hort.* 35:217–226.

North, C. J. 1971. The use of the starch-iodine staining test for assessing the picking date for pears. Rep. E. Malling Res. Sta. for 1970:149–151.

Olsen, K. L. 1982. Picking maturity of apples. Fruit and Fruit Technology Research Institute Information Bulletin 496. Stellenbosch, Republic of South Africa.

Olsen, K. L., R. D. Bartram, and M. W. Williams. 1986. 'Golden Delicious' apples: Productivity and quality with adequate but modest leaf nitrogen levels. p. 11–16. In: Apple maturity program. Wenatchee, WA.

Olsen, K. L., and G. C. Martin. 1980. Influence of apple bloom date on maturity and storage quality of 'Starking Delicious' apples. *J. Am. Soc. Hort. Sci.* 105:183–186.

Olsen, K. L., H. A. Schomer, and R. D. Bartram. 1967. Segregation of 'Golden Delicious' apples for quality by light transmission. *Proc. Am. Soc. Hort. Sci.* 91:821–828.

Park, D. M., W. S. Kim, and W. C. Kim. 1984. The fruit growth curves and changes in carbohydrates of the fruit of four pear cultivars (*Pyrus pyrifolia*) (in Korean, English summary). *J. Kor. Soc. Hort. Sci.* 25:45–49.

Phillips, W. R. 1939. Respiration curve for 'McIntosh' apple. *Sci. Agric.* 19:505–509. [*Hort. Abstr.* 9:1452; 1939.]

Phillips, W. R., and P. A. Poapst. 1952. Storage of apples. Horticultural Div. Exptal. Farm Service. Publication 776. Canada Dept. Agr.

Poapst, P. A., G. M. Ward, and W. R. Phillips. 1959. Maturation of 'McIntosh' apples in relation to starch loss and abscission. *Can. J. Plant. Sci.* 39:257–263.

Quast, P. 1987. Fruit development and ripening in 'Jamba', 'Gravenstein', 'James Grieve', and 'Horneburger' in 1985 and 1986 on the lower Elbe (in German). *Mitteilungen des Obstbauversuchringes des Alten Landes.* 42:292–298. [*Hort. Abstr.* 59(3):1844; 1989.]

Reid, M. S., C. A. S. Padfield, C. B. Watkins, and J. E. Harman. 1982. Starch iodine pattern as a maturity index for 'Granny Smith' apples. 1. Comparison with flesh firmness and soluble solids content. *New Zealand J. Agr. Res.* 25:229–237.

Reid, M. S., M. J. C. Rhodes, A. C. Hulme. 1973. Changes in ethylene and CO_2 during the ripening of apples. *J. Sci. Food Agr.* 24:971–979.

Robinson, T. L., E. J. Seeley, and B. H. Barritt. 1983. Effect of light environment and spur age on 'Delicious' apple fruit size and quality. *J. Am. Soc. Hort. Sci.* 108:855–861.

Rouchard, J., C. Moons, and J. A. Meyer. 1983. Effects of selected fungicide treatments on the sugar content of 'Golden Delicious' and 'Jonagold' apples. *Scientia Hort.* 20:161–168.

Ryall, A. L., E. Smith, and W. T. Pentzer. 1941. The elapsed period from full bloom as an index of harvest maturity of pears. *Proc. Am. Soc. Hort. Sci.* 38:273–281.

Saltveit, M. E. 1983. Relationship between ethylene production and taste panel preference of 'Starkrimson Red Delicious' apples. *Can. J. Plant Sci.* 63:303–306.

Sfakiotakis, E. M., and D. R. Dilley. 1973. Internal ethylene concentrations in apple fruits attached to or detached from the tree. *J. Am. Soc. Hort. Sci.* 98:501–503.

Shaw, S. J., and R. N. Rowe. 1982. Sources of variability on apple maturity index data. *New Zealand Agr. Sci.* 16:51–53.

Smith, R. B., E. C. Lougheed, and E. W.Franklin. 1969. Ethylene production as an index of maturity for apple fruits. *Can. J. Plant Sci.* 49:805–807.

Smith, R. B., E. C. Lougheed, E. W. Franklin, and I. McMillan. 1979. The starch iodine test for determining stage of maturation in apples. *Can. J. Plant Sci.* 59:725–735.

Smith, S. M. 1985. Agriculture. Measurements of the quality of apples. Recommendations of an EEC working group. Commission of European Communities, Luxembourg.

Smock, R. M. 1948. A study of maturity indices for 'McIntosh' apples. *Proc. Am. Soc. Hort. Sci.* 52:176–182.

Smock, R. M., and D. Boynton. 1944. The effects of differential nitrogen treatments in the orchard on the keeping quality of 'McIntosh' apples. *Proc. Am. Soc. Hort. Sci.* 45:77–86.

Stow, J. R. 1988. The effect of cooling rate and harvest date on the storage behavior of 'Conference' pears. *J. Hort. Sci.* 63:59–67.

Tripathi, S. N., T. R. Chadha, S. K. Chopra, and J. N. Bhargava. 1986. Determination of maturity indices of 'Red Delicious' apple (*Malus domestica* Borkh). p. 301–305. In: T. R. Chadha, V. P. Bhutani, J. L. Kaul (eds), *Advances in Research on Temperate Fruits.* Dr. Y. S. Parmar Univ. of Horticulture and Forestry, Solan, India.

Truter, A. B., and R. F. Hurndall. 1988a. New findings on determining maturity of 'Starking', 'Topred' and 'Starkrimson' apples. *Decid. Fruit Grower* 38:26–29.

———. 1988b. Experimental and commercial maturity studies with 'Granny Smith' apples. *Decid. Fruit Grower* 38:364–367.

Tukey, H. B. 1942. Time interval between full bloom and fruit maturity for several varieties of apples, pears, peaches and cherries. *Proc. Am. Soc. Hort. Sci.* 40:133–140.

Tustin, D. S., P. M. Hirst, and I. J. Warrington. 1988. Influence of orientation and position of fruiting laterals on canopy light penetration, yield, and fruit quality of 'Granny Smith' apple. *J. Am. Soc. Hort. Sci.* 113:693–699.

Vangdal, E. 1982. Eating quality of pears. *Acta Agr. Scand.* 33:135–139.

———. 1985. Quality criteria for fruit for fresh consumption. *Acta Agr. Scand.* 35:41–47.

Walsh, C. S. 1977. The relationship between endogenous ethylene and abscission of mature apple fruits. *J. Am. Soc. Hort. Sci.* 102:615–619.

Wang, C. Y. 1982. Pear fruit maturity, harvesting, storage and ripening. p. 431–443. In: T. van der Zwet and N. F. Childers (eds), *The Pear.* Horticultural Publishers, Gainesville, FL.

Wang, S. Y., P. C. Wang, and M. Faust. 1988. Non-destructive detection of watercore in apple with nuclear magnetic resonance imaging. *Scientia Hort.* 35:227–234.

Warrington, I. J., C. J. Stanley, R. Volz, and D. C. Morgan. 1984. Effects of summer pruning on 'Gala' apple quality. *New Zealand Orchardist* 57:518–52.

Watada, A. E., R. C. Herner, A. A. Kader, R. J. Romani, G. L. Staby. 1984. Terminology for the description of developmental stages of horticultural crops. *HortScience* 19:20–21.

Watkins, C. B., M. S. Reid, J. E. Harman, and C. A. S. Padfield. 1982. Starch iodine pattern as a maturity index for 'Granny Smith' apples. 2. Differences between districts and relationship to storage disorders and yield. *New Zealand J. Agr. Res.* 25:587–592.

Wills, R. B. H., P. A. Bambridge, and K. J. Scott. 1980. The use of flesh firmness and other objective tests to determine consumer acceptability of 'Delicious' apples. *Austral. J. Expt. Agr. Anim. Husb.* 20:252–256.

Wills, R. B. H., W. B. McGlasson, D. Graham, T. H. Lee, and E. G. Hall. 1989. *Postharvest: An Introduction To the Physiology and Handling of Fruit and Vegetables* (3rd rev. ed.). BSP Professional Books, Oxford, England.

Worthington, J. T. 1974. A light-transmittance technique of determining tomato ripening rate and quality. *Acta Hort.* 38I:193–215.

Worthington, J. T., T. van der Zwet, and H. L. Keil. 1977. Reflectance and light transmittance technique for measuring maturity and ripening of 'Eldorado' and 'Bartlett' pears. *Acta Hort.* 69:327–331.

Yamaki, S., I. Kajiura, N. Kakiuchi. 1979. Changes in sugars and their related enzymes during development and ripening of Japanese pear fruit (*Pyrus serotina* Rehder var. *culta* Rehder). *Bull. Fruit Tree Res. Sta.* A 6:15–26.

Subject Index

Cumulative Subject Index (Volumes 1–13)

Cumulative Contributor Index
(Volumes 1–13)